NORTHWEST EUROPEAN MICROPALAEONTOLOGY AND PALYNOLOGY

ELLIS HORWOOD SERIES IN GEOLOGY

Editors: D. T. DONOVAN, Professor of Geology, University College London, and J. W. MURRAY, Professor of Geology, University of Exeter

This series aims to build up a library of books on geology which will include student texts and also more advanced works of interest to professional geologists and to industry. The series will include translation of important books recently published in Europe, and also books specially commissioned.

RADIOACTIVITY IN GEOLOGY: Principles and Applications
E. M. DURRANCE, Department of Geology, University of Exeter
FAULT AND FOLD TECTONICS
W. JAROSZEWSKI, Faculty of Geology, University of Warsaw
A GUIDE TO CLASSIFICATION IN GEOLOGY
J. W. MURRAY, Department of Geology, University of Exeter
THE CENOZOIC ERA: Tertiary and Quaternary
C. POMEROL, University of Paris VI
Translated by D. W. HUMPHRIES, Department of Geology, University of Sheffield, and E. E. HUMPHRIES
Edited by D. CURRY and D. T. DONOVAN, Department of Geology, University College London

BRITISH MICROPALAEONTOLOGICAL SOCIETY SERIES

This series, published by Ellis Horwood Limited for the British Micropalaeontological Society, aims to gather together knowledge of a particular faunal group for specialist and non-specialist geologists alike. The original series of Stratigraphic Atlas or Index volumes ultimately will cover all groups and will describe and illustrate the common elements of the microfauna through time (whether index or long-ranging species) thus enabling the reader to identify characteristic species in addition to those of restricted stratigraphic range. The series has now been enlarged to include the reports of conferences, organized by the Society, and collected essays on specialist themes.
The synthesis of knowledge presented in the series will reveal its strengths and prove its usefulness to the practising micropalaeontologist, and to those teaching and learning the subject. By identifying some of the gaps in this knowledge, the series will, it is believed, promote and stimulate further active research and investigation.

PALAEOBIOLOGY OF CONODONTS
Editor: R. J. ALDRIDGE, Department of Geology, University of Nottingham
CONODONTS: Investigative Techniques and Applications
Editor: R. L. AUSTIN, Department of Geology, University of Southampton
FOSSIL AND RECENT OSTRACODS
Editors: R. H. BATE, British Museum Natural History, London, and Stratigraphic Services International, Guildford, E. ROBINSON, Department of Geology, University College London, and L. SHEPPARD, British Museum Natural History, London, and Stratigraphic Services International, Guildford
NANNOFOSSILS AND THEIR APPLICATIONS
Editor: J. A. CRUX, British Petroleum Research Centre, Sunbury-on-Thames, and S. E. VAN HECK, Shell Internationale Petroleum, Maatschappij
A STRATIGRAPHICAL INDEX OF THE PALAEOZOIC ACRITARCHS AND OTHER MARINE MICROFLORA
Editors: K. J. DORNING, Pallab Research, Sheffield, and S. G. MOLYNEUX, British Geological Survey, Nottingham
MICROPALAEONTOLOGY OF CARBONATE ENVIRONMENTS
Editor: M. B. HART, Department of Geological Studies, Plymouth Polytechnic
A STRATIGRAPHICAL INDEX OF CONODONTS
Editors: A. C. HIGGINS, Geological Survey of Canada, Calgary, and R. L. AUSTIN, Department of Geology, University of Southampton
STRATIGRAPHICAL ATLAS OF FOSSIL FORAMINIFERA, 2nd Edition
Editors: D. G. JENKINS, The Open University, and J. W. MURRAY, Department of Geology, University of Exeter
A STRATIGRAPHICAL INDEX OF CALCAREOUS NANNOFOSSILS
Editor: A. R. LORD, Department of Geology, University College London
MICROFOSSILS FROM RECENT AND FOSSIL SHELF SEAS
Editors: J. W. NEALE, and M. D. BRASIER, Department of Geology, University of Hull
THE STRATIGRAPHIC DISTRIBUTION OF DINOFLAGELLATE CYSTS
Editor: A. J. POWELL, British Petroleum Research Centre, Sunbury-on-Thames
OSTRACODA
Editors: R. C. WHATLEY and C. MAYBURY, Department of Geology, University College of Wales

ELLIS HORWOOD SERIES IN APPLIED GEOLOGY

Published by Ellis Horwood Limited for the Institution of Geologists, Burlington House, Piccadilly London W1V 9AG

The books listed below are motivated by the up-to-date applications of geology to a wide range of industrial and environmental factors; they are practical, for use by the professional and practising geologist or engineer in the field, for study, and for reference.

A GUIDE TO PUMPING TESTS
F. C. BRASSINGTON, North West Water Authority
QUATERNARY GEOLOGY: Processes and Products
JOHN A. CATT, Rothamsted Experimental Station, Harpenden
TUNNELLING GEOLOGY AND GEOTECHNICS
Editors: M. C. KNIGHTS and T. W. MELLORS, Consulting Engineers, W. S. Atkins & Partners
PRACTICAL PEDOLOGY: Manual of Soil Formation, Description and Mapping
S. G. McRAE, Department of Environmental Studies and Countryside Planning, Wye College (University of London)
MODERN STANDARDS FOR AGGREGATES
D. C. PIKE, Consultant in Aggregates, Reading
LASER HOLOGRAPHY IN GEOPHYSICS
S. TAKEMOTO, Disaster Prevention Research Unit, Kyoto University

NORTHWEST EUROPEAN MICROPALAEONTOLOGY AND PALYNOLOGY

Editors:
DAVID J. BATTEN, Ph.D., B.A., B.Sc., M.Sc., D.U.C.
Reader, Department of Geology and Petroleum Geology, Aberdeen University
MICHAEL C. KEEN, Ph.D.
Lecturer, Department of Geology, Glasgow University

Published by
ELLIS HORWOOD LIMITED
Publishers · Chichester
for
THE BRITISH MICROPALAEONTOLOGICAL SOCIETY

First published in 1989
ELLIS HORWOOD LIMITED
Market Cross House, Cooper Street,
Chichester, West Sussex, PO19 1EB, England
The publisher's colophon is reproduced from James Gillison's drawing of the ancient Market Cross, Chichester.

Distributors:
Australia and New Zealand:
JACARANDA WILEY LIMITED
GPO Box 859, Brisbane, Queensland 4001, Australia
Canada:
JOHN WILEY & SONS CANADA LIMITED
22 Worcester Road, Rexdale, Ontario, Canada
Europe and Africa:
JOHN WILEY & SONS LIMITED
Baffins Lane, Chichester, West Sussex, England
North and South America and the rest of the world:
Halsted Press: a division of
JOHN WILEY & SONS
605 Third Avenue, New York, NY 10158, USA
South-East Asia
JOHN WILEY & SONS (SEA) PTE LIMITED
37 Jalan Pemimpin # 05–04
Block B, Union Industrial Building, Singapore 2057
Indian Subcontinent
WILEY EASTERN LIMITED
4835/24 Ansari Road
Daryaganj, New Delhi 110002, India

British Library Cataloguing in Publication Data
Northwest European micropalaeontology and palynology
1. Oceans. Bed. Micropalaeontology
I. Batten, D. J. II. Keen, M.C.
560.92

Library of Congress data available

ISBN 0–7458–0498–5 (Ellis Horwood Limited)
ISBN 0–470–21487–2 (Halsted Press)

Printed in Great Britain by Hartnolls, Bodmin

List of contributors

N. R. Ainsworth
Department of Geology, Trinity College, Dublin 2, Ireland

H. F. Barron
Biostratigraphy Research Group, British Geological Survey, Keyworth, Nottingham NG12 5GG, UK

S. Braley
Department of Geological Sciences, Plymouth Polytechnic, Plymouth PL4 8AA, UK

P. Copestake
B.P. Exploration, 301 St Vincent St, Glasgow G2 5DD, UK

R. Dyer
GeoStrat Ltd, Unit 23A, Motherwell Business Centre, Dalziel St, Motherwell ML1 9PJ, UK

K. M. Farr
Department of Environmental Sciences, Plymouth Polytechnic, Plymouth PL4 8AA, UK

F. J. Gregory
Department of Geology, University of Hull, Cottingham Road, Kingston-upon-Hull, HU6 7RX, UK

D. Guy-Ohlson
Section of Palaeobotany, Swedish Museum of Natural History, Box 50007, S-10405 Stockholm, Sweden

M. B. Hart
Department of Geological Sciences, Plymouth Polytechnic, Plymouth PL4 8AA, UK

M. C. Keen
Department of Geology, University of Glasgow, Glasgow G12 8QQ, UK

A. M. MacLennan
3 Little Spott Cottages, Dunbar, East Lothian EH42 1RH, UK

H. Nøhr-Hansen
The Geological Survey of Greenland, Øster Voldgade 10, DK-1350 København K, Denmark

M. O'Neill
Department of Geology, Trinity College, Dublin 2, Ireland

S. Packer
Paleoservices Ltd, Unit 15, Paramount Industrial Estate, Sandown Rd, Watford WD2 4XA, UK

M. M. Rutherford
British Petroleum Company plc, Stratigraphy Branch, Chertsey Rd, Sunbury-on-Thames, Middlesex TW16 7LN, UK

B. A. Tocher
Institute of Earth Studies, Llandinam Building, University College of Wales, Aberystwyth, Dyfed SY23 3DB, UK

N. H. Trewin
Department of Geology and Petroleum Geology, Marischal College, University of Aberdeen, Aberdeen AB9 1AS, UK

R. V. Tyson
Department of Geological Sciences, University College London, Gower Street, London WC1E 6BT, UK

Table of contents

Preface

Oil and gas exploration has now been concentrated on the Northwest European Continental Shelf, and in particular on the North Sea, for the past 20 years. Throughout this period, numerous structural and petroleum geologists, sedimentologists, geophysicists, geochemists, micropalaeontologists and others who specialize in the many subdisciplines of the Earth Sciences have been engaged in research on the region. They have perused the logs of hundreds of wells, studied seismic sections covering thousands of kilometres, worked on numerous core and cuttings samples, and written countless reports. Although most of the data accumulated remain confidential, the amount of information that has been made available for publication has increased substantially over the years, and quite a few large volumes of collected works on the geology and hydrocarbon exploration and production of the continental shelf and its borderlands have appeared in print. Some of the contributions contain significant amounts of palaeontological discussion but this is mostly in connection with dating and zonations. Papers that are primarily concerned with fossils are comparatively rare.

In view of the importance of micropalaeontology *sensu lato* (i.e. including palynology) in oil and gas exploration, the Council of the British Micropalaeontological Society (BMS) decided in 1986 that there should be a petroleum-orientated meeting in Aberdeen, the main operational centre of the North Sea oil industry in the UK. It was also agreed that the proceedings should, if possible, be published in the BMS Micropalaeontology Series. The meeting eventually took place at Aberdeen University on 2–3 April 1987, under the joint auspices of the BMS and the Petroleum Group of the Geological Society, London. It was followed by a 'long-weekend' field-trip to examine the sedimentology of some of the Palaeozoic and Mesozoic coastal outcrops bordering the Moray Firth. The excursion was led by Dr N. H. Trewin, a senior lecturer on the staff of the University and a contributor to this volume.

As local organizer (with much help from colleagues), I had high hopes of a significant contribution to the proceedings by oil company personnel. In the event, however, the majority of offers of papers came from individuals working in universities and service companies. The resulting meeting proved to be interesting and worthwhile, but the intended theme of applications of micropalaeontology and palynology to petroleum exploration, was not fully realized. The main points of ten of the papers in this volume were presented orally and hence comprised an important part of the formal sessions. The other two (by Guy-Ohlson and Tyson) were slightly later offerings considered suitable for inclusion in the volume. Although the majority are not primarily contributions to 'applied micropalaeontology', all have some, if indirect, relevance to hydrocarbon exploration, both on land and offshore in Northwest Europe. Most are biased towards palaeoenvironments and/or the Jurassic period. Overall there is also an emphasis on palynology, but non-palynological microfossils form an equally important component of four of the papers and are the main subject of two others. It was appropriate, in view of the aims of the subsequent field-trip, that the Jurassic rocks of the Moray Firth area received special attention at the meeting; three of the five contributions are included here.

There seems to be a tendency among some of our colleagues in the Earth Science fraternity to regard micropalaeontologists as rather narrow-minded specialists who isolate themselves from 'mainstream' geology. Although there may be an element of truth in such an opinion, it is, nevertheless, an over-generalization as I think this collection of papers demonstrates. Several are collaborative endeavours and none is so devoted to microfossils that there is no geological perspective.

I am very grateful to Michael Keen, who agreed to take on the responsibility of handling the papers in which non-palynological microfossils play an important part. This helped considerably to lighten the editorial burden at a difficult time when major changes of staff and other adjustments were being made in our respective Departments in response to the 'rationalization' of provision for Earth Sciences in UK universities. He joins me in thanking those of our colleagues who kindly reviewed the manuscripts and made valuable suggestions for improvements. We do not necessarily concur with all of the interpretations presented and opinions expressed; indeed, a few run contrary to our own views and experience. As editors we adopted a neutral stance on points of contention which is only fair on the authors, without whose efforts this volume could not have been published.

D. J. Batten

1

Jurassic and Upper Triassic biostratigraphy of the North Celtic Sea and Fastnet Basins

N. R. Ainsworth, M. O'Neill and M. M. Rutherford

ABSTRACT

A detailed biostratigraphy is proposed for the Jurassic and Upper Triassic of the North Celtic Sea and Fastnet Basins, offshore southwest Ireland. Within this interval, 25 associations can be recognized: six in the Upper Jurassic, nine in the Middle Jurassic, nine in the Lower Jurassic and one in the Upper Triassic. These associations are based on the tops of ranges of ostracods, foraminifera, dinoflagellate cysts and miospores from approximately 3500 ditch-cutting samples and sidewall cores from 30 wells. All stages from the top of the Upper Jurassic through to the Upper Triassic can be recognized by microfaunas and/or microfloras; however, precise dating of the Callovian–Oxfordian interval is difficult. This is due to the very poor microfossil recovery, because of the extensive development of continental red beds. The Jurassic–Cretaceous boundary can be distinguished throughout both basins using the zonation schemes described by Anderson in 1971 and Colin *et al.* in 1981 for non-marine ostracods. The Triassic–Jurassic boundary is more difficult to locate, owing to long-ranging and relatively sparse microfossil assemblages transgressing the boundary.

INTRODUCTION

The North Celtic Sea Basin offshore southwest Ireland is an Upper Palaeozoic to Recent depositional downfaulted trough into which approximately 6000 m of sediment has been deposited. The North Celtic Sea Basin is aligned in an ENE–WSW orientation and extends from south of Ireland into Cardigan Bay, offshore west of Wales (Fig. 1). Robinson *et al.* (1981) have shown from seismic profiles that the Celtic Sea covers a pair of parallel fault-bounded northeast–southwest-trending grabens (the North and South Celtic Sea Basins) separated by a narrow basement high (the Pembroke Ridge). At its northeasterly end, the North Celtic Sea Basin is linked to basins of the Irish Sea, while its southwesterly extension is the Fastnet Basin.

Previous studies of the microfaunas and microfloras of the North Celtic Sea and Fastnet Basins began with Colin *et al.* (1981) who described the Cretaceous and Upper Jurassic microfossils recovered from 11 Esso–Marathon wells in the North Celtic Sea, producing a useful biostratigraphical analysis. This study has been followed by eight papers concentrating upon the microfaunas from the more southwesterly

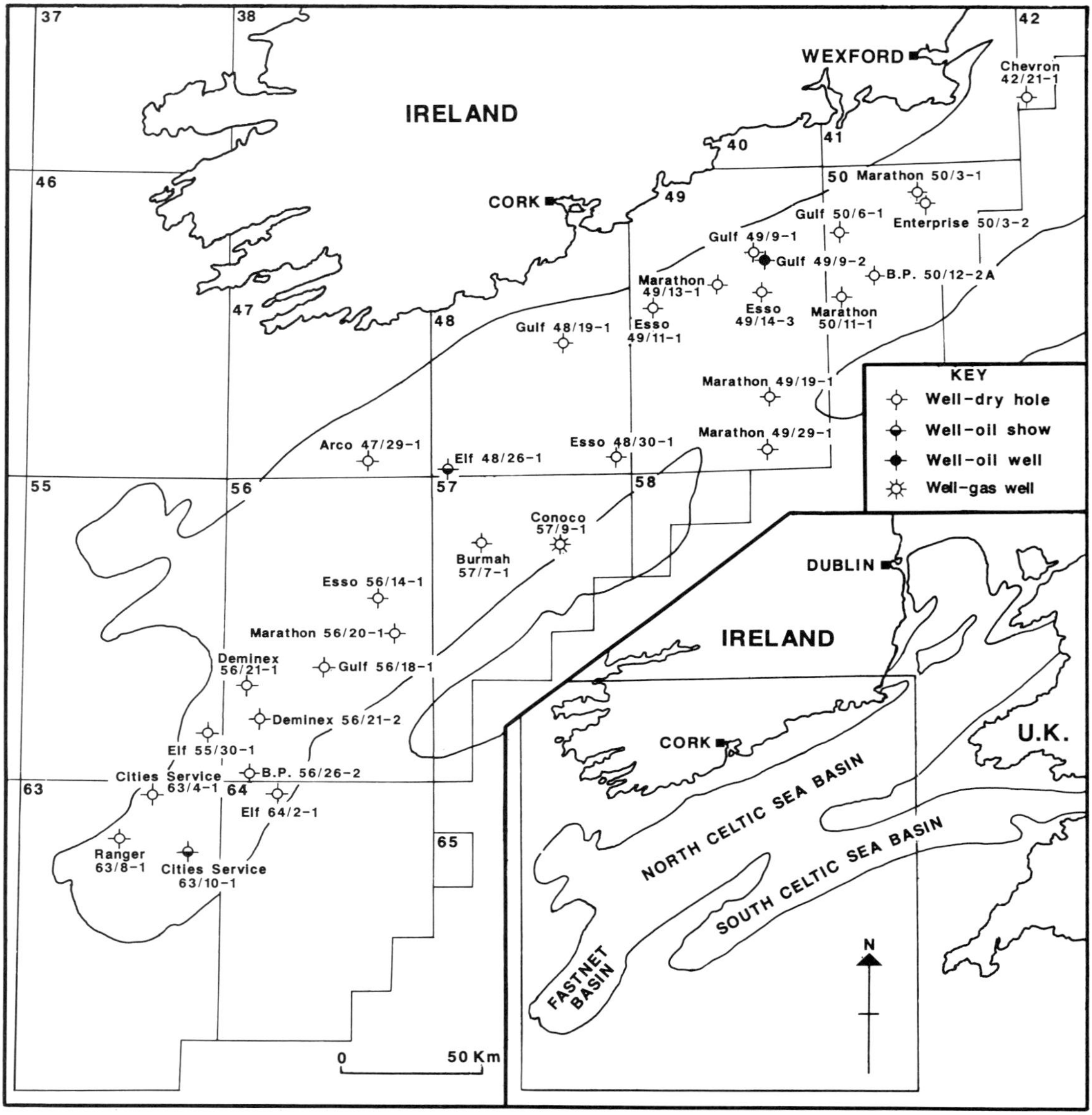

Fig. 1 — Location of wells investigated.

Fastnet Basin. Ainsworth *et al.* (1985) described 12 microfaunal and microfloral associations recognized in the Lower Cretaceous and lowermost Upper Cretaceous of the Fastnet Basin; Ainsworth & Horton (1986) discussed the microfaunas recovered from the Upper Triassic, Lower Jurassic and Lower Cretaceous in Elf 55/30-1. The ostracod faunas from the Lower Cretaceous through to the Pliensbachian have been extensively described by Ainsworth (1985, 1986a, b, c, 1987a, b). These studies have been further continued in a paper by Ainsworth *et al.* (1987) describing a composite microfaunal and palynomorph biostratigraphy, in terms of 38 biostratigraphic 'events', each recognized by the tops of ranges of one or more taxa, from the Lower Cretaceous, Jurassic and uppermost Triassic of the North Celtic Sea and Fastnet Basins.

In 1971, Marathon began the first drilling with exploration well 48/25-1. Since that time, 68 wells have been drilled to date (June 1988), with some commercial success. This present study utilizes micropalaeontological and palynological data from 30 wells from the North Celtic Sea and Fastnet Basin. From the 30 wells studied, over 3500 ditch-cutting samples and sidewall cores were analysed for ostracods, foraminifera, dinocysts and miospores between the top of the Upper Jurassic and the base of the Rhaetian (Upper Triassic). It was not possible to recognize fossil assemblage zones in the formal sense owing to heavy dependence on ditch-cutting samples, which contain inevitable mixing and caving. For this reason, the term 'association' is used. 25 micropalaeontological–palynological associations have been recognized based on the study of these wells. In all cases, the age of the association is determined on the basis of the first downhole occurrence of selected microfossils (ostracods, foraminifera or palynomorphs).

Those species which have well-recorded ranges in Europe confined to a particular stage or stages in an association are termed diagnostic, while those species which have ranges beyond an inferred stage or stages are termed characteristic. In the marine intervals, mainly between the Bathonian through to the Rhaetian, all fossil groups could be used for dating and correlation; in the non-marine to marine facies of the Portlandian to Callovian, ostracods and palynomorphs proved to be the most useful.

BIOSTRATIGRAPHY

The following section gives a brief description of the 25 microfaunal and microfloral associations recognized in the Jurassic and Upper Triassic of the North Celtic Sea and Fastnet Basins, emphasizing the taxa used in dating (see Fig. 2 later in this chapter). Lithostratigraphic units enclosed by inverted commas in the text denote British onshore units which have been widely used in the Irish and British offshore in an informal sense, by workers including Penn (1987) and Ainsworth *et al.* (1987).

UPPER JURASSIC

Six biostratigraphical associations (Associations 1–6) are recognized in the Upper Jurassic, mostly in the North Celtic Sea Basin. The upper Portlandian is characterized by rich ostracod assemblages and abundant saccate pollen grains. The lower Portlandian–upper Kimmeridgian interval also yields abundant ostracods; however, dinocysts are generally rare. The lowermost upper–lower Kimmeridgian is marked by moderately common dinocysts, abundant ostracods and small agglutinating foraminifera. Rare to common ostracod faunas are observed in the lowermost Kimmeridgian–upper Oxfordian. In the lower and lowermost Oxfordian there are only very rare occurrences of short-ranging dinocysts.

Association 1 (upper Portlandian)

This association is recognized throughout the North Celtic Sea by a diverse and very abundant ostracod fauna, in conjunction with long-ranging miospores.

Ostracoda

Diagnostic species

Bisulcocypris aff. *B. verrucosa* (Jones) *sensu* Ainsworth *et al.*, *Cetacella* cf. *C. striata* (Helmdach) *sensu* Colin *et al.*, *Cypridea dunkeri carinata* Jones, *C. granulosa granulosa* (Sowerby), *C.* sp. A Ainsworth *et al.* and *Klieana alata* Martin.

Characteristic species

Bisulcocypris? dilatata Anderson, *Cypridea tumenscens praecursor* Oertli, *C.* cf. *C. peltoides peltoides* Anderson *sensu* Ainsworth *et al.*,

Damonella ellipsoidea (Wolburg), *D. pygmaea* (Anderson), *Darwinula leguminella* (Forbes), *D. oblonga* (Roemer), *Rhinocypris jurassica* (Martin) and *Scabriculocypris trapezoides* Anderson.

Upper Portlandian ostracod assemblages are dominated by *Cypridea* spp., with alternations of *Damonella*, *Darwinula* and *Rhinocypris*. The top of the upper Portlandian is placed at the first occurrence of generally common *Cypridea tumenscens praecursor* Oertli, in association with *C.* cf. *C. peltoides peltoides* Anderson *sensu* Ainsworth *et al.* In the uppermost horizons of a few North Celtic Sea wells, rare *Cypridea granulosa granulosa* (Sowerby) and *C.* sp. A Ainsworth are recovered. The index taxon *Cypridea dunkeri carinata* Jones is also very rare in the North Celtic Sea occurring towards the bottom of this interval, before the rich influxes of *Damonella* and *Darwinula* from the lower Portlandian–upper Kimmeridgian interval.

Many of the taxa observed in this interval have been described from similar stratigraphical horizons in southern England (Anderson 1966, 1971, 1985, Barker 1966a, b), France (Oertli 1963), Germany (Martin 1940, Wolburg 1971) and Sweden (Christensen 1968).

Palynomorphs

Diagnostic species

Classopollis echinatus Burger is used locally to indicate uppermost Portlandian age sediments (Colin *et al.* 1981).

Characteristic species

Araucariacites australis Cookson, *Baculatisporites comaumensis* (Cookson) Potonié, *Callialasporites* spp., *Cerebropollenites mesozoicus* (Couper) Nilsson, *Chomotriletes minor* (Kedves) Pocock, *Cibotiumsporites juriensis* (Balme) Filatoff, *Cicatricosisporites abacus* (Burger) Fensome, *C. australiensis* (Cookson) Potonié, *C. purbeckensis* Norris, *Classopollis classoides* (Pflug) Pocock & Jansonius, *Couperisporites complexus* (Couper) Pocock, *Cyathidites australis* Couper, *C. minor* Couper, *Densoisporites velatus* Weyland & Krieger, *Dictyophyllidites equiexinus* (Couper) Dettmann, *D. harrisii* Couper, *Duplexisporites problematicus* (Couper) Playford & Dettmann, *Foraminisporis wonthaggiensis* (Cookson & Dettmann), *Foveosporites subtriangularis* (Brenner) Döring, *Gleicheniidites senonicus* (Ross) Scarby, *Ischyosporites pseudoreticulatus* (Couper) Döring, *Januasporites spinulosus* Dettmann, *J. tumulosus* Norris, *Osmundacidites wellmanii* Couper, *Perinopollenites elatoides* Couper, *Pilosisporites trichopapillosus* (Thiergart) Delcourt & Sprumont, *Retitriletes austroclavatidites* (Cookson) Döring *et al.*, *Sestrosporites pseudoalveolatus* (Couper) Dettmann, *Staplinisporites caminus* Balme, *Trilobosporites apiverrucatus* Couper, *T. aequiverrucosus* Dörhöfer, *T.* spp., *Uvaesporites argenteaeformis* (Bolkhovitina) Schulz, and disaccate pollen grains.

Decreasing diversity of spores downhole coincides with increasing numbers of saccate pollen grains. Palynomorph assemblages close to the Berriasian–Portlandian boundary often comprise abundant amorphous kerogen and woody material. The freshwater alga, *Botryococcus* Kützing, is present and the brackish-water alga, *Celyphus rallus* Batten, has been recorded in abundance close to this boundary. The assemblage described here corresponds well with palynomorph assemblages from southern England (Couper 1958, Norris 1969) including Dorset (Lord & Bown 1987).

Association 2 (lower Portlandian–upper Kimmeridgian)

This association is observed throughout the North Celtic Sea and is marked by abundant ostracods and dinocysts.

Ostracods

Diagnostic species

Cetacella armata Martin, *Damonella* cf. *D. ellipsoidea* (Wolburg) *sensu* Ainsworth *et al.* (large form) and *Mantelliana* cf. *M. purbeckensis* (Forbes) *sensu* Ainsworth *et al.*

Characteristic species

Bisulcocypris? dilatata Anderson, *Cypridea tumenscens praecursor* Oertli, *C.* cf. *C. peltoides peltoides* Anderson *sensu* Ainsworth *et al.*,

Damonella ellipsoidea (Wolburg), *D. pygmaea* (Anderson), *Darwinula leguminella* (Forbes), *D. oblonga* (Roemer), *Rhinocypris jurassica* (Martin) and *Scabriculocypris trapezoides* Anderson.

Association 2 ostracod faunas are dominated by *Damonella* spp., *Rhinocypris jurassica* (Martin) and *Mantelliana* cf. *M. purbeckensis* (Forbes) *sensu* Ainsworth *et al.* Many of these species also occur in the overlying upper Portlandian sediments, but in lesser proportions with respect to *Cypridea*. The *Cypridea* faunas are still observed, albeit in fewer horizons and numbers. Many of these taxa have been reported from rocks of a similar age throughout northwest Europe. These include southern England (Anderson 1966, 1971, 1985, Barker 1966a, b), France (Oertli 1963), Germany (Martin 1940, Wolburg, 1971), Sweden (Christensen 1968) and northeast Spain (Brenner 1976).

Palynomorphs

Characteristic species

Cribroperidinium globatum (Gitmez & Sarjeant) Helenes, *C. longicorne* (Downie) Lentin & Williams, *Mendicodinium groenlandicum* (Pocock & Sarjeant) Davey, *Omatia montgomeryi* Cookson & Eisenack, *Sentusidinium* sp. Sarjeant & Stover, *Araucariacites australis* Cookson, *Callialasporites* spp., *Cerebropollenites mesozoicus* (Couper), *Cicatricosisporites* spp., *Classopollis classoides* (Pflug) Pocock & Jansonius, *Couperisporites complexus* (Couper) Pocock, *Cyathidites minor* Couper, *Densoisporites perinatus* Couper, *D. velatus* (Weyland & Krieger) Bharadwaj & Kumar, *Distalanulisporites verrucosus* Pocock, *Gleicheniidites senonicus* (Ross) Scarby, *Ischyosporites pseudoreticulatus* (Couper) Döring, *Leptolepidites* spp. (Couper) Norris, *Osmundacidites wellmanii* Couper, *Perinopollenites elatoides* Couper, *Pilosisporites trichopapillosus* (Thiergart) Delcourt & Sprumont, *Retitriletes austroclavatidites* (Cookson) Döring *et al.*, *Staplinisporites caminus* (Balme) Pocock, and disaccate pollen.

Palynomorph recovery is poor to moderate, while preservation is normally poor. The occurrence of species of *Cribroperidinium globatum* is characteristic of assemblages of Kimmeridgian age (Gitmez 1970, Gitmez & Sarjeant 1972), although these species do appear downhole in the lower Portlandian.

Association 3 (lowermost upper–lower Kimmeridgian)

This interval in the North Celtic Sea is characterized by diverse ostracod faunas and diverse palynomorph assemblages, whereas the foraminifera are generally rare and of a limited diversity.

Ostracoda

Diagnostic species

Amphicythere confundens Oertli, *Cytherella* sp. B, *Damonella* aff. *D. ellipsoidea* (Wolburg) *sensu* Ainsworth *et al.*, *D.* cf. *D. pygmaea* (Anderson) *sensu* Ainsworth *et al.*, *Fabanella boloniensis* (Jones), *F.* aff. *F. ansata* (Jones), *Galliaecytheridea dorsetensis* Christensen & Kilenyi, *G.* sp. Dépêche, *G.* sp. A, *Macrodentina* (*M.*) *annulata* Malz, *M.* (*M.*) cf. *M.* (*M.*) *cicatriosa* Malz *sensu* Colin *et al.*, *M.* (*M.*) cf. *M.* (*M.*) *klingleri* Malz *sensu* Dépêche, *Macrodentina* (*Polydentina*) *steghausi steghausi* (Klingler), *Paranotacythere* (*Unicosta*) *extendata* Bassiouni, *Sabriculocypris?* sp. A Ainsworth *et al.*, *Schuleridea triebeli* (Steghaus) and *Theriosynoecum* sp. A, *T.* sp. B.

Characteristic species

Bisulcocypris? dilatata Anderson, *Cetacella inermis* Martin, *Damonella ellipsoidea* (Wolburg), *D. pygmaea* (Anderson), *Darwinula leguminella* (Forbes), *D. oblonga* (Roemer), *Leiria paucistriata* Helmdach and *Rhinocypris jurassica* (Martin).

The lowermost upper Kimmeridgian–lower Kimmeridgian contains both diverse marine and non-marine ostracod faunas. The top of this interval is characterized by the first downhole occurrence of rare *Macrodentina* (*M.*) and *Macrodentina* (*P.*) spp., often in association with *Fabanella boloniensis* (Jones). Where the 'Kimmeridgian Limestone' is thickly developed, *Fabanella boloniensis* (Jones) is numerically abundant, often forming a near monospecific assemblage both within and slightly above the development of the oolitic–pelloidal limestone. The more marine ostracods (*Amphicythere*, *Galliaecytheridea*, *Macrodentina*, *Paranotacythere* and *Schuleridea*) are generally rare within the limestone but may form a near-exclusive fauna surrounding the limestone. Other new notable occurrences are *Theriosynoecum* sp. A from within the limestone and *T.* sp. B from below the limestone. This genus occurs in abundance in association with the large numbers of *Damonella* spp., with rarer *Leiria* and *Rhinocypris* spp., all of which are observed throughout the interval. Towards the lower horizons of the lower Kimmeridgian, ostracod faunas decrease.

The lowermost upper–lower Kimmeridgian ostracods of the North Celtic Sea show close comparisons with the faunas of onshore Europe. The marine ostracods recovered by the present authors are similar to those described by Cox *et al.* (1987), Ahmed (1987), Wilkinson (1983), Kilenyi (1969), Dépêche (1985) and Malz (1958) from the Kimmeridgian of the southern North Sea, England, France and Germany. The non-marine faunas, however, are only similar to those reported by Brenner (1976) and Helmdach (1971) from northeast Spain and Portugal.

Foraminifera

Characteristic species

Ammobaculites sp., *Ammodiscus* sp., *Eoguttulina* sp., *Lenticulina subalata* (Reuss), *L.* sp., *Spirillina infima* (Strickland) and *Spirolina* sp.

Palynomorphs

Diagnostic species

Aldorfia dictyota (Cookson & Eisenack) Davey, *Glossodinium dimorphum* Ioannides, Stavrinos & Downie, *Gochteodinia mutabilis* (Riley) Fisher & Riley, *Occisucysta balia* Gitmez, *Subtilisphaera inaffectum* (Drugg) Bujak & Davies and *S. paeminosum* (Drugg) Bujak & Davies.

Characteristic species

Adnatosphaeridium caulleryi (Deflandre) Williams & Downie, *Chytroeisphaeridia chyroeides* (Sarjeant) Downie & Sarjeant emend. Davey, *Cribroperidinium globatum* (Gitmez & Sarjeant) Helenes, *C. longicorne* (Downie) Lentin & Williams, *Dingodinium tuberosum* (Gitmez) Fisher & Riley, *Ellipsoidictyum cinctum* Klement/*Valensiella ovula* (Deflandre) Eisenack, *Escharisphaeridium pocockii* (Sarjeant) Erkmen & Sarjeant, *Mendicodinium groenlandicum* Pocock & Sarjeant) Davey, *Polystephanephorus sarjeantii* Gitmez, *Prolixosphaeridium granulosum* (Deflandre) Davey *et al.*, *Sentusidinium echinatum* (Gitmez & Sarjeant) Sarjeant & Stover, *Araucariacites australis* Cookson, *Callialasporites* spp., *Classopollis classoides* (Pflug) Pocock & Jansonius, *Cerebropollenites mesozoicus* (Couper), *Cicatricosisporites* spp., *Couperisporites complexus* (Couper) Pocock, *Cyathidites* spp., *Dictyophyllidites equiexinus* (Couper) Dettmann, *D. harrisii* Couper, *Distalanulisporites verrucosus* Pocock, *Densoisporites perinatus* Couper, *D. velatus* (Weyland & Krieger) Bharadwaj & Kumar, *Gleicheniidites senonicus* (Ross) Scarby, *Osmundacidites wellmanii* Couper, *Ischyosporites pseudoreticulatus* Couper) Döring, *Pilosisporites trichopapillosus* (Thiergart) Delcourt & Sprumont, *Retitriletes austroclavatidites* (Cookson) Döring and *Staplinisporites caminus* (Balme) Pocock.

Although the preservation is quite poor, the assemblage is moderately rich and dominated by *Subtilisphaera* spp. and *Aldorfia dictyota.* The assemblage is comparable with Kimmeridgian assemblages from onshore UK, notably those described by Gitmez (1970), Gitmez & Sarjeant (1972) and Riding (1987). The North Celtic Sea, however, differs in the absence of such dinocysts as *Gonyaulacysta jurassica* (Deflandre) Norris & Sarjeant, *Ctenidodinium panneum* (Norris) Lentin & Williams, and only rare and poorly preserved specimens of *Glossodinium dimorphum* Ioannides, Stavrinos & Downie are present. Perhaps this can be explained by provincialism of dinoflagellates within the Kimmeridgian where species and stratigraphical ranges have shown local presence or absence rather than cosmopolitan characteristics (Norris 1975, Riding 1987).

Association 4 (lowermost Kimmeridgian–upper Oxfordian)
This association is marked throughout the North Celtic Sea by the intermittent recovery of ostracod faunas. Rare to common charophyte oogonia are seen throughout this sequence.

Ostracoda
Diagnostic species
Bisulcocypris fluxans fluxans Helmdach, *B. fluxans spiculata* Helmdach, *B. levis* Helmdach, *Cetacella* sp. A, *Leiria striata* Helmdach, *L.* sp. A, *Klieana* sp. A, Ainsworth *et al.*, *Theriosynoecum hemigymnon* Helmdach and *T. wyomingenis* (Branson).
Characteristic species
Cetacella inermis Martin, *Darwinula leguminella* (Forbes), *D. oblonga* (Roemer) and *Leiria paucistriata* Helmdach.

This ostracod assemblage is poorly defined throughout the North Celtic Sea because of the limited number of productive horizons. Where the various taxa are recovered they are generally abundant. The top of the interval is defined by the first downhole occurrence of either or both *Leiria striata* Helmdach and *Bisulcocypris* spp. The very distinctive *Theriosynoecum hemigymnon* Helmdach and *T. wyomingensis* (Branson) occur further downhole, often in association with abundant *Cetacella* and *Leiria* spp.

Very few close faunal comparisons can be made to those described from onshore Europe. Only the taxa discussed by Brenner (1976) and Helmdach (1971, 1972) from the Upper Jurassic of northeast Spain and Portugal possess similar faunas.

Association 5 (lower Oxfordian)
The recovery of the short-ranging dinocyst (*Acanthaulax scarburghensis* (Sarjeant) Lentin & Williams) dates the lower Oxfordian interval.

Ostracoda
Diagnostic species
Darwinula sp. A.

Only a new species of *Darwinula* was sporadically recorded within this interval.

Palynomorphs
Diagnostic species
Acanthaulax scarburghensis (Sarjeant) Lentin & Williams.
Characteristic species
Chytroeisphaeridia chytroeides (Sarjeant) Downie & Sarjeant emend. Davey, *Escharisphaeridium pocockii* (Sarjeant) Erkmen & Sarjeant, *Gonyaulacysta?* spp., *Mendicodinium groenlandicum* (Pocock & Sarjeant) Davey, *Cerebropollenites mesozoicus* (Couper), species of *Callialasporites* and *Classopollis*, and bisaccate pollen grains.

The above index species for the top lower Oxfordian appears to have a wide geographic distribution (Davies 1983) and is relatively commonly recorded from the North Celtic Sea Basin. The overall assemblage ranges through to the upper Callovian. Dinocyst diversity is poor, in comparison with the abundant lower Oxfordian assemblages described from the onshore UK (Riding 1984a, 1987).

Association 6 (lowermost Oxfordian)
This association is based solely on the rare recovery of the lowermost Oxfordian dinocyst *Wanaea fimbriata* Sarjeant.

Palynomorphs
Diagnostic species
Wanaea fimbriata Sarjeant.
Characteristic species
Cerebropollenites mesozoicus (Couper), *Ischyosporites crateris* Blame, *Duplexisporites problematicus* (Couper) Playford & Dettmann, *Gleicheniidites senonicus* Ross, *Densoisporites velatus* Weyland & Krieger, *Classopollis* spp. and *Neoraistrickia* spp.

This diagnostic dinocyst species has an established range restricted to the lowermost Oxfordian. It is extremely rare, recorded from only two wells. The interval between the lowermost Oxfordian and the uppermost Bathonian contains very sparse non-marine palynomorphs.

These are largely attributable to long-ranging Oxfordian–Bathonian taxa.

MIDDLE JURASSIC

Nine biostratigraphical associations (Associations 7–15) are recognized in the Middle Jurassic. These are best developed in the northern North Celtic Sea. No stratigraphically diagnostic palynomorphs or microfossils are recovered from presumed Callovian sediments. Spore assemblages, typically Middle Jurassic in aspect, occur throughout and are associated with a continental facies. The uppermost Bathonian yields common ostracod microfaunas. This is replaced downhole by rare upper Bathonian microfaunas extracted from the thin argillaceous horizons within the 'Bathonian Limestone'. Microfloras, however, are commonly recovered. The middle–lower Bathonian palynomorphs and microfaunas are generally diverse and abundant, becoming rare towards the lower intervals. Throughout the lowermost Bathonian and Bajocian, both ostracods and foraminifera are impoverished, whereas dinocysts are abundant.

Association 7 (Callovian)

Throughout much of the North Celtic Sea, the Callovian–Bathonian boundary is delineated by a sharp unconformity or disconformity. This junction places the continental sediments of the Callovian upon the Bathonian. In many of the wells studied by the present authors, the top Bathonian constitutes the upper Bathonian limestones, a hard cemented sequence, which acted as a non-erosional surface during the mid-Cimmerian tectonic event. The uppermost Bathonian (Association 8) and part of or all the Callovian are often missing owing to variable erosion.

At present, because of the lithologies developed, it has not been possible to prove the presence of Callovian sediments, because only rare non-age-diagnostic assemblages have been recovered. However, often thick sequences of Callovian sediments can be inferred by the presence of well-dated rocks of lowermost Oxfordian age above, and strata of upper Bathonian age below.

Palynomorphs
Characteristic species
Cerebropollenites mesozoicus (Couper), *Ischyosporites pseudoreticulatus* (Couper) Döring, *Duplexisporites problematicus* (Couper) Playford & Dettmann, *Gleicheniidites senonicus* Ross, *Densoisporites velatus* Weyland & Krieger, and species of *Classopollis* and *Neoraistrickia.*

Association 8 (uppermost Bathonian)

This association is based on a rich ostracod assemblage. This characteristic fauna has only been recorded in the more northerly North Celtic Sea wells and occurs between the continental Callovian and the more fully marine Bathonian.

Ostracoda
Diagnostic species
Belekocytheridea cf. *B. punctata* Bate *sensu* Ainsworth *et al.*, *Darwinula incurva* Bate, *Glyptocythere oscillum* (Jones & Sherborn), *G. guembeliana* (Jones), *G.* sp. A., *Klieana levis* Oertli, *Micropneumatocythere falcata* Sheppard, *M. subconcentrica* (Jones) and *Pichottia muris* Oertli.

The top horizons of the uppermost Bathonian contain rare specimens of *Darwinula* and common specimens of *Belekocytheridea*, *Klieana* and *Pichottia.* The lower levels yield rarer numbers of the brackish genera in conjunction with the marine genera *Glyptocythere* and *Micropneumatocythere*, suggesting a fluctuating marine–non-marine environment.

These ostracods bear strong comparisons with those faunas described from the Forest Marble of southern England (Sheppard 1981, Bate 1967) and northern France (Oertli 1957, 1959). This assemblage can be placed within the *Micropneumatocythere falcata* Ostracod Biozone (Sheppard 1981), indicating uppermost Bathonian sediments.

Palynomorphs

Characteristic species

Cerebropollenites mesozoicus (Couper), *Ischyosporites pseudoreticulatus* (Couper) Döring, *Duplexisporites problematicus* (Couper) Playford & Dettmann, *Gleicheniidites senonicus* Ross, *Densoisporites velatus* Weyland & Krieger, and species of *Classopollis* and *Neoraistrickia*.

Association 9 (upper Bathonian)

This distinctive upper Bathonian association is recovered from the thin argillaceous horizons between the thicker bedded intervals of well-cemented oolitic peloidal limestones and sandstones.

Ostracoda

Diagnostic species

Acanthocythere sp., *Caytonidea* aff. *C. terraefullonicae* (Jones & Sherborn) *sensu* Ainsworth *et al.*, *Fastigatocythere juglandia* (Jones), *Fossaterquemula blakeana* (Jones), *Glabellacythere dolabra* (Jones & Sherborn), *Lophocythere ostreata* (Jones & Sherborn), *Micropneumatocythere quadrata* Bate, *M.* cf. *M. crassa* Peterson *sensu* Dépêche, *Monoceratina vulsa* Jones & Sherborn, *Oligocythereis* cf. *O. fullonica* (Jones & Sherborn) *sensu* Sheppard, *Pichottia magnamuris* Bate, *Praeschuleridea quadrata* Bate, *Progonocythere polonica* Blaszyk, *P. stilla* Sylvester-Bradley, *Pseudoprotocythere? bessinensis* Dépêche & Oertli, *Schuleridea* (*Eoschuleridea*) *bathonica* Bate and *S.* (*E.*) cf. *S.* (*E.*) *trigonalis* (Jones) *sensu* Ainsworth *et al.*

Characteristic species

Bairdia hilda Jones.

The marine ostracods of Association 9 are often very rare and poorly preserved because of the difficulty of extraction from the well-cemented sandy limestones and limestones. Diversity, however, is high. Many of these taxa have been described from sediments belonging to the Great Oolite, Upper Fullers Earth and Upper Estuarine Series (upper Bathonian) of England by Bate (1967, 1978), Sheppard (1981) and from the upper Bathonian of France (Dépêche 1985).

Foraminifera

Diagnostic species

Vaginulina barnardi Gordon.

Characteristic species

Epistomina ex. gr. *bireticulata* Pazdro, *E. stelligera* (Reuss), *L. muensteri* Roemer plex. and *Spirillina infima* (Strickland).

Foraminiferal assemblages are very impoverished, poorly preserved and dominated by *Epistomina* spp. in association with *Lenticulina muensteri.*

Palynomorphs

Characteristic species

Ctenidodinium combazii Dupin, *C. continuum* Gocht, *C. sellwoodii* (Sarjeant) Stover & Evitt, *Diacanthum filapicatum* (Gocht) Stover & Evitt, *Endoscrinium luridum* (Deflandre) Gocht, *Escharisphaeridium* spp., *Ellipsoidictyum/Valensiella* spp., *Gonyaulacysta* spp., *Hystrichogonyaulax pectinigera* (Gocht) Fensome, *H. regalis* (Gocht) Stover & Evitt, *Korystocysta gochtii* (Sarjeant) Woollam, *Meiourogonyaulax* spp., *Mendicodinium groenlandicum* (Pocock & Sarjeant) Davey, *Pareodinia ceratophora* (Deflandre) Gocht, *Sentusidinium* spp., *Tubotuberella dangeardii* (Sarjeant) Stover & Evitt, *Cerebropollenites mesozoicus* (Couper), *Cyathidites australis* Couper, *C. minor* Couper, *Dictyophyllidites equiexinus* (Couper) Dettmann, *D. harrisii* Couper, *Duplexisporites problematicus* (Couper) Playford & Dettmann, *Ischyosporites pseudoreticulatus* (Couper) Döring, *Leptolepidites major* Couper, *Neoraistrickia* spp., *Osmundacidites wellmanii* Couper, *Quadraeculina anellaeformis* Maljavkina, *Retitriletes austroclavatidites* (Cookson) Döring *et al.*, *Staplinisporites caminus* (Balme) Pocock, and disaccate pollen.

This assemblage persists through the Bathonian and into the upper Bajocian and is dominated by the *Ctenidodinium/Dichadogonyaulax* group. *Escharisphaeridium* spp. and *Ellipsoidictyum/Valensiella* spp. are common. Abundance and diversity of *Ctenidodinium* spp. are characteristic of the Bathonian (Riding 1984a). Abundance of *Ctenidodinium* spp. in the

North Celtic Sea and Fastnet Basins occurs in downhole succession to the non-marine ostracod assemblages of uppermost Bathonian age.

Association 10 (middle–lower Bathonian)

Middle–lower Bathonian microfaunas and microfloras are diverse and abundant and are commonly recorded from calcareous argillaceous lithologies, below the sandstones and limestones of the upper Bathonian. Towards the base of the interval, microfaunas decrease in diversity and abundance.

Ostracoda

Diagnostic species

Bairdia sp. *sensu* Sheppard, *Cytherella* cf. *C. fullonica* Jones & Sherborn *sensu* Ainsworth *et al.*, *Cytherelloidea catenulata* (Jones & Sherborn), *Mandocythere?* sp. A Ainsworth *et al.*, *Marslatourella globulosa* Dépêche, *Merocythere postangusta* Sheppard, *Morkhovenicythereis bouvadensis* (Dépêche), *Monoceratina* sp., *Nophrecythere bessinensis* Dépêche, *N. rimosa* Dépêche, *N.* sp. A Ainsworth *et al.*, *Oligocythereis? lacunosa* Dépêche. *Oligocythereis* sp., *Pontocyprella* sp. cf. *P. aureola* Ljubimova *sensu* Dépêche, *Praeschuleridea confossa* Sheppard, *P.* cf. *P. subtrigonia* (Jones & Sherborn) *sensu* Ainsworth *et al.*, *Progonocythere* sp. A Ainsworth *et al.*, *Polycope* sp. B, *Pseudomacrocypris atypica* Sheppard, *Schuleridea (Eoschuleridea) batei* Dépêche, *S.* (*E.*) sp. A Ainsworth *et al.* and *Tethysia bathonica* Sheppard.

Characteristic species

Bairdia hilda Jones.

Middle–lower Bathonian marine ostracod faunas are generally diverse and well preserved although not usually very abundant. The recorded faunas show strong similarities to those of Middle and Lower Bathonian Lower Fullers Earth and Fullers Earth Rock of Dorset and Bath, southern England, as described by Bate (1978), Sheppard (1981), Bate & Sheppard (1982) and from northern France (Dépêche 1985).

Foraminifera

Diagnostic species

Ammobaculites agglutinans (d'Orbigny), *A. coprolithiformis* (Schwager), *A. fontinensis* (Terquem), *Astacolus* cf. *A. limata* (Schwager), *Citherina serratocostata* (Gümbel), *Eoguttulina oolithica* (Terquem), *Epistomina* cf. *E. callovica* Kaptarenco-Chernousova, *E.* cf. *E. decorata* Kaptarenco-Chernousova, *E.* cf. *E. elshankaensis poltavica* Kaptarenco-Chernousova, *E.* cf. *E peregrina* Kaptarenco-Chernousova, *E. regularis* Terquem, *E. tenuicostata* Bartenstein & Brand, *E. turgidula* Pazdro, *E.* sp. A, *Falsopamula obliqua* (Terquem), *Garantella ampasindavaensis* Espitalé & Sigal, *Haplophragmoides canariensis* (d'Orbigny), *Haplophragmoides* sp., *Lagena globosa* (Montagu), *Lenticulina dictyoides* (Deeke), *L. galeata* (Terquem), *L. major* (Bornemann), *L. volubilis* Dain, *Lingulina longiscata* (Terquem), *Massilina dorsetensis* Cifelli, *Nodosaria obscura* Reuss, *Nubeculinella infraoolithica* (Terquem), *Marginulinopsis* sp., *Ophthalmidium carinatum* Kübler & Zwingli, O. sp. A Coleman, *Planularia eugenii* (Terquem), *Saracenaria oxfordiana* Tappan, *S.* aff. *S. oxfordiana* Tappan, *S.* aff. *S. segmentata* Tappan, *S.* sp. A, *Tetrataxis?* sp. Coleman, *Vaginulina clathrata eypensa* Cifelli and *V. flabelloides* (Terquem).

Characteristic species

Cyclogyra liassina (Terquem), *Eoguttulina liassica* (Strickland), *E. polygona* (Terquem), *Epistomina* ex. gr. *bireticulata* Pazdro, *E. mosquensis* Uhlig, *E. stellicostata* Bielecka & Pozaryski, *E. stelligera* (Reuss), *Frondicularia nympha* Kopik, *Garantella floscula* (Kaptarenco-Chernousova), *Lenticulina ectypa* (Loeblich & Tappan), *L. muensteri* (Roemer) plex., *L. quenstedti* (Gümbel), *L. tricarinella* (Reuss), *L. varians* (Bornemann) plex., *Nodosaria fontinensis* Terquem, *O. strumosum* (Gümbel), *Paalzowella feifeli* (Paalzow), *Planularia beierana* (Gümbel), *P. cordiformis* (Terquem), *P. crepidularis* Roemer, *Reinholdella cebra* Pazdro, *R. crebra alta* Pazdro, *R. media* (Kaptarenco-Chernousova), *Spirillina infima* (Strickland), *S. tenuissima* (Gümbel) and *Spirophthal-*

midium concentricum (Terquem & Berthelin).

Flood occurrences of epistominids dominate the assemblage, in particular *E. mosquensis*, *E. regularis* and *E. stelligera*. Influxes of *Reinholdella crebra*, *R. media* and *Garantella ampasandavaensis* are also common, as are *Lenticulina* spp. This fauna is very similar to that described from the Winterborne Kingston Borehole in Dorset (Coleman 1982) and from the Dorset and Bath areas (Cifelli 1959). It can also be correlated with northwest Europe (Bartenstein & Brand 1937).

Palynomorphs

Diagnostic species

Ctenidodinium aff. *sellwoodii* (Sarjeant) Stover & Evitt, *Hapsidaulax margarethae* Sarjeant and *Meiourogonyaulax reticulata* Dodekova.

Characteristic species

Aldorfia aldorfensis (Gocht) Stover & Evitt, *Carpathodinium predae* (Beju) Drugg, *Chytroeisphaeridia chytroeides* (Sarjeant) Davey, *Ctenidodinium combazii* Dupin, *C. continuum* Gocht, *C. sellwoodii* (Sarjeant) Stover & Evitt, *Diacanthum filapicatum* (Gocht) Stover & Evitt, *Ellipsoidictyum/Valensiella* spp., *Energlynia acollaris* (Dodekova) Sarjeant, *Escharisphaeridium pocockii* (Sarjeant) Erkmen & Sarjeant, *Escharisphaeridium* spp., *Hystrichogonyaulax regalis* (Gocht) Stover & Evitt, *Meiourogonyaulax valensii* Sarjeant, *Mendicodinium groenlandicum* (Pocock & Sarjeant) Davey, *Nannoceratopsis pellucida* (Deflandre) Evitt, *Pareodinia ceratophora* Sarjeant, *P. prolongata* Deflandre, *Sentusidinium rioulti* (Sarjeant) Stover & Evitt, *S.* spp., *Caddesphaera halosa* (Filatoff) Fenton, Neves & Piel, *Callialasporites* spp., *Cerebropollenites mesozoicus* (Couper), *Cibotiumsporites juriensis* (Balme) Filatoff, *Cyathidites australis* Couper, *C. minor* Couper, *Duplexisporites problematicus* (Couper) Playford & Dettmann, *Ischyosporites pseudoreticulatus* (Couper) Döring, *Leptolepidites major* Couper, *Osmundacidites wellmanii* Couper, *Neoraistrickia* spp., *Quadraeculina anellaeformis* Maljavkina, *Botryococcus* Kützing, and disaccate pollen.

A difference between the compositions of the lower Bathonian palynomorph assemblages can be observed across the North Celtic Sea Basin. The lower Bathonian assemblage from the south central part of the basin is dominated by *Ctenidodinium combazii* and species of the miospore genus *Callialasporites*, whereas the assemblage from the northern and eastern parts of the basin is dominated by spores and proximate–proximo-chorate cysts with very few species of *Ctenidodinium* recorded. This may indicate some Tethyan affinites in the southern and central parts of the North Celtic Sea Basin and some Boreal influence further towards the northeastern part, similar to assemblages described by Fenton & Fisher (1978) from southern Europe, Greenland and Arctic Canada.

Association 11 (lowermost Bathonian)

Associations 11 and 12 (lowermost Bathonian and uppermost Bajocian, respectively) are characterized by the occurrences of short-ranging dinocysts. Calcareous microfaunas are generally rare with the exception of influxes of the foraminiferal genus *Reinholdella*.

Ostracoda

Characteristic species

Cardobairdia sp. C Ainsworth *et al.*

This short interval is composed of a single species of *Cardobairdia*, occurring rarely to abundantly throughout the North Celtic Sea. The genus is observed above the relatively rich faunas of the middle–lower Bathonian (Association 10).

Foraminifera

Characteristic species

Cyclogyra liasina (Terquem), *Eoguttulina polygona* (Terquem), *Epistomina mosquensis* Uhlig, *E. regularis* Terquem, *E. stellicostata* (Gümbel), *E. stelligera* (Reuss), *Frondicularia nympha* Kopik, *Lenticulina muensteri* (Roemer) plex., *L.*

quenstedti (Gümbel), *L. tricarinella* (Reuss), *Nodosaria fontinensis* Terquem, *Ophthalmidium strumosum* (Gümbel), *Paalzowella feifeli* (Paalzow), *Planularia crepidularis* Roemer, *Reinholdella crebra alta* Pazdro, *Spirillina tenuissima* Gümbel and *Spirophthalmidium concentricum* (Terquem & Berthelin).

Diversity and abundance of foraminifera decrease from this point and there are no new species recorded. The fauna is dominated by occassional influxes of *Reinholdella crebra alta.*

Palynomorphs
Diagnostic species
Kylindrocysta spinosa Fenton, Neves & Piel.
Characteristic species
Aldorfia aldorfensis (Gocht) Stover & Evitt, *Chytroeisphaeridia chytroeides* (Sarjeant) Davey, *Ctenidinium sellwoodii* (Sarjeant) Stover & Evitt, *Diacanthum filapicatum* (Gocht) Stover & Evitt, *Ellipsoidictyum/Valensiella* spp., *Endoscrinium luridum* (Deflandre) Gocht, *Energlynia acollaris* (Dodekova) Sarjeant, *Escharisphaeridium* spp., *Hystrichogonyaulax pectinigera* (Gocht) Fensome, *H. regalis* (Gocht) Stover & Evitt, *Meiourogonyaulax valensii* Sarjeant, *Nannoceratopsis gracilis* Alberti emend. Evitt, *N. pellucida* (Deflandre) Evitt, *N. spiculata* Stover, *Pareodinia ceratophora* (Deflandre) Gocht, *Caddesphaera halosa* (Filatoff) Fenton, Neves & Piel, *Callialasporites* spp., *Cerebropollenites mesozoicus* (Couper), *Baculatisporites comaumensis* (Cookson) Potonié, *Cyathidites australis* Couper, *C. minor* Couper, *Duplexisporites problematicus* (Couper) Playford & Dettmann, *Ischyosporites pseudoreticulatus* (Couper) Döring, *Leptolepidites major* Couper, *Osmundacidites wellmanii* Couper, *Neoraistrickia* spp., *Botryococcus* Kützing, and disaccate pollen grains.

A transition from ctenidodiniod cysts to those referable to the genus *Nannoceratopsis* marks the incoming of assemblages of lower Middle–Lower Jurassic aspect.

Association 12 (uppermost Bajocian)
This marine association of uppermost Bajocian age is defined on first downhole occurrence of the uppermost Bajocian dinocyst *Acanthaulax crispa* (Wetzel) Lentin & Williams. Ostracods are very rare.

Ostracoda
Characteristic species
Cardobairdia sp. C Ainsworth *et al.*

No new ostracod taxa occur in this interval.

Palynomorphs
Diagnostic species
Acanthaulax crispa (Wetzel) Lentin & Williams.
Characteristic species
Aldorfia aldorfensis (Gocht) Stovert & Evitt, *Carpathodinium predae* (Beju) Drugg, *Ctenidodinium combazii* Dupin, *C. continuum* Gocht, *C. sellwoodii* (Sarjeant) Stover & Evitt, *Diacanthum filapicatum* (Gocht) Stover & Evitt, *Ellipsoidictyum/Valensiella* spp., *Escharisphaeridium* spp., *Endoscrinium luridum* (Deflandre) Gocht, *Energlynia acollaris* (Dodekova) Sarjeant, *Hystrichogonyaulax pectinigera* (Gocht) Fensome, *H. regalis* (Gocht) Stover & Evitt, *Kalyptea stegasta* (Sarjeant) Wiggins, *Korystocysta gochtii* (Sarjeant) Woollam, *Lithodinia jurassica* (Eisenack) Gocht, *Meiourogonyaulax valensii* Sarjeant, *Nannoceratopsis gracilis* (Alberti) Evitt, *N. pellucida* Deflandre, *N. spiculata* Stover, *Pareodinia ceratophora* (Deflandre) Gocht, *Sentusidinium* spp., *Caddesphaera halosa* (Filatoff) Fenton, Neves & Piel, *Callialasporites* spp., *Cerebropollenites mesozoicus* (Couper), *Baculatisporites comaumensis* (Cookson) Potonié, *Cyathidites australis* Couper, *C. minor* Couper, *Duplexisporites problematicus* (Couper) Playford & Dettmann, *Ischyosporites pseudoreticulatus* (Couper) Döring, *Leptolepidites major* Couper, *Neoraistrickia* spp., *Osmundacidites wellmanii* Couper, *Retitriletes austroclavatidites* (Cookson) Döring *et al.*, *Staplinisporites caminus* (Balme) Pocock, *Uvaesporites argenteaeformis* (Bolkhovitina) Schulz, *Botryococcus* Kützing, and disaccate pollen grains.

The diagnostic dinocyst taxon, *Acan-*

thaulax crispa (Wetzel) Lentin & Williams, marking the top appearance downhole of uppermost Bajocian sediments is relatively common in the North Celtic Sea and occurs within a diverse dinocyst assemblage. This species has been previously described in sediments of upper Bajocian age from England and southwest Germany (Fenton 1981). Long-ranging gymnospermous pollen contributes to over 50% of the palynomorph assemblage. The diversity of dinoflagellate cysts in the North Celtic Sea Basin compares well with assemblages described from the British onshore (Riding 1984a).

Association 13 (upper Bajocian)
This association is poorly defined throughout the North Celtic Sea and is based on sparse microfaunal evidence. It occurs between the rich uppermost Bajocian–Bathonian assemblages above and the incoming assemblages of the Aalenian–upper Toarcian.

Ostracoda
Characteristic Species
Cardobairdia sp. C Ainsworth *et al.*, *Cytherella* cf. *C. catenulata* (Jones & Sherborn) *sensu* Ainsworth *et al.*, and *Praeschuleridea* sp. B Ainsworth *et al.*

Throughout the North Celtic Sea the Bajocian yields only rare specimens of *Praeschuleridea* sp. B Ainsworth *et al.* and *Cardobairdia* sp. C Ainsworth *et al.*; the latter extends down from the lowermost Bathonian interval above.

Foraminifera
Diagnostic species
Garantella stellata Kaptarenco-Chernousova and *Guttulina pygmaea* (Schwager).
Characteristic species
Eoguttulina liassica (Strickland), *E. polygona* (Terquem), *Epistomina mosquensis* Uhlig, *Garantella floscula* (Kaptarenco-Chernousova), *Lenticulina ectypa* (Loeblich & Tappan), *L. muensteri* (Roemer) plex., *Reinholdella crebra* Pazdro and *R. media* (Kaptarenco-Chernousova).

There is a decrease in diversity of foraminifera into the Upper Bajocian, with specimens poorly preserved and small. Assemblages are usually dominated by floods of *Reinholdella* spp. and *Garantella* spp. associated with small influxes of *Lenticulina* spp.

Palynomorphs
Characteristic species
Ctenidodinium combazii Dupin, *C. continuum* Gocht, *C. sellwoodii* (Sarjeant) Stover & Evitt, *Diacanthum filapicatum* (Gocht) Stover & Evitt, *Ellipsoidictyum/Valensiella* spp., *Escharisphaeridium* spp., *Endoscrinium luridum* (Deflandre) Gocht, *Energlynia acollaris* (Dodekova) Sarjeant, *Korystocysta gochtii* (Sarjeant) Woollam, *Lithodinia jurassica* (Eisenack) Gocht, *Meiourogonyaulax valensii* Sarjeant, *Nannoceratopsis gracilis* (Alberti) Evitt, *Pareodinia ceratophora* (Deflandre) Gocht, *Sentusidinium* spp., *Caddesphaera halosa* (Filatoff) Fenton, Neves & Piel, *Micrhystridium* spp., *Tasmanites* spp., *Callialasporites* spp., *Cerebropollenites mesozoicus* (Couper), *Baculatisporites comaumensis* (Cookson) Potonié, *Cyathidites australis* Couper, *C. minor* Couper, *Duplexisporites problematicus* (Couper) Playford & Dettmann, *Ischyosporites pseudoreticulatus* (Couper) Döring, *Leptolepidites major* Couper, *Neoraistrickia* spp., *Osmundacidites wellmanii* Couper, *Quadraeculina anellaeformis* Maljavkina, *Retitriletes austroclavatidites* (Cookson) Döring *et al.*, *Staplinisporites caminus* (Balme) Pocock, *Uvaesporites argenteaeformis* (Bolkhovitina) Schulz, *Botryococcus* Kützing, and disaccate pollen grains.

Although no new taxa appear downhole, abundances of the dinocyst genus *Nannoceratopsis* become important constituents of the assemblage.

Association 14 (lower Bajocian)
This marine association is defined by a suite of commonly occurring dinoflagellate cysts, in conjuction with rare non-age-diagnostic pyritized microfaunas. Abundant miospores

and gymnospermous pollen indicate a strong terrestrial input.

Ostracoda
Characteristic species
Cytherella cf. *C. catenulata* (Jones & Sherborn) *sensu* Ainsworth *et al.* and *Praeschuleridea* sp. B Ainsworth *et al.*

The ostracod fauna is similar to that in the overlying upper Bajocian interval—sparse and poorly diverse.

Foraminifera
Diagnostic species
Planularia cordiformis Bartenstein.
Characteristic species
Epistomina mosquensis Uhlig, *Eoguttulina liassica* (Strickland) and *L. muensteri* (Roemer) plex.

Faunas are impoverished and poorly preserved.

Palynomorphs
Diagnostic species
Pareodinia evitti (Pocock) Wiggins.
Characteristic species
Mancodinium semitabulatum Morgenroth, *Nannoceratopsis dictyambonis* (Drugg) Riding, *N. gracilis* (Alberti) Evitt, *N. spiculata* Stover, *Pareodinia ceratophora* (Deflandre) Gocht, *P. prolongata* Sarjeant, *Scriniocassis weberi* Gocht, *Sentusidinium* spp., *Wallodinium cylindricum* (Habib) Duxbury, *Caddesphaera halosa* (Filatoff) Fenton, Neves & Piel, *Cymatiosphaera* spp., *Micrhystridium* spp., *Araucariacites australis* Cookson, *Callialasporites* spp., *Cerebropollenites mesozoicus* (Couper), *Classopollis* spp., *Duplexisporites problematicus* (Couper) Playford & Dettmann, *Ischyosporites pseudoreticulatus* (Couper) Döring, *Leoptolepidites major* Couper, *Neoraistrickia* spp., *Osmundacidites wellmanii* Couper, *Retitriletes austroclavatidites* (Cookson) Döring *et al.*, *Staplinisporites caminus* (Balme) Pocock, *Uvaesporites argenteaeformis* (Bolkhovitina) Schulz, and disaccate pollen grains.

Miospores contribute significantly to the assemblage with an overall dominance of gymnospermous pollen. The association compares well with lower Bajocian–Aalenian assemblages from Britain. *Wallodinium cylindricum* has been described by Riding (1984b, 1987) as a good marker for the lower Middle Jurassic and uppermost Lower Jurassic in England, and this appears to be true also in the Irish offshore. The occurrence of *Nannoceratopsis dictyambonis* associated with *N. gracilis/senex* and *N. spiculata*, is typical of the lower Bajocian in Europe (Riding 1987).

Association 15 (Aalenian–upper Toarcian)
Throughout the North Celtic Sea and Fastnet Basins, a very distinctive Aalenian–upper Toarcian association is recognized by diverse and abundant calcareous microfaunas; however, dinocysts are rare.

Ostracoda
Diagnostic species
Aphelocythere tenuicostata Ainsworth, *Bairdia? eirensis* Ainsworth, *B. ohmerti* Knitter, *Camptocythere mediofoveolata* Ainsworth, *Cytherella? depressum* Ainsworth, *Cytherelloidea lordi* Ainsworth, *Cytheropteron alafastigatum* Fisher, *C. bispinosum cribrum* Fisher, *Ernstella anterocostata* (Boomer), *E. lanceolata* (Boomer), *Liasina? cylindrica* Ainsworth, *Monoceratina striata* Triebel & Bartenstein, *Nudacythere bizoni* (Ainsworth), *Otocythere callosa* Triebel & Klingler, *Paracypris* cf. *P. bajociana* Bate *sensu* Ainsworth, *Platella toarcensis* Ainsworth, *Polycope pelta* Fisher, *P. transversicostata* Ainsworth, *Praeschuleridea arguta arguta* Ainsworth, *P. arguta magna* Ainsworth, *P. ellipsoidea* Ainsworth, *P. foveolata* Ainsworth, *P. whatleyi* Ainsworth, *P.* cf. *P. pseudokinkelinella* Bate & Coleman *sensu* Ainsworth, *P.* sp. A Ainsworth, *Procytherura celtica* Ainsworth, *P. eugyphea* Ainsworth, *P. multicostata* Ainsworth, *P. reticulata* Ainsworth, *P.? costata* Ainsworth, *P.? exquista* Ainsworth, *P.? liassica* Ainsworth, *Pseudomacrocypris* sp. A Ainsworth, *Rutlandella batei* Ainsworth, *R. transversiplicata* Bate & Coleman and *Wellandia ornata* Ainsworth, Indet. Gen. 1 sp. A Ainsworth, Indet. Gen. 2 sp. A Ainsworth, Indet. Gen. 3 sp. A Ainsworth.

Characteristic species
Bairdiacypris rectangularis Ainsworth, *B. triangularis* Ainsworth, *Cardobairdia fastnetensis* Ainsworth, *C. toarcensis* Ainsworth, *Cytherella apostolescui* Ainsworth, *Ernstella* cf. *E. intrepida* (Bate & Coleman *sensu* Ainsworth), *Kinkelinella sermoisensis* Apostolescu and *K. costata* Knitter, Indet. Gen. 4 sp. A Ainsworth.

This interval yields an extremely diverse ostracod fauna, particularly with regard to the Cytheruridae and Schulerideidae. Specimens are generally well preserved and sometimes abundant. In the Fastnet Basin, both diversity and abundance are greater than in the North Celtic Sea and this is most likely because of the less-indurated pyritic dark-grey argillaceous sequences.

The ostracods could not be used to place the Aalenian–Toarcian boundary owing to their long stratigraphical ranges, the rare occurrences of short-ranging taxa, and the fact that much of fauna is new. Of the 52 taxa recorded, 38 are new, with the remainder described from the Aalenian and Toarcian of England (Lord 1978, Bate & Coleman 1975, Coleman 1980), France (Apostolescu 1959, 1961), northern Germany (Plumhoff 1963, Knitter 1983, Riegraf 1985) and Portugal (Exton 1979).

Foraminifera
Diagnostic species
Cytherina cf. *C.* sp. A Copestake & Johnson, *Dentalina ensis* Wisniowski, *D. sinemurensis* Terquem, *D. ventricosa* Franke, *Frondicularia brizaeformis* Bornemann, *Lenticulina bocardi* (Terquem), *Lenticulina dorbignyi* (Roemer), *Nodosaria regularis regularis* Terquem, *N. regularis* subsp. A Copestake & Johnson, *Nubeculinella tibia* Parker & Jones, *Spiroloculina inopinata* Wood & Barnard, *Vaginulina clathrata* (Terquem), *V. clausa* (Terquem) and *V. hechti* Bartenstein.
Characteristic species
Ammodiscus asper (Terquem), *Astacolus minuta* Bornemann, *Dentalina paucicurvata* Franke, *D. pseudocommunis* Franke, *Eoguttulina bilocularis* (Terquem), *E. liassica* (Strickland), *Lenticulina gottingensis* Bornemann, *L. muensteri* (Roemer) plex., *L. polygonata* Franke, *L. varians* (Bornemann) plex., *Lingulina cernua* (Berthelin), *L. nodosaria* (Terquem), *Nodobacularia* sp., *Nodosaria dispar* Franke gr., *Ophthalmidium orbiculare* Burbach, *Planularia crepidularis* Roemer, *Pseudonodosaria vulgata* (Bornemann) gr., *R. dreheri* (Bartenstein) and *R. macfadyeni* (Ten Dam).

Compared with the overlying lower Bajocian, foraminifera are much more diverse and numerous. Foraminiferal assemblages are dominated by *Lenticulina* and *Reinholdella* spp. There is some similarity between the assemblage found in these two basins and that described from the Jurassic of the southern Jura mountains (France) (Wernli & Septfontaine 1971) and the Prealpes Medianes Plastiques romandes (Switzerland) (Wernli 1971), and the Lower Inferior Oolite of the Cotswolds as described by Morris (1982) although the fauna in the North Celtic Sea and Fastnet Basins is less diverse and abundant than the fauna of Morris.

Palynomorphs
Diagnostic species
Eyachia prisca Gocht, *Nannoceratopsis plegas* Drugg, *Parvocysta barbata* Bjaerke, *P. bullula* Bjaerke, *Maumia verrucata* Pocock, *Neoraistrickia gristhorpensis* (Couper) Tralau.
Characteristic species
Escharisphaeridium spp., *Mancodinium semitabulatum* Morgenroth, *Nannoceratopsis dictyambonis* (Drugg) Riding, *N. gracilis/senex* (Alberti) Evitt & van Helden, *N. triceras* Drugg, *Scriniocassis weberi* Gocht, *Sentusidinium* spp., *Wallodinium cylindricum* (Habib) Duxbury, *Caddesphaera halosa* (Fenton, Neves & Piel), *Micrhystridium* spp., *Tasmanites* spp., *Araucariacites australis* Cookson, *Callialasporites* spp., *Cerebropollenites mesozoicus* Couper, *Classopollis* spp., *Duplexisporites problematicus* (Couper) Playford & Dettmann, *Osmundacidites wellmanii* Couper, *Retitriletes austroclavatidites* (Cookson) Döring *et al.*, *Staplinisporites caminus* (Balme) Pocock, disaccate pollen and *Botryococcus* sp. Kützing.

The Aalenian Stage may be distinguished in a limited number of wells in the southern part of the North Celtic Sea Basin, based on recovery of the dinocyst taxon *Nannoceratopsis plegas*. Elsewhere an Aalenian association cannot be separated from an Aalenian–upper Toarcian association. Dinocyst assemblages of Aalenian age in northwest Europe show strong similarity to assemblages from the above Bajocian and underlying Toarcian.

LOWER JURASSIC

Nine biostratigraphical associations (Associations 16–24) can be recognized within the Lower Jurassic of the North Celtic Sea and Fastnet Basins. The lower Toarcian is characterized by influxes of foraminiferal genus *Reinholdella*, in association with *Tasmanites* (green algae). The upper Pliensbachian is marked by abundant microfaunas and dinocysts, whereas the lower Pliensbachian generally yields only rare assemblages. The Sinemurian microfaunas are diverse and abundant in the central and southern regions of the Fastnet Basin; however, they are generally very impoverished in the North Celtic Sea. The lowermost Sinemurian–uppermost Hettangian interval yields very abundant microfaunas. This is replaced in the Hettangian by rare marine but common non-marine ostracods and palynomorphs. The lowermost Hettangian-upper Rhaetian is characterized by an abundant ostracod fauna and palynomorphs.

Association 16 (lower Toarcian)

The lower Toarcian is characterized by rare ostracods, an abundance of the foraminiferal genus *Reinholdella*, in association with influxes of *Tasmanites* spp. (green algae).

Ostracoda

Characteristic species

Bairdiacypris rectangularis Ainsworth, *B. triangularis* Ainsworth, *Cardobairdia fastnetensis* Ainsworth, *C. toarcensis* Ainsworth, *Cytherella apostolescui* Ainsworth, *Ernstella* cf. *E. intrepida* (Bate & Coleman *sensu* Ainsworth), *Kinkelinella sermoisensis* Apostolescu and *K. costata* Knitter, Indet. Gen. 4 sp. A Ainsworth.

The lower Toarcian is similar to the overlying Bajocian interval and is characterized by a very impoverished ostracod fauna, commonly composed of juvenile pyritized *Bairdiacypris*, *Cardobairdia* and *Cytherella* spp. Species of these genera occur in larger numbers in the overlying Aalenian–upper Toarcian interval.

Foraminifera

Diagnostic species

Astacolus pediacus Tappan, *Nodosaria denticulatacostata* Franke and *Saracenaria subleavis subleavis* Franke.

Characteristic species

Dentalina paucicurvata Franke, *Eoguttulina bilocularis* (Terquem), *E. liassica* (Strickland), *Haplophragmoides kingensis* Tappan, *Lenticulina muensteri* (Roemer) plex., *L. polygonata* (Franke), *L. varians* (Bornemann) plex., *Lingulina nodosaria* (Terquem), *Marginulina prima prima* d'Orbigny, *Nodobacularia* sp., *Nodosaria dispar* Franke gr., *Planularia crepidularis* Roemer, *P. pauperata* Jones & Parker, *Pseudonodosaria oviformis* (Terquem), *P. sexcostata* (Bornemann), *P. vulgata* (Bornemann) gr., *Reinholdella dreheri* (Bartenstein) and *R. macfadyeni* (Ten Dam).

Foraminiferal assemblages are characterized by influxes of a *Nodobacularia* sp. and are again dominated by floods of *Reinholdella* spp. and *Lenticulina* spp.

Palynomorphs

Diagnostic species

Kekryphalospora distincta Fenton & Riding.

Characteristic species

Luehndea spinosa Morgenroth, *Mancodinium semitabulatum* Morgenroth, *Nannoceratopsis gracilis/senex* (Alberti) Evitt & van Helden, *Crassosphaera hexagonalis* Wall, *Cymatiosphaera* spp., *Micrhystridium* spp., *Tasmanites newtoni* Wall, *T. suevicus* (Eisenack) Wall, *Sphaeromorphitae* Downie, Evitt & Sarjeant,

Cerebropollenites mesozoicus (Couper) Nilsson, *Chasmatosporites magnolioides* (Erdmann) Nilsson, *Gliscopollis meyeriana* (Klaus) Venkatachala, *Ischyosporites pseudoreticulatus* (Couper) Döring, *Osmundacidites wellmanii* Couper, and disaccate pollen grains.

An assemblage dominated by the green algae (Prasinophyceae) characteristically occurs within the lower Toarcian sediments, similar to assemblages described from the Posidonia Shale of Germany (Wille 1982) and the Lower Toarcian shales of Yorkshire (Wall 1965). *L. spinosa* is rare within assemblages of lower Toarcian age.

Association 17 (lowermost Toarcian)
This association is based on the very rare occurrence of the foraminifera *Ophthalmidium* sp. 2 (Ruget & Sigal).

Foraminifera
Diagnostic species
Ophthalmidium sp. 2 (Ruget & Sigal).
Characteristic species
Involutina liassica (Jones), *Lenticulina muensteri* (Roemer) plex. and *L. varians* (Bornemann) plex.

Association 18 (upper Pliensbachian)
This interval is characterized by well-preserved, very abundant microfaunas and dinocysts throughout the North Celtic Sea and Fastnet Basins.

Ostracoda
Diagnostic species
Bairdia parva Ainsworth, *B.* aff. *B. molesta* Apostolescu *sensu* Ainsworth, *B.* sp. A *sensu* Ainsworth, *Cardobairdia* sp. A *sensu* Ainsworth, *Hermiella inflata* Ainsworth, *Isobythocypris dorsoconvexa* Ainsworth, *I. pliensbachensis* Ainsworth, *I. tumida* Ainsworth, *Ledahia septenaria* (Gründel), *Liasina vestibulifera* Gramann, *Ogmoconcha* aff. *O. rotunda* Dreyer *sensu* Ainsworth, *Ogmoconchella propinqua* Malz, *O.* aff. *O. aequalis* Herrig *sensu* Ainsworth, *Pseudohealdia etaulensis* (Apostolescu), *P.* cf. *P. etaulensis* (Apostolescu) *sensu* Ainsworth and *P.* sp. A *sensu* Ainsworth.
Characteristic species
Cardobairdia posteroprolata Ainsworth, *C.* sp. K (Apostolescu), *Liasina lanceolata* (Apostolescu), *Ogmoconcha* aff. *O. contractula* Triebel *sensu* Ainsworth, *O.* aff. *O. amalthei* (Quenstedt) *sensu* Ainsworth, *Paracypris* cf. *P. redcarensis* (Blake) *sensu* Ainsworth and *Polycope* aff. *P. cerasia* Blake *sensu* Ainsworth.

The upper Pliensbachian ostracods of Association 18 are diverse and often very abundant throughout both basins. The assemblage is generally dominated by the Healdiidae and/or the Bairdiidae. In the topmost horizons, the genus *Hermiella* is observed. It is quickly replaced downhole by *Ogmoconchella propinqua* Malz. Towards the base of the upper Pliensbachian, faunal diversity and abundance rapidly decrease.

Many of the ostracods are new, occurring in only these two basins. The few species with a wider geophraphical distribution have been recorded from the Pliensbachian of England (Lord 1971), France (Donze 1985), Germany (Gründel 1964, Malz 1971), Denmark (Michelsen 1975) and Portugal (Exton 1979).

Foraminifera
Diagnostic species
Astacolus cf. *A. denticula-carinata* (Franke), *Frondicularia terquemi* subsp. B Copestake & Johnson, *Planularia pseudocrepidularia filosa* (Terquem).
Characteristic species
Astacolus minuta Bornemann. *Dentalina terquemi* d'Orbigny, *Eoguttulina liassica* (Strickland), *Frondicularia terquemi bicostata* d' Orbigny, *F. terquemi sulcata* Bornemann, *F. terquemi terquemi* d'Orbigny, *Haplophragmoides kingensis* Tappan, *Involutina liassica* (Jones), *Lenticulina gottingensis* Bornemann, *L. muensteri* (Roemer) plex., *L. varians* (Bornemann) plex., *Lingulina nodosaria* (Terquem), *L. tenera pupa* (Terquem), *L. tenera tenera* (Bornemann), *Marginulina prima interrupta* (Terquem), *M. prima praerugosa* Nørvang, *M. prima prima*

d'Orbigny, *Marginulinopsis vetusta* (d'Orbigny), *Ophthalmidium orbiculare* Burbach, *Pseudonodosaria sexcostata* (Bornemann), *P. vulgata* (Bornemann) gr., *Reinholdella macfadyeni* (Ten Dam) and *Vaginulina listi* (Bornemann).

There is quite a rich fauna of typically Middle Liassic foraminifera, similar to that described from the Mochras Borehole (Copestake & Johnson 1981), Britain (Barnard 1950, 1956, 1957), western Scania, Sweden (Norling 1972), Jutland (Nørvang 1957) and from northwest Germany (Bartenstein & Brand 1937).

Palynomorphs
Characteristic species
Luehndea spinosa Morgenroth, *Mancodinium semitabulatum* Morgenroth, *Nannoceratopsis gracilis/senex* (Alberti) Evitt & van Helden, *Micrhystridium* spp., *Cerebropollenites mesozoicus* (Couper) Nilsson, *Chasmatosporites magnolioides* (Erdmann) Nilsson, *Gliscopollis meyeriana* (Klaus) Venkatachala, *Ischyosporites pseudoreticulatus* (Couper) Döring, *Osmundacidites wellmanii* Couper, *Quadraeculina anellaeformis*, and disaccate pollen grains.

The upper Pliensbachian association is commonly recorded in most wells throughout both basins. It is easily recognized by the occurrence in large numbers of the dinocyst taxon *Luehndea spinosa*, often comprising up to 50% of the dinocyst component of the palynomorph assemblage. The upper Pliensbachian association is very similar to UK assemblages, but of very low diversity in comparison with German upper Pliensbachian dinocyst assemblages (Morgenroth 1970).

Association 19 (lower Pliensbachian)
This association is generally marked by rare foraminifera and ostracods throughout both basins. In the Cities Service 63/4-1 well situated in the Fastnet Basin, however, foraminifera are abundant and diverse.

Ostracoda
Characteristic species
Cardobairdia posteroprolata Ainsworth, *C.* sp. K (Apostolescu), *Liasina lanceolata* (Apostolescu), *Ogmoconcha* aff. *O. contractula* Triebel *sensu* Ainsworth, *O.* aff. *amalthei* (Quenstedt) *sensu* Ainsworth, *Paracypris* cf. *P. redcarensis* (Blake) *sensu* Ainsworth and *Polycope* aff. *P. cerasia* Blake *sensu* Ainsworth.

In total contrast with the lower Pliensbachian of northwest Europe, the ostracod faunas are very impoverished in both diversity and abundance, with all the recorded species extending down from the upper Pliensbachian interval.

Foraminifera
Diagnostic species
Conicospirillina trochoides (Berthelin), *Lingulina* cf. *L. gottingensis* Franke, *Marginulina lamellosa* (Terquem & Berthelin) and *Spiroloculina aspera* Terquem & Berthelin.
Characteristic species
Ammodiscus asper (Terquem), *Astacolus minuta* Bornemann, *Eoguttulina bilocularis* (Terquem), *E. liassica* (Strickland), *Frondicularia brizaeformis* Bornemann form C Barnard, *F. paradoxa* Berthelin, *F. terquemi bicostata* d'Orbigny, *F. terquemi sulcata* Bornemann, *F. terquemi terquemi* d'Orbigny, *Involutina liassica* (Jones), *Lenticulina muensteri* (Roemer) plex., *L. polygonata* Franke, *L. varians* (Bornemann) plex., *L. tenera pupa* (Terquem), *L. tenera pupoides* (Nørvang), *L. tenera tenuistriata* (Nørvang), *Marginulina prima praerugosa* Nørvang, *M. prima prima* d'Orbigny, *Marginulinopsis matutina* (d'Orbigny), *M. vetusta* (d'Orbigny), *Nodosaria dispar* Franke gr., *N. fontinensis* Terquem, *Pseudonodosaria multicostata* (Bornemann), *P. sexcostata* (Bornemann), *P. vulgata* (Bornemann) gr., *Reinholdella macfadyeni* (Ten Dam), *Vaginulina listi* (Bornemann), *Vaginulinopsis exarata* (Terquem) and *Verneuilinoides mauritii* (Terquem).

In most wells in the North Celtic Sea and Fastnet Basins the lower Pliensbachian yields impoverished assemblages. The poor recovery of benthonic foraminifera is an indication of stagnant conditions after upper Sinemurian shallowing and is recorded throughout the Lias of Britain and northwestern Europe (Copestake & Johnson 1981). In the Cities Service 63/4-1 well the assemblage is diverse and abundant, owing to less restricted marine conditions. This fauna is similar to that described by Barnard (1950, 1956, 1957), from the Lias of Britain by Nørvang (1957) from Jutland and by Norling (1972) from western Scania, Sweden.

Association 20 (upper Sinemurian)

Microfaunas belonging to Association 20 are generally very impoverished in the North Celtic Sea; however, within the central and southern regions of the Fastnet Basin, microfaunas are diverse and abundant. Rare dinocysts diagnostic of Sinemurian sediments are recovered.

Ostracods

Dignostic species

Acrocythere? cf. *A.? tricostata* Michelsen, *Bairdia carinate* Drexler, *Klinglerella exiloreticulata* Ainsworth, MS., *K. herrigi* Ainsworth, MS., *Lophodentina?* cf. *L.? puchella* Apostolescu *sensu* Donze, *L.* sp. A, *Ogmoconchella danica* Michelsen and *O. mouherensis* (Apostolescu).

Characteristic species

Bairdia molesta Apostolescu, *Cardobairdia* sp. B Ainsworth *et al.*, *Klinglerella lacunosa* Ainsworth, MS., *K. sinemuriana* Ainsworth, MS., *Lophodentina pumicosa* Apostolescu, *Nanacythere (Goniocythere) minor* Michelsen, *Ogmoconchella celticensis* Ainsworth, MS., *Paracypris* aff. *P. redcarensis* Blake *sensu* Ainsworth *et al.*, *Polycope cerasia* Blake, *Pseudomacrocypris* cf. *P. subtriangularis* Michelsen *sensu* Ainsworth *et al.* and *Ruchholzella frequens* Ainsworth, MS.

The top of the upper Sinemurian is marked by the incoming of *Klinglerella* spp. and *Ogmoconchella celticensis* Ainsworth, MS. The mid-Sinemurian is well defined in the Fastnet Basin by the tophole occurrence of *Ruchholzella frequens* Ainsworth, MS., a species possessing an ornament similar to *Cristacythere betzi* (Klingler & Neuweiler 1959) which occurs throughout northwest Europe at this level. Although the upper Sinemurian ostracod fauna contains many new species, the remainder have been recorded and described from the Sinemurian of England (Lord 1978), France (Apostolescu 1959; Donze 1985), northwest Germany (Klingler & Neuweiler 1959), southern Sweden (Sivhed 1977, 1980) and Denmark (Michelsen 1975).

Foraminifera

Diagnostic species

Berthelinella involutina subsp. A, *Dentalina primaeva* (d'Orbigny), *Frondicularia terquemi muelensis* Ruget & Sigal, *Guttulina* sp., *Lingulina longiscata* (Terquem), *L.* aff. *L. nodosaria* (Terquem), *L. tenera occidentalis* (Berthelin), *L. tenera praepupa* (Nørvang), *L.* cf. *L. tenera substriata* (Nørvang), *Marginulina prima spinata* (Terquem) and *Saracenaria* aff. *S. segmenta* Tappan.

Characteristic species

Ammodiscus asper (Terquem), *Berthelinella involutina involutina* (Terquem), *Brizalina liassica* (Terquem), *Dentalina paucicurvata* Franke, *D. matutina* d'Orbigny, *D. terquemi* d'Orbigny, *Eoguttulina liassica* (Strickland), *Frondicularia brizaeformis* form C Barnard, *F. paradoxa* (Berthelin), *F. terquemi bicostata* d'Orbigny, *F. terquemi sulcata* Bornemann, *F. terquemi* form F Barnard, *Involutina liassica* (Jones), *Lagena aphela* Tappan, *Lenticulina muensteri* (Roemer) plex., *L. polygonata* Franke, *L. varians* (Bornemann) plex., *Lingulina nodosaria* (Terquem), *L. tenera pupa* (Terquem), *L. tenera tenera* (Bornemann), *L. tenera tenuistriata* (Nørvang), *L. tenera pupoides* (Nørvang), *Marginulina prima interrupta* (Terquem), *M. prima prima* d'Orbigny, *M. prima praerugosa* Nørvang, *M. radiata* Terquem, *Marginulinopsis vetusta* (d'Orbigny), *Nodosaria columnaris* Franke, *N. dispar* Franke gr., *N. fontinensis* Terquem, *N. mitis* (Terquem & Berthelin), *Planularia inaequistriata* (Terquem), *P. pauperata* Jones & Parker, *Pseudonodosaria multicostata* (Bornemann), *P. oviformis* (Terquem), *P. sexcostata* (Bornemann), *P. vulgata* (Bornemann), *Spirillina infima* (Strickland), *Vaginulina listi* (Bornemann), *Vaginulinopsis exarata*

(Terquem) and *Verneuilinoides mauritii* (Terquem).

The fauna is abundant and diverse and includes many long-ranging species. It is dominated by the *Lingulina tenera* plexus in association with an abundance of *Vaginulina listi*, *Marginulina radiata* and *Marginulinopsis vetusta*. The assemblage can be compared with the Lias of Britain (Barnard 1949, 1956, 1957, Copestake & Johnson 1981), Jutland (Nørvang 1957) and western Scania, Sweden (Norling 1972).

Palynomorphs

Diagnostic species

Liasidium variabile Drugg.

Characteristic species

Crassosphaera hexagonalis Wall, *Cymatiosphaera* spp., *Tasmanites newtonii* Wall, *T. suevicus* (Eisenack) Wall, *Veryhachium* var. *reductum* (Deunff) Jekhowsky, and disaccate pollen.

Characteristically, assemblages are impoverished and usually contain non-diagnostic long-ranging miospore taxa. *L. variable* Drugg, however, is a good marker species for the upper Sinemurian (Bujak & Williams 1985, Riding 1987). This species has not been recorded outside Europe (Riding 1987).

Association 21 (lower Sinemurian)

This association has similar characteristics to the overlying upper Sinemurian: impoverished microfaunas in the North Celtic Sea, and diverse and abundant microfaunas in the Fastnet Basin. Within this interval the foraminiferal species *Reinholdella margarita margarita* and *Involutina liassica* are abundant.

Ostracoda

Diagnostic species

Acrocythere? cf. *A.? gassumensis* Michelsen, *A.?* sp. A, *Donzocythere* sp. A, *Klinglerella retia* Ainsworth, MS., *K.?* cf. *K.? undata* Apostolescu *sensu* Ainsworth *et al.*, *Ljubimovella? frequens* Donze, *Ruchholzella luxuriosa* (Apostolescu) and *R. moorei* (Jones).

Characteristic species

Bairdia molesta Apostolescu, *B.* cf. *B. carinata undulata* Herrig *sensu* Ainsworth *et al.*, *Cardobairdia* sp. B Ainsworth *et al.*, *Cristacythere costata* Michelsen, *Klinglerella lacunosa* Ainsworth, MS., *K. sinemuriana* Ainsworth, MS., *Lophodentina pumicosa* Apostolescu, *Nanacythere* (*Goniocythere*) *minor* Michelsen, *Ogmoconchella celticensis* Ainsworth, MS., *Paracypris* aff. *P. redcarensis* Blake *sensu* Ainsworth *et al.*, *Polycope cerasia* Blake, *Pseudomacrocypris* cf. *P. subtriangularis* Michelsen *sensu* Ainsworth *et al.*, and *Ruchholzella frequens* Ainsworth, MS.

As in the overlying upper Sinemurian, ostracods are more commonly recovered in the Fastnet Basin than in the North Celtic Sea, with their occurrences limited to the interbedded argillaceous horizons between the sandstones.

The ostracod faunas bear closest affinities to those of the onshore Sinemurian of northern and central France (Donze 1985). Fewer similarities are seen with those described from England (Lord 1971, 1978), southern Sweden (Sivhed 1977, 1980) and Denmark (Michelsen 1975).

Foraminifera

Diagnostic species

Astacolus semireticulata (Fuchs), *Bullopora globulata minima* Adams, *Frondicularia brizaeformis* form E Barnard, *Marginulina/Saracenaria hamus* (Terquem), *Neobulimina* sp. 2 Bang, *Nodosaria metensis* Terquem and *Spirillina tenuissima* Gümbel.

Characteristic species

Ammodiscus asper (Terquem), *Berthelinella involutina involutina* (Terquem), *Brizalina liassica* (Terquem), *Dentalina matutina* d'Orbigny, *D. pseudocommunis* Franke, *Eoguttulina liassica* (Strickland), *Frondicularia brizaeformis* form C Barnard, *F. brizaeformis* form F Barnard, *F. terquemi sulcata* Bornemann, *F. paradoxa* (Berthelin), *Involutina liassica* (Jones), *Lagena aphela* Tappan, *Lenticulina muensteri* (Roemer) plex., *Lingulina cernua* (Berthelin), *L. tenera pupa* (Terquem), *L. tenera subprismatica* Franke, *L. tenera tenuistriata* Nørvang, *Marginulina radiata* Terquem, *Marginulinopsis vetusta* (d'Orbigny), *Nodosaria columnaris* Franke, *N. dispar* Franke gr., *N. fontinensis* Terquem, *N. mitis* (Terquem & Berthelin), *N. nitidana* Brand, *Oberhausella mesotriassica* (Oberhauser), *Planularia inaequistriata* (Terquem), *Pseudonodosaria vulgata* (Bornemann), *Reinholdella margarita margarita* (Terquem), *Spirillina infima* (Strickland), *Vaginulina listi* (Bornemann) and *Vaginulinopsis exarata* (Terquem).

The foraminiferal assemblage is similar to that described from Jutland (Nørvang 1957) and from western Scania, Sweden (Norling 1972). The influxes of *Involutina liassica* are important events in the lower Sinemurian and can be correlated with similar intervals in Britain (Copestake & Johnson 1981), where a similar fauna is also described.

Association 22 (lowermost Sinemurian–uppermost Hettangian)

This very distinctive association occurs throughout both the North Celtic Sea and the Fastnet Basins. It is dated by moderately diverse but very abundant marine ostracods, in conjunction with generally long-ranging foraminifera.

Ostracoda

Diagnostic species

Donzocythere cf. *D. convergens* (Donze *sensu* Donze), *Isobythocypris* cf. *I. cylindrica* Herrig *sensu* Ainsworth *et al.*, *Ogmoconcha hagenowi* Drexler, *Ogmoconchella* cf. *O. ellipsoidea* (Jones) *sensu* Ainsworth *et al.* and *O. michelseni* Ainsworth.

Characteristic species

Bairdia molesta Apostolescu, *B.* cf. *B. carinata undulata* Herrig *sensu* Ainsworth *et al.*, *Cardobairdia* sp. *B sensu* Ainsworth *et al.*, *Klinglerella translucens* (Blake), *Ogmoconchella ellipsoidea* (Jones), *Paracypris* aff. *P. redcarensis* Blake *sensu* Ainsworth *et al.*, *Polycope cerasia* Blake and *Pseudomacrocypris* cf. *P. subtriangularis* Michelsen *sensu* Ainsworth *et al.*

This interval is dominated by flood occurrences of *Ogmoconcha* and in particular *Ogmoconchella* throughout both basins. Many of the above species have been well described throughout the Hettangian and lower Sinemurian of northern Europe (Lord 1978, Donze 1985, Michelsen 1975, Malz 1971, Drexler 1958).

Foraminifera

Diagnostic species

Lingulina cf. *L. tenera pupoides* (Nørvang), *Planularia nucleata* (Terquem).

Characteristic species

Involutina liassica (Jones), *Lagena aphela* Tappan, *Lenticulina muensteri* (Roemer) plex., *L. polygonata* (Franke), *L. varians* (Bornemann) plex., *Lingulina tenera substriata* (Nørvang), *L. tenera tenera* (Bornemann), *Oberhausella mesotriassica* (Oberhauser), *Reinholdella margarita margarita* (Terquem) and *Vaginulinopsis exarata* (Terquem).

Palynomorphs

Characteristic species

Cerebropollenites mesozoicus (Couper) Nilsson, *Chasmatosporites* spp., *Classopollis* spp., *Deltoidospora* spp., *Duplexisporites problematicus* (Couper) Playford & Dettmann, *Gliscopollis meyeriana* (Klaus) Venkatachala, *Kraeuselisporites reissingeri* (Harris) Morbey, *Lycopodiacidites rugulatus* (Couper) Schulz, *Microreticulatisporites fuscus* (Nilsson) Morbey, *Polycingulatisporites circulus* (Simoncsics & Kedves), *P. liassicus* Schulz, *Quadraeculina anellaeformis* Maljavkina, *Uvaesporites argenteaeformis* (Bolkhovitina) Schulz, disaccate pollen grains, *Crassosphaera hexagonalis* Wall, *Leiosphaeridia* spp., *Micrhystridium* spp., *Veryhachium* spp., and foraminiferal test linings.

This association is characterized by poor recovery of palynomorphs; this interval is devoid of dinocysts. The association can be correlated with early Jurassic assemblages from northwest Europe, described by Morbey & Dunay (in: Thusu 1978).

Association 23 (Hettangian)

This fluctuating non-marine–marine association is of Hettangian age and extends throughout the North Celtic Sea and Fastnet Basins. It is characterized by freshwater and brackish ostracods in the lower and middle levels of the 'Liassic Limestone', with rare marine ostracod faunas occurring in the upper horizons.

Ostracoda

Diagnostic species

Darwinula hettangiana Ainsworth, *D. rara* Ainsworth, *Lutkevichinella fastigata* Ainsworth and *L. hortonae* Ainsworth.

Characteristic species

Klinglerella translucens (Blake).

Within many of the wells studied, ostracod abundance is high; however, diversity is low (fewer than five species), with generally poorly preserved carapaces. Faunas are most commonly recorded from the middle and lower horizons of the 'Liassic Limestone', from within the softer 'marly' horizons rather than from the thicker well-cemented oolitic, pelloidal and micritic limestones. The faunas in the lower and middle horizons are dominated by *Darwinula* and *Lutkevichinella*, whereas the upper sections of the 'Liassic Limestone' yield rare to moderately common specimens of *Klingerella translucens* (Blake).

Of the five species seen in this interval, only two species (*Lutkevichinella hortonae* Ainsworth and *Klingerella translucens* (Blake)) occur outside this region, the former from the Hettangian of the Paris Basin and central France (Apostolescu 1959, 1961) and the latter from the Hettangian of northern England (Lord 1971) and central France (Donze 1985).

Palynomorphs

Diagnostic species

Dapcodinium priscum Evitt.

Characteristic species

Cerebropollenites spp., *Chasmatosporites magnoloides* (Erdmann) Nilsson, *Classopollis* spp., *Deltoidospora* spp., *Duplexisporites problematicus* (Couper) Playford & Dettmann, *Gliscopollis meyeriana* (Klaus) Venkatachala, *Kraeuselisporites reissingeri* (Harris) Morbey, *Lycopodiacidites rugulatus* (Couper) Schulz, *Microreticulatisporites fuscus* (Nilsson) Morbey, *Polycingulatisporites circulus* Simoncsics & Kedves, *P. liassicus* Schulz, *Porcellispora longdonensis* (Clarke) Morbey, *Quadraeculina anellaeformis* Maljavkina and species of *Leiosphaeridia* and *Micrhystridium*.

This association is of low diversity with respect to dinocyst recovery. This is in agreement with Hettangian palynofloral diversity throughout northwest Europe (Morbey 1978). The cosmopolitan marker species for the Hettangian, *Dapcodinium priscum*, is a rare component of the Irish offshore assemblages.

Association 24 (lowermost Hettangian–upper Rhaetian)

This marine association consists of abundant ostracods and acritarch dominated palynomorph assemblages of a lowermost Hettangian–upper Rhaetian age.

Ostracoda

Diagnostic species

Ogmoconchella serratostriata Ainsworth, *Klinglerella* sp. and *Cytherella* sp. A Ainsworth & Horton.

Characteristic species

Ogmoconchella ellipsoidea (Jones)

Ostracods of Association 24 are numerically abundant and yet poorly diverse, and they are dominated by *Ogmoconchella serratostriata* Ainsworth, with rare specimens of *Klinglerella* sp. and *Cytherella* sp. A. These ostracods are extracted from shale horizons sandwiched between the 'Langport Member' and the thickly bedded limestones pertaining to the 'Laissic Limestone'. Where the shales have been compacted, ostracod abundance is low. At present no ostracods have been recovered from the 'Langport Member' because of the difficulty of extraction from the well-cemented limestone lithologies.

Palynomorphs

Characteristic species

Acanthotriletes ovalis Nilsson, *A. varius* Nilsson,

Carnisporites megaspiniger Morbey, *C. spiniger* (Leschik) Morbey, *Cerebropollenites* spp., *Chasmatosporites* spp., *Deltoidospora* spp., *Kraeuselisporites reissingeri* (Harris) Morbey, *Microreticulatisporites fuscus* (Nilsson) Morbey, *Polycingulatisporites circulus* Simoncsics & Kedves, *P. liassicus* Schulz, *Lunatisporites rhaeticus* (Schulz) Warrington, *Porcellispora longdonensis* (Clarke) Scheuring emend. Morbey, *Gliscopollis meyriana* (Klaus) Venkatachala, *Classopollis* spp., *Quadraeculina anellaeformis* Maljavkina, *Ricciisporites tuberculatus* Lundblad, *Stereisporites* spp., *Uvaesporites argenteaeformis* (Bolkhovitina) Schulz, *Micrhystridium* spp. and *Leiosphaeridia* spp.

Assemblages described from the Hettangian–Rhaetian of onshore Britain by Warrington (1978) are richer and more diverse than those encountered in this study.

TRIASSIC

A single association (Association 25) is described from the Triassic.

Association 25 (Rhaetian)

The Rhaetian association is dominated by dinocysts and miospores, with one very rare ostracod species.

Ostracoda

Diagnostic species

Ogmoconchella martini (Anderson)

A single well (Arco 47/29-1) has yielded rare ostracods, the sole species being *Ogmoconchella martini* Anderson, 1964); this has been recorded from the middle and upper Rhaetian of both England and northern Germany (Bate 1978). This stage is generally devoid of microfauna throughout the North Celtic Sea and Fastnet Basins.

Palynomorphs

Diagnostic species

Rhaetogonyaulax rhaetica (Sarjeant) Loeblich & Loeblich emend. Harland *et al.*, *Granuloperculatipollis rudis* (Venkatachala & Goczan) Morbey, *Ovalipollis pseudoalatus* (Thiergart) Schuurman, *Rhaetipollis germanicus* Schulz, *Tsugaepollenites? pseudomassulae* (Madler) Morbey and *Vesicaspora fuscus* (Pautsch) Morbey.

Characteristic species

Acanthotriletes ovalis Nilsson, *A. varius* Nilsson, *Carnisporites megaspiniger* Morbey, *C. spiniger* (Leschik) Morbey, *Cerebropollenites* spp., *Deltoidospora* spp., *Kraeuselisporites reissingeri* (Harris) Morbey, *Microreticulatisporites fuscus* (Nilsson) Morbey, *Lunatisporites rhaeticus* (Schulz) Warrington, *Porcellispora longdonensis* (Clarke) Scheuring emend. Morbey, *Classopollis* spp., *Ricciisporites tuberculatus* Lundblad and *Uvaesporites argenteaeformis* (Bolkhovitina) Schulz.

This assemblage corresponds well with assemblages from northwest Europe. Dinocyst diversity in the Rhaetian of Europe is very low; however, *Rhaetogonyaulax rhaetica* is often recovered in abundance (Riding 1984a).

DEPOSITIONAL ENVIRONMENTS

Many of the faunas encountered in the Upper Triassic and Jurassic of both the North Celtic Sea and the Fastnet Basins are similar to those described elsewhere from onshore and offshore northwest Europe. The microfossils described above reflect these facies changes, as seen in their relative abundances, diversities and the preservation modes of their constituent species.

Conformably overlying the dominantly fluvial, sabkha and playa-lake red-bed and evaporite sequences of the 'Mercia Mudstone Group' is the Rhaetian 'Penarth Group'. This group comprises a lower sequence of alternating thin micritic limestones and grey to black fissile shales with evaporites ('Westbury Formation–Cotham Member'), above which there is a pale-coloured well-cemented limestone ('Langport Member') (Fig. 2). The lower sequence dated by rare dinocysts and often abundant pollen (Association 25), indicates the earliest onset of the marine transgression, in conjunction with

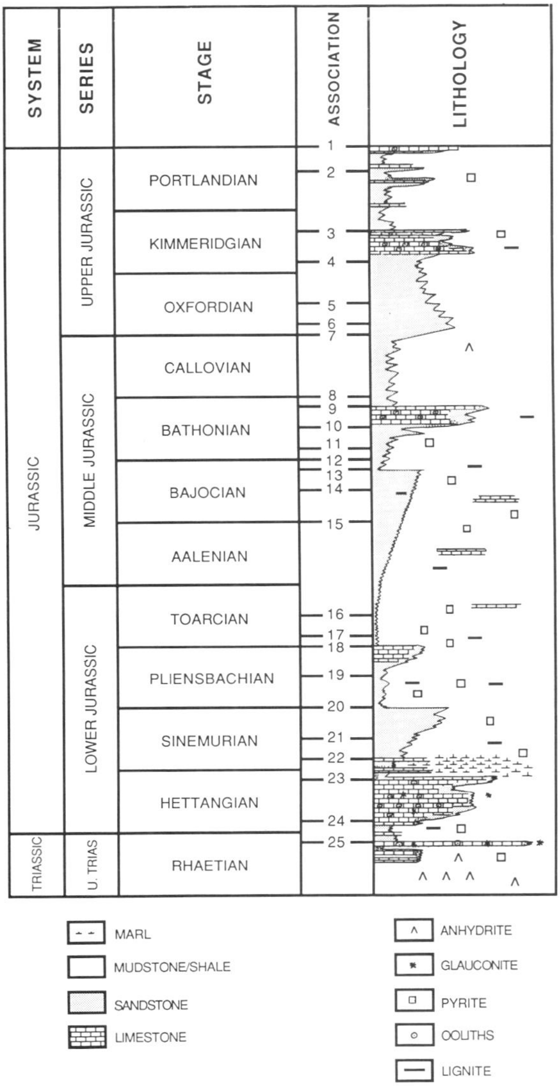

Fig. 2 — Jurassic and Upper Triassic biostratigraphy and generalized lithofacies of the North Celtic Sea and Fastnet Basins.

the gradual downwarping of both basins. Lagoonal conditions are suggested by both the lithologies and the microfloras, with intermittent connections to a shallow open sea, alternating with a drying of the region for the evaporites to form. During the formation of the 'Langport Member', a generally low-energy, warm, very shallow marine environment prevailed. Short higher-energy conditions are suggested by the recovery of rare dinocysts, acanthormorph acritarchs, in association with the rare occurrences of glauconite, oolites and peloids. The absence of microfauna is mainly due to the difficulty of fossil extraction from the well-cemented nature of the limestones. Shallow marine conditions continued into the lowermost Hettangian with the deposition of a moderately thick (less than 89 m) grey to dark-grey mudrock, yielding abundant but often poorly preserved ostracods, normally comprising a single taxon (Association 24) (Fig. 2).

In the Hettangian, a thick (less than 270 m) carbonate succession, the 'Liassic Limestone' (Hettangian Limestone' *sensu* Robinson *et al.* 1981) was deposited. In the North Celtic Sea, this formation possesses much thicker argillaceous units (Fig. 2). The oolitic, pelloidal and micritic limestones yield only very rare dinocysts, suggesting marine turbid carbonate-rich environments. In some of the lower and middle horizons throughout the succession, within the marl lithologies, however, abundant freshwater and brackish ostracod faunas are recorded (Association 23). The occurrence of these faunas indicates an unstable marginal marine platform at this time. Towards the top of the 'Liassic Limestone' in the upper Hettangian, the brackish aspect of the ostracod faunas disappears and is replaced by rare marine ostracods, because the lower Jurassic marine trangression becomes well established throughout the two basins.

Overlying these carbonates is the 'Liassic Marl', a sequence of moderately soft marls and thin limestones of uppermost Hettangian–lowermost Sinemurian age (Fig. 2). Flood abundances of the ostracod genera *Ogmoconcha* and *Ogmoconchella* (Association 22) occur throughout the region. Both lithology and faunas most probably represent a slightly deeper tranquil inner-shelf depositional environment.

These marls are commonly succeeded by a very thick sequence of argillaceous sediments, typical of 'Liassic' lithologies throughout northern Europe (Fig. 2). Within the North Celtic Sea this sequence extends up into the Bajocian. Generally throughout the North Celtic Sea and in the northern region of the Fastnet Basin, the Sinemurian marine inner shelf microfaunas (Associations 21 and 22) are impoverished and poorly preserved, with the exception of the foraminifera *Involutina liassica* (Jones) and *Reinholdella* spp. which occur in abundance within the lower Sinemurian. Microfossil recovery is often poor owing to the indurated nature of the sediments, in association with the dark-grey, often pyritic shales, representing restricted marine conditions. In the south and central regions of the Fastnet Basin, well-preserved, diverse and moderately abundant marine microfaunas occur within a thick (less than 390 m) sequence of sandstones with interbedded mudstones and shales representing delta fringe deposits (Robinson *et al.* 1981, Naylor & Shannon 1982) (Fig. 2). These sandstones are strongly diachronous, thinning rapidly towards the north of the basin.

In all wells excluding the Fastnet Cities Service 63/4-1 well, the lower Pliensbachian is represented by infrequent, often pyritized microfaunas (Association 19), extracted from dark-grey, often pyritic shales and micritic mudstones (Fig. 2). Many of the specimens observed in this sequence are small and/or juvenile forms and extend up into the upper Pliensbachian. This restricted fauna in conjunction with the sediments is thought to represent a moderately deeper-shelf environment, with poor water circulation, forming 'stagnant pools', similar to that described from the Lias of Britain (Copestake & Johnson 1981). In the Cities Service 63/4-1 well, foraminiferal faunas are abundant, indicating a more open marine environment. With the gradual shallowing of the basins in the upper Pliensbachian, marine microfaunas of Association 18 become very abundant, quite diverse and moderately to well-preserved, increasing in all aspects towards the top levels of this interval. Within the topmost intervals, the dinocyst taxon *Luehndea spinosa* Morgenroth is also abundant. The argillaceous sediments are less pyritized and generally more calcareous, with some arenaceous and carbonate stringers, and this sequence is thought to represent a more open marine environment, with moderate to good water circulation.

The lower Toarcian is characterized by a distinct lithological change from grey and dark-grey calcareous shales to pyritic, dark-grey to black shales (Fig. 2). This coincides with an abrupt microfaunal turnover, especially with respect to the ostracod family Healdiidae, which becomes extinct near the base of the lower Toarcian. This interval is dominated by influxes of the foraminiferal genus *Reinholdella*, in association with rarer juvenile specimens of the ostracod genera *Bairdiacypris*, *Cytherella* and *Cardobairdia*, the species of which extend up into the upper Toarcian–Aalenian interval. This faunal turnover is associated with the lower Toarcian eustatic event, with a change from the shallow conditions at the end of the Pliensbachian to the transgressive seas of the Toarcian. During the lower Toarcian, restricted water circulation within 'basinal deeps' led to stagnant bottom waters. Similar associations occur throughout northwest Europe, examples include the Posidonia Shale of southwest Germany (Wille 1982) and the Jet Rock and Bituminous Shale of Yorkshire, northeast England (Lord 1978).

More normal marine conditions gradually spread throughout both basins during the upper Toarcian, continuing into the Aalenian. Rich, diverse and generally well-preserved foraminifera, ostracods and dinocysts (Association 15) are recovered from the commonly calcareous to very calcareous grey shales and claystones, indicating inner-shelf environments throughout the region (Fig. 2). In general, the microfaunas are more diverse and abundant in the Fastnet Basin, where conditions were probably less restricted and/or shallower.

In the Bajocian, sediments predominantly consist of sandstones, pyritic and calcareous mudrocks with argillaceous micrite sequences (Fig. 2). Shallow marine environments, often with long periods of marine stagnation are suggested by the pyritic nature and poor recovery of both foraminifera and ostracods (Associations 13 and 14). The abundant miospores indicate a strong terrestial input at this time. Those wells situated in the most northerly regions of the North Celtic Sea possess a thick siliciclastic dominated interval, again yielding rare faunas and floras. Their consistuent microfossils are less pyritized and more abraded, suggesting less-restricted higher-energy shallow marine environments.

Overlying these sediments is a soft to moderately undurated grey-coloured claystone–shale succession of lower–middle Bathonian age, generally yielding diverse, abundant and well-preserved microfaunas and microfloras (Association 10) (Fig. 2). Abundance and diversity increase up through this sequence. The foraminifera are dominated by flood influxes of the genus *Epistomina*, indicating high-energy shallow-water conditions.

The marls and shales are succeeded by a sequence of carbonates and sandy carbonates of upper Bathonian age. The lower units often comprise coarsening-upward cycles of oolitic and bioclastic sandstones. Above, lie a succession of oolitic sandy micrites, sandy oosparites and oolitic biosparites and micrites. The intercalated marls are thinly developed (Fig. 2). Microfaunas are rare and poorly preserved, because of the difficulty of extraction from the well-cemented limestones. Microfloras are dominated by diverse dinocysts (Association 9). These sediments comprise a series of shallow-shelf marine limestones, indicating the onset of a regressive episode during the Middle Jurassic similar to those of the Great Oolite and Forest Marble of southern England.

The uppermost Bathonian interval is marked by a distinct change in facies and microfaunas. The lower horizons yield sparse brackish and marine ostracods, whereas the upper intervals contain rich brackish and freshwater ostracod assemblages in association with common charophyte oogonia. The occurrence of the freshwater algae and the presence of freshwater and brackish ostracods extracted from darkish-grey claystones indicates the continued regressive trend throughout the Bathonian interval, resembling that described from the upper Bathonian of southern England (Fig. 2).

Throughout much of the Callovian and lower Oxfordian intervals, argillaceous sequences are dominated by red lithologies, with multicoloured sandstones (Fig. 2). Siderite nodules, evaporites and lignitic debris are generally rare. These overall lithologies, together with the often common influxes of charophyte oogonia, suggest continental, freshwater and fluviatile environments. Very rare dinocysts of lowermost Oxfordian and lower Oxfordian age (Associations 5 and 6) are recorded in some of the wells studied, indicating very brief shallow marine incursions within the dominantly non-marine environment.

The upper Oxfordian–lowermost Kimmeridgian interval is characterized by infrequent, poorly diverse but locally abundant ostracod faunas (Association 4). The sediments comprise an alternating sequence of red, yellow, grey and brown mudrocks and sandstones, with stringers of limestone (Fig. 2). As in the underlying Callovian and Oxfordian, rare to common charophyte oogonia are recorded. Throughout the North Celtic Sea, lacustrine through to brackish lagoonal environments are indicated by both lithologies and ostracod faunas, with periodic drying of the lagoonal areas, suggested by the absence of continuous faunas throughout the section.

Throughout the North Celtic Sea, the lower Kimmeridgian–lowermost upper Kimmeridgian comprises a complex suite of microfossils and lithologies. The lowermost intervals are dominated by grey-brown argillaceous sequences, with minor light-coloured sandstones and white to grey, sparry, micritic and sandy limestones (Fig. 2). These generally yield sparse ostracod faunas, mainly comprising of *Theriosynoecum* spp., indicating short freshwater–brackish conditions within a lagoonal environment. These alternate with longer periods of lacustrine deposition, suggested by the absence of microfossils. Ascending the succession, ostracods increase in both diversity and abundance, dominated by *Damonella*, *Darwinula* and *Theriosynoecum*. These genera, likewise, suggest freshwater–brackish salinities within a lagoonal environment. In the middle and upper levels of the lower Kimmeridgian–lowermost upper Kimmeridgian interval, a variety of microfossils are observed, indicating a rapidly fluctuating environment. Through the latter section, diverse and often very abundant freshwater–brackish ostracods occur in association with abundant agglutinating foraminifera, suggesting similar ecological conditions to that of the lower intervals of the lower Kimmeridigian. In between these faunas, are the rare, more fully marine foraminifera (*Lenticulina* spp.), ostracods belonging to the genera *Macrodentina*, *Galliaecytheridea* and *Schuleridea* and common to rare, poorly diverse dinocysts including *Subtilisphaera* spp. These imply minor marine influences within the lagoonal environment. This is further indicated by the presence of echinoderm fragments. In the top sections, the ostracod genus *Fabanella*, a genus with a wide-ranging salinity tolerence, is often very common within the oolitic, pelloidal, micritic, sandy limestones and sandstones, suggesting marginal through to near-normal marine salanities within the lagoon at this time. Towards the basin margins, these limestones are poorly developed and are replaced by fine to coarse clastics, yielding sparse microfaunas and microfloras (Fig. 2).

The basinal limestones and marginal clastics of the lower Kimmeridgian are succeeded by a thick sequence of light to dark mudstones of upper Kimmeridgian to lower Portlandian age, together with thin beds of pale-coloured limestones and fine- to medium-grained sandstones. Thin coals and sphaerosiderites are rare throughout (Fig. 2). These argillaceous sequences yield diverse and abundant ostracods, dominated by the freshwater to brackish marine faunas, many species of which occur below in the lower Kimmeridgian and also above in the upper Portlandian. No true marine ostracod genera occur within this interval. Poorly preserved marine dinocysts are recorded; however, their recovery is variable and may in part indicate that they have been washed into the region. The presence of large numbers of *Damonella* and *Mantelliana* spp., in conjunction with rarer *Cypridea* faunas suggest marginal to near-normal marine salinities within a lagoon.

The overlying light- to dark-grey mudrocks of the upper Portlandian, the 'Lower–Middle Purbeck' equivalent, with subordinate thin to medium argillaceous micrites and oolitic carbonates, yield a very diverse and abundant ostracod fauna throughout the North Celtic Sea (Association 1) (Fig. 2). In the lowermost interval, ostracod faunas represented by the genus *Cypridea* are not common, the interval being dominated by *Damonella* and *Rhinocypris* spp. Increasingly up through this interval, however, well-preserved specimens of *Cypridea* occur in very large numbers and represented mainly by two species: *Cypridea tumenscens praecursor* and *C. peltoides peltoides*. Towards the topmost levels of the upper Portlandian, faunal abundance reaches a peak. Interspersed with the *Cypridea* faunas are the more rarer associations of the genera *Damonella*, *Bisulcocypris*, *Klieana* and *Rhinocypris*. This upper Portlandian sequence is thought to represent a shallowing of the region progressively throughout the interval, with the deposition of faunas, floras and sediments in freshwater and brackish-water lakes and lagoons. The occurrences of the *Damonella*, *Bisulcocypris*, *Klieana* and *Rhinocypris* faunas most probably indicates only a slight increase in salinities within these lakes and lagoons.

Similar faunas and lithologies extend into the lower Berriasian, suggesting comparable environments, before the onset of alluvial freshwater and marginal environments within the upper Berriasian.

ACKNOWLEDGEMENTS

The research summarized in this paper was carried out under contract to the Petroleum

Affairs Division of the Department of Energy, Dublin, Ireland, and is published by permission of the Minister for Energy. The authors wish to thank Dr K. Robinson and Mr N. Murphy for their constructive criticism and Professor C. H. Holland and Dr G. Clayton for advice and comments.

Plate 1 —

Fig. 1 — *Bisulcocypris? dilatata* Anderson, right view of carapace, × 81, upper Portlandian.

Fig. 2 — *Bisulcocypris* aff. *B. verrucosa* (Jones) *sensu* Ainsworth, left view of carapace, × 49, upper Portlandian.

Fig. 3 — *Cetacella* cf. *C. striata* (Helmdach) *sensu* Colin *et al.*, right view of carapace, × 53, upper Portlandian.

Fig. 4 — *Cypridea tumenscens praecursor* Oertli, right view of carapace, × 38, upper Portlandian.

Fig. 5 — *Cypridea* cf. *C. peltoides peltoides* Anderson *sensu* Ainsworth *et al.*, left view of carapace, × 38, upper Portlandian.

Fig. 6 — *Cypridea* sp. A Ainsworth *et al.*, right view of carapace, × 38, upper Portlandian.

Fig. 7 — *Damonella pygmaea* (Anderson), right view of carapace, × 68, upper Portlandian.

Fig. 8 — *Klieana alata* Martin, right view of carapace, × 68, upper Portlandian.

Fig. 9 — *Cetacella armata* Martin, right view of carapace, × 55, lower Portlandian–upper Kimmeridgian.

Fig. 10 — *Damonella* cf. *D. ellipsoidea* (Wolburg) *sensu* Ainsworth *et al.*, right view of carapace, × 54, lower Portlandian–upper Kimmeridgian.

Fig. 11 — *Mantelliana* cf. *M. purbeckensis* (Forbes) *sensu* Ainsworth *et al.*, right view of carapace, × 48, lower Portlandian–upper Kimmeridgian.

Fig. 12 — *Fabanella boloniensis* (Jones), right view of male carapace, × 43, lowermost upper–lower Kimmeridgian.

Fig. 13 — *Macrodentina* (*M.*) cf. *M.*(*M.*) *cicatriosa* Malz *sensu* Colin, Lehmann & Morgan, left view of female carapace × 72, lowermost upper–lower Kimmeridgian.

Fig. 14 — *Scabriculocypris?* sp. A Ainsworth, right view of carapace, × 58, lowermost upper–lower Kimmeridgian.

Fig. 15 — *Theriosynoecum* sp. A Ainsworth *et al.*, right view of female carapace, × 48, lowermost upper–lower Kimmeridgian.

Fig. 16 — *Bisulcocypris levis* Helmdach, right view of male carapace, × 48, lowermost Kimmeridgian–upper Oxfordian.

Fig. 17 — *Cetacella inermis* Martin, right view of carapace, × 62, lowermost Kimmeridgian–upper Oxfordian.

Fig. 18 — *Theriosynoecum hemigymnon* Helmdach, right view of male carapace, × 40, lowermost Kimmeridgian–upper Oxfordian.

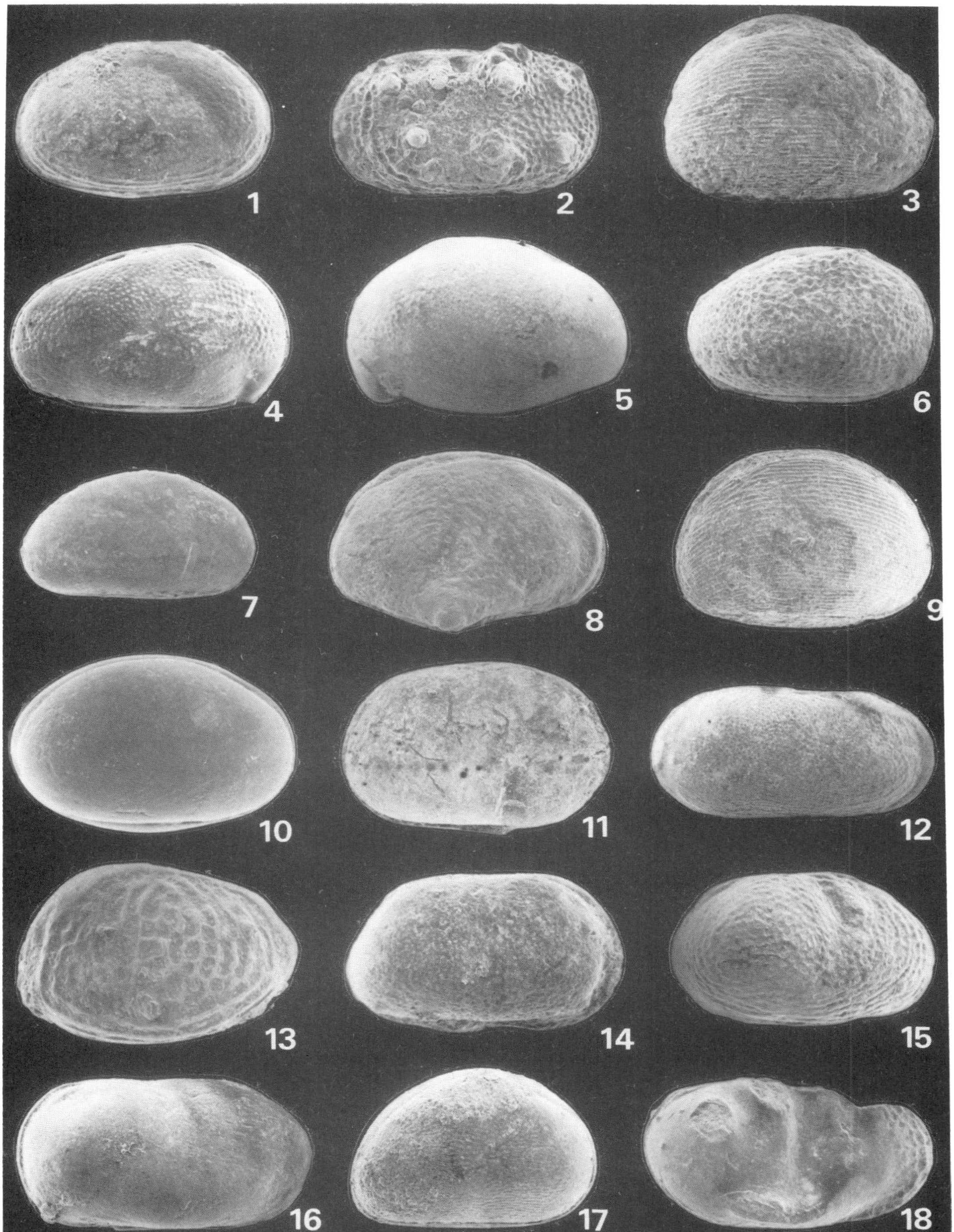
1
2
3
4
5
6
7
8
9
10
11
12
13
14
15
16
17
18

Plate 2 —

Fig. 1 — *Belekocytheridea* cf. *B. punctata* Bate *sensu* Ainsworth *et al.*, left view of carapace, × 93, uppermost Bathonian.

Fig. 2 — *Glyptocythere oscillum* (Jones & Sherborn), left view of female carapace, × 65, uppermost Bathonian.

Fig. 3 — *Klieana levis* Oertli, left view of female carapace, × 56, uppermost Bathonian.

Fig. 4 — *Pichottia muris* Oertli, left view of male carapace, × 56, uppermost Bathonian.

Fig. 5 — *Fastigatocythere juglandia* (Jones), external view of male right valve, × 59, upper Bathonian.

Fig. 6 — *Progonocythere stilla* Sylvester-Bradley, external view of male left valve, × 60, upper Bathonian.

Fig. 7 — *Schuleridea* (*Eoschuleridea*) cf. *S.* (*E.*) *trigonalis* (Jones) *sensu* Ainsworth *et al.*, right view of female carapace, × 72, upper Bathonian.

Fig. 8 — *Marslatourella globulosa* Dépêche, right view of carapace, × 74, middle–lower Bathonian.

Fig. 9 — *Merocythere postangusta* Sheppard, external view of female right valve, × 76, middle–lower Bathonian.

Fig. 10 — *Nophrecythere rimosa* Dépêche, left view of female carapace, × 62, middle–lower Bathonian.

Fig. 11 — *Nophrecythere* sp. A Ainsworth *et al.*, right view of female carapace, × 62, middle–lower Bathonian.

Fig. 12 — *Praeschuleridea confossa* Sheppard, right view of male carapace, × 66, middle–lower Bathonian.

Fig. 13 — *Progonocythere* sp. A Ainsworth *et al.*, right view of male carapace, × 67, middle–lower Bathonian.

Fig. 14 — *Mandocythere?* sp. A Ainsworth *et al.*, right view of carapace, × 62, middle–lower Bathonian.

Fig. 15 — *Schuleridea* (*Eoschuleridea*) *batei* Dépêche, right view of female carapace, × 66, middle–lower Bathonian.

Fig. 16 — *Cardobairdia* sp. C Ainsworth *et al.*, right view of carapace, × 93, uppermost Bathonian.

Fig. 17 — *Praeschuleridea* sp. B Ainsworth *et al.*, left view of carapace, × 71, upper Bajocian.

Fig. 18 — *Cytherelloidea* cf. *C. catenulata* (Jones & Sherborn) *sensu* Ainsworth *et al.*, left view of male carapace, × 87, upper Bajocian.

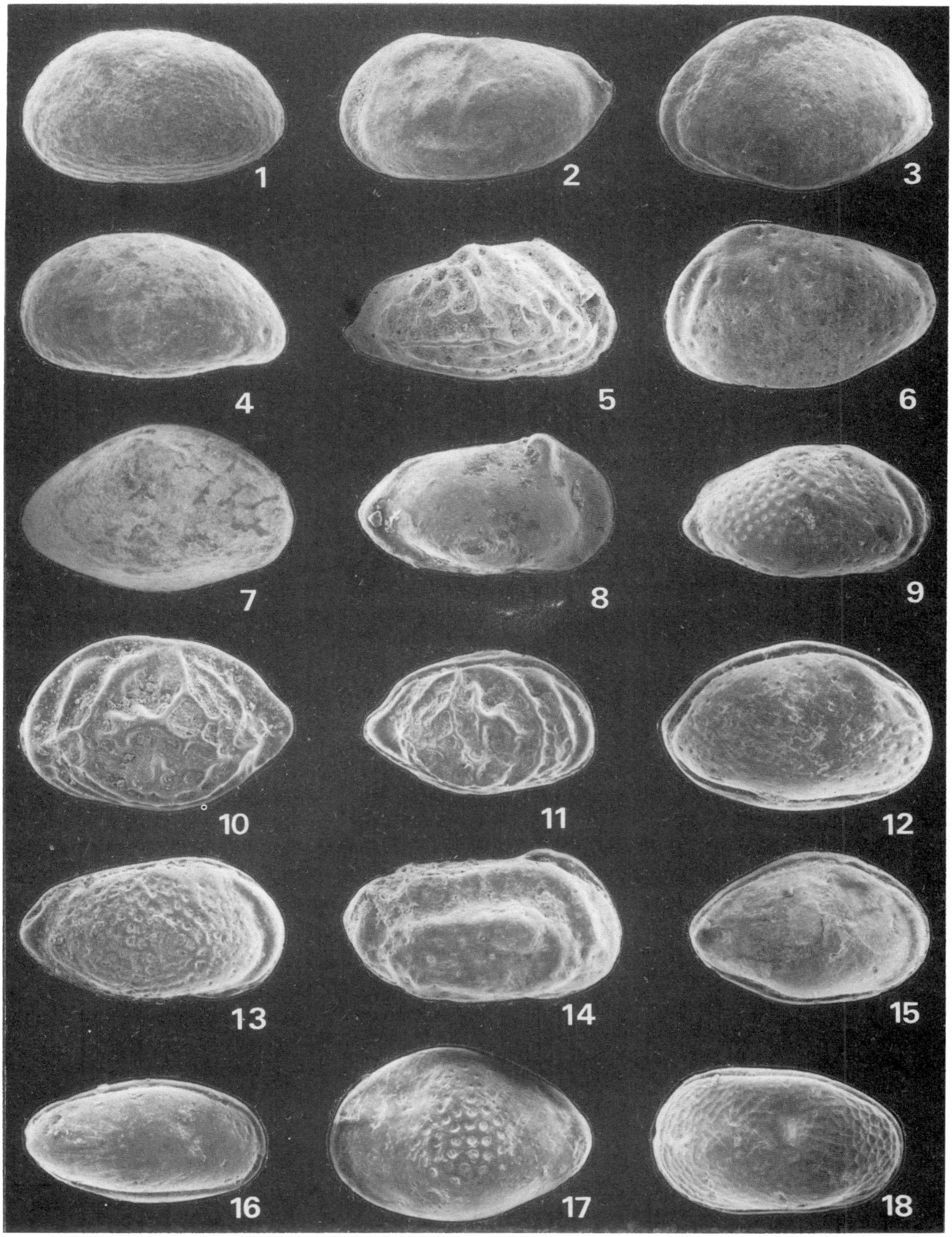
1
2
3
4
5
6
7
8
9
10
11
12
13
14
15
16
17
18

Plate 3

Fig. 1 — *Camptocythere mediofoveolata* Ainsworth, left view of female carapace, × 88, Aalenian–upper Toarcian.

Fig. 2 — *Kinkelinella sermoisensis* (Apostolescu), right view of female carapace, × 103, Aalenian–upper Toarcian.

Fig. 3 — *Praeschuleridea ellipsoidea* Ainsworth, left view of female carapace, × 66, Aalenian–upper Toarcian.

Fig. 4 — *Rutlandella batei* Ainsworth, right view of carapace, × 130, Aalenian–upper Toarcian.

Fig. 5 — *Bairdia parva* Ainsworth, left view of carapace, × 67, upper Pliensbachian.

Fig. 6 — *Isobythocypris pliensbachensis* Ainsworth, right view of carapace, × 55, upper Pliensbachian.

Fig. 7 — *Ledahia septenaria* (Gründel), right view of carapace, × 80, upper Pliensbachian.

Fig. 8 — *Ogmoconcha* aff. *O. rotunda* Dreyer *sensu* Ainsworth, right view of carapace, × 52, upper Pliensbachian.

Fig. 9 — *Ruchholzella frequens* Ainsworth, MS, right view of male carapace, × 60, upper Sinemurian.

Fig. 10 — *Ogmoconchella celticensis* Ainsworth, MS, right view of carapace, × 56, upper Sinemurian.

Fig. 11 — *Cristacythere costata* Michelsen, right view of female carapace, × 65, lower Sinemurian.

Fig. 12 — *Ljubimovella? frequens* Donze, right view of male carapace, × 56, lower Sinemurian.

Fig. 13 — *Acrocythere?* cf. *A.? gassumensis* Michelsen *sensu* Ainsworth *et al.*, left view of carapace, × 129, lower Sinemurian.

Fig. 14 — *Klinglerella translucens* (Blake), external view of female left valve, × 69, lowermost Sinemurian–uppermost Hettangian.

Fig. 15 — *Ogmoconchella ellipsoidea* (Jones), right view of carapace, × 72, lowermost Sinemurian–uppermost Hettangian.

Fig. 16 — *Darwinula rara* Ainsworth, right view of carapace, × 56, Hettangian.

Fig. 17 — *Lutkevichinella hortonae* Ainsworth, right view of female carapace, × 59, Hettangian.

Fig. 18 — *Ogmoconchella martini* Anderson, right view of carapace, × 68, middle–lower Rhaetian.

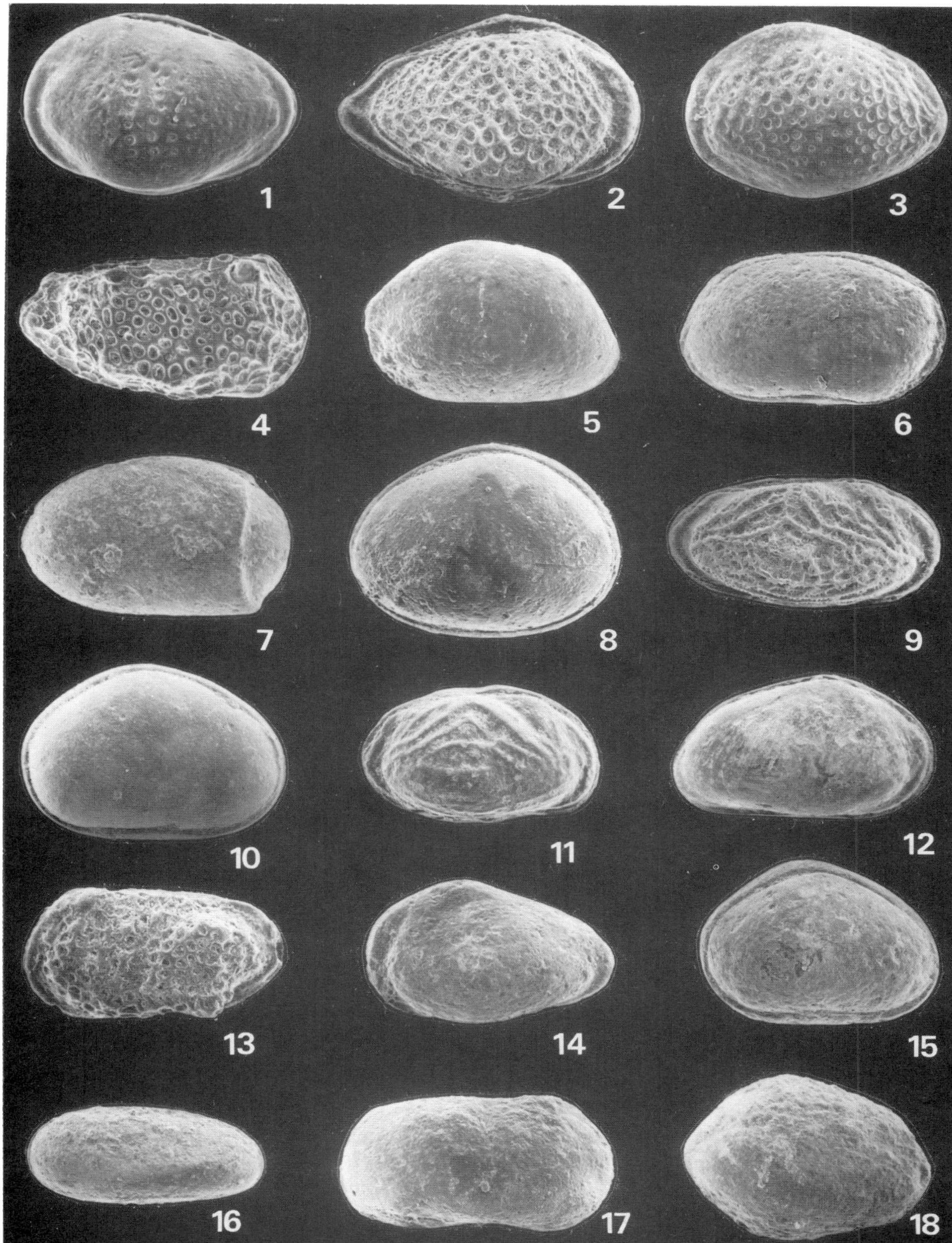
1
2
3
4
5
6
7
8
9
10
11
12
13
14
15
16
17
18

Plate 4 —

Fig. 1 — *Lenticulina galeata* (Terquem), × 66, middle–lower Bathonian.
Fig. 2 — *Lenticulina ectypa* (Loeblich & Tappan), × 78, middle–lower Bathonian.
Fig. 3 — *Lenticulina tricarinella* (Reuss), × 60, middle–lower Bathonian.
Fig. 4 — *Lenticulina volubilis* Dain, × 63, middle–lower Bathonian.
Fig. 5 — *Lenticulina dictyoides* (Deeke), × 66, middle–lower Bathonian.
Fig. 6 — *Astacolus* cf. *A. limata* (Schwager), × 92, middle–lower Bathonian.
Fig. 7 — *Nubeculinella infraoolithica* (Terquem), × 35, middle–lower Bathonian.
Fig. 8 — *Saracenaria oxfordiana* Tappan, × 92, middle–lower Bathonian.
Fig. 9 — *Epistomina stelligera* (Reuss), dorsal view, × 100, middle–lower Bathonian.
Fig. 10 — *Frondicularia nympha* Kopik, × 88, middle–lower Bathonian.
Fig. 11 — *Ophthamidium strumosum* (Gümbel), × 78, lowermost Bathonian.
Fig. 12 — *Epistomina* ex. gr. *bireticulata* Pazdro, ventral view, × 109, middle–lower Bathonian.
Fig. 13 — *Epistomina regularis* Terquem, ventral view, × 58, middle–lower Bathonian.
Fig. 14 — *Massilina dorsetensis* Cifelli, × 100, middle–lower Bathonian.
Fig. 15 — *Cyclogyra liasina* (Terquem), × 40, lowermost Bathonian.
Fig. 16 — *Ammobaculites coprolithiformis* (Schwager), × 44, middle–lower Bathonian.
Fig. 17 — *Vaginulina flabelloides* (Terquem), × 39, middle–lower Bathonian.
Fig. 18 — *Reinholdella crebra* Pazdro, dorsal view, × 92, middle–lower Bathonian.
Fig. 19 — *Paalzowella feifeli* (Paalzow), ventral view, × 152, middle–lower Bathonian.
Fig. 20 — *Nodosaria obscura* Reuss, × 60, middle–lower Bathonian.
Fig. 21 — *Garantella floscula* Kaptarenco-Chernousova, ventral view × 70, upper Bajocian.
Fig. 22 — *Epistomina* cf. *E. callovica* Kaptarenco-Chernousova, dorsal view × 81, middle–lower Bathonian.

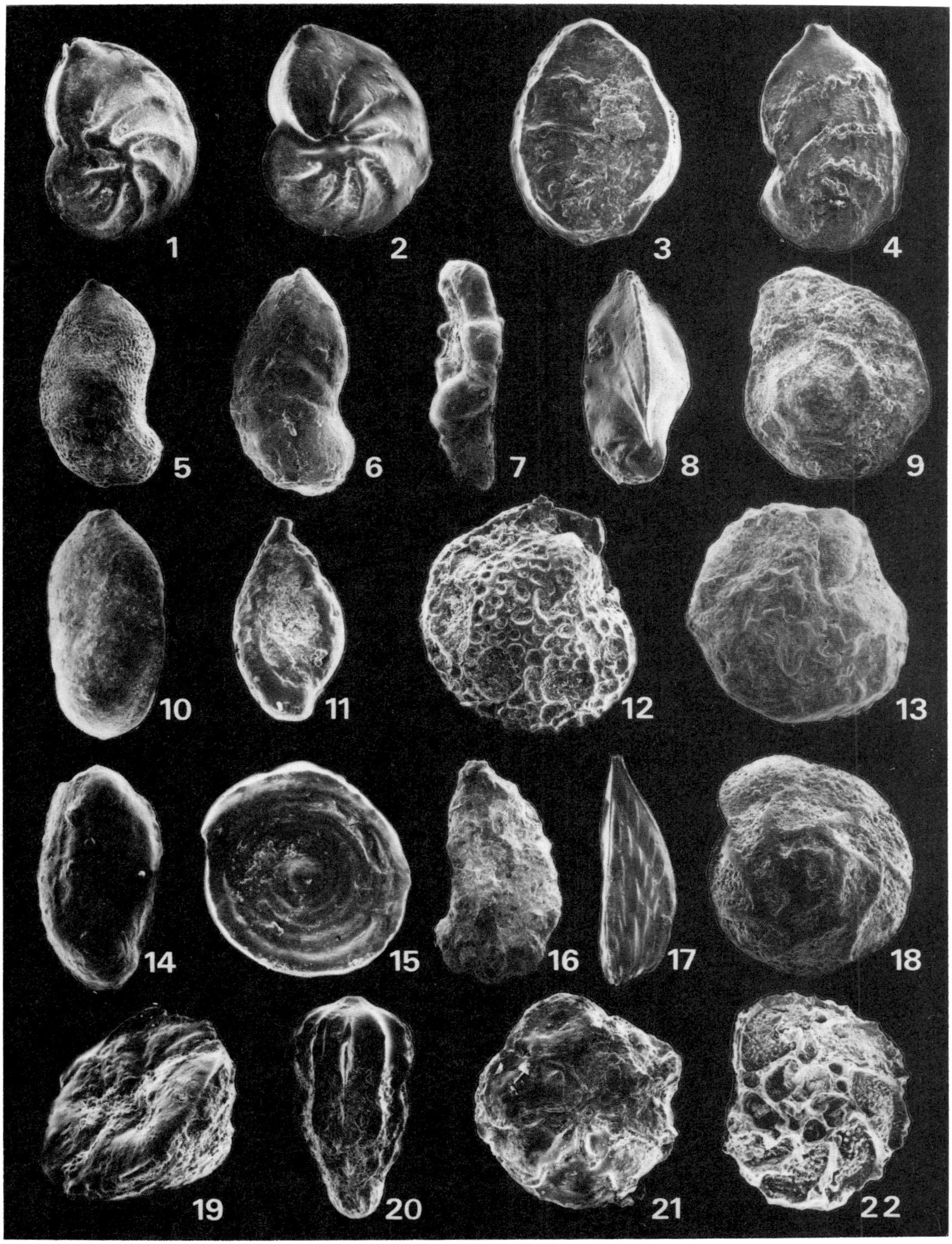
1
2
3
4
5
6
7
8
9
10
11
12
13
14
15
16
17
18
19
20
21
22

Plate 5 —

Fig. 1 — *Lenticulina muensteri* (Roemer), × 50, Aalenian–upper Toarcian.
Fig. 2 — *Lenticulina polygonata* Franke, × 99, Aalenian–upper Toarcian.
Fig. 3 — *Lenticulina dorbignyi* (Roemer), × 70, Aalenian–upper Toarcian.
Fig. 4 — *Nodosaria regularis* subsp. A Copestake & Johnson, × 73, Aalenian–upper Toarcian.
Fig. 5 — *Dentalina ensis* Wisniowski, × 70, Aalenian–upper Toarcian.
Fig. 6 — *Nodobacularia* sp., × 70, lower Toarcian.
Fig. 7 — *Pseudonodosaria oviformis* (Terquem), × 109, lower Toarcian.
Fig. 8 — *Vaginulina clathrata* (Terquem), × 27, Aalenian–upper Toarcian.
Fig. 9 — *Ophthalmidium* sp. 2 (Ruget & Sigal), × 81, lowermost Toarcian.
Fig. 10 — *Saracenaria subleavis subleavis* (Franke), × 70, lower Toarcian.
Fig. 11 — *Lingulina tenera pupa* (Terquem), × 90, lower Pliensbachian.
Fig. 12 — *Lingulina* cf. *L. tenera substriata* (Nørvang), × 73, upper Sinemurian.
Fig. 13 — *Frondicularia terquemi* subsp. B Copestake & Johnson, × 50, upper Pliensbachian.
Fig. 14 and 15 — *Reinholdella dreheri* (Bartenstein), Fig. 14, dorsal view, × 117, lower Toarcian; Fig. 15, ventral view, × 130, lower Toarcian.
Fig. 16 — *Marginulina prima praerugosa* Nørvang, × 58, upper Pliensbachian.
Fig. 17 — *Eoguttulina bilocularis* (Terquem), × 117, lower Pliensbachian.
Fig. 18 and 19 — *Reinholdella margarita margarita* (Terquem), Fig. 18, dorsal view, × 92, lower Sinemurian; Fig. 19, side view, × 92, lower Sinemurian.
Fig. 20 — *Involutina liassica* (Jones), × 70, lower Sinemurian.
Fig. 21 — *Astacolus semireticulata* (Fuchs), × 63, lower Sinemurian.
Fig. 22 — *Planularia inaequistrata* (Terquem), × 85, lower Sinemurian.
Fig. 23 — *Reinholdella macfadyeni* (Ten Dam), × 77, lower Toarcian.

1
2
3
4
5
6
7
8
9
10
11
12
13
14
15
16
17
18
19
20
21
22
23

Plate 6 —

Fig. 1 — *Cribroperidinium globatum* (Gitmez & Sarjeant) Helenes, lower Portlandian–upper Kimmeridgian.
Fig. 2 — *Chytroeisphaeridia chytroeides* (Sarjeant) Downie & Sarjeant emend. Davey, Bathonian.
Fig. 3 — *Carpathodinium predae* (Beju) Drugg, lower Bathonian.
Fig. 4 — *Aldorfia dictyota* (Cookson & Eisenack) Davey, lower Kimmeridgian.
Fig. 5 — *Mendicodinium groenlandicum* (Pocock & Sarjeant) Davey, lowermost upper–lower Kimmeridgian.
Fig. 6 — *Pareodinia prolongata* Sarjeant, lower Bathonian.
Fig. 7 — *Kylindrocysta spinosa* Fenton, Neves & Piel, lowermost Bathonian.
Fig. 8 — *Luehndea spinosa* Morgenroth, upper Pliensbachian.
Fig. 9 — *Diacanthum filapicatum* (Gocht) Stover & Evitt, uppermost Bajocian.
Fig. 10 — *Endoscrinium luridum* (Deflandre) Gocht, lowermost Bathonian.
Fig. 11 — *Ctenidodinium combazii* Dupin, upper Bathonian.
Fig. 12 — *Energlynia acollaris* (Dodekova) Sarjeant, lower Bathonian.
Fig. 13 — *Meiourogonyaulax reticulata* Dodekova, lower Bathonian.
Fig. 14 — *Korystocysta gochtii* (Sarjeant) Woollam, uppermost Bajocian.

All magnifications approximately × 500.

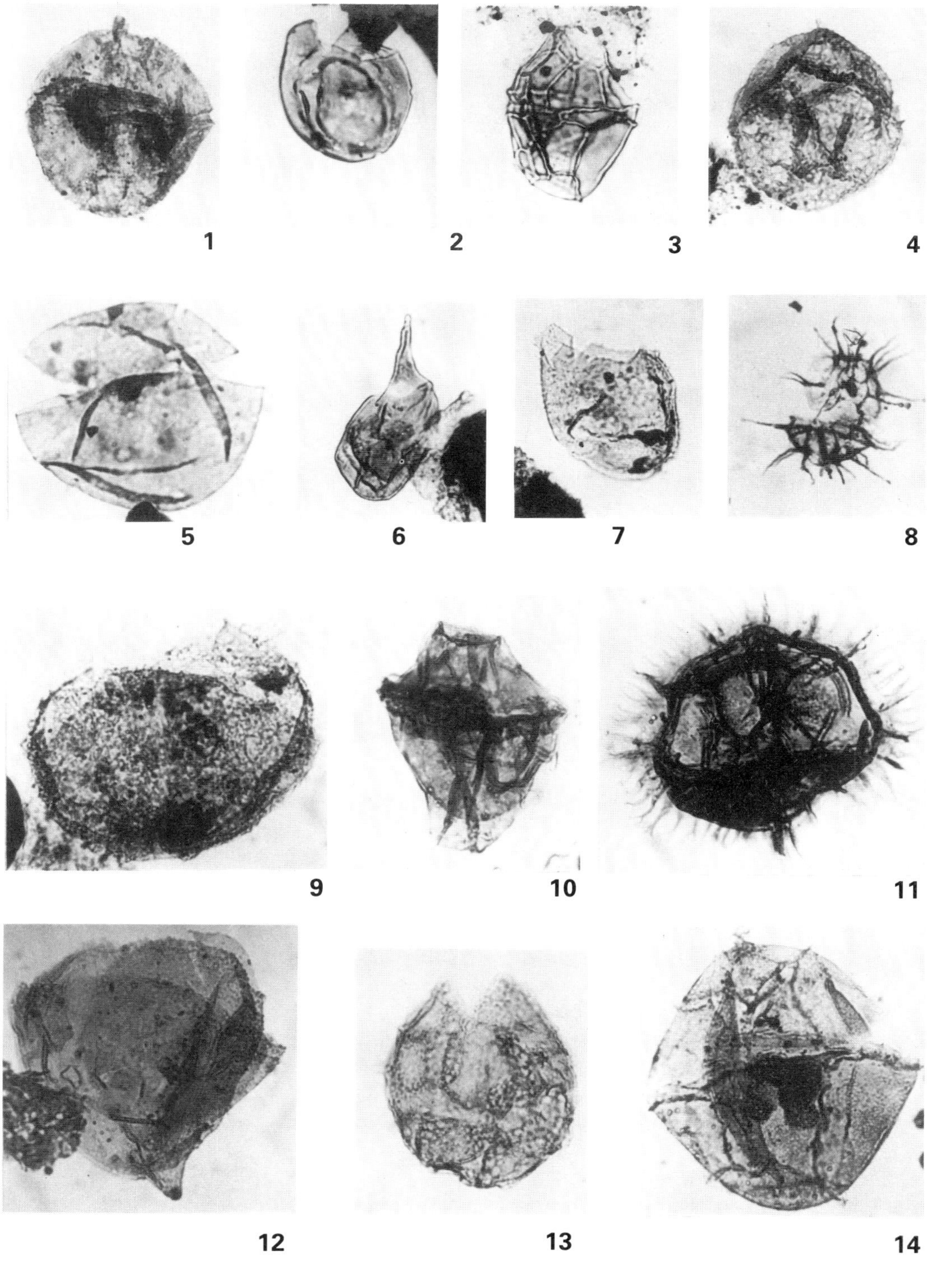
1
2
3
4
5
6
7
8
9
10
11
12
13
14

Plate 7 —

Fig. 1 — *Hapsidaulax margarethe* Sarjeant, lower Bathonian.
Fig. 2 — *Mancodinium semitabulatum* Morgenroth, Aalenian–upper Toarcian.
Fig. 3 — *Valensiella/Ellipsoidictyum* sp. Eisenack & Klement, lowermost upper-lower Kimmeridgian.
Fig. 4 — *Parvocysta bullula* Bjaerke, Aalenian–upper Toarcian.
Fig. 5 — *Ischyosporites pseudoreticulatus* (Couper) Döring, upper Portlandian.
Fig. 6 — *Kraeuselisporites reissingeri* (Harris) Morbey, lower Sinemurian.
Fig. 7 — *Neoraistrickia gristhorpensis* (Couper) Traulau, Aalenian–upper Toarcian.
Fig. 8 — *Rhaetipollis germanicus* Schulz, Rhaetian.
Fig. 9 — *Pilosisporites trichopapillosus* (Thiergart) Delcourt & Sprumont, upper Portlandian.
Fig. 10 — *Retitriletes austroclavatidites* (Cookson) Döring *et al.*, upper Portlandian.
Fig. 11 — *Microreticulatisporites fuscus* (Nilsson) Morbey, Rhaetian.
Fig. 12 — *Sestrosporites pseudoalveolatus* (Couper) Dettmann, upper Portlandian.
Fig. 13 — *Quadraeculina anellaeformis* Maljavkina, middle–lower Bathonian.
Fig. 14 — *Classopollis echinatus* Burger, upper Portlandian.
Fig. 15 — *Couperisporites complexus* (Couper) Pocock, upper Portlandian.

All magnifications approximately × 500.

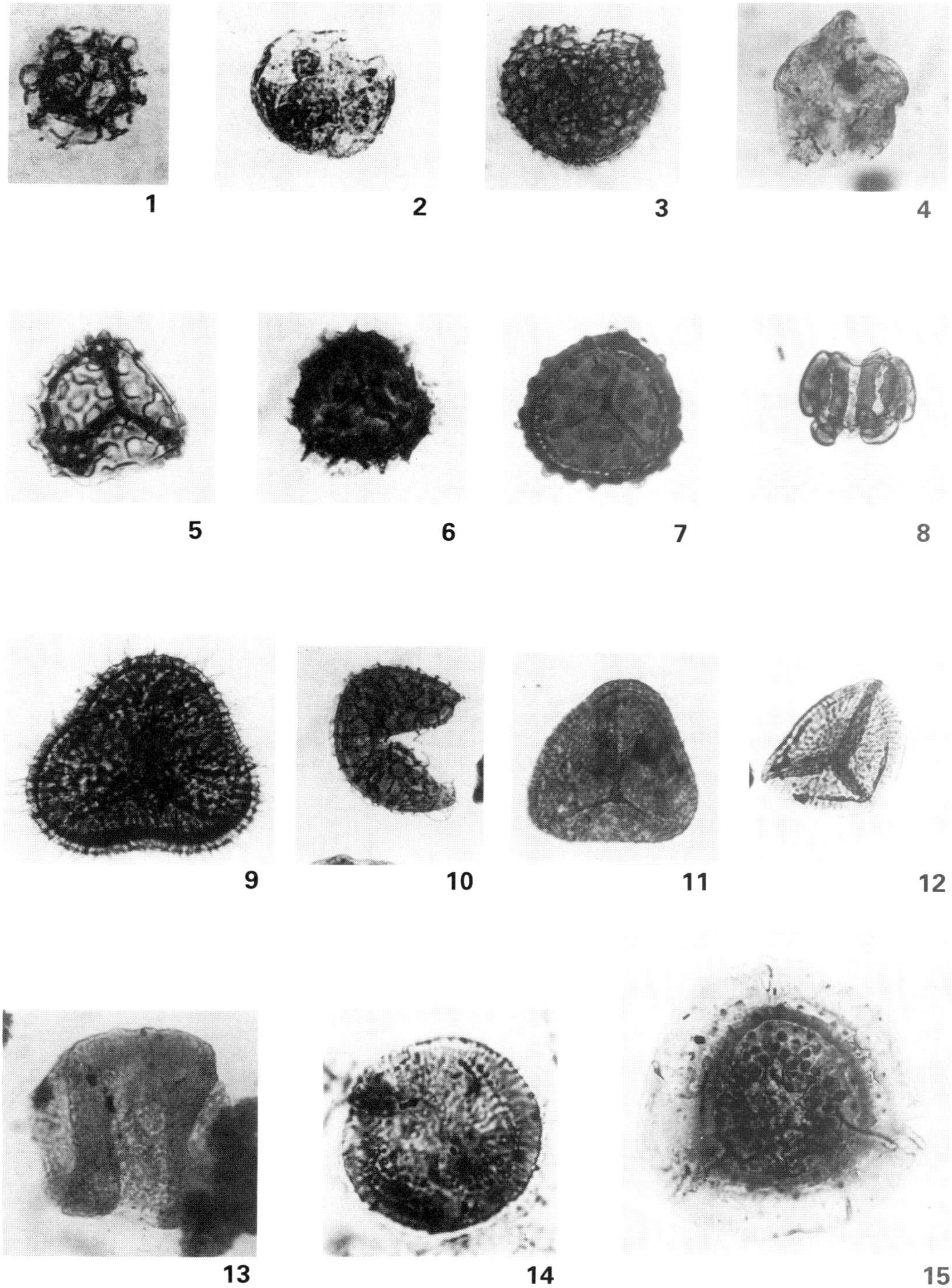
1
2
3
4
5
6
7
8
9
10
11
12
13
14
15

REFERENCES

Ahmed, S. T. (1987). Upper Oxfordian and lower Kimmeridgian ostracods from the South Ferriby Quarry, south Humberside. *Proceedings of the Yorkshire Geological Society* **46** 267–274.

Ainsworth, N. R. (1985). Upper Jurassic and Lower Cretaceous Ostracoda from the Fastnet Basin, offshore southwest Ireland. *Irish Journal of Earth Sciences* **7** 15–33.

Ainsworth, N. R. (1986a). Upper Jurassic and Lower Cretaceous Ostracoda from the Fastnet Basin, offshore southwest Ireland. *Irish Journal of Earth Sciences* **7** 145–168.

Ainsworth, N. R. (1986b). Upper Jurassic and Lower Cretaceous Ostracoda from the Fastnet Basin, offshore southwest Ireland. *Irish Journal of Earth Sciences* **8** 53–72.

Ainsworth, N. R. (1986c). Toarcian and Aalenian Ostracoda from the Fastnet Basin, offshore southwest Ireland. *Geological Survey of Ireland Bulletin* **3** 277–336.

Ainsworth, N. R. (1987a). Upper Jurassic and Lower Cretaceous Ostracoda from the Fastnet Basin, offshore southwest Ireland. *Irish Journal of Earth Sciences* **8** 139–153.

Ainsworth, N. R. (1987b). Pliensbachian Ostracoda from the Fastnet Basin, offshore southwest Ireland. *Geological Survey Ireland Bulletin* **4** 41–62.

Ainsworth, N. R. & Horton, N. F. (1986). Mesozoic micropalaeontology of exploration well Elf 55/30-1 from the Fastnet Basin, offshore southwest Ireland. *Journal of Micropalaeontology* **5** 19–29.

Ainsworth. N. R., Horton, N. F. & Penney, R. A. (1985). Lower Cretaceous micropalaeontology of the Fastnet Basin, offshore southwest Ireland. *Marine and Petroleum Geology* **2** 341–349.

Ainsworth, N. R., O'Neill, M., Rutherford, M. M., Clayton, G., Horton, N. F. & Penney, R.A. (1987). Biostratigraphy of the Lower Cretaceous, Jurassic and Late Triassic of the North Celtic Sea and Fastnet Basins. In: Brooks, J. & Glennie, K. (eds), *Petroleum geology of north west Europe*. Graham & Trotman, London, pp. 611–622.

Anderson, F. W. (1966). New genera of Purbeck and Wealden Ostracoda. *Bulletin of the British Museum (Natural History), Geology* **11** 435–446.

Anderson, F. W. (1971). The Ostracods. In: Anderson, F. W. & Bazley, R. A. B. (eds), The Purbeck Beds of the Weald (England). *Bulletin of the Geological Survey of Great Britain* **34** 27–174.

Anderson, F. W. (1985). Ostracod faunas in the Purbeck and Wealden of England. *Journal of Micropalaeontology* **4** 1–68.

Apostolescu, V. (1959). Ostracodes du Lias du Bassin de Paris. *Revue de l'Institut Français du Pétrole* **14** 795–826.

Apostolescu, V. (1961). *In*: Cousin, N., Espitalie, J., Signal, J. & Apostolescu, V. Sud du Bassin, région d'Argenton-sur-Creuse et de la Chatre (Départment du Cher et de l'Indre). Ostracodes. In: Cousin, N., Espotalie, J., Signal, J. & Apostolescu V. (eds), *Colloque sur le Lias Français. Mémoire du Bureau de Recherches Géologiques Minières* **4** 449.

Barker, D. (1966a). Ostracods from the Portland Beds of Dorset. *Bulletin of the British Museum (Natural History), Geology* **11** 447–457.

Barker, D. (1966b). Ostracods from the Portland and Purbeck Beds of the Aylesbury district. *Bulletin of the British Museum (Natural History), Geology* **11** 459–489.

Barnard, T. (1949). Foraminifera from the Lower Lias of the Dorset coast. *Quarterly Journal of the Geological Society, London* **105** 347–391.

Barnard, T. (1950). Foraminifera from the Upper Lias of Byfield, Northamptonshire. *Quarterly Journal of the Geological Society, London* **106** 1–36.

Barnard, T. (1956). Some Lingulinae from the Lias of England. *Micropaleontology* **2** 271–282.

Barnard, T. (1957). *Frondicularia* from the Lower Lias of England. *Micropaleontology* **3** 171–181.

Bartenstein, H. & Brand, E. (1937). Mikropalaontologische Untersuchungen zur Stratigraphie des nordwestdeutschen Lias und Doggers. *Abhandlungen Herausgegeben von der Senckenbergischen Naturforschenden Gesellschaft* **439** 1–224.

Bate, R. H. (1967). The Bathonian Upper Estuarine Series of eastern England, Part 1: Ostracods. *Bulletin of the British Museum (Natural History), Geology* **14** 21–66.

Bate, R. H. (1978). The Jurassic Part II—Aalenian to Bathonian. In: Bate, R. & Robinson, E. (eds), *A stratigraphical index of British Ostracoda. Geological Journal, Special Issue* **8** 213–258.

Bate, R. H. (1978). The Trias. In: Bate, R. & Robinson, E. (eds), *A stratigraphical index of British Ostracoda. Geological Journal, Special Issue* **8** 175–187.

Bate, R. H. & Coleman, B. E. (1975). Upper Lias Ostracoda from Rutland and Huntingdonshire. *Bulletin of the Geological Survey of Great Britain* **55** 1–42.

Bate, R. H. & Sheppard, L. M. (1982). Bathonian Ostracoda in the Winterborne Kingston Borehole, Dorset. In: Rhys, G. H., Lott, G. K. & Calver, M. A. (eds), *The Winterborne Kingston Borehole, Dorset, England. Report, Institute of Geological Sciences, London* **81** 77–81.

Bizon, G. (1961). Contributions a l'etude micropaleontologique du Lias du Bassin de Paris, Deuxieme Partie: Lorraine, region de Nancy et Thionville. In: Cousin Espitalie, J., Signal, J. & Apostolescu, V. (eds), *Colloque sur le Lias Français. Mémoire du Bureau de Recherches Géologiques Minières* **4** 433–436.

Bizon, G. & Oertli, H. (1961). Contributions a l'etude micropaleontologique (foraminiferes–ostracodes) du Lias du Basin de Paris, Septieme Partie: Conclusions. In: Cousin, N., Espitalie, J., Signal, J. & Apostolescu, V. (eds), *Colloque sur le Lias Français. Mémoire du Bureau de Recherches Geologiques Minières* **4** 107–119.

Bjaerke, T. (1980). Mesozoic palynology of Svalbard IV. Toarcian dinoflagellates from Spitzbergen. *Palynology* **4** 57–77.

Brenner, P. (1976). Ostracoden und Charophyten des Spanischen Wealden (Systematik, Ökologie, Stratigraphie, Paläogeographie). *Palaeontographica* **152** 113–201.

Brouwer, J. (1969). Foraminiferal assemblages from the Lias of N.-W. Europe. *Verhandelingen der Koninklinjke Nederlandsche (Nederlandse) Akademie van Wetenschappen, Amsterdam, Afdeeling Natuurkunde* **25** 1–48.

Bujak, J. P. & Williams, G. L. (1985). Mesozoic and

Cenozoic dinoflagellates. In: Bolli, H. M., Saunders, J. B. & Perch-Nielsen, K. (eds), *Plankton stratigraphy*. Cambridge University Press, Cambridge, pp. 847–965.

Cifelli, R. (1959). Bathonian Foraminifera of England. *Bulletin of the Museum of Comparative Zoology* **121** 265–368.

Christensen, O. B. (1968). Some deposits and microfaunas from the Upper Jurassic in Scania. *Sveriges Geologiska Undersökning, Afhandlingar* **62** 3–46.

Coleman, B. E. (1980). In: Penn, I. E., Dingwall, R. G. & Knox, R. W. O'B. (eds), *The Inferior Oolite (Bajocian) sequence from a borehole in Lyme Bay, Dorset. Report, Institute of Geological Sciences, London* **79** 1–27.

Coleman, B. E. (1982). In: Rhys, G. H., Lott, G. K. & Calver, M. A. (eds), *Lower and Middle Jurassic foraminifera in the Winterborne Kingston Borehole, Dorset. Report, Institute of Geological Sciences, London* **81** 82–88.

Colin, J. P., Lehmann, R. A. & Morgan, B. E. (1981). Cretaceous and Late Jurassic biostratigraphy of the North Celtic Sea Basin, offshore southern Ireland. In: Neale, J. W. & Brasier, M. D. (eds), *Microfossils from recent and fossil shelf seas*. Ellis Horwood, Chichester, pp. 122–155.

Copestake, P. & Johnson, B. (1981). Jurassic. In: Jenkins D. G. & Murray, J. W. (eds), *Stratigraphical atlas of fossil foraminfera*. Ellis Horwood, Chichester, pp. 81–105.

Couper, R. A. (1958). British Mesozoic microspores and pollen grains. *Palaeontographica* **103** 75–179.

Cox, B. M., Lott, G. K., Thomas, J. E. & Wilkinson, I. P. (1987). Upper Jurassic stratigraphy of four shallow cored boreholes in the UK sector of the southern North Sea. *Proceedings of the Yorkshire Geological Society* **46** 97–109.

Davies, E. H. (1983). The dinoflagellate Oppel-Zonation for the Jurassic–Lower Cretaceous sequence in the Sverdrup Basin, Arctic Canada. *Geological Survey of Canada Bulletin* **359** 1–59.

Dépêche, F. (1985). Lias Supérieur, Dogger, Malm. In: Oertli, H. J. (ed.), *Atlas des ostracodes de France. Bulletin Centres Recherches Exploration–Prodution Elf-Aquitaine, Mémoires* **9** 119–145.

Donze, P. (1985). Lias inferieur et moyen. In: Oertli, H. J. (ed), *Atlas des ostracodes de France. Bulletin Centres Recherches Exploration–Prodution Elf-Aquitaine, Mémoires* **9** 101–117.

Drexler, E. (1958). Foraminiferen und Ostracoden aus dem Lias von Siebeldingen/Pfalz. *Geologisches Jahrbuch* **75** 475–554.

Dreyer, E. (1965). Mikrofossilen des Rät und Lias von SW-Brandenburg. *Jahrbuch für Geologie* **1** 491–531.

Drugg, W. S. (1978). Some Jurassic dinoflagellate cysts from England, France and Germany. *Paläontologie B* **168** 61–79.

Exton, J. (1979). Pliensbachian and Toarcian microfauna of Zambujal, Portugal: systematic palaeontology. *Geological Paper, Carleton University* **79** 1–104.

Fenton, J. P. G. (1981). Taxonomic revision of selected dinoflagellate cysts from the late Bajocian (Middle Jurassic) of northwest Germany. *Review of Palaeobotany and Palynology* **31** 249–260.

Fenton, J. P. G. & Fisher, M. J. (1978). Regional distribution of marine microplankton in the Bajocian and Bathonian of northwest Europe. *Palinologia, Número Extraordinario* **1** 233–243.

Fenton, J. P. G. & Riding, J. B. (1987). *Kekryphalospora distincta* gen. et sp. nov., a trilete spore from the Lower and Middle Jurassic of North-West Europe. *Pollen et Spores* **29** 427–435.

Gitmez, G. U. (1970). Dinoflagellate cysts and acritarchs from the basal Kimmeridgian (Upper Jurassic) of England, France and Scotland. *Neues Jahrbuch für Geologie und Paläontologie, Abhandlungen* **119** 113–133.

Gitmez, G. U. & Sarjeant, W. A. S. (1972). Dinoflagellate cysts and acritarchs from the Kimmeridgian (Upper Jurassic) of England, Scotland and France. *Bulletin of the British Museum (Natural History), Geology* **21** 171–257.

Gründel, J. (1964). Zur Gattung *Healdia* (Ostracoda) und zu einiger verwandten Formen aus dem unteren Jura. *Geologie* **13** 456–477.

Helmdach, F. F. (1971). Zur Glienderung limnisch-brackischer Sedimente des portugiensischen Oberjura (ob. Callovien-Kimmeridge) mit Hilfe von Ostrakoden. *Neues Jahrbuch für Mineralogie. Geologie und Paläontologie Monatshefte* **11** 546–662.

Helmdach, F. F. (1972). Ontogenie und Ornamentation einiger oberjurassischer Arten von *Bisulcocypris* (Ostracoda). *Neues Jahrbuch für Mineralogie, Geologie und Paläontologie, Abhandlungen* **141** 286–300.

Herrig, E. (1969). Ostracoden aus dem Ober-Domérien von Grimmen westlich Griefswald (Teil 1). *Geologie* **18** 446–471.

Kilenyi, T. I. (1969). The Ostracoda of the Dorset Kimmeridge Clay. *Palaeontology* **12** 112–160.

Klingler, W. & Neuweiler, F. (1959). Leitende Ostracoden aus dem deutschen Lias B. *Geologisches Jahrbuch* **76** 373–410.

Knitter, H. (1983). Biostratigraphie Untersuchungen mit Ostracoden im Toarcian Süddeutschlands. *Facies* **8** 213–262.

Lord, A. (1971). Revision of some Lower Lias Ostracoda from Yorkshire. *Palaeontology* **14** 642–665.

Lord, A. (1978). The Jurassic Part 1—Hettangian–Toarcian. In: Bate, R. & Robinson, E. (eds), *A stratigraphical index of British Ostracoda. Geological Journal, Special Issue* **8** 189–212.

Lord, A. R. & Bown, P. R. (eds) (1987). Mesozoic and Cenozoic stratigraphical micropalaeontology of the Dorset coast and Isle of Wight, southern England. *Field Guide for the 20th European Micropalaeontological Colloquium. British Micropalaeontological Society Guide Book* **1**, pp. 1–183.

Malz, H. (1958). Die Gattung *Macrodentina* und einige andere Ostracoden-Arten aus dem Oberen Jura von NW-Deutschland, England und Frankreich. *Ablandlungen der Senckenbergischen Naturforschenden Gesellschaft* **497** 1–67.

Malz, H. (1971). Zur Taxonomie 'glattschaliger' Lias-Ostracoden. *Senckenbergiana Lethaea* **52** 433–455.

Martin, G. P. R. (1940). Ostracoden des norddeutschen Purbeck und Wealden. *Senckenbergiana* **22** 275–361.

Michelsen, O. (1975). Lower Jurassic biostratigraphy and Ostracods of the Danish Embayment. *Danmarks Geologiske Undersogelse* **104** 1–289.

Morbey, S. J. (1975). The palynostratigraphy of the Rhaetian Stage, Upper Triassic in the Kendelbachgraben, Austria. *Palaeontographica B* **152** 1–75.

Morbey, S. J. (1978). Late Triassic and Early Jurassic subsurface palynostratigraphy in north west Europe. *Palinologia, Número Extraordinario* **1** 355–365.

Morgenroth, P. (1970). Dinoflagellate cysts from the Lias Delta of Lühnde, Germany. *Neues Jahrbuch für Geologie und Paläontologie, Abhandlungen* **136** 345–359.

Morris, P. H. (1982). Distribution and paleoecology of Middle Jurassic foraminifera from the Lower Inferior Oolite of the Cotswolds. *Palaeogeography, Palaeoclimatology, Palaeoecology* **37** 319–364.

Naylor, D. & Shannon, P. M. (1982). Fastnet Basin. In: Naylor, D. & Shannon, P. M. (eds), *The geology of offshore Ireland and west Britain.* Graham & Trotman, London, pp. 49–58.

Norling, E. (1972). Jurassic stratigraphy and foraminifera of western Scania, southern Sweden. *Sveriges Geologiska Undersokning, Afhandlingar och Uppsatser, Serie Ca* **47** 1–120.

Norris, G. (1969). Miospores from the Purbeck Beds and marine Upper Jurassic of southern England. *Palaeontology* **12** 574–620.

Norris, G. (1975). Provincialiam of Callovian–Neocomian dinoflagelate cysts in the northern and southern hemispheres. *American Association of Stratigraphic Palynologists Contribution Series* **4** 29–35.

Nørvang, A. (1957). The foraminifera of the Lias series in Jutland, Denmark. *Mededelingen Dansk, Geolisch Forening* **13** 1–135.

Oertli, H. J. (1957). In: Bernard, F. Bizon, J.-J. & Oertli, H. J. (eds) *Ostracodes lacustres du Bathonian du Poitou (Bassin de Paris). Bulletin de la Société Géologique de France* **6** 753–770.

Oertli, H. J. (1959). Les ostracodes du Bathonian du Boulonnais 1. Les ⟨micro-ostracodes⟩. *Revue de Micropaléontology* **2** 115–126.

Oertli, H. J. (1963). Ostracodes du 'Purbeckian' du Bassin Parisien. *Revue de l'Institut Français du Pétrole* **18** 5–40.

Penn, I. E. (1987). Geophysical logs in the stratigraphy of Wales and adjacent offshore and onshore areas. *Proceedings of the Geologists' Association* **98** 275–315.

Plumhoff, F. (1963). Die Ostracoden des Oberaalenium und tiefen Unterbajocium (Jura) des Gifhorner Troger, Nordwestdeutschlands. *Abhandlungen der Senckenbergischen Naturforschenden Gesellschaft* **503** 1–100.

Riding, J. B. (1984a). Dinoflagellate cysts range-top biostratigraphy of the uppermost Triassic to lowermost Cretaceous of northwest Europe. *Palynology* **8** 195–210.

Riding, J. B.(1984b). A palynological investigation of Toarcian to early Aalenian strata from the Blea Wyke area, Ravenscar, North Yorkshire. *Proceedings of the Yorkshire Geological Society* **45** 109–122.

Riding, J. B. (1987). Dinoflagellate cyst stratigraphy of the Nettleton Bottom Borehole (Jurassic: Hettangian to Kimmeridgian), Lincolnshire, England. *Proceedings of the Yorkshire Geological Society* **46** 231–266.

Riegraf, W. (1985). Mikrofauna, Biostratigraphie und Fazies im Unteren Toarcian Südwestdeutschlands und Vergleiche mit benachbarten Gebieten. *Tübinger Mikropaläontogische Mitteilungen* **3** 1–232.

Robinson, K. W., Shannon, P. M. & Young, D. G. G. (1981). The Fastnet Basin: an integrated analysis. In: Illing, L. V. & Hobson, G. D. (eds), *The Geology of the continental shelf of north-west Europe.* Heyden and Son, London, pp. 444–454.

Sheppard, L. M. (1981). Bathonian ostracod correlation north and south of the English Channel with the description of two new Bathonian ostracods. In: Neale, J. W. & Brasier, M. D. (eds), *Microfossils from recent and fossil shelf seas.* Ellis and Horwood, Chichester, pp. 73–89.

Sivhed, U. (1977). A Lower Jurassic ostracode fauna in the Gantofta Brick Pit, Skane, southern Sweden. *Sveriges Geologiska Undersoeking* **71** 1–31.

Sivhed, U. (1980). Lower Jurassic ostracodes and stratigraphy of western Skane, southern Sweden. *Sveriges Geologiska Undersoekning* **50** 1–85.

Thusu, B. (ed.) (1978). Distribution of stratigraphically diagnostic dinoflagellate cysts and miospores from the northwest European continental shelf and adjacent areas. *Continental Shelf Institute (Trondheim, Norway) Publication* **100** 1–111.

Wall, D. (1965). Microplankton, pollen and spores from the Lower Jurassic in Britain. *Micropaleontology* **11** 151–190.

Warrington, G. (1977). Appendix 3: Palynology of the White Lias, Cotham and Westbury Beds and the Keuper Marl of the Steeple Aston Borehole. In: Poole, E. G. (ed.), *Stratigraphy of the Steeple Aston Borehole, Oxfordshire. Bulletin of the Geological Survey of Great Britain* **57** 40–43.

Warrington, G. (1978). Palynology of the Keuper, Westbury and Cothan Beds and the White Lias of the Withycombe Farm Borehole. *Bulletin of the Geological Survey of Great Britain* **68** 22–28.

Warrington, G. *et al* (1980). A correlation of Triassic rocks in the British Isles. *Special Report, Geological Society, London* **13** 1–78.

Wernli, R. (1971). Les Foraminifères du Dogger du Jura Meridional (France). *Archives des Sciences* **24** 305–364.

Wernli, R. & Septfontaine, M. (1971). Micropaleontologie comparee du Dogger du Jura meridional (France) et des Prealpes Medianes Plastiques romandes (Suisse). *Ecologae Geologicae Helvetiae* **64** 437–458.

Wilkinson, I. P. (1983). Kimmeridge Clay Ostracoda of the North Wotton Borehole, Norfolk, England. *Journal of Micropalaeontology* **2** 17–29.

Wille, W. (1982). Evolution and ecology of Upper Liassic dinoflagellates from SW Germany. *Neues Jahrbuch für Geologie und Paläontologie, Abhandlungen* **164** 74–82.

Wolburg, J. (1971). Zur taxonomie und Nomenklatur einiger Wealden-Ostracoden. *Senckenbergiana Lethaea* **52** 39–51.

2

Micropalaeontological and stratigraphical recognition of the Triassic–Jurassic boundary in the North Celtic Sea and Fastnet Basins

M. M. Rutherford and N. R. Ainsworth

ABSTRACT

The Triassic–Jurassic boundary in the North Celtic Sea and Fastnet Basins is difficult to recognize because of variable poor recovery of short-ranging palynomorphs and ostracod taxa. There are no clear lithological breaks coinciding with the system boundary. Micropalaeontological and palynological analysis of ditch-cutting samples from 10 wells has allowed a composite workable biostratigraphy to be established, based on range tops and abundances of selected taxa. Petrophysical log interpretation has been used to identify lithological units by comparison with those described from the British onshore. Four associations of ostracod and palynomorph taxa have been identified. These have been used to recognize the Hettangian and Rhaetian stages, and to distinguish lowermost Sinemurian–upper most Hettangian and lowermost Hettangian–upper Rhaetian intervals. Correlation of the proposed succession with micropalaeontological and stratigraphical successions of the Triassic–Jurassic stages of onshore Britain and Europe is discussed.

INTRODUCTION

The North Celtic Sea and Fastnet Basins accommodate sediments of Upper Palaeozoic to Recent age. Exploration drilling began in 1971 and over 60 wells have now been drilled. The present account is based upon a micropalaeontological study of 10 of these wells (Fig. 1).

The lower Sinemurian–Rhaetian succession of the North Celtic Sea and Fastnet Basins has previously been described by Ainsworth & Horton (1986), who discuss microfaunas from Upper Triassic, Lower Jurassic and Lower Cretaceous rocks in Elf exploration well 55/30-1, and by Ainsworth *et al.* (1987) in a study of the micropalaeontology and palynology of uppermost Triassic, Jurassic and Lower Cretaceous rocks from 25 wells.

In the present study, approximately 600 samples were examined from 10 wells. A generalized lithostratigraphic succession for the area has been developed by correlation of well logs with British onshore successions and biostratigraphic results have been considered in relation to this succession. Palaeoecology and compar-

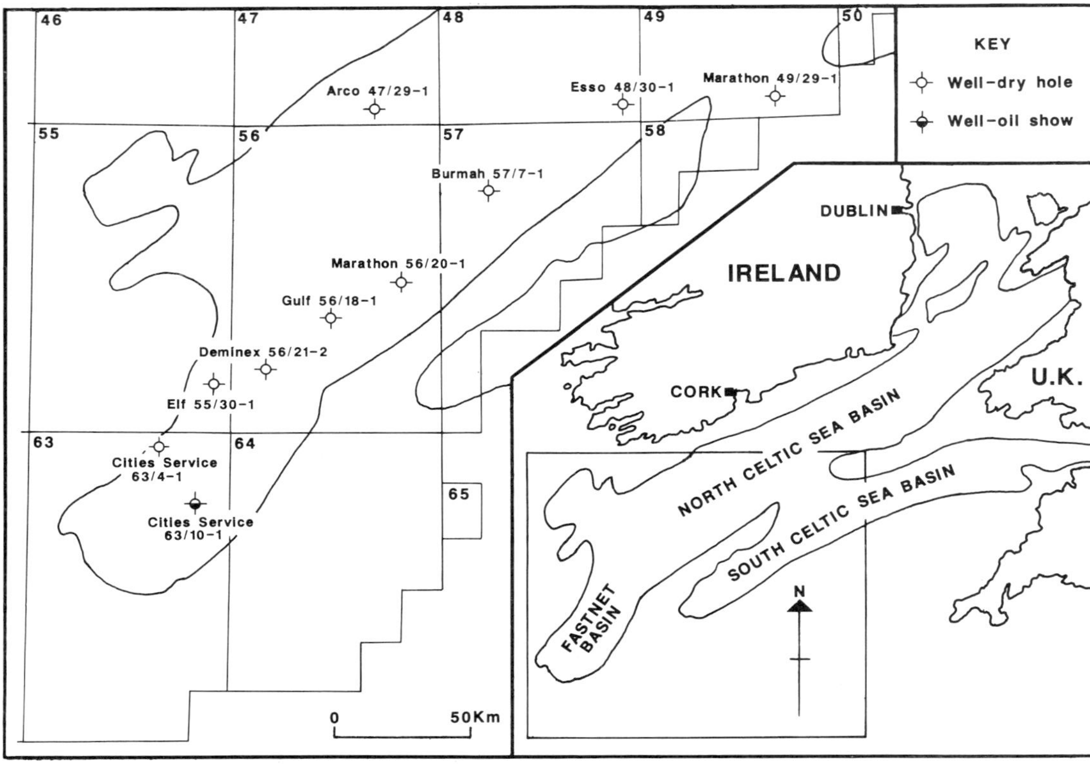

Fig. 1 — Location of North Celtic Sea and Fastnet Basins, and of wells studied.

isons with micropalaeontological associations from north west Europe are discussed. Ostracod and palynomorph taxa are illustrated in Plates 1, 2 and 3.

In Britain, the Triassic–Jurassic system boundary has been defined by Cope *et al.* (1980) and Warrington *et al.* (1980) at the appearance of ammonites of the genus *Psiloceras*; this is usually a few metres above the base of the Lias sequence, the basal beds of which (pre-*Planorbis* Beds) are therefore assigned to the Triassic. The use of ditch-cutting samples and lack of cored material precludes the recognition of the system boundary on this criterion in the Irish offshore.

LITHOSTRATIGRAPHY

The Triassic sediments occurring in both the North Celtic Sea and the Fastnet Basins can be correlated with the standard British onshore sections, suggesting continuation of the lithofacies developed within the British sections to the Irish offshore. This cannot be proved because of lack of outcrop, but available information from geophysical logs, lithologies and biostratigraphic dating indicate strong correlation. To avoid confusion, the nomenclature used to describe the British Upper Triassic sections has therefore been applied to the Irish sections; consequently the classifications employed by Warrington *et al.* (1980), Whittaker *et al.* (1985) and Penn (1987) have been used. The Jurassic sediments, however, differ from the British onshore as discussed by Cope *et al.* (1980); therefore existing terminology cannot be used.

The Penarth Group

This group comprises the Westbury Formation and the overlying Lilstock Formation; it is conformable upon the Mercia Mudstone Group. The Westbury Formation consists of interbedded shales, limestones and dolomites. The shales are brown, medium to dark grey, black, occasionally grey-green, locally calcareous, subfissile to fissile and often contain pyrite. The thin limestone stringers are white, grey to brown, and often micritic or argillaceous. The dolomites are dark brown, grey, hard and generally fine grained and often replace the limestones. Up to three stringers of dolomite and/or limestone occur throughout this sequence. Anhydrite is common, occurring as discrete crystals within the shale beds. In the North Celtic Sea and Fastnet Basins, the lower boundary is defined as the base of the subfissile–fissile shales marked by a decrease in the gamma-ray values and a more irregularly serrated sonic motif. This boundary is also recognized by a distinct colour change from red or brown to white, grey or black. The upper boundary is placed below an increase in gamma-ray values at the base of the incoming Cotham Member (Figs. 2–4).

The Lilstock Formation comprises the Cot-

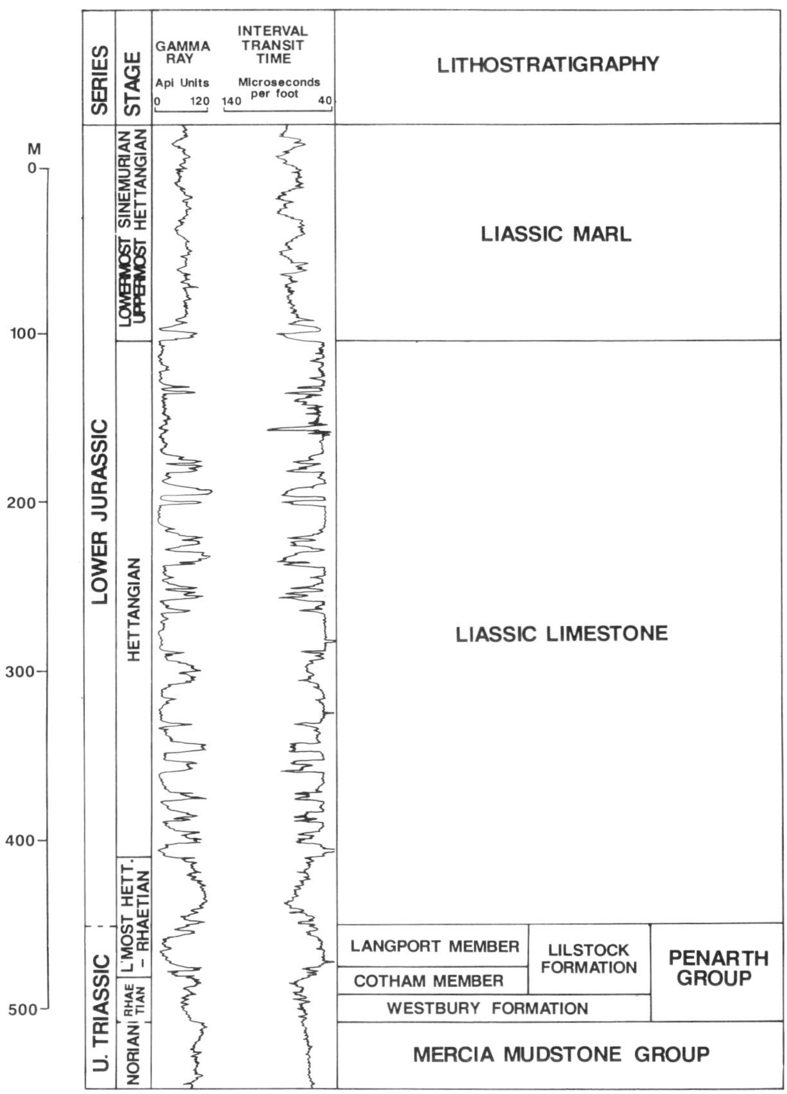

Fig. 2 — Summary lithostratigraphy of the Lower Jurassic and Upper Triassic.

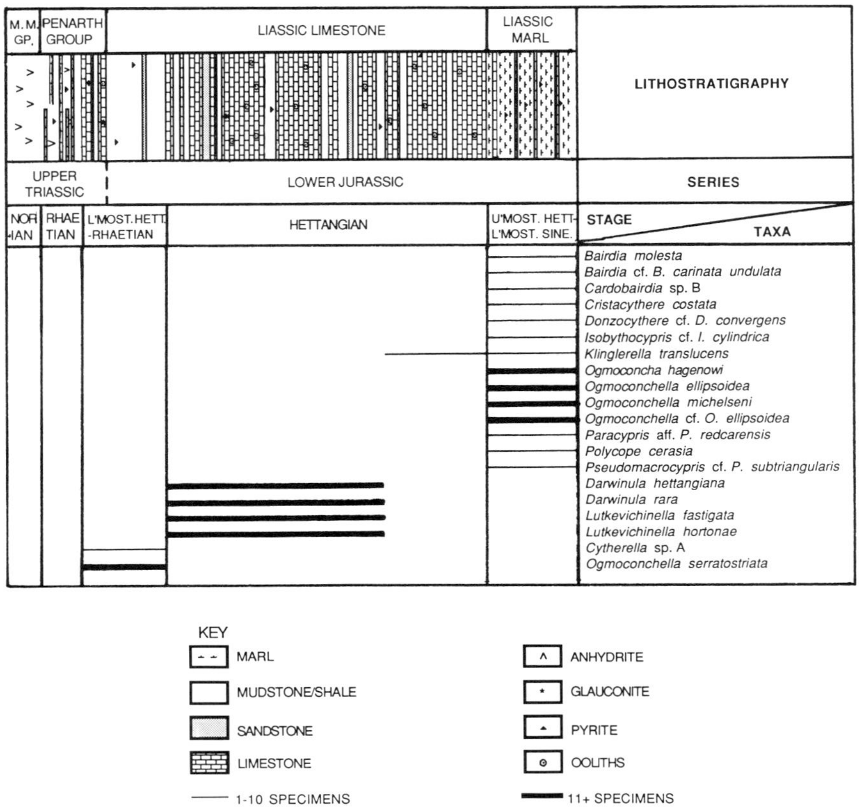

Fig. 3 — The stratigraphic distribution of ostracods in Lower Jurassic and Upper Triassic deposits from the North Celtic Sea and Fastnet Basins.

ham Member and overlying Langport Member. In the North Celtic Sea and Fastnet Basins the Cotham Member comprises medium to dark grey, green, moderately firm to hard, calcareous shales and mudstones with some anhydrite and limestones as in the Westbury Formation. The lower boundary is defined at a gamma-ray peak and the upper boundary at a decrease in the gamma-ray, spontaneous potential (SP) and sonic responses, and an increase in resistivity values which coincide with the incoming of the limestone lithologies of the Langport Member (Figs. 2–4).

The Langport Member in the North Celtic Sea and Fastnet Basins consists largely of sparitic or micritic limestone which is off-white, light to grey–brown and firm to hard; locally it is dolomitic or formed of a soft micrite. Ooliths, peloids and echinoderm fragments occur sporadically within a sparitic or micritic matrix; glauconite occurs very rarely. Thin shale–mudstone bands, generally light grey to brown grey, slightly calcareous and in part dolomitic, are present throughout the member. The lower boundary is placed at the base of a decrease in the gamma-ray, SP and sonic log responses, and a coincident increase in resistivity values. The upper boundary is placed at the base of an increase in gamma–ray and sonic and a decrease in resistivity values that reflect the incoming of the argillaceous sediments of the Lower Liassic Limestone member (Figs. 2–4).

The Liassic Limestone

The Liassic Limestone (Hettangian Limestone

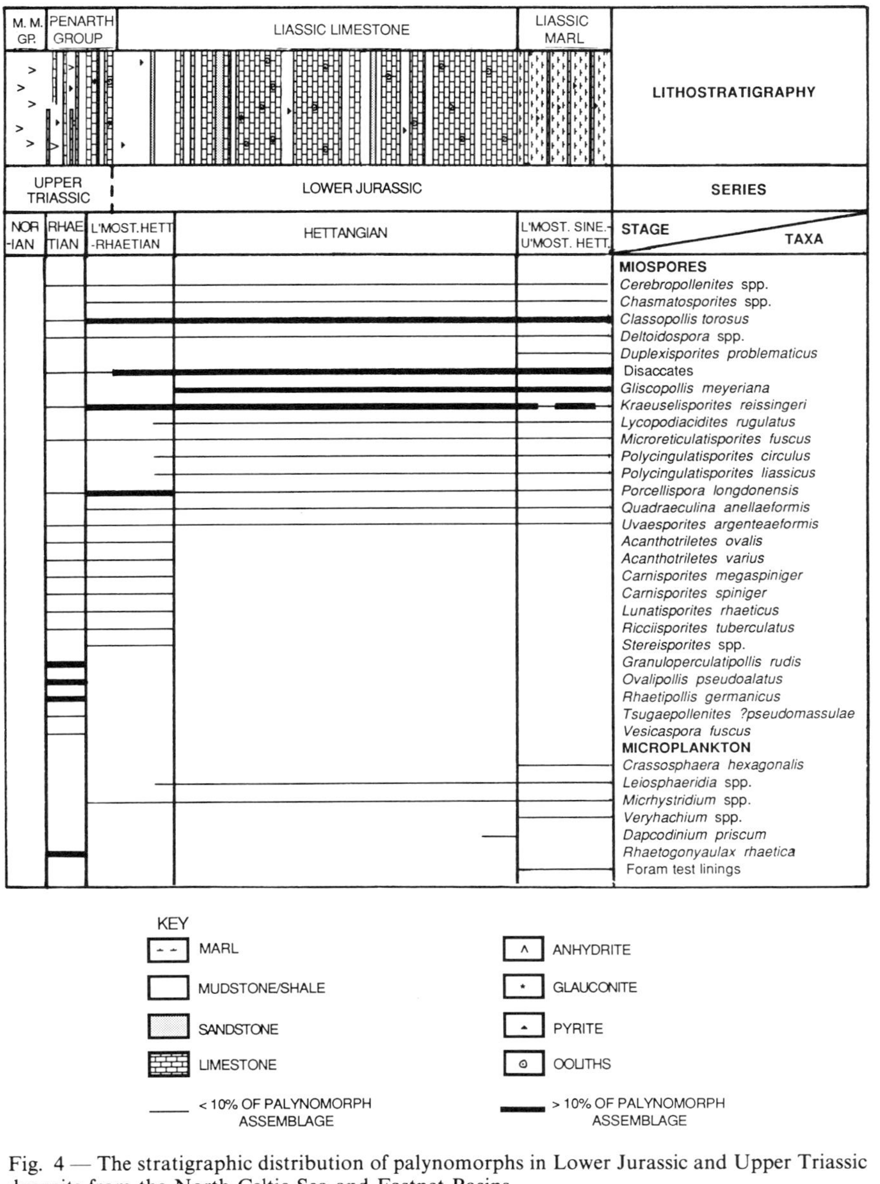

Fig. 4 — The stratigraphic distribution of palynomorphs in Lower Jurassic and Upper Triassic deposits from the North Celtic Sea and Fastnet Basins.

sensu Robinson *et al.* (1981) is an alternating sequence of carbonate and argillaceous sediments, often exceeding 200 m in thickness; it consists of a lower shale–mudstone member and an upper limestone or limestone–shale sequence. The lower member comprises a mudstone or shale, light to medium grey, grey–green, moderately calcareous to very calcareous, generally moderately hard and locally grading to siltstone. Some very thin micritic limestone, which is medium to dark brown, and/or fine-grained argillaceous sandstone stringers, which are white to grey and pyritic, also occur. The lower boundary throughout the North Celtic Sea and Fastnet Basins is recognized in borehole petrophysical logs by an increase to peak values in the gamma-ray log and an increase in the sonic and decrease in the resistivity values that reflect the abrupt lithological change from the limestones of the underlying Langport Member. The upper boundary is placed at the base of a sharp decrease in the SP, gamma-ray and sonic log responses and an increase in resistivity values, which mark the incoming of the thickly bedded units of limestone of the upper member of the Liassic Limestone (Figs. 2–4).

The upper member comprises a limestone–shale sequence, in which carbonates are generally dominant. In wells in the Fastnet Basin, limestone units are commonly thicker in relation to the argillaceous units whereas, in the southern part of the North Celtic Sea, argillaceous units are thicker and dominate the sequences. Three major groups of limestones are observed: firstly, white, off-white and light to dark-grey, moderately soft argillaceous limestones; secondly, light to dark-grey and dark-brown, often hard micrites; thirdly, white and light to dark-brown, hard oolitic and sparry limestones. The oolitic limestones in both the North Celtic Sea and the Fastnet Basins are generally less common than the micrites. The former are best developed on the western edge of the Fastnet Basin. Local dolomite replacement of the limestones occurs within some levels. Interbedded throughout the limestones are shales, siltstones and sandstones. The shales are similar to those in the underlying lower mudstone member; the siltstones generally comprise light to dark-grey, hard calcareous sediments, locally grading to sandy calcilutites. The sandstones are white, grey, brown and yellow, fine to coarse-grained, moderately well sorted, angular to subangular, usually with a calcareous cement and grade through to sandy limestones. Lignitic debris occurs randomly throughout these rocks. The base of the sequence is placed at the incoming of the thick limestone beds above the lower shale–mudstone member. This coincides with a decrease in the SP, gamma-ray and sonic log responses and an increase in resistivity values. The top of this sequence is placed at the base of the Liassic Marl marked by an increase in gamma-ray and SP log values and a decrease in sonic log response. Throughout this sequence, the wireline logs show irregular serrate motifs that reflect rapid alternation between the carbonates and argillaceous and arenaceous rocks (Figs. 2–4).

The Liassic Marl

The Liassic Marl overlies the Liassic Limestone in all wells studied, where it generally exceeds 60 m in thickness. It consists largely of calcareous clay or shale, which is commonly light to dark grey, soft to firm and subfissile to fissile, locally grading to grey or brown, soft to firm calcareous siltstones. In some sections, stringers of white to grey, hard, sparry micritic argillaceous limestones occur, generally within the lower parts of the sequence. In both the North Celtic Sea and the Fastnet Basins the base of this unit is taken at the first occurrence of marls without thickly bedded limestones and is associated with a distinct log motif; a high gamma-ray and sonic response, and a low resistivity value. The top boundary is at the base of the incoming sandstones or darker grey, usually less calcareous shales, coinciding in the latter case with a more regularly even gamma-ray motif. In all wells studied by the present authors the gamma-ray and sonic log responses show a distinct cyclic funnel-shaped motif, with up to three or four cycles repeated throughout the unit. This cyclicity reflects an increase in carbonate content, with each cycle capped by a thin limestone (Figs. 2–4).

BIOSTRATIGRAPHY

Range tops and abundances of palynomorphs and ostracod taxa used to describe four biostratigraphic associations have been determined from analysis of 10 wells within the Fastnet and North Celtic Sea Basins, allowing establishment of a composite biostratigraphy. The lowermost Sinemurian–uppermost Hettangian association (Association 1) is extracted from the Liassic Marl overlying the Liassic Limestone. A Hettangian microfaunal and palynomorph association is described from the upper Liassic Limestone unit (Association 2). The lowermost Hettangian–upper Rhaetian association (Association 3) is described from the Langport Member of the Lilstock Formation, and from the mudstone–shale member below the upper Liassic Limestone unit. The oldest (Rhaetian) association is extracted from an interbedded mudstone–shale–limestone sequence, probably equivalent to the Westbury Formation, and in part, the Cotham Member of the Lilstock Formation (Association 4).

Composite range charts for palynomorphs and ostracod taxa show top appearances and abundances of all species encountered in this study (Figs. 3 and 4), and selected taxa are illustrated in Plates 1–3.

Association 1 (lowermost Sinemurian–uppermost Hettangian)

Poorly diverse palynomorph assemblages are obtained from the siltstone–claystone sequences of the Liassic Marl. There are no palynomorph taxa with range tops within this interval. No dinocysts are recorded. Pollen grains dominate the palynomorph assemblages. *Classopollis torosus* (Reissinger) Balme 1957 attains proportions of up to 75% of the overall palynomorph assemblage in sampling intervals regularly spaced throughout the Liassic Marl within the Fastnet Basin well Elf 55/30-1. Abundances of *C. torosus* have been recorded by Aldorf & Doubinger (1978) from the Sinemurian–Hettangian of Portugal and by Wall (1965) from the Sinemurian–Hettangian of Britain.

In Cities Service 63/10-1 and Texaco 56/22-1, also located in the Fastnet Basin, *Gliscopollis meyeriana* (Klaus) Venkatachala 1966 is the most abundant miospore recorded, reaching up to 70% of the overall palynomorph assemblage in horizons throughout the Liassic Marl interval. Disaccate pollen grains are consistently present in the Liassic Marl assemblages, generally comprising 10–20% of the assemblage. The most common spore (more than 5% palynomorph assemblage) is *Kraeuselisporites reissingeri* (Harris) Morbey 1975. Miospore taxa also present include *Microreticulatisporites fuscus* (Nilsson) Morbey 1975, *Cerebropollenites* spp. (Nilsson) Singh & Kumar 1969, *Polycingulatisporites liassicus* Schulz 1967, *Porcellispora longdonensis* (Clarke) Scheuring emend. Morbey 1975, *Deltoidospora* spp. Miner 1935, *Duplexisporites problematicus* (Couper) Playford & Dettmann 1965, *Quadraeculina anellaeformis* Maljavkina 1949 and *Lycopodiumsporites rugulatus* (Couper) Schulz 1967. Low numbers (less than 1%) of the acanthomorph acritarchs *Micrhystridium* spp. (Deflandre) Lister 1970 and *Veryhachium* spp. Downie & Sarjeant 1963 are encountered. More frequently, the prasinophycean algae including *Tasmanites* spp. Wall 1965 and *Crassosphaera* spp. Cookson & Manum 1960 are present.

The lowermost Sinemurian–uppermost Hettangian association is defined by the occurrence of rich ostracod faunas. The ostracod fauna is dominated by flood occurrences of the family Healdiidae, throughout both of the basins. The marine ostracods are dominated by four species; *Ogmoconcha hagenowi* Drexler 1958, *Ogmoconchella ellipsoidea* (Jones 1872), *O.* cf. *O. ellipsoidea* (Jones 1872) *sensu* Ainsworth *et al.* 1987 and *O. michelseni* sp nov. *Klinglerella translucens* (Blake 1876) is generally sparsely recorded in this interval, in similar numbers to the underlying topmost horizons of the Hettangian interval. The other characteristic species seen within this interval are generally rarer and occur mainly within the Fastnet Basin. These include *Bairdia molesta* Apostolescu 1959, *B.* cf. *B. carinata undulata* Herrig 1981, *Cardobairdia* sp. B Ainsworth *et al.* 1987, *Cristacythere costata* Michelsen 1975, *Donzocythere* cf. *D. convergens* (Donze 1968 *sensu* Donze 1985), *Isobythocypris* cf. *I. cylindrica* Herrig 1981 *sensu* Ainsworth *et al.* 1987, *Paracypris* aff. *P. redcarensis* (Blake 1876) *sensu* Ainsworth *et al.* 1987, *Polycope cerasia* Blake 1876 and *Pseudomacrocypris* cf. *P. subtriangularis* Michelsen 1975 *sensu* Ainsworth *et al.* 1987. These species are all extracted from the Liassic Marl, with many of the above species occurring throughout the Hettangian and lower Sinemurian of northern Europe (Lord 1978, Donze 1985, Michelsen 1975, Drexler 1958, Herrig 1979a, b, c, Malz 1971).

Association 2 (Hettangian)

The Liassic Limestone contains shale–marl horizons, from which moderate palynomorph assemblages have been obtained. In comparison with European Hettangian microplankton assemblages, dinocyst occurrence in the Irish offshore basins within this interval is rare and poorly diverse. *Dapcodinium priscum* Evitt 1961 is recorded within the top metre of the Hettangian Limestone unit in the Fastnet Basin well (Elf 55/30-1). Apart from the top appearance downhole of this taxon, palynomorphs extracted from the Hettangian Limestone yield miospore and acritarch assemblages similar in composition and abundance to the overlying lowermost Sinemurian–uppermost Hettangian (Association 1). *Classopollis torosus* and/or *Gliscopollis meyeriana* variably dominate the assemblage and *Kraeuselisporites reissingeri* is abundant (more than 10%), these taxa occurring as single grains and in tetrads. Miospore taxa which are present but not contributing more than between 1% and 5% of the assemblage include *Polycingulatisporites liassicus*, *Microreticulatisporites fuscus*, *Porcellispora longdonensis*, *Quadraeculina anellaeformis*, *Uvaesporites argenteaeformis* (Bolkhovitina) Schulz 1967, *Duplexisporites problematicus* and *Cerebropollenites* spp. prasinophycean algae including species of *Tasmanites* and *Crassosphaera* are also recorded. Hettangian miospore assemblages from the North Celtic Sea are normally poor in diversity and preservation. This is due to overall high organic maturity, to deep burial and/or to proximity to intrusions. Palynomorphs are consequently degraded and darkened.

Both marine and non-marine ostracods are recovered from within the Liassic Limestone. Generally, a non-marine ostracod fauna, comprising no more than four species, is observed in the lower and middle horizons of the Liassic Limestone, while a marine fauna may be present in the upper horizons. Within the lower and middle horizons, the association is dominated by two genera (*Lutkevichinella* and *Darwinula*) each with two species. Where recovered, abundance is generally high, with both genera being extracted from the softer argillaceous limestone horizons rather than from the well-cemented micrites, pelsparites or oosparites.

Of the four non-marine species, only *Lutkevichinella hortonae* sp.nov. has been recorded outside the North Celtic Sea and Fastnet Basins, by Apostolescu (1959, 1961) from the Hettangian of the Paris Basin and central France, respectively. In the top horizons of the Liassic Limestone, *Klinglerella translucens* is often recovered, a species which extends up into the Liassic Marl in many of the wells studied. Throughout the Liassic Limestone, the specimes are poorly preserved and, like the non-marine faunas below, are extracted from the softer limestone lithologies.

Association 3 (lowermost Hettangian–upper Rhaetian)

The carbonate-rich mudstone–shale sequence above the Langport Member and underlying the upper Liassic Limestone unit represents the lithological unit within which the Jurassic–Triassic (Hettangian–Rhaetian) boundary exists. It is not possible to place this boundary accurately without recovery of ammonite faunas (see Lithostratigraphy.)

Both faunas and floras show no change across this boundary, and therefore Association 3 is assigned a lowermost Hettangian–upper Rhaetian age. Included within this association are faunas and floras from the Langport Member (by lithological comparisons, Rhaetian in age) and the mudstone–shale sequence overlying this (lowermost Hettangian–uppermost Rhaetian). Recovery of microfaunas and palynomorphs from the Langport Member is very poor, owing to the calcareous nature of the lithologies. New occurrences downhole of *Acanthotriletes ovalis* Nilsson 1958, *A. varius* Nilsson 1958, *Carnisporites megaspiniger* Morbey 1975, *C. spiniger* (Leschik) Morbey 1975, *Lunatisporites rhaeticus* (Schulz) Warrington 1974 and *Ricciisporites tuberculatus* Lundblad 1954 are recorded from within the calcareous shale–mudstone sequence just above the Langport Member; therefore an age of lowermost Hettangian–upper Rhaetian must be assigned to this association. Of local significance at least is the abundant occurrence of

Porcellispora longdonensis (up to 11% of the palynomorph assemblage). *Kraeuselisporites reissingeri* is still abundant, comprising greater than 30% of overall palynomorph assemblage in the Elf 55/30-1 well. The ?green algae–sphaeromorph group, *Leiosphaeridia* (Timofeev) Mädler 1963, is represented sporadically throughout the Lower Jurassic and Upper Triassic sections. Leiospheres have been recorded in abundance by Wall (1965) from the Lower Lias of Britain.

The marine ostracod faunas occurring in the lowermost Hettangian–Rhaetian interval are generally abundant, but poorly diverse (less than four species). They are dominated by *Ogmoconchella serratostriata* sp.nov., with rare *O. ellipsoidea* (Jones), *Cytherella* sp. A Ainsworth & Horton 1986, and a poorly preserved *Klinglerella* sp. This assemblage is extracted from the lower shale–mudstone member of the Liassic Limestone, above that of the Langport Member. The faunas are best observed in the Fastnet Basin, within the Cities Service 63/4-1 and Elf 55/30-1 wells. In the other wells examined by the present authors, the shales were indurated, and extraction of ostracod faunas proved difficult. Many of the extracted specimens were fragmented. In these wells, numerical abundance was also very low. No faunas have been recovered from the Langport Member, because of difficulty of extraction from the well-cemented limestone lithologies.

Association 4 (?Middle–lower Rhaetian)

Rhaetian palynomorph assemblages vary in composition, abundance and preservation in the North Celtic Sea and Fastnet Basins. They are indicated by the top downhole appearance of abundant *Rhaetogonyaulax rhaetica* (Sarjeant) Loeblich & Loeblich emend. Harland, Morbey & Sarjeant 1975, *Ovalipollis pseudoalatus* (Thiergart) Schuurman 1976 and by the first downhole occurrence of *Rhaetipollis germanicus* Schulz 1967, *Tsugaepollenites? pseudomassulae* (Mädler) Morbey 1975 and *Vesicaspora fuscus* (Pautsch) Morbey 1975.

In the North Celtic Sea Basin, often poorly preserved assemblages are dominated by *Ovalipollis pseudoalatus* which may attain percentages in excess of 60% of the total palynomorph assemblage and *Granuloperculatipollis rudis* (Venkatachala & Goczan) Morbey 1975 (more than 18% palynomorph assemblage), a taxon occurring in tetrads and individual grains and having a range top within the siltstone–shale unit underlying the Langport Member. The dinoflagellate cyst *Rhaetogonyaulax rhaetica* is also recorded in abundance within this shale–siltstone unit, often attaining percentages as high as 14% of the total palynomorph assemblage; *Kraeuselisporites reissingeri*, *Porcellispora longdonensis*, *Quadraeculina anellaeformis*, *Uvaesporites argenteaeformis*, *Microreticulatisporites fuscus* and the green algae *Tasmanites* and *Cymatiosphaera* spp. (Wetzel) Deflandre 1954 each comprise less than 1% of the palynomorph assemblage. The first appearance downhole of *Tsugaepollenites? pseudomassulae* is also at this level, this taxon accounting for up to 4% of the palynomorphs present. Morbey (1978) documents the incoming of abundant *Granuloperulatipollis rudis* towards the base of the *Rhaetipollis germanicus* miospore Biozone, which he recognized throughout northwest Europe. He also suggested, however, that abundant *Granuloperculatipollis rudis* may be indicative of a lower Rhaetian age.

In the Fastnet Basin, the Elf 55/30-1, Cities Service 63/10-1 and Deminex 56/21-2 wells yield abundant *Rhaetipollis germanicus* (20 % assemblage), *Ovalipollis pseudoalatus* (more than 60% assemblage), with lesser amounts of *Gliscopollis meyeriana*, *Classopollis* spp., *Granuloperculatipollis rudis*, *Uvaesporites argenteaeformis*, *Microreticulatisporites fuscus*, *Kraeuselisporites reissingeri*, *Tsugaepollenites? pseudomassulae*, *Vesicaspora fuscus* and *Lunatisporites rhaeticus*. The dinocyst taxon *Rhaetogonyaulax rhaetica* may account for up to 15% of the palynomorph assemblage.

This association is thought to be equivalent in age to assemblages described from the Lower Westbury to Upper Cotham Beds (Warrington 1978). This correlation is supported by the abundant recovery of *Ovalipollis ovalis* and the presence of

Tsugaepollenites? pseudomassulae, *Rhaetipollis germanicus*, *Lunatisporites rhaeticus* and *Vesicaspora fuscus*. Warrington (1978) found that, within the upper part of the Cotham Beds, there was a decline in the numbers of *Ovalipollis ovalis* and disappearance of *Tsugaepollenites? pseudomassulae*, *Rhaetípollis germanicus*, *Vesicaspora fuscus* and *Rhaetogonyaulax rhaetica*. This association is also closely comparable with assemblages described from the Penarth Group of the Winterbourne Kingston Borehole, Dorset (Warrington 1982), the Withycombe Farm Borehole (Warrington 1978), the Steeple Aston Borehole (Warrington 1977) and from upper Triassic palynomorph assemblages from the western English Channel and neighbouring areas (Warrington 1983).

Only a single well at present has yielded a very sparse marine ostracod fauna, comprising a single species: *Ogmoconchella martini* (Anderson 1964). This species has been recorded from the middle and upper Rhaetian sediments of England and northern Germany (Anderson 1964). The paucity of fauna is unlike that of northern Europe as described by workers including Anderson (1964), Christensen (1962), Dreyer (1965), Kozur (1968, 1970) and Will (1969).

PALAEOECOLOGY AND DEPOSITIONAL HISTORY

The Rhaetian sequence of sediments, often containing evaporites, conformably overlie the generally red-coloured, fluvial, sabhka and playa-lake deposits of the Mercia Mudstone Group. During the deposition of the Westbury Formation and Cotham Member, a lagoonal environment, with intermittent connections to a shallow open sea, is suggested by the lithologies and the microfossils. The limited diversity of the dinocyst component of the palynomorph association reflects deposition in an environment of reduced salinity (Wall 1965, Warrington 1978). Drying of the region is reflected by the accumulation of evaporites.

A low-energy warm carbonate-rich, very shallow marine environment prevailed during the deposition of the Langport Member. The rare glauconite, ooliths and peloids within the carbonates and the recovery of acanthomorph acritarchs suggest short intermittent higher-energy environments. These shallow marine conditions continued into the lowermost Hettangian, when a moderately thin grey to dark-grey mudrock was deposited, yielding abundant, usually poorly preserved marine ostracods, often comprising a single taxon.

The Hettangian sequence is dominated by limestones, some of which are oolitic or peloidal, suggesting warm shallow marine high-energy carbonate-rich enviroments. However, the thin marl and soft micritic limestone lithologies in the lower and middle parts of the Liassic Limestone yield abundant freshwater (*Darwinula*) and brackish-water oligohaline (*Lutkevichnella*) ostracods. The abundant occurrence of the pollen *Classopollis* within the Hettangian is thought to indicate relatively arid or semi-arid conditions (Vakhrameev 1981). Plants producing this pollen type appear to have thrived in lowlands bordering lagoons (Sladen & Batten 1984). The floras and lithologies indicate an unstable marine environment, perhaps reflecting a rapidly migrating shoreline due to the commencement of the Rhaetian–Hettangian marine transgression (Robinson *et al.* 1981, Naylor & Shannon 1982). Higher in the Liassic Limestone, only rare marine ostracods are recovered, suggesting that the Jurassic marine transgression had become more established throughout both of the basins by upper Hettangian times.

In the uppermost Hettangian–lowermost Sinemurian interval, carbonate deposition was replaced by cyclic deposition of marls and thin micritic limestones. The ostracods, although not very diverse, are abundant and may reflect a slightly deeper, more tranquil inner-shelf marine environment.

DISCUSSION

Association 1 (lowermost Sinemurian–uppermost Hettangian) consists of poorly diverse miospore assemblages characteristically dominated by the pollen *Classopollis*. Such dominance has been previously described from lower Sinemurian sections in Britain by Wall (1965). Abundance of *Kraeuselisporites reissingeri* is characteristic of lower Sinemurian assemblages from northwest Europe (Morbey 1978). The presence of green algae, acanthomorph acritarchs and foram test linings is also considered typical of lower Sinemurian assemblages (Riding 1987). The rich ostracod faunas in this association are dominated by *Ogmoconcha hagenowi*, *O. ellipsoidea* and *O. michelseni*. The former two species are commonly recorded throughout lower Sinemurian and Hettangian sediments from northwest Europe (Lord 1971, 1978, Donze 1985, Drexler 1958, Malz 1971, Michelsen 1975). The remaining fauna discussed in the biostratigraphic section is also well described from France (Apostolescu 1959, Donze 1968, 1985), East Germany (Herrig 1979, a,b,c) and Denmark (Michelsen 1975).

The Hettangian interval (Association 2), is similar to the overlying lower Sinemurian–upper Hettangian interval, characterized by monotonous miospore assemblages. The dominance of *Classopollis* spp. and *Gliscopollis meyeriana* has been described from the Lias and Rhaetian of the Larne Borehole, Northern Ireland (Warrington & Harland 1975) and is analogous to assemblages described by Wall (1965) from English Hettangian deposits. Morbey (1978) stated that the dinocyst *Dapcodinium priscum* occurs characteristically within the Hettangian Stage all over northwest Europe; the taxon has, however, a recorded range from the Rhaetian to lower Sinemurian (Riding 1987). To date, it has only been recovered from the Hettangian interval within the Fastnet Basin. Of the four non-marine ostracods recovered from the Hettangian only *Lutkevichinella hortonae* occurs outside the two basins, from the Hettangian of the Paris Basin and central France (Apostolescu 1959, 1961). These three regions (the North Celtic Sea, Fastnet Basin and central France) possessed similar ecological conditions at this time, all of which differ from the deeper marine faunas and environments prevailing in the rest of northern Europe. Within the top horizons of the Liassic Limestone, sparse numbers of *Klinglerella translucens* occur. This taxon has a previously recorded occurrence from the marine Hettangian and lower Sinemurian of England (Lord 1971) and France (Donze 1985).

The palynomorph assemblages of lowermost Hettangian–upper Rhaetian (Association 3) age are poorly diverse and contrast markedly to those of the underlying Rhaetian association. Clear abundances of *Classopollis torosus*, *Kraeuselisporites reissingeri* and *Porcellispora longdonensis* dominate the assemblages. An abundance of the latter taxon has been documented by Warrington (1978) from the Upper Cotham Member of the Penarth Group, who also noted within this interval an increase in the relative abundances of several taxa of trilete miospores coinciding with the disappearance of rich Rhaetian miospore assemblages. Association 3 spans the Triassic–Jurassic system boundary and it is now commonly agreed that there are no major palynofloral breaks that could be correlated precisely with that boundary. The basal Jurassic palynofloras appear to reflect an extension of the ranges of the upper Rhaetian assemblages across the Triassic–Jurassic boundary (Fisher & Dunay 1981, Warrington 1983). Ostracod faunas from Association 3, except for *Ogmoconchella ellipsoidea*, have not yet been described from outside the North Celtic Sea and Fastnet Basins.

Rhaetogonyaulax rhaetica is commonly recorded in large numbers from many British onshore sections and is especially abundant within the Lower Cotham Member of the Lilstock Formation and the Westbury Formation, both formations included in the Penarth Group (Warrington 1974, 1977, 1978). Miospores characteristic of Association 4 (Rhaetian) are comparable in composition with those documented from the Westbury Formation and Cotham Member succession elsewhere in onshore Britain (Orbell 1973, Warrington 1974, 1977, 1978, 1982) and include *Rhaetipollis germanicus*, *Granuloperculatipollis rudis*, *Tsugaepollenites? pseudomassulae* and *Ovalipollis pseudoalatus*, and are indicative of a Rhaetian age. Morbey (1978) has used a zonation scheme based on miospores and microplankton to propose a palynostratigraphy for the Upper Triassic and Lower Jurassic in northwest Europe. He recognized a *Rhaetipollis germanicus* upper Rhaetian Zone, characterized by the association of that taxon with common–abundant *Ovalipollis pseudoalatus* and *Gliscopollis meyeriana*. The dinoflagellate cyst zone equivalent, the *Rhaetogonyaulax rhaetica* Biozone is assigned an age of lowermost Hettangian–Rhaetian, the incoming of *R. rhaetica* close to the base of the Hettangian stage. He also states that *R. rhaetica* becomes more abundant and is characteristically developed in marine sequences of Rhaetian age.

Recovery of Rhaetian ostracods from the North Celtic Sea and Fastnet Basins is very sparse to non-existent, in comparison with rich and diverse faunas described from Great Britain (Anderson 1964), France (Colin *et al.* 1985), Germany (Wicher 1951, Dreyer 1965, Kozur 1968, 1970, Will 1969) and Denmark (Christensen 1962). The faunal paucity is most probably due to two reasons: firstly, the sediments in the Irish offshore basins contain varying quantities of evaporites, indicating a more continental to marginal marine environment, alternating with very short marine conditions; secondly, the dolomitization of the limestone may indicate percolation of ground water, leading to decalcification of the benthonic faunas.

SYSTEMATIC DESCRIPTIONS OF NEW OSTRACOD SPECIES

N. R. Ainsworth

The ostracods described in this paper are deposited in the collections of the Museum of the Geology Department, Trinity College, Dublin (prefix TCD.). All measurements are in millimetres.

Subclass Ostracoda Latreille 1806
Order Podocopida Müller 1894
Suborder Podocopina Sars 1866
Superfamily Darwinulacea Brady & Norman 1889
Family Darwinulidae Brady & Norman 1889

Genus *Darwinula* Brady & Robertson 1885

Darwinula hettangiana sp. nov.
(Plate 1, Figs. 7, 8 and 11)

1986 *Darwinula* sp. 2; Ainsworth & Horton, p. 28.
1987 *Darwinula* sp. B; Ainsworth, *et al.*, p. 620.

Derivato nominis. From the occurrence of this species from Hettangian strata within the Fastnet and North Celtic Sea Basins.
Diagnosis. Carapace large to very large, subrectangular. Anterior margin asymmetrically rounded, extremity situated at one-third height. Posterior margin broadly rounded, symmetrical. Dorsal margin arched, with greatest height in posterior third. Ventral margin straight, with weak oral concavity slightly in front of mid-height. Left valve larger than right valve. Valve surface smooth.
Holotype. Carapace, TCD. 29032; from the Hettangian of Elf 55/30-1; depth, 1955 m; latitude, 59°09′09.630″ N; longitude, 10°05′54.190″ W; Fastnet Basin.
Material. Nine valves, 153 carapaces.
Distribution. Recovered from the Hettangian of Cities Service 63/4-1, Cities Service 63/10-1, Deminex 56/21-2, Elf 55/30-1 and Gulf 56/18-1.
Dimensions.

	Length (mm)	Height (mm)
Holotype		
Carapace TCD. 29032	1.01	0.50
Paratypes. Same sample as holotype.		
Carapace TCD. 29033	0.93	0.48
Carapace TCD. 29034	0.95	0.46
Carapace TCD. 29035	0.98	0.46
Carapace TCD. 29036	1.05	0.48

Description. Carapace large to very large, subrectangular. Anterior margin asymmetrically rounded, extremity at one-third height. Posterior margin broadly rounded, symmetrical. Dorsal margin arched, greatest height in posterior third.

Ventral margin straight, with weakly formed oral concavity slightly in front of mid-length. Dorsal view subpyriform. Maximum length at one-third height, maximum height and width in posterior third. Left valve larger than right valve, overlapping right valve along dorsal and ventral margins. Sexual dimorphism not observed. Moderately thinly calcified. Valve surface smooth. Internal details only very poorly observed owing to poor preservation. Hinge adont. Muscle scar pattern viewed externally consists of 14 elongate scars arranged in a rosette pattern.

Remarks. *Darwinula hettangiana* sp. nov. possesses similarities to *D. major* Jones 1894 from the Rhaetian of northwest Europe. It is readily distinguished by its more narrowly rounded anterior, more strongly arched dorsal margin and more subpyriform outline in dorsal view, rather than the more subrectangular view as in *Darwinula major* Jones 1894. *Darwinula globulosa* (Jones 1862) differs in that its greatest height is situated in mid-length, with a more regular dorsal arching, and the ventral margin is concave rather than straight. *Darwinula oblonga* (Roemer 1839) is also similar but can be distinguished by its larger size, less arched dorsal margin and more asymmetrically rounded anterior.

Generally occurring in large numbers in association with *Darwinula rara* sp. nov. and *Lutkevichinella* spp. in the soft marl horizons within the Liassic Limestone of Hettangian age. This species is the more common of the two *Darwinula* species within the Liassic Limestone.

Darwinula rara sp. nov.
(Plate 1, Figs. 9, 10 and 12)

1986 *Darwinula* sp. 1; Ainsworth & Horton, p. 28.
1987 *Darwinula* sp. A; Ainsworth *et al*., p. 620.

Derivato nominis. Latin *rarus*, infrequent, with reference to the less common occurrence of this species with the Liassic Limestone.

Diagnosis. Carapace of medium size, subcylindrical. Anterior margin narrowly rounded, extremity at two-fifths height. Posterior margin asymmetrically rounded, extremity slightly above mid-height. Dorsal margin very weakly arched to straight. Ventral margin straight, with weak oral concavity.

Holotype. Carapace, TCD. 29037; from the Hettangian of Elf 55/30-1; depth, 1950 m; latitude, 50°09′09.630″ N, longitude, 10°05′54.190″ W; Fastnet Basin.

Material. Two valves, 46 carapaces.

Distribution. Recorded in the Hettangian of Cities Service 63/4-1, Cities Service 63/10-1 and Elf 55/30-1.

Dimensions.

	Length (mm)	Height (mm)
Holotype		
Carapace TCD. 29037	0.61	0.28
Paratypes. Same sample as holotype.		
Carapace TCD. 29038	0.61	0.28
Carapace TCD. 29039	0.60	0.30
Carapace TCD. 29040	0.53	0.28
Carapace TCD. 29041	0.57	0.25

Description. Carapace of medium size, subcylindrical. Anterior margin narrowly rounded, slightly asymmetrical extremity at two-fifths height. Posterior margin asymmetrically rounded, extremity slightly above mid-height. Dorsal margin very weakly arched to straight, weakly tapers to anterior, curve continuous into antero-dorsal margin. Ventral margin straight, with weakly defined oral concavity. Dorsal view elongate subpyriform. Maximum length at mid-height, maximum height immediately posterior of mid-length, maximum width posteriorly. Left valve larger than right valve, overlapping right valve dorsally and ventrally. Sexual dimorphism not observed. Thinly calcified. Valve surface smooth. Internal details only very poorly observed owing to the poorly preserved material. Hinge adont.

Remarks. A species that closely resembles both *Darwinula liassica* Jones 1894 and *D. leguminella* (Forbes 1855). It can be distinguished from the type material of the former by its smaller size and the more asymmetrically rounded posterior margin and from the latter by its smaller size, less inclined dorsal margin and less narrowly rounded anterior.

Occurring with *Darwinula hettangiana* sp. nov., but in fewer numbers and horizons, within the Hettangian Liassic Limestone.

Superfamily Cytheracea Baird 1850
Family Limnocytheridae Klie 1938
Genus *Lutkevichinella* Snejder 1956

Lutkevichinella fastigata sp. nov.
(Plate 1, Figs. 13–18)

1986 Indet. Gen. B sp. 1; Ainsworth & Horton, p. 28, Fig. 3.
1987 *Lutkevichinella?* sp. A; Ainsworth *et al.*, p. 620.

Derivato nominis. Latin *fastigatus*, sloping up or down, with reference to the dorsal margin which slopes down to the posterior margin.
Diagnosis. Carapace of medium size, subrectangular. Anterior margin broadly rounded. Posterior margin strongly asymmetrically rounded, extremity at one-third height. Dorsal margin straight, tapers at posterior. Ventral margin concave. Ornament of weak reticulation; on ventral surface of inflation longitudinal riblets, with weak cross-ribbing. Median sulcus only moderately developed in front of mid-length.
Holotype. Female carapace, TCD. 29042; from the Hettangian of Elf 55/30-1; depth, 2025 m; latitude, 50°09′09.630″ N; longitude, 10°05′54.190″ W; Fasnet Basin.
Material. One valve, 137 carapaces.
Distribution. Recovered from the Hettangian of Cities Service 63/4-1, Cities Service 63/10-1, Deminex 56/21-2 and Elf 55/30-1.
Dimensions.

		Length (mm)	Height (mm)
Holotype			
Female carapace	TCD. 29042	0.60	0.31

Paratypes. Carapaces from Elf 55/30-1, depth, 2025 m, carapace juv. from Cities Service 63/4-1, depth, 1213 m.

Female carapace	TCD. 29043	0.63	0.28
Male carapace	TCD. 29044	0.64	0.29
Male carapace	TCD. 29045	0.65	0.31
Carapace juv.	TCD. 29046	0.53	0.28

Description. Carapace of medium size, subrectangular. Anterior margin broadly rounded. Posterior margin asymmetrically rounded, extremity at one-third height. Dorsal margin straight, tapering to posterior, cardinal angles prominent. Posterior of mid-length, margin obscured by weak dorsal inflation and short dorsal rib. Ventral margin concave, behind mid-length curve tapers up to posterior, margin obscured by inflation. Dorsal view fusiform. Maximum length at mid-height, maximum height at anterior cardinal angle, maximum width behind mid-length, towards ventral margin. Left valve larger than right valve, overlapping dorsally and ventrally. Sexually dimorphic, male dimorph more elongate and less wide than female dimorph. Moderately thinly calcified. A weak sulcus extends down from anterior of mid-dorsal margin, terminating above mid-height. Valve surface moderately weakly reticulate, reticulae polygonal, extend over whole valve surface, except for compressed anterior and posterior marginal rims. On ventral surface of inflation, reticulae aligned subparallel to margin, forming longitudinal riblets. Posterior of mid-length, a short, weakly formed curved dorsal rib is situated above inflation, rib extends slightly above margin. Internal details not observed, owing to poor preservation of valve.
Remarks. *Lutkevichinella fastigata* sp. nov. is somewhat similar to *L.? gruendeli* Kozur 1970, differing in possessing only a single median sulcus, the greatest width in the female dimorph being situated more towards the middle of the valve and in the more compressed anterior and posterior margins. It can be distinguished from *Lutkevichinella hortonae* sp. nov. by its smaller size, more narrowly and asymmetrically rounded posterior margin, and more pronounced reticulate ornament.

Recorded in association with *Lutkevichinella hortonae* sp. nov. and the *Darwinula* spp. from the Hettangian of the Fastnet Basin.

Lutkevichinella hortonae sp. nov.
(Plate 2, Figs. 1–6)

1959 Ostracoda ∞; Apostolescu, p. 816, Plate 4, Fig. 78.
1961 Ostracoda ∞; Apostolescu, Apostolescu, Table 4.
1986 *Limnocythere* sp. 1; Ainsworth & Horton, p. 28, Plate 1, Fig. 19.
1987 *Limnocythere* sp. A; Ainsworth *et al.*, p. 620.

Derivato nominis. In honour of Nicola Horton, who with one of the present authors illustrated this species from the Fastnet Basin.

Diagnosis. Carapace large, subrectangular. Anterior margin bluntly rounded. Posterior margin asymmetrically rounded, extremity below mid-height. Dorsal margin straight. Ventral margin concave. Sulcus extends from mid-dorsal margin to central part of valve. Valve surface very weakly reticulate. A very weak dorsal rib extends along dorsal margin.

Holotype. Female carapace, TCD. 29047; from the Hettangian of Elf 55/30-1, depth, 2015 m; latitude, 50°09′09.630″ N, longitude, 10°05′54.190″ W; Fastnet Basin.

Material. 1 valve, 228 carapaces.

Distribution. Recovered from the Hettangian of Cities Service 63/4-1, Cities Service 63/10-1, Deminex 56/21-2 and Elf 55/30-1.

Dimensions.

		Length (mm)	Height (mm)
Holotype			
Female carapace	TCD. 29047	0.75	0.35
Paratypes. Same sample as holotype.			
Female carapace	TCD. 29048	0.73	0.35
Female carapace	TCD. 29049	0.74	0.37
Male carapace	TCD. 29050	0.71	0.37
Male carapace	TCD. 29051	0.73	0.41

Description. Carapace large, subrectangular. Anterior margin bluntly rounded. Posterior margin asymmetrically rounded, extremity slightly below mid-height. Dorsal margin straight, with prominent cardinal angles, slightly obscured by weak dorsal rib. Ventral margin concave at mid-length, strongly convex antero-ventrally, less so postero-ventrally. Dorsal view males elongate subovate, with compressed anterior margin, females subpyriform, also with compressed anterior margin. Maximum length slightly below mid-height, maximum height at anterior cardinal angle, maximum width slightly posterior of mid-length in male dimorph, in posterior third in female dimorph. Sexually dimorphic, female dimorph less high, with greatest width situated more towards posterior margin. Moderately thinly calcified. A moderately developed sulcus extends down from mid-dorsal margin to central region of valve. Valves moderately inflated, with greatest inflation extending down to mid-ventral margin. In female carapaces, inflation extends to posterior margin. Valve surface very weakly reticulate, reticulae polygonal. A very weak dorsal rib extends along dorsal margin. Internal details not observed.

Remarks. Owing to the generally poorly preserved nature of the material, many of the specimens appear to possess a smooth valve surface. It is only within a few specimens that the very weakly reticulate ornament can be observed.

This very distinct species cannot be easily confused with any other species. It does, however, show some similarities to both *Lutkevichinella keupera* (Will 1969) and *L.? gruendeli* Kozur 1970. *Lutkevichinella hortonae* sp. nov. can be distinguished from the former by its larger size, more concave ventral margin and less deep median sulcus, and from the latter by possessing only a single sulcus, a weakly reticulate valve surface and a more convex ventral margin. *Lutkevichinella* sp. (Bizon & Oertli 1961) differs from the present author's species by its compressed posterior margin and concave dorsal margin.

First illustrated by Apostolescu (1959, 1961) from the Hettangian of both the Paris Basin and central France. In the Fastnet Basin, it is recorded in abundance in similar aged strata, often in association with both the *Darwinula* spp. and *Lutkevichinella fastigata* sp. nov.

Suborder Metacopina Sylvester-Bradley 1961
Superfamily Healdiacea Harlton 1933
Family Healdiidae Harlton 1933
Genus *Ogmoconchella* Gründel 1964

Ogmoconchella michelseni sp. nov.
(Plate 2, Figs. 13, 14 and 17)

1987 *Ogmoconchella* cf. *O. mouherensis* (Apostolescu); Ainsworth *et al.*, p. 620.

Derivato nominis. In honour of Dr O. Michelsen; in recognition of his work on the Lower Jurassic Ostracoda of the Danish Embayment.

Diagnosis. Carapace of medium size, subovate in lateral view, fusiform in dorsal view. Anterior margin more narrowly rounded than posterior margin. Dorsal margin strongly arched, with greatest convexity slightly behind mid-length. Ventral margin weakly concave in right valve, straight in left valve. Dorsal view; in posterior third, a distinct bend towards posterior, coinciding with maximum width. Valve surface smooth.

Holotype. Carapace, TCD. 29052; from the uppermost Hettangian–lowermost Sinemurian of Elf 55/30-1; depth, 1800 m; latitude, 50°09′09.630″ N; longitude, 10°05′54.190″ W; Fastnet Basin.

Material. Nine valves, 369 carapaces.

Distribution. Recorded from the uppermost Hettangian-lowermost Sinemurian of Burmah 57/7-1, Cities Service 63/4-1, Cities Service 63/10-1, Deminex 56/21-2, Elf 55/30-1, Esso-Marathon 48/30-1, Gulf 56/18-1, Marathon 49/29-1 and Marathon 56/20-1.

Dimensions.

		Length (mm)	Height (mm)
Holotype			
Carapace	TCD. 29052	0.63	0.45
Paratypes. Same sample as holotype.			
LV	TCD. 29053	0.65	0.43
Carapace	TCD. 29054	0.63	0.48
Carapace	TCD. 29055	0.65	0.46
Carapace	TCD. 29056	0.65	0.45

Description. Carapace of medium size, subovate. Right valve; anterior margin narrowly rounded, extremity at one-third height. Posterior margin broadly rounded. Dorsal margin strongly arched behind mid-length, curve tapering towards anterior margin. Ventral margin weakly concave. The transitions between dorsal and posterior margins, and between posterior and ventral margins are marked by moderately sharp curves. Left valve; anterior margin narrowly rounded, extremity at one-third height. Posterior margin broadly rounded. Dorsal margin strongly arched slightly further behind mid-length than in right valve, curve strongly tapers towards anterior margin. Ventral margin straight to very weakly convex. Transitions between dorsal and posterior margins, and between posterior and ventral margins marked by sharp curves. Dorsal view fusiform, with a well-defined bend towards posterior in posterior third. Maximum length below mid-height, maximum height behind mid-length, maximum width in posterior third. Left valve larger than right valve, overlapping right valve dorsally and ventrally. Sexual dimorphism not observed. Moderately strongly calcified. Valve surface smooth. In some specimens, the anterior margin of the right valve has a thin external lamella. Inner lamella moderately narrow. Hinge prionodont; crenulated selvage of smaller right valve fits into prominent peripheral groove of larger left valve. Ventrally, posteriorly and anteriorly, hinge groove is very narrow. Muscle scar pattern not observed.

Remarks. *Ogmoconchella michelseni* sp. nov. shows strong affinities to *O. ellipsoidea* (Jones 1872), but differs from the latter in the maximum width being situated in the posterior third rather than at mid-point or slightly posterior of mid-length. Both *Ogmoconchella propinqua* Malz 1971 and *O. adenticulata* (Pietrzenuk 1961) have similarities to *O. michelseni* sp. nov. but can be distinguished by their larger sizes, the maximum widths being situated towards mid-length, the more regularly rounded lateral outlines and the fine fingerprint ornament. *Ogmoconchella* sp. 1 (Donze 1967) differs in its larger size, the more regularly rounded lateral outline, with the maximum width situated at mid-length. *Ogmoconchella* *mouherensis* (Apostolescu 1959) also resembles *O. michelseni* sp. nov. but can be distinguished by its characteristic vertical swelling postero-medianly and/or very distinct vertical bend posteriorly on the lateral surface.

This species occurs in abundance in many of the wells in association with both *Ogmoconcha hagenowi* Drexler 1958 and *Ogmoconchella ellipsoidea* (Jones 1872) within the Liassic Marl of uppermost Hettangian–lowermost Sinemurian age.

Ogmoconchella serratostriata sp. nov.
(Plate 2, Figs. 15, 16 and 18)

1987 *Ogmoconchella* sp. A (reticulate form); Ainsworth *et al.*, p. 621.

Derivato nominis. Latin *serratus*, serrated, *striata*, ridged, with reference to the concentric ribbed ornament, extending subparallel to margins.
Diagnosis. Carapace of medium size, subovate. Ornament of fine bifurcating ribbing which extends subparallel to all margins.
Holotype. Carapace, TCD. 29057; from the upper Rhaetian–lowermost Hettangian of Elf 55/30-1; depth, 2060 m; latitude, 50°09′09.630″ N; longitude, 10°05′54.190″ W; Fastnet Basin.
Material. Seven valves, 273 carapaces.
Distribution. Recorded from the upper Rhaetian–lowermost Hettangian of Cities Service 63/4-1, Cities Service 63/10-1, Elf 55/30-1, Esso-Marathon 56/20-1 and Gulf 56/18-1.
Dimensions.

		Length (mm)	Height (mm)
Holotype			
Carapace	TCD. 29057	0.55	0.38
Paratypes. Same sample as holotype.			
Carapace	TCD. 29058	0.53	0.36
Carapace	TCD. 29059	0.53	0.35
Carapace	TCD. 29060	0.53	0.36
Carapace juv.	TCD. 29061	0.48	0.31

Description. Carapace of medium size, subovate. Right valve; anterior margin slightly asymmetrically rounded, extremity below mid-height. Posterior margin broadly rounded. Dorsal margin arched, with greatest convexity at mid-length. Margin convex in front of mid-length, straight to very weakly convex behind mid-length. Ventral margin straight to weakly concave. Left valve; anterior margin asymmetrically rounded, extremity at two-fifths height. Posterior margin broadly rounded. Dorsal margin arched, highest at mid-length, in front of mid-length curve continuous into anterior margin. Ventral margin straight to very weakly convex. Transitions between dorsal and posterior margins and between posterior and ventral margins marked by sharp curves. Dorsal view subovate. Maximum length slightly below mid-length, maximum height at mid-length, maximum width slightly posterior of mid-length. Left valve larger than right valve, overlapping right valve along entire periphery. Sexual dimorphism not observed. Moderately strongly calcified. Ornament of fine bifurcating concentric ribbing, extending subparallel to all margins. Inner lamella moderately narrow. Hinge prionodont; crenulated selvage of smaller right valve fits into prominent peripheral groove of larger left valve. Ventrally, posteriorly and anteriorly, hinge groove is narrow. Muscle scar pattern not observed.
Remarks. A very distinct species due to its ornament of fine bifurcating concentric ribbing. When the ornament is abraded, this species looks very similar to *Ogmoconchella ellipsoidea* (Jones 1872) but can be distinguished by its smaller size, less strongly arched dorsal margin, with the greatest convexity situated more towards mid-length, while the posterior margin is less bluntly rounded. *Ogmoconchella michelsensi* sp. nov. differs in its more strongly arched dorsal margin, the maximum width being situated in the posterior third, the smooth valve surface and larger size. This species also shows affinities to *Ogmoconchella owthorpensis* (Anderson 1964) and *O. moorei* (Jones 1894). It is readily distinguished from the former by its smaller size, the less well-rounded dorsal margin, less strongly overlapping left valve and ribbed ornament. It differs from the latter in the straight ventral margin, less regularly arched dorsal margin and ribbed

ornament. *Ogmoconchella adenticulata* (Pietrzenuk 1961) also possesses a ribbed ornament but differs in its larger size, more regularly rounded lateral outline and more flattened subovate outline in dorsal view.

In many of the observed specimens, the concentric ribbing is poorly preserved owing to the drilling techniques used. It commonly occurs in abundance within the shales and mudstones sandwiched between the Langport Member of Triassic age and the thick bedded limestones belonging to the Liassic Limestone of Hettangian age. Specimens have also been recorded in the uppermost Hettangian–lowermost Sinemurian interval, but these have been reworked owing to their often abraded appearance and different carapace colour compared with the main assemblage.

ACKNOWLEDGEMENTS

The research summarized in this paper was carried out under contract to the Petroleum Affairs Division of the Department of Energy, Dublin, Ireland, and is published by permission of the Minister for Energy. The authors wish to thank Dr K. Robinson and Mr N. Murphy for their constructive criticism of the manuscript and Professor C. H. Holland and Dr G. Clayton for advice and comments.

Plate 1—

Fig. 1 — *Polycope cerasia* Blake 1876, right view of carapace, TCD. 29183, × 100, Deminex 56/21-2, 2860 m, lowermost Sinemurian–uppermost Hettangian.

Fig. 2 — *Cytherella* sp. A, left view of female carapace, TCD. 29184, × 58, Elf 55/30-1, 2070 m, lowermost Hettangian–Rhaetian.

Fig. 3 — *Bairdia molesta* Apostolescu 1959, external view of right valve, TCD. 29186, × 64, Cities Service 63/4-1, 1096 m, lowermost Sinemurian–uppermost Hettangian.

Fig. 4 — *Bairdia* cf. *B. carinata undulata* Herrig 1981 *sensu* Ainsworth *et al.* 1987, external view of left valve, TCD. 29188, × 55, Cities Service 63/4-1, 1096 m, lowermost Sinemurian–uppermost Hettangian.

Fig. 5 — *Isobythocypris* cf. *I. cylindrica* Herrig 1981 *sensu* Ainsworth *et al.* 1987, right view of carapace, TCD. 29189, × 52, Deminex 56/21-2, 2860 m, lowermost Sinemurian–uppermost Hettangian.

Fig. 6 — *Pseudomacrocypris* cf. *P. subtriangularis* Michelsen 1975 *sensu* Ainsworth *et al.* 1987, left view of carapace, TCD. 29190, × 82, Cities Service 63/4-1, 1113 m, lowermost Sinemurian–uppermost Hettangian.

Fig. 7 — *Darwinula hettangiana* sp. nov., left view of carapace, TCD. 29032, holotype, × 51, Elf 55/30-1, 1955 m, Hettangian.

Fig. 8 — *Darwinula hettangiana* sp. nov., right view of carapace, TCD. 29033, paratype, × 51, Elf 55/30-1, 1955 m, Hettangian.

Fig. 9 — *Darwinula rara* sp. nov., left view of carapace, TCD. 29037, holotype, × 58, Elf 55/30-1, 1950 m, Hettangian.

Fig. 10 — *Darwinula rara* sp. nov., right view of carapace, TCD. 29037, holotype, × 58, Elf 55/30-1, 1950 m, Hettangian.

Fig. 11 — *Darwinula hettangiana* sp. nov., dorsal view of carapace, TCD. 29032, holotype, × 51, Elf 55/30-1, 1955 m, Hettangian.

Fig. 12 — *Darwinula rara* sp. nov., dorsal view of carapace, TCD. 29037, holotype, × 58, Elf 55/30-1, 1950 m, Hettangian.

Fig. 13 — *Lutkevichinella fastigata* sp. nov., left view of female carapace, TCD. 29042, holotype, × 63, Elf 55/30-1, 2025 m, Hettangian.

Fig. 14 — *Lutkevichinella fastigata* sp. nov., right view of female carapace, TCD. 29043, paratype, × 63, Elf 55/30-1, 2025 m, Hettangian.

Fig. 15 — *Lutkevichinella fastigata* sp. nov., left view of male carapace, TCD. 29044, paratype, × 63, Elf 55/30-1, 2025 m, Hettangian.

Fig. 16 — *Lutkevichinella fastigata* sp. nov., right view of male carapace, TCD. 29044, paratype, × 63, Elf 55/30-1, 2025 m, Hettangian.

Fig. 17 — *Lutkevichinella fastigata* sp. nov., dorsal view of female carapace, TCD. 29042, holotype, × 63, Elf 55/30-1, 2025 m, Hettangian.

Fig. 18 — *Lutkevichinella fastigata* sp. nov., dorsal view of male carapace, TCD. 29044, paratype, × 63, Elf 55/30-1, 2025 m, Hettangian.

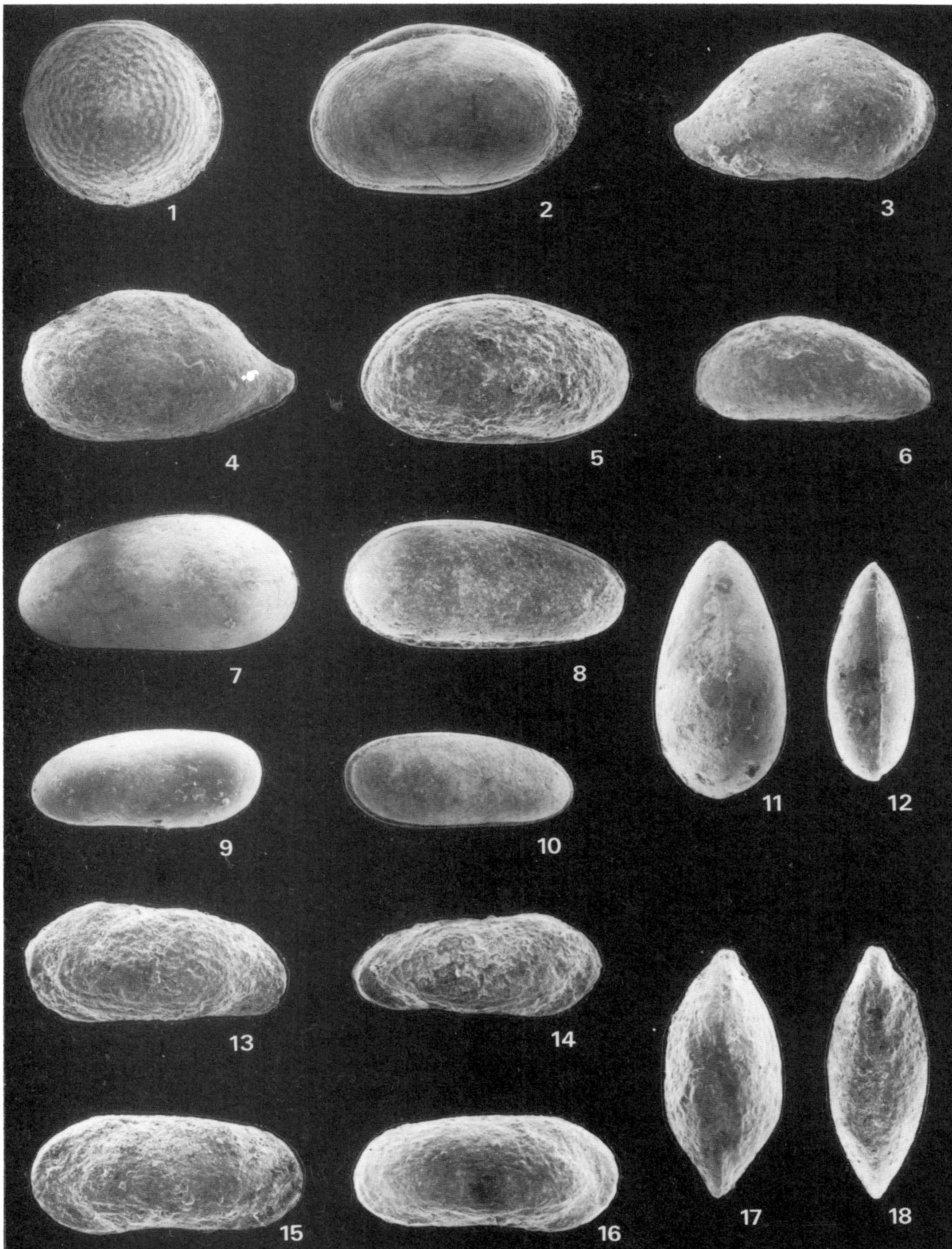
1
2
3
4
5
6
7
8
11
12
9
10
13
14
17
18
15
16

Plate 2—

Fig. 1 — *Lutkevichinella hortonae* sp. nov., left view of female carapace, TCD. 29048, paratype, × 55, Elf 55/30-1, 2015 m, Hettangian.

Fig. 2 — *Lutkevichinella hortonae* sp. nov., right view of female carapace, TCD. 29047, holotype, × 55, Elf 55/30-1, 2015 m, Hettangian.

Fig. 3 — *Lutkevichinella hortonae* sp. nov., left view of male carapace, TCD. 29051, paratype, × 55, Elf 55/30-1, 2015 m, Hettangian.

Fig. 4 — *Lutkevichinella hortonae* sp. nov., right view of male carapace, TCD. 29051, paratype, × 55, Elf 55/30-1, 2015 m, Hettangian.

Fig. 5 — *Lutkevichinella hortonae* sp. nov., dorsal view of female carapace, TCD. 29047, holotype, × 55, Elf 55/30-1, 2015 m, Hettangian.

Fig. 6 — *Lutkevichinella hortonae* sp. nov., dorsal view of male carapace, TCD. 29051, paratype, × 55, Elf 55/30-1, 2015 m, Hettangian.

Fig. 7 — *Donzocythere* cf. *D. convergens* (Donze 1985 *sensu* Donze 1985), right view of carapace, TCD. 29198, x 76, Cities Service 63/4-1, 1123 m, lowermost Sinemurian–uppermost Hettangian.

Fig. 8 — *Klinglerella translucens* (Blake 1876), right view of male carapace, TCD. 29069, × 57, Elf 55/30-1, 1800 m, lowermost Sinemurian–uppermost Hettangian.

Fig. 9 — *Ogmoconcha hagenowi* Drexler 1958, right view of carapace, TCD. 29251, × 59, Elf 55/30-1, 1790 m, lowermost Sinemurian–uppermost Hettangian.

Fig. 10 — *Ogmoconchella ellipsoidea* (Jones 1872), right view of carapace, TCD. 29071, × 63, Elf 55/30-1, 1800 m, lowermost Sinemurian–uppermost Hettangian.

Fig. 11 — *Ogmoconchella* cf. *O. ellipsoidea* (Jones 1872) *sensu* Ainsworth *et al.* 1987, right view of carapace, TCD. 29072, × 58, Cities Service 63/4-1, 1123 m, lowermost Sinemurian–uppermost Hettangian.

Fig. 12 — *Ogmoconchella martini* (Anderson 1964), right view of carapace, TCD. 29073, × 68, Arco 47/29-1, 3204 m, Rhaetian.

Fig. 13 — *Ogmoconchella michelseni* sp. nov., external view of left valve, TCD. 29053, paratype, × 60, Elf 55/30-1, 1800 m, lowermost Sinemurian–uppermost Hettangian.

Fig. 14 — *Ogmoconchella michelseni* sp. nov., right view of carapace, TCD. 29052, holotype, × 60, Elf 55/30-1, 1800 m, lowermost Sinemurian–uppermost Hettangian.

Fig. 15 — *Ogmoconchella serratostriata* sp. nov., left view of carapace, TCD. 29058, paratype, × 64, Elf 55/30-1, 2045 m, lowermost Hettangian–upper Rhaetian.

Fig. 16 — *Ogmoconchella serratostriata* sp. nov., right view of carapace, TCD. 29057, holotype, × 64, Elf 55/30-1, 2045 m, lowermost Hettangian–upper Rhaetian.

Fig. 17 — *Ogmoconchella michelseni* sp. nov., dorsal view of carapace, TCD. 29052, holotype, × 60, Elf 55/30-1, 1800 m, lowermost Sinemurian–uppermost Hettangian.

Fig. 18 — *Ogmoconchella serratostriata* sp. nov., dorsal view of carapace, TCD. 29057, holotype, × 64, Elf 55/30-1, 2045 m, lowermost Hettangian–upper Rhaetian.

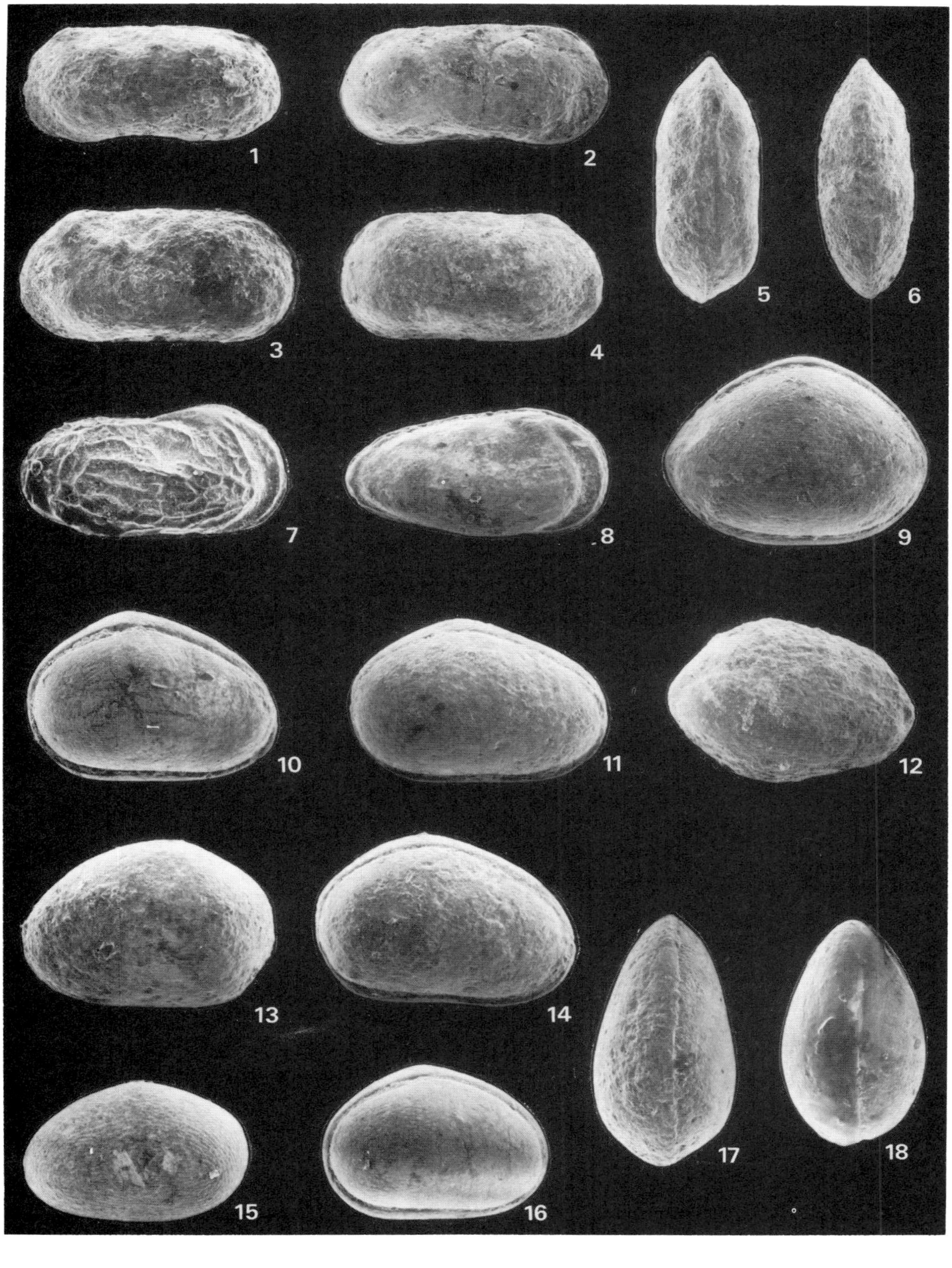
1
2
3
4
5
6
7
8
9
10
11
12
13
14
15
16
17
18

Plate 3—

Fig. 1 — *Rhaetogonyaulax rhaetica* (Sarjeant) Harland, Morbey & Sarjeant 1975, Esso-Marathon 48/30-1, 7,900′, Rhaetian.

Fig. 2 — *Porcellispora longdonensis* (Clarke) Morbey 1975, TCD. 28201, Elf 55/30-1, 2030 m, lowermost Hettangian-upper Rhaetian.

Fig. 3 — *Ovalipollis pseudoalatus* (Thiergart) Schuurman 1976, TCD. 28202, Elf 55/30-1, 2100 m, Rhaetian.

Figs. 4 and 5 — *Rhaetipollis germanicus* Schulz 1967, TCD. 28203 and TCD. 28204, Elf 55/30-1, 2100 m, Rhaetian.

Fig. 6 — *Kraeuselisporites reissingeri* (Harris) Morbey 1975, TCD. 28205, Elf 55/30-1, 2100 m, Rhaetian.

Fig. 7 — *Carnisporites spiniger* (Leschik) Morbey 1975, Esso-Marathon 48/30-1, 7,900′, Rhaetian.

Fig. 8 — *Acanthotriletes ovalis* Nilsson 1958, TCD. 28206, Cities Service 63/10-1, 3276 m, lowermost Hettangian–upper Rhaetian.

Fig. 9 — *Microreticulatisporites fuscus* (Nilsson) Morbey 1975, TCD. 28207, Elf 55/30-1, 1820 m, Hettangian.

Fig. 10 — *Granuloperculatipollis rudis* (Venkatachala & Góczán) Morbey 1975, TCD. 28208, Elf 55/30-1, 2100 m, Rhaetian.

Figs. 11 and 12 — *Quadraeculina anellaeformis* Maljavkina, 1949, TCD. 28209 and TCD. 28210, Elf 55/30-1, 2030 m, lowermost Hettangian–upper Rhaetian; 1820 m, Hettangian.

Figs. 13 and 14 — *Tsugaepollenites? pseudomassulae* (Mädler) Morbey 1975, TCD. 28211 and TCD. 28212, Elf 55/30-1, 2100 m, Rhaetian.

Fig. 15 — *Dapcodinium priscum* Evitt 1961, TCD. 28213, Elf 55/30-1, 1820 m, Hettangian.

All magnifications approximately × 500.

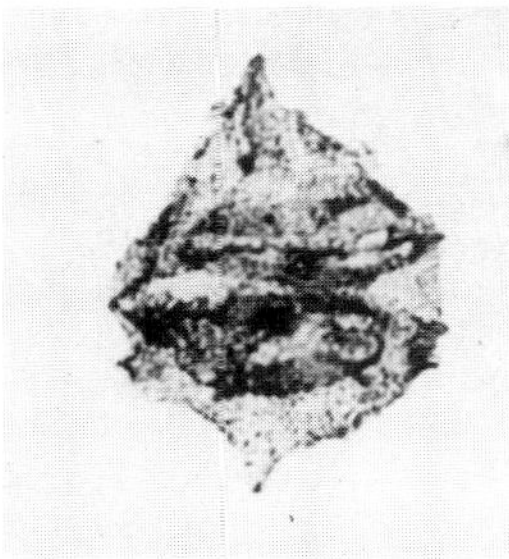
1

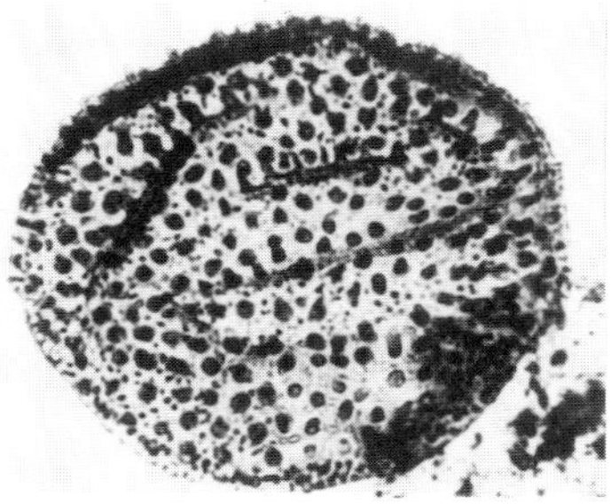
2

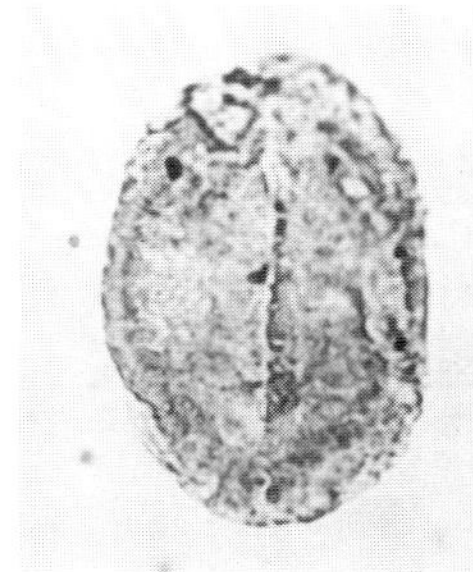
3

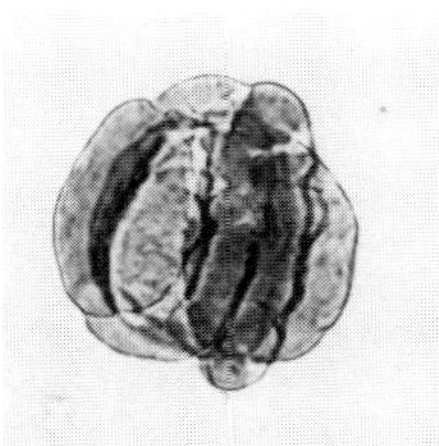
4

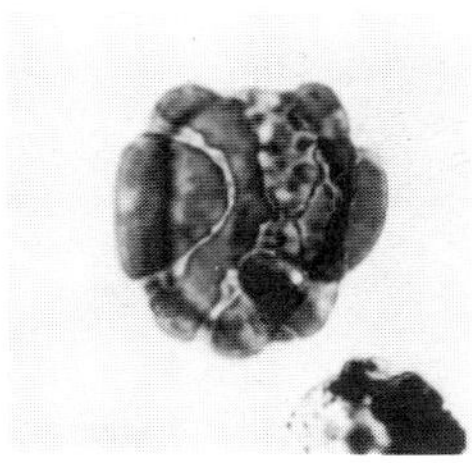
5

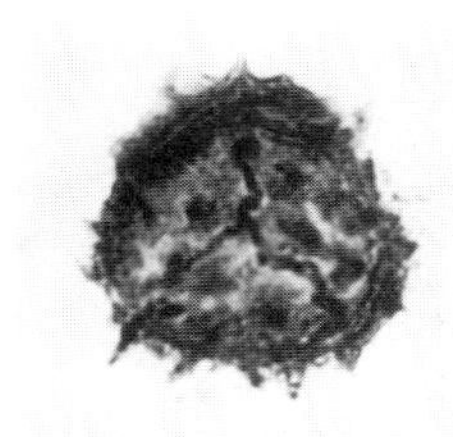
6

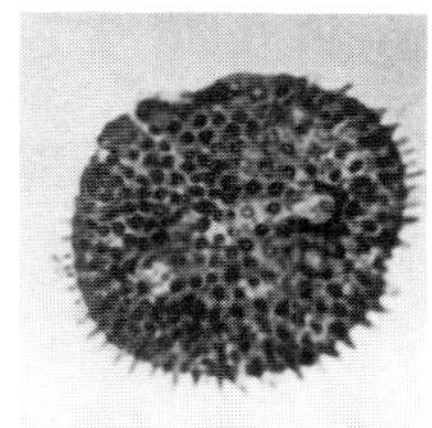
7

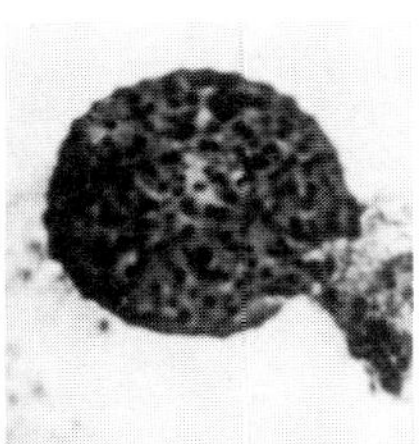
8

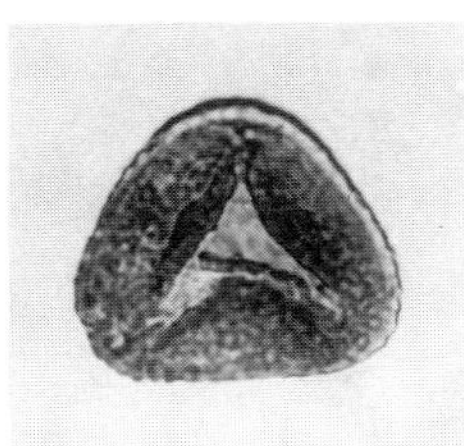
9

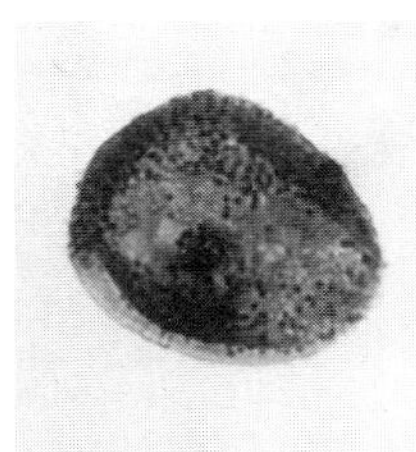
10

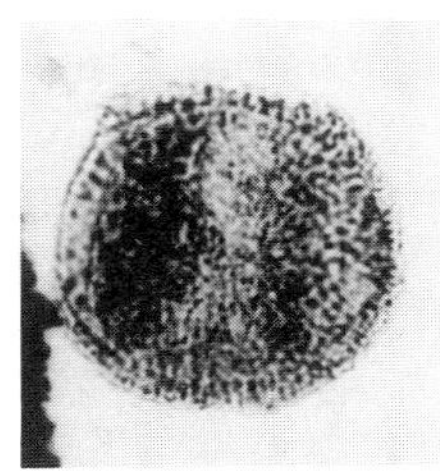
11

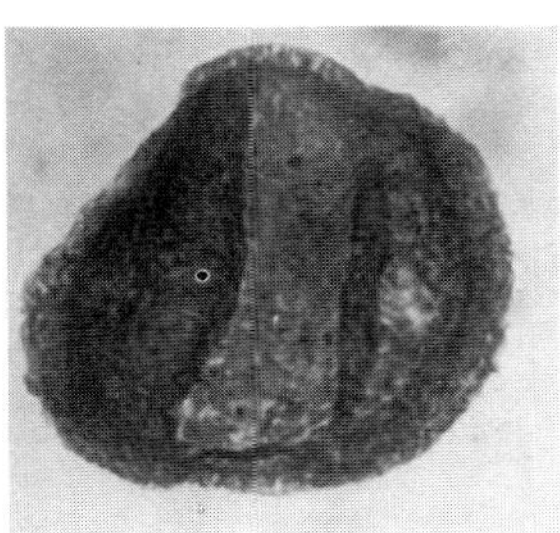
12

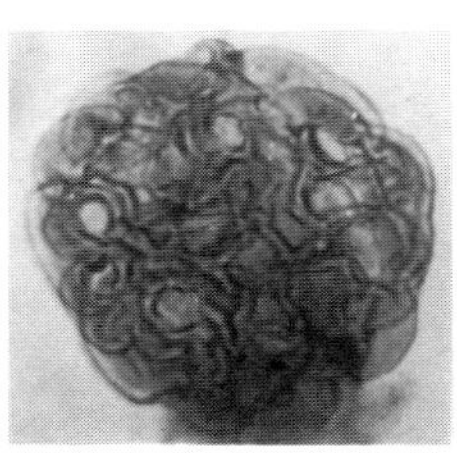
13

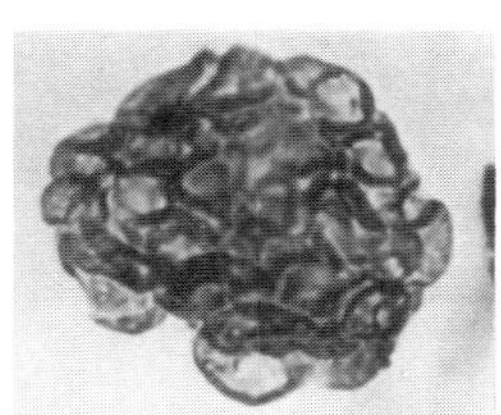
14

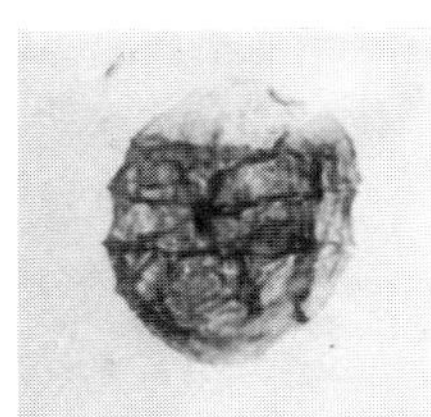
15

REFERENCES

Ainsworth, N. R. & Horton, N. F. (1986). Mesozoic micropalaeontology of exploration well Elf 55/30-1 from the Fastnet Basin, offshore southwest Ireland. *Journal of Micropalaeontology* **5** 19–29.

Ainsworth, N. R., O'Neill, M., Rutherford, M. M., Clayton G. Horton, N. F. & Penney, R. A. (1987). Biostratigraphy of the Lower Cretaceous, Jurassic and uppermost Triassic of the North Celtic Sea and Fastnet Basins. In: Brooks, J. & Glennie, K. (eds), *Petroleum geology of north west Europe*. Graham & Trotman, London, pp. 611–622.

Aldorf, M. C. & Doubinger, J. (1978). Associations palynologiques du Trias et du Lias inférieur de l'Quest de l'Europe (Luxembourg, France, Portugal). *Palynologia, Número Extraordinario* **1** 21–25.

Anderson, F. W. (1964). Rhaetic Ostracoda. *Bulletin of the Geological Survey of Great Britain* **21** 133–174.

Apostolescu, V. (1959). Ostracodes du Lias du Bassin de Paris. *Revue de l'Institut Français du Pétrole* **14** 795–826.

Apostolescu, V. (1961). Sud du Bassin, région D'Argenton-sur-Creuse et de la Chatre (Départment du Cher et de l'Indre). Ostracodes. In: (Cousin, N., Espitalie, J., Signal, J. & Apostolescu, V. (eds),) *Colloque sur le Lias Français. Mémoire du Bureau de Recherches Géologiques Minières* **4** 449.

Christensen, O. B. (1962). Ostracodtyper fra Keuper-Rhaet lagserian i dybdeboringerne ved Harte og Ullerslev. *Meddelelser Dansk Geologisk Forening* **15** 90–98.

Colin, J. P., Babinot, J. F. & Grékoff, N. (1985). Trias. In: Oertli, H. J. (ed.) *Atlas des ostracodes de France. Bulletin Centres Recherches Exploration–Production Elf-Aquitaine, Mémoires* **9** 89–99.

Cope, J. C. W. *et al.* (1980). A correlation of Jurassic rocks in the British Isles. Part One: Introduction and Lower Jurassic. *Special Report Geological Society*, London **14** 1–73.

Donze, P. (1967). Les Ostracodes du sontage de Laneuveville-devant-Nancy (Lotharingien de la région type). *Sciences Terre* **12** 71–92.

Donze, P. (1968). Espèces nouvelles d'Ostracodes du Lias inférieur normand. *Bulletin de la Société Linnéenne du Nord de la France* **9** 78–87.

Donze, P. (1985). Lias inferieur et moyen. In: Oertli, H. J. (ed.), *Atlas des Ostracodes de France. Bulletin Centres Recherches Exploration–Production Elf-Aquitaine, Mémoires* **9** 101–117.

Drexler, E. (1958). Foraminiferen und Ostracoden aus dem Lias von Siebeldingen/Pfalz. *Geologisches Jahrbuch* **75** 475–554.

Dreyer, E. (1965). Mikrofossilen des Rät und Lias von SW-Brandenburg. *Jahrbuch für Geologie* **1** 491–531.

Fisher, M. J. (1972). The Triassic palynofloral succession in England. *Geoscience and Man* **4** 101–109.

Fisher, M. J. & Dunay, R. (1981). Palynology and the Triassic/Jurassic boundry. *Review of Palaeobotany and Palynology* **34** 129–135.

Herrig, E. (1979a). Die Gattung *Bairdia* (Ostracoda, Crustacea) im Lias von Thüringen. Teil 1. *Zeitschrift für Geologie Wissenschaften* **7** 641–661.

Herrig, E. (1979b). Ostrakoden aus dem Lias von Thüringen: Die Gattungen *Bairdia* (Teil II), *Fabalicypris* und *Bairdiacypris*. *Zeitschrift für Geologie Wissenschaften* **7** 763–782.

Herrig, E. (1979c). Weitere glattschalige Ostrakoden aus dem Lias von Thüringen. *Zeitschrift für Geologie Wissenschaften* **11** 1343–1361.

Kozur, H. (1968). Neue Ostracoden aus dem Röt und Muschelkalk des germanischen Binnenbeckens. *Monatsberichte Akademie Wissenschaften* **10** 498–519.

Kozur, H. (1970). Neue Ostracoden der germanischen Mittel und Obertrias. *Geologie* **19** 434–447.

Lord, A. (1971). Revision of some Lower Lias Ostracoda from Yorkshire. *Palaeontology* **14** 642–665.

Lord, A. (1978). The Jurassic Part 1 (Hettangian–Toarcian). In: Bate, R. & Robinson, E. (eds), *A stratigraphical index of British Ostracoda. Geological Journal, Special Issue* **8** 189–212.

Malz, H. (1971). Zur Taxonomie 'glattschaliger' Lias-Ostracoden. *Senckenbergiana Lethaea* **52** 433–455.

Michelsen, O. (1975). Lower Jurassic biostratigraphy and ostracods of the Danish Embayment. *Danmarks Geologiske Undersogelse* **104** 1–289.

Morbey, S. J. (1975). The Palynostratigraphy of the Rhaetian Stage, Upper Triassic in the Kendelbachgraben, Austria. *Palaeontographica* **152** 1–75.

Morbey, S. J. (1978). Late Triassic and Early Jurassic subsurface palynostratigraphy in north-west Europe. *Palinologia, Número Extraordinario* **1** 355–365.

Naylor, D. & Shannon, P. M. (1982). Fastnet Basin. In: Naylor, D. & Shannon, P. M. (eds), *The geology of offshore Ireland and west Britain*. Graham & Trotman, London, pp. 49–58.

Orbell, G. (1973). Palynology of the British Rhaeto-Liassic. *Bulletin of the Geological Survey of Great Britain* **44** 1–44.

Penn, I. E. (1987). Geophysical logs in the stratigraphy of Wales and adjacent offshore and onshore areas. *Proceedings of the Geologists' Association* **98** 275–314.

Poole, E. G. (1979). The Triassic–Jurassic boundary in Great Britain. *Geological Magazine* **116** 303–311.

Riding, J. B. (1987). Dinoflagellate cyst stratigraphy of the Nettleton Bottom Borehole (Jurassic: Hettangian to Kimmeridgian), Lincolnshire, England. *Proceedings of the Yorkshire Geological Society* **46** 231–266.

Robinson, K. W., Shannon, P. M. & Young, D. G. G. (1981). The Fastnet Basin: an integrated analysis. In: L. V. Illing & G. D. Hobson (eds), *The geology of the continental shelf of north-west Europe*. Heyden and Son, London, pp. 444–454.

Sladen, C. P. & Batten, D. J. (1984). Source-area environments of Late Jurassic and Early Cretaceous sediments in southeast England. *Proceedings of the Geologists' Association* **95** 149–163.

Vakhrameev, V. A. (1981). Pollen *Classopollis* indicator of Jurassic and Cretaceous climates. *Palaeobotanist* **28–29** 301–307.

Wall, D. (1965). Microplankton, pollen and spores from the Lower Jurassic in Britain. *Micropaleontology* **11** 151–190.

Warrington, G. (1971). Palynology of the Upper Triassic strata of the Llanbedr (Mochras Farm) Borehole. *Report, Institute of Geological Sciences London* **71** 75–88.

Warrington, G. (1974). Studies in the palynological biostratigraphy of the British Trias. 1. Reference sections in west Lancashire and north Somerset. *Review of Palaeobotany and Palynology* **17** 133–147.

Warrington, G. (1977). Appendix 3: Palynology of the White Lias, Cotham and Westbury Beds and the Keuper Marl of the Steeple Aston Borehole. In: Poole, E. G. (ed.),

Stratigraphy of the Steeple Aston Borehole, Oxfordshire. Bulletin of the Geological Survey of Great Britain **57** 40–43.

Warrington, G. (1978). Appendix 1. Palynology of the Keuper, Westbury and Cotham Beds and the White Lias of the Withycombe Farm Borehole. In: Poole, E. G. (ed.), *Stratigraphy of the Withycombe Farm Borehole, near Banbury, Oxfordshire. Bulletin of the Geological Survey of Great Britain* **68** 22–28.

Warrington, G. (1982). Palynology of cores from the basal Lias and the Permian (?)–Triassic sequence of the Winterborne Kingston Borehole, Dorset. In: Rhys, G. H., Lott, G. K. & Calver, M. A. (eds), *The Winterborne Kingston Borehole, Dorset, England. Report, Institute of Geological Sciences, London* **81** 122–126.

Warrington, G. (1983). Late Triassic and earliest Jurassic palynomorph assemblages from the western English Channel and neighbouring areas. *Proceedings of the Ussher Society* **5** 473–476.

Warrington, G. & Harland, R. (1975). Appendix 5: Palynology of the Trias and Lower Lias of the Larne Borehole. In: Manning, P. I. & Wilson, H. E. (eds), *The stratigraphy of the Larne Borehole. Bulletin of the Geological Survey of Great Britain* **50** 37–50.

Warrington, G. *et al.* (1980). A correlation of Triassic rocks in the British Isles. *Special Report. Geological Society, London* **13** 1–78.

Whittaker, A., Holliday, D. W. & Penn, I. E. (1985). Geophysical logs in British stratigraphy. *Special Reports, Geological Society, London* **18** 1–73.

Wicher, C. A. (1951). Zur mikropalaontologische Gliederung des nichtmarinen Rat. *Erdol und Kohle* **4** 755–760.

Will, H. J. (1969). Untersuchungen zur stratigraphie und genese des Oberkeupers in Nordwestdeutschland. *Beiheftezum Geologischen Jahrbuch* **54** 1–240.

3

Spore and pollen assemblage zonation of Swedish Bajocian and Bathonian sediments

Dorothy Guy-Ohlson

ABSTRACT

19 samples from a previously unknown Middle Jurassic sequence penetrated by Humlarp Borehole No. 363 have yielded a well-preserved palynoflora consisting mainly of spores and pollen grains. On the basis of their distribution within the borehole and the known stratigraphic ranges of the 86 species identified, four assemblages zones have been proposed. These are dated as early Bajocian, middle Bajocian, late Bajocian–early Bathonian and early Bathonian, and they permit precise correlation with other Middle Jurassic palynological assemblages from several boreholes in northwest Scania. General correlations of the zones are also made with other northwest European spore and pollen-dominated palynofloras. The results of both biostratigraphic and palynofacies analyses reflect transgressions and regressions of the Middle Jurassic sea at the Humlarp site of deposition.

INTRODUCTION

The Humlarp Borehole No. 363 (Fig. 1) is situated in northwest Scania, the southernmost province of Sweden, in an area known as the Ängelholm Trough. It was drilled in 1962 for reconnaissance purposes by the Höganäs AB Company in connection with coal and fire clay exploration. As a result of the economic interest, the geology of the area is fairly well known; details can be found in the papers by Norling (1972, 1978, 1981).

Palynological investigations of several boreholes in the area have been carried out. Palynofloras have been described from the Toarcian and Middle Jurassic of Vilhelmsfält No. 1 (Guy 1971, Guy-Ohlson 1976, 1986) and from the Middle Jurassic of Härninge Nos. 64 and 159 and Rosenhäll No. 62 (Guy-Ohlson 1978), and assemblages of Sinemurian and late Liassic age have been reported from Oppegård No. 1 (Guy-Ohlson & Malmquist 1985). From Åstorp No. 20 and Karindal No. 1 palynofloras of the late Middle Jurassic and Late Jurassic respectively have been recorded (Guy-Ohlson 1985, Guy-Ohlson & Norling 1988).

In this paper on the Humlarp Borehole the palynomorph content is documented, the core is dated and assemblage zones intended for use in further biostratigraphic correlations are established. The changes in palynofacies through the

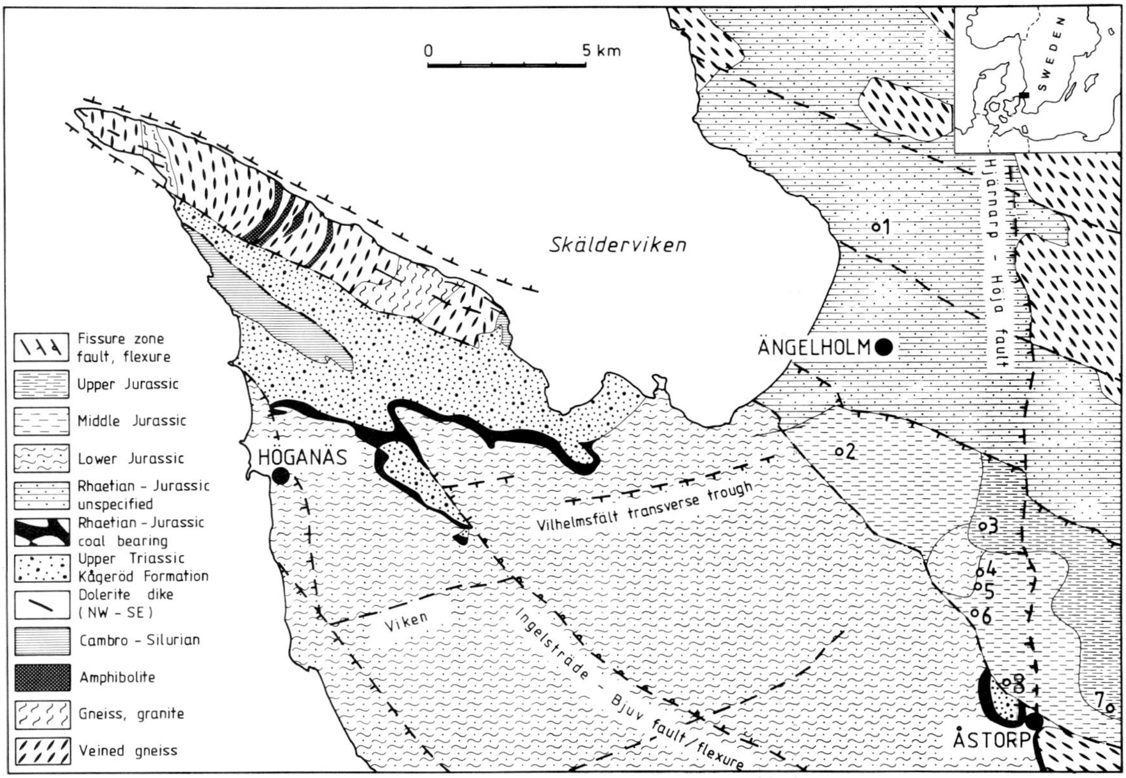

Fig. 1 — Geological sketch map of northwest Scania. The numbers refer to the following boreholes: 1, Valhall No. 1; 2, Vilhelmsfält No. 1; 3, Karindal No. 1; 4, Härninge Nos. 159 and 64; 5, Oppegård No. 1; 6, Rosenhäll No. 62; 7, Åstorp No. 20; 8, Humlarp HB No. 363. (After Wikman *et al.* 1981.)

core are assessed in order to determine environments of deposition, and the palynomorph species content is compared with that of other assemblages of similar age.

MATERIALS AND METHODS

Sampling and Lithology

A brief lithological description of the vertical core from which the 19 samples were taken is given in Fig. 2. The dip of the sedimentary sequence varies markedly in different parts of the core. Two intersecting faults in the vicinity of the borehole may explain, at least in part, the wide spectrum of dip angles encountered (noted in Fig. 2 where known). An alternative explanation is that the erosion of basal sediments of a cliff face causes the overlying beds to slump. This might account for the range of dip angles from 15° to vertical for the interval 85.50–65.75 m.

Preparation and Examination

The samples were prepared using well-known techniques, and two standard slide preparations were routinely examined (Batten 1981, Batten & Morrison 1981). Palynofacies studies were carried out using the second slide as a basis for comparison between the samples. Fields of view approximately in the middle of each slide and considered to be representative of the palynofacies were photographed in transmitted light (same reference coordinates for all). These are illustrated on Plates 1 and 2, Figs. 1–16.

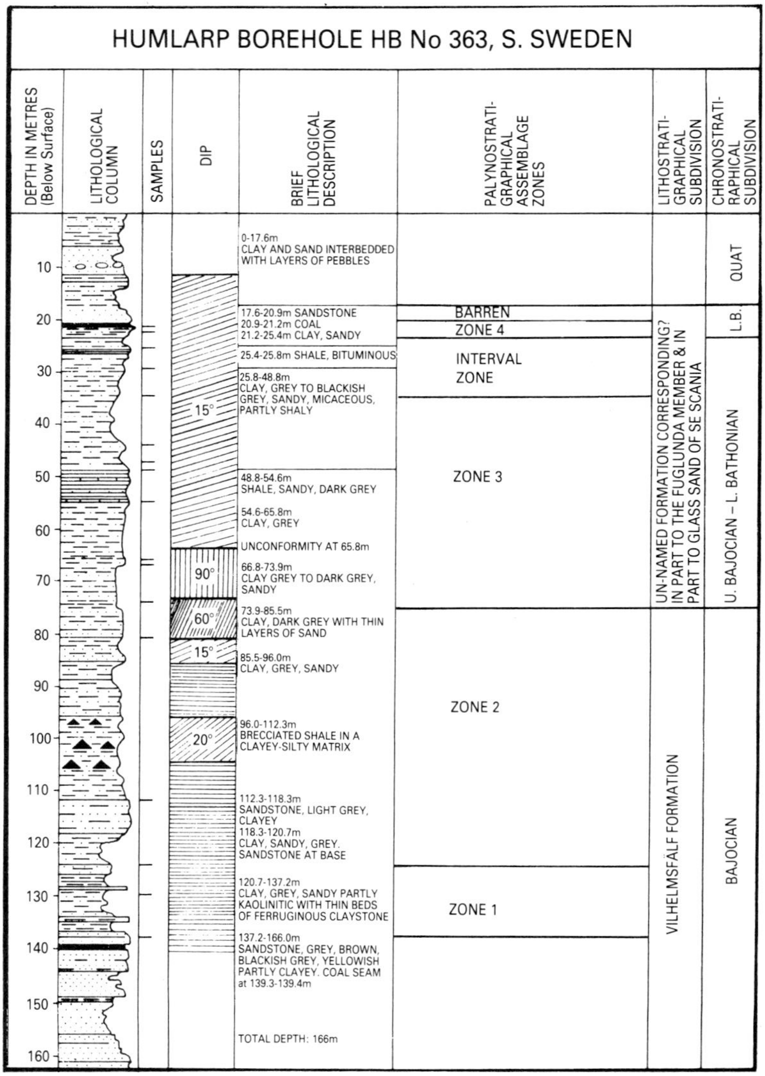

Fig. 2 — Well log (compiled by Erik Norling, Geological Survey of Sweden) and palynostratigraphical zonation of the Humlarp Borehole HB No. 363. The 19 sampling horizons are indicated at 20.90 m, 22.35 m, 22.85 m, 25.45 m, 29.45 m, 34.65 m, 43.10 m, 44.10 m, 47.30 m, 48.75 m, 54.60 m, 65.75 m, 66.65 m, 73.90 m, 80.55 m, 112.30 m, 124.45 m, 130.00 m and 137.80 m.

For the biostratigraphic analysis all the palynomorphs found during examination of the second slide in the transmitted-light microscope were recorded. In addition, routine scanning of one stub bearing strewn residue from each sample was carried out using a JEOL 35 C scanning electron microscope. Micrographs of selected palynomorphs are illustrated on Plates 3–5, Figs. 17–50.

Repository

The illustrated material is retained for reference

purposes in the Section of Paleobotany, Swedish Museum of Natural History, Stockholm, Sweden.

ASSEMBLAGE COMPOSITION

Numerous well-preserved palynomorphs occur, mainly spores and pollen grains but also in a few slides (Fig. 3) dinoflagellates, acritarchs and other organic walled microplankton. 86 species, three of which are reworked Carboniferous spores, were identified; a complete species list in alphabetical order with names of authors and dates is given in Table 1. In Fig. 3 the species are ordered according to their known published stratigraphic ranges and the number of specimens in each sample is recorded.

It can been seen from this figure that species with long stratigraphic ranges dominate the palynoflora. These are placed in one of three groups. Species in group 1 have ranges commencing at or below the base of the Jurassic and terminating in the Middle Jurassic. Those in group 2 (by far the largest) first appear in pre-Jurassic rocks and are present throughout the Jurassic and in many cases also occur in the Cretaceous. Group 3 contains species which have their inceptions in Middle Jurassic rocks and range up to the uppermost Jurassic or even into the Cretaceous. The remaining species, referred to as group 4, have ranges that are restricted to part of the Middle Jurassic or extend from the Lower–Middle Jurassic transition into the Middle Jurassic. The group to which each species is considered to belong is given in parentheses after each name in Fig. 3, which also shows the qualitative changes in palynofloral composition of the samples studied and the quantitative differences in the number of species and individuals. From these data, four assemblage zones have been distinguished.

Assemblage zone 1 is delineated by the incoming of 24 species (Table 2) and encompasses samples from 130.00 m and 124.45 m. Zone 2 is characterized by the first appearance of a further eight species and comprises samples from 112.30 m and 80.55 m. Six species first appear in zone 3, 11 occur only as single specimens and the ranges of a further 11 are terminated within the zone, which encompasses samples from 73.90 m to 34.65 m. An interval zone containing very few palynomorphs is recognized for samples from 29.45 m and 24.45 m. Zone 4 above 25.45 m is characterized by the incoming of four species and a pronounced increase in the number of *Classopollis classoides*.

BIOSTRATIGRAPHICAL DATING

Accurate dating of palynomorph assemblages is often impossible without the aid of dinoflagellates, but in this investigation the presence of these algae helps little because their stratigraphic ranges are comparatively long. Consideration of the total assemblage composition of each zone in the light of the published ranges of the species (Fig. 3) provides some indication of age and allows tentative determinations to be made.

The general composition and diversity of the species within the form genus *Callialasporites* are consistent with a Bajocian age for zones 1 and 2. More specifically, the incoming in zone 2 of *Neoraistrickia gristhorpensis*, *Leptolepidites major* and *L. paverus* suggests a mid-Bajocian age (Couper 1958, Levet-Carette 1964) whereas a lack of these and other Middle Jurassic indicators may imply an early Bajocian age for zone 1.

Assemblage zone 3 consists of samples which have been taken from the core interval which contains an unconformity (Fig. 3). The general palynomorph composition of samples from the overlying and underlying strata are, however, very similar—hence the recognition of only a single assemblage zone for this interval. The new incoming species and the last occurrences of species such as *Lycopodiacidites pseudoreticulatus*, *Callialasporites trilobatus*, *Leptolepidites paverus*, *Lycopodiumsporites vilhelmii* and *Leptolepidites rotundus* suggest a late Bajocian age for the assemblage, whereas single occurrences of species such as *Pinuspollenites*

Table 1 — List of taxa in alphabetical order.

Alisporites robustus Nilsson 1958
Araucariacites australis Cookson 1947
Baculatisporites comaumensis (Cookson) Potonié 1956
Brachysaccus microsaccus (Couper) Mädler 1964
Botryococcus sp.
Calamospora mesozoica Couper 1958
Callialasporites dampieri (Balme) Dev 1961
Callialasporites microvelatus Schulz 1966
Callialasporites minus (Tralau) Guy 1971
Callialasporites trilobatus (Balme) Dev 1961
Callialasporites turbatus (Balme) Schulz 1967
Canningia sp.
Ceratosporites spinosus Schulz 1967
Cerebropollenites mesozoicus (Couper) Nilsson 1958
Chasmatosporites apertus (Rogalska) Nilsson 1958
Chasmatosporites major (Nilsson) Pocock & Jansonius 1958
Chasmatosporites sp.
Cibotiumspora jurienensis (Balme) Filatoff 1975
Cingutriletes clavus (Balme) Dettmann 1963
Classopollis classoides (Pflug) Pocock & Jansonius 1961
Contignisporites dunrobinensis (Couper) Schulz 1967
Contignisporites problematicus (Couper) Döring 1965
Cyathidites australis Couper 1953
Cyathidites concavus (Bolkhovitina) Dettmann 1963
Cyathidites minor Couper 1953
Converrucosisporites utriculosus (Krasnova) Pocock 1970
Converrucosisporites variverrucatus (Couper) Norris 1969
Converrucosisporites sp.
Densoisporites crassus Tralau 1968
Densoisporites erdtmanii Guy-Ohlson 1979
Densoisporites scanicus Tralau 1968
Densoisporites velatus Weyland & Krieger 1953
Densoisporites sp.
Densosporites sp. (reworked)
Dictyophyllidites equiexinus (Couper) Dettmann 1963
Eucommiidites troedssonii Erdtman 1948
Exesipollenites tumulus Balme 1957
Ginkgocycadophytus spp.
Gleicheniidites senonicus Ross 1949
Ischyosporites variegatus (Couper) Schulz 1967
Leptolepidites equatibossus (Couper) Tralau 1968
Leptolepidites major Couper 1958
Leptolepidites paverus Levet-Carette 1964
Leptolepidites rotundus Tralau 1968
Leptolepidites sp.
Lycopodiacidites rugulatus (Couper) Schulz 1967
Lycopodiacidites pseudoreticulatus Guy 1971
Lycopodiumsporites clavatoides (Couper) Tralau 1968
Lycopodiumsporites reticulumsporites (Rouse) Dettmann 1963
Lycopodiumsporites semimuris Danzé-Corsin & Laveine 1963

Lycopodiumsporites vilhelmii Guy 1971
Marattisporites scabratus Couper 1958
Micrhystridium spp.
Monolites couperi Tralau 1968
Murospora sp. (reworked)
Nannoceratopsis gracilis (Alberti) Evitt 1961
Nannoceratopsis spiculata Stover 1966
Neoraistrickia gristhorpensis (Couper) Tralau 1968
Neoraistrickia truncatus (Cookson) Potonie 1956
Osmundacidites wellmanii Couper 1953
Parvisaccites enigmatus Couper 1958
Perinopollenites elatoides Couper 1958
Pinuspollenites globosaccus Filatoff 1975
Pityosporites scaurus (Nilsson) Schulz 1967
Podocarpidites ellipticus Cookson 1947
Podocarpidites sp.
Polycingulatisporites triangularis (Bolkhovitina) Playford & Dettmann 1963
Protopinus scanicus Nilsson 1958
Pterosperma sp.
Sestrosporites pseudoalveolatus (Couper) Dettmann 1963
Schizosporis sp.
Species indet.
Spheripollenites sp.
Spore tetrad (reworked)
Staplinisporites caminus (Balme) Pocock 1962
Stereisporites cicatricosus (Rogalska) Danzé-Corsin & Laveine 1956
Tasmanites sp.
Todisporites major Couper 1958
Todisporites minor Couper 1958
Todisporites sp.
Trachysporites sp.
Tripartites sp. (reworked)
Uvaesporites argentaeformis (Bolkhovitina) Schulz 1967
Uvaesporites cerebralis Tralau 1968
Uvaesporites puzzlei Guy 1971
Uvaesporites sp.
Vitreisporites pallidus (Reissinger) Nilsson 1958
Zebrasporites sp.

---

globosaccus, *Cingutriletes clavus* and *Uvaesporites cerebralis* point more to an early Bathonian age. A late Bajocian to early Bathonian age is therefore suggested for zone 3. This means that, if the vertical beds between 73.90 m and 66.65 m were the result of tectonic movements, these must have occurred some time during this period (Fig. 2).

Assemblage zone 4 lacks several of the stratigraphically restricted Bajocian species mentioned above. Instead species frequently found in Bathonian assemblages (Couper 1958, Tralau 1968, Schulz 1966, Riding 1987b), such as *Monolites couperi* and *Callialasporites microvelatus*, are present. This suggests that the zone is early Bathonian in age. The uppermost sample at 20.90 m is barren of palynomorphs and Quaternary deposits are present at 17.6 m.

The assemblages at Humlarp differ in character from the older assemblages of Aalenian and Toarcian age at Vilhelmsfält (Guy-Ohlson 1986) and from the younger assemblages at

Table 2 — Important zonal species at Humlarp.

Assemblage zone	*Important zonal species at Humlarp (in alphabetical order)*
Zone 4 (above 25.45 m)	Incoming species: *Callialasporites microvelatus*, *Chasmatosporites major*, *Monolites couperi* and *Zebrasporites* sp. Prominent increase in number of *Classopollis classoides*
Interval zone	Almost barren of palynomorphs
Zone 3 (73.90–34.65 m)	Incoming species: *Callialasporites trilobatus*, *Contignisporites dunrobinensis*, *Dictyophyllidites equiexinus*, *Leptolepidites equatibossus*, *Nannoceratopsis gracilis* and *Podocarpidites* sp. Only occurrence of single specimens: *Chasmatosporites* sp., *Cingutriletes clavus*, *Converrucosisporites utriculosus*, *Densoisporites erdtmanii*, *Densoisporites* sp., *Pinuspollenites globosaccus*, *Pterosperma* sp., *Todisporites* sp., *Uvaesporites puzzlei*, *Uvaesporites cerebralis*, *Uvaesporites* sp. Outgoing species: *Callialasporites trilobatus*, *Densoisporites crassus*, *Leptolepidites equatibossus*, *Leptolepidites major*, *Leptolepidites paverus*, *Leptolepidites rotundus*, *Lycopodiacidites pseudoreticulatus*, *Lycopodiumsporites vilhelmii*, *Nannoceratopsis gracilis*, *Nannoceratopsis spiculata*, *Protopinus scanicus*
Zone 2 (112.30–80.55 m)	Incoming species: *Calamospora mesozoicus*, *Densoisporites crassus*, *Densoisporites scanicus*, *Leptolepidites major*, *Leptolepidites paverus*, *Neoraistrickia gristhorpensis*, *Neoraistrickia truncatus*, *Polycingulatisporites triangularis*
Zone 1 (130.00–124.45 m)	Incoming species: *Callialasporites minus*, *Callialasporites turbatus*, *Ceratosporites spinosus*, *Chasmatosporites apertus*, *Cibotiumspora jurienensis*, *Contignisporites problematicus*, *Converrucosisporites* sp., *Cyathidites concavus*, *Densoisporites velatus*, *Eucommiidites troedssonii*, *Leptolepidites rotundus*, *Leptolepidites* sp., *Lycopodiacidites pseudoreticulatus*, *Lycopodiumsporites clavatoides*, *Lycopodiumsporites semimuris*, *Lycopodiumsporites vilhelmii*, *Pityosporites scaurus*, *Protopinus scanicus*, *Sestrosporites pseudoalveolatus*, *Spheripollenites* sp., *Stereisporites cicatricosus*, *Tasmanites* sp., *Todisporites major*, *Todisporites minor*

Åstorp and Karindal (Guy-Ohlson & Norling 1988 in terms of both the variety of species and the number of individuals of the form genus *Callialasporites*.

PALYNOFACIES ANALYSIS AND INTERPRETATION

The results of the palynofacies examination are presented pictorially in Plates 1 and 2. Observations are provided in the explanations of the plates and are interpreted in conjunction with the results of the palynostratigraphic analysis in Fig. 3. Together these results indicate changes in depositional environment at Humlarp from fresh to fully marine. Deposition in freshwater at 137.80 m was followed by a marine transgression giving brackish conditions from 130.00 m to 80.55 m and a fully marine environment at 73.90 m and 66.65 m. A regression is assumed at 65.75 m as freshwater conditions are indicated at that level. There was then a return to brackish conditions at 54.60 m and 48.75 m, followed by fully marine conditions at 47.30 m and 44.10 m and a return to a brackish environment at 43.10 m and 34.65 m. The depositional environment could not be determined for samples 29.45 m and 25.45 m but brackish conditions are likely at 22.85 m and a marine environment at 22.35 m.

COMPARISON AND CORRELATION WITH RELEVANT PALYNOFLORAS

The composition of the Humlarp assemblages is similar to that of other palynofloras already described from boreholes in the same region. Of the 86 species identified, 67 were also recorded from Vilhelmsfält. Comparison of the data in Figs. 3 and 4 in this paper with those from Vilhelmsfält (Guy 1971, Table 6, pp. 82–86, and Fig. 11, p. 87) reveals a great similarity in palynomorph content, except with respect to *Araucariacites australis* and *Perinopollenites elatoides*, between the Humlarp core samples below 73.50 m and the Vilhelmsfält interval 75–140 m, previously dated as Bajocian–Bathonian

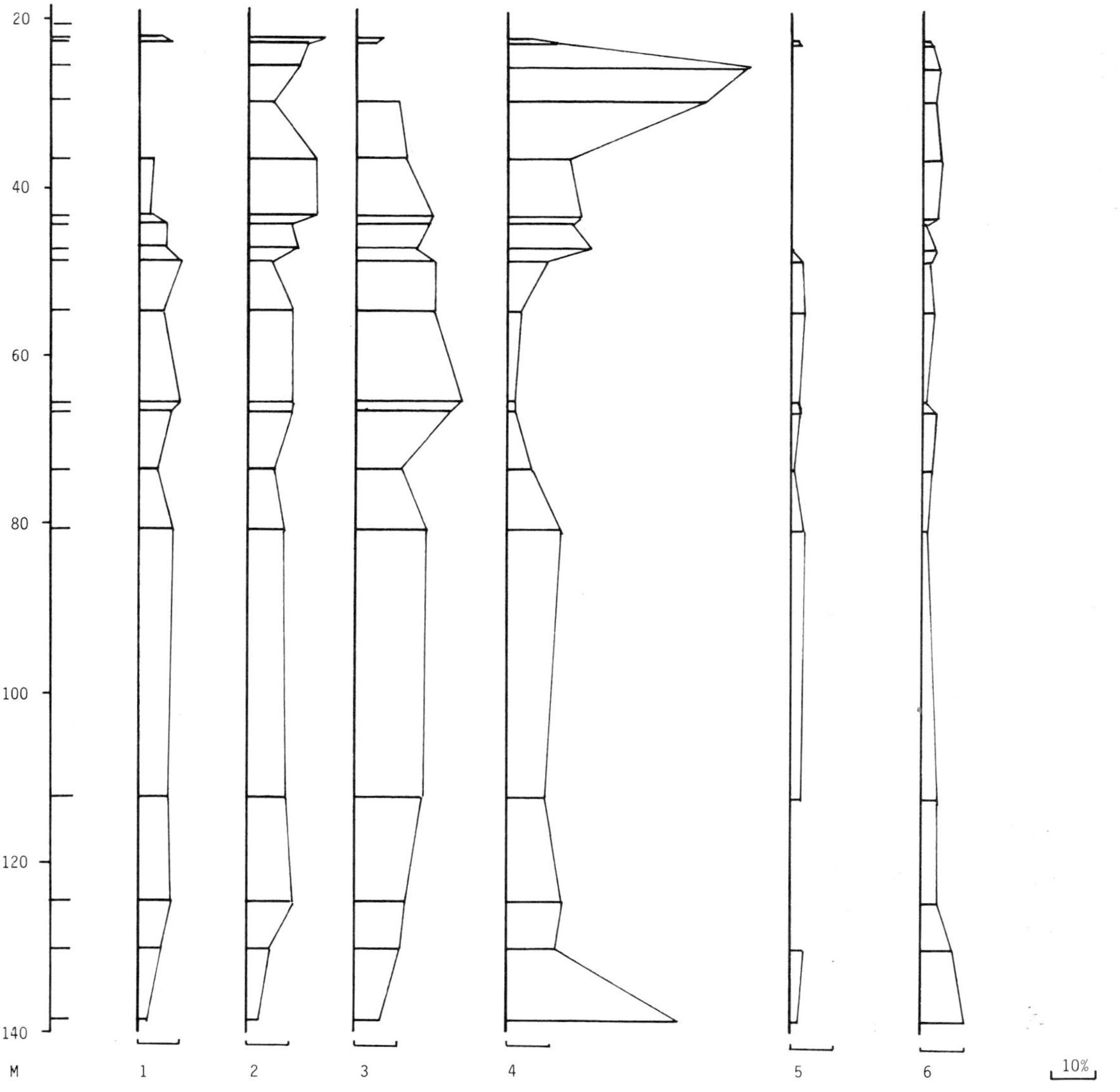

Fig. 4 — Changes in relative abundance at Humlarp of six representative species. 1, *Cerebropollenites mesozoicus*; 2, *Classopollis classoides*; 3, *Araucariacites australis*; 4, *Perinopollenites elatoides*; 5, *Osmundacidites wellmanii*; 6, *Cyathidites minor*.

(Guy 1971). Humlarp zones 1 and 2 can in fact be correlated with two equivalent zones at Vilhelmsfält, that from 134.30 m to 102.00 m corresponding to Humlarp zone 1 and that from 102 m to 75 m corresponding to zone 2.

The assemblages are also recognizable in the boreholes Härninge Nos. 159 and 64 and Rosenhäll No. 62 (Guy-Ohlson 1978). Härninge No. 159 bears the closest resemblance and two palynomorph zones can be detected; zone 2 from 108 m to 91 m and zone 1 from 135.5 m to 108 m. Despite the fact that the samples from Härninge No. 64 and Rosenhäll No. 62 are cable tool drill cuttings rather than cores, the same zones are recognizable. In Härninge No. 64, zone 2 encompasses 110 m to 85 m and zone 1 extends from 140 m to 113.5 m, but in the Rosenhäll No. 62 borehole only zone 2 can be detected, from 90 m to 65.5 m. All these correlations are presented in diagrammatic form in Fig. 5.

Middle Jurassic (Bajocian–Bathonian) palynofloras have been described by Tralau (1968) from Eriksdal in southeast Scania. Comparison of the palynological recovery from Humlarp with that from Eriksdal shows that 46 of the 86 species (53%) occur at Eriksdal. If only palynomorph assemblage zone 4 is compared, the figures change to 36 species out of 52 (69%). Thus the youngest assemblage at Humlarp bears the greatest resemblance to the Eriksdal palynoflora. Unfortunately, Tralau (1968) did not publish any quantitative results, so that a detailed comparison which might permit accurate correlation is not possible. From the systematic descriptions of the dispersed pollen grains and spores it is, however, possible to obtain the exact sampling horizon for Tralau's illustrated specimens and hence a qualitative estimate of which species came from which sampling horizon. The uncertain age of the horizons, however, presents

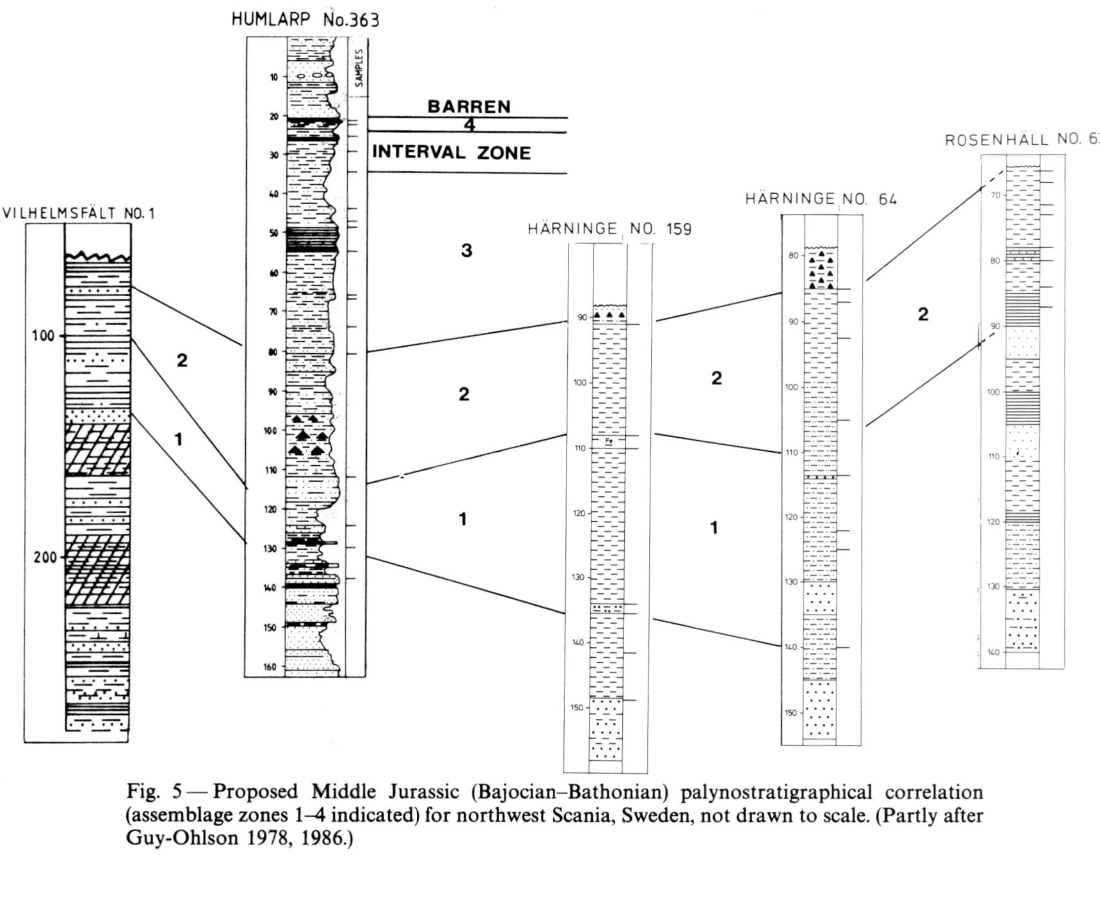

Fig. 5 — Proposed Middle Jurassic (Bajocian–Bathonian) palynostratigraphical correlation (assemblage zones 1–4 indicated) for northwest Scania, Sweden, not drawn to scale. (Partly after Guy-Ohlson 1978, 1986.)

another difficulty. Tralau (1968, p. 104) claimed that no Bathonian elements such 'as *Lycopodiacidites rugulatus*, *Calamospora mesozoica*, *Triletes rariverrucatus*, *Eucommiidites granulosus* and *Leptolepidites equatibossus* are found above layer 41, whereas Bathonian elements continue upwards'. A check of the illustrated specimens reveals that the Bathonian elements come from layers 41, 45 and 53. If the statement 'Bathonian elements continue upwards' refers to the layer numbers then this is quite in order, but the top layers 12–43 (i.e. 100 m) are labelled Bathonian in his Fig. 10 (p. 105).

The Eriksdal samples clearly need to be reinvestigated in order to refine the age determination of the palynoflora. Since the composition of the Humlarp assemblages from zones 3 and 4 bears a greater resemblance to those from Eriksdal than that of zones 1 and 2, it is, however, suggested on purely qualitative grounds that the Eriksdal palynoflora is younger than the Bajocian assemblages of Vilhelmsfält, Humlarp and Rosenhäll.

No stratigraphical correlations can be attempted with Lilla Altona and Öresund No. 2 (Tralau & Artursson 1971) and Höllviken (Mahin 1968) because of a lack of adequate data. Sediments of Middle Jurassic (Bajocian) age were recorded by Norling (1972) in the Rydebäck-Fortuna Borehole No. 4. Samples from these beds have been prepared for palynological investigation but were found to be devoid of palynomorphs.

Middle Jurassic assemblages that are dominated by land-derived palynomorphs and contain very few or no dinoflagellates have always been difficult to date with accuracy because the palynomorphs comprising the assemblages have relatively long stratigraphical ranges compared with those of dinoflagellates. Although descriptions of many well-dated Middle Jurassic assemblages are published, comparisons here are limited to those that consist mainly of spores and pollen grains. The Jurassic strata of the Danish island of Bornholm have been investigated palynologically by Hoelstad (1985) who defined three biozones, one of which, namely zone III, he dated as Middle Jurassic. The palynomorph composition of this zone is most closely similar to that of Humlarp assemblage zone 2. It is also similar to Tralau's Eriksdal palynoflora (33 species of 51 recorded from the Bornholm zone having been found at Eriksdal) but, when quantitative aspects and species in common are considered, the greater similarity is perhaps to Humlarp. Thus, it is suggested that Hoelstad's zone III may be of late Bajocian or Bajocian–Bathonian age.

Other Jurassic palynofloras of similar age from, for example, the northwest European Continental Shelf (Batten 1978), Greenland (Lund & Pedersen 1985), Britain (Couper 1958, Lam & Porter 1977, Riding 1987a, b), Ireland (Higgs & Beese 1986), Holland (Herngreen & De Boer 1974), France (Levet-Carette 1964, Schmidt 1987, Srivastava 1987), Germany (Schulz 1966, Weiss in: Schweitzer *et al.* 1987), Italy (Del Rio 1984), Bulgaria (Černjavska 1986), Siberia (Ilyina 1985, 1986), Canada (Pocock 1970), Argentina (Volkheimer 1968, 1969), Australia (Filatoff 1975), Iran (Arjang 1975, Achilles *et al.* 1984) and Afghanistan (Ashraf 1977) are clearly comparable as far as species content and compositional characters are concerned, but no detailed correlation is possible because of the incomplete presentation of results in many of the publications. A very generalized correlation of spore- and pollen-grain-dominated assemblages of Bajocian and Bathonian age compiled from the northwest European studies referred to above is presented in Fig. 6.

CONCLUSION

The palynostratigraphic investigation of sediments from Humlarp Borehole No. 363 in northwest Scania has resulted in the defining of four assemblage zones in a hitherto unknown Middle Jurassic sequence. These are based on a total spore and pollen grain content of 86 species. They are considered to range from early Bajocian to early Bathonian in age, and they permit detailed correlation with other Middle Jurassic palynological assemblages from several boreholes in northwest Scania. Even though only a very generalized correlation of the Swedish zones

	SWEDEN (NW Scania) 1	DENMARK (Bornholm) 2	HOLLAND (Eastern) 3	ENGLAND (Yorkshire) 4	SCOTLAND (Brora) 5	IRELAND (Co. Cork) 6	GREENLAND (James Land) 7	NORWAY (Andøya, Trøndelag) 8
BATHONIAN					BRORA COAL FORMATION	COLBOND CLAY OF CLOYNE MICROFLORA	SUB ASSEMBLAGE C 3 (in part)	ASSEMBLAGE 1
BATHONIAN	ZONE 4		POLLEN ZONE VII	UPPER DELTAIC				
BATHONIAN	INTERVAL							
BAJOCIAN	ZONE 3	ZONE III					SUB ASSEMBLAGE C 1 - C 2	
BAJOCIAN	ZONE 2		POLLEN ZONE VI	MIDDLE DELTAIC				
BAJOCIAN	ZONE 1			LOWER DELTAIC				

Fig. 6 — Tentative correlation of selected Bajocian—Bathonian spore- and pollen-dominated palynological assemblages for northwest Europe based respectively on the following: 1, the present paper; 2, Hoelstad (1985); 3, Herngreen & De Boer (1974); 4, Couper (1958); 5, Lam & Porter (1977); 6, Higgs & Beese (1986); 7, Lund & Pedersen (1985); 8, Thusu & Vigran (1975).

with the northwest European spore- and pollen-grain-dominated palynofloras is presented, it is hoped that this will encourage others to persevere with the non-dinoflagellate Middle Jurassic assemblages with the aim of arriving at more refined and detailed correlations of such assemblages in the future.

ACKNOWLEDGEMENTS

For help in many different ways I wish to thank the following people, namely Yvonne Arremo, Lisbeth Brevig, Kamlesh Khullar, Olof Haglund, Lars Imby, Peter Kvarnestam, Bengt Lindqvist and Erik Norling.

Plates 1–5 — Photographs illustrating palynofacies of the samples taken from different levels (Figs. 1–16) and micrographs of selected palynomorphs (Figs. 17–50).

Plate 1 —

Fig. 1 — 22.35 m, a shallow near-shore environment of deposition is envisaged for this grey clay sample containing some sand. The palynofacies is dominated by land plant debris, abundant well-preserved spores and pollen grains and microscopic algae (tasmanitids).

Fig. 2 — 22.85 m, brackish sedimentary conditions are suggested for this mixed clay – sandstone sample. The palynofacies consists of a reduced amount of well-preserved land-derived plant material which is also smaller in size by comparison with that in the overlying sample. In addition, there are fewer palynomorphs, but those present (*Botryococcus* colonies occur in sample but not shown on photograph) indicate freshwater influence on the marine depositional environment, the site of which is assumed to be further away from the source of the miospores.

Figs. 3 and 4 — 34.55 m and 43.10 m, both palynofacies and the presence of *Tasmanites* and *Botryococcus* indicate a brackish depositional environment. The sample at 34.55 m is a greenish clay with laminae of fine sand and mica and that from 43.10 m is a dark-grey clay with some light-coloured sand.

Figs. 5 and 6 — 44.10 m and 47.30 m, these palynofacies, in which *Tasmanites* and *Micrhystridium* spp. occur, indicate a decrease in the number of land-derived palynomorphs in a quiet marine environment; both samples are grey clays (partly sandy and shaley).

Fig. 7 — 48.75 m, the palynofacies indicates an increase in land-derived palynomorphs compared with the previous sample (47.30 m) which, together with the presence of *Botryococcus* colonies (not in photographs), indicate a freshwater influence on the marine environment; this is in turn indicated by such genera as *Nannoceratopsis*, *Pterosperma* and *Tasmanites* (see Plate 1, Fig. 3). The sample is an unlaminated grey clay containing sand and unorientated fragments of slate.

Fig. 8 — 54.60 m, the palynofacies contains well-preserved palynomorphs; deposition in a brackish environment is inferred from the presence of both *Botryococcus* and *Tasmanites*. The sample is a dark-grey shaley clay with light-grey sandy and micaceous laminae.

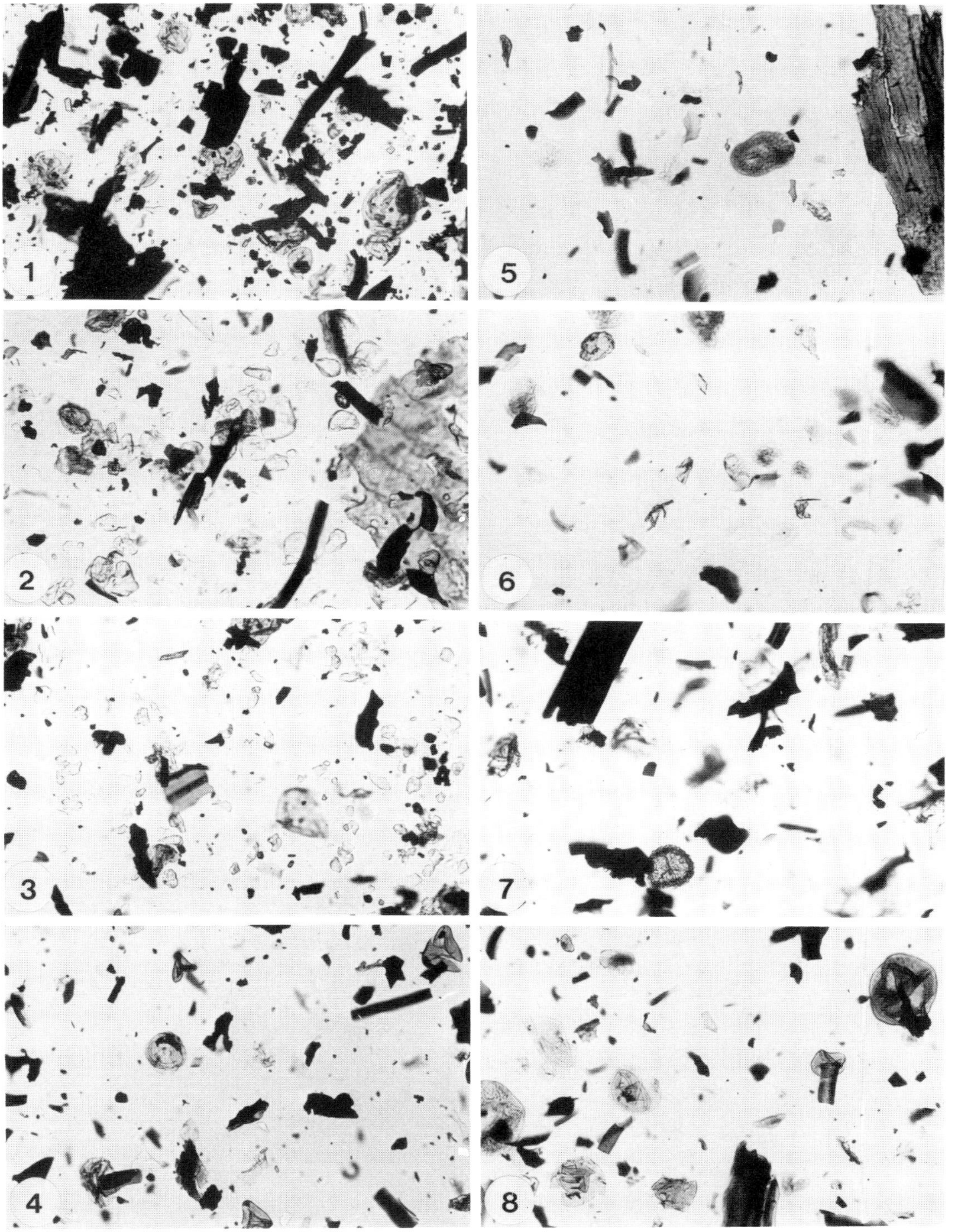

Plate 2 —

Fig. 9 — 65.75 m, palynofacies containing spores, pollen grains and the alga *Botryococcus*. No dinoflagellates or acritarchs were recovered from this unsorted sandy grey clay sample which is presumed to have been deposited in a freshwater environment.

Figs. 10 and 11 — 66.65 m and 73.90 m, the palynofacies are dominated by gymnospermous pollen and this, together with the presence of *Tasmanites* (not in photographs), suggests a marine environment not particularly close to land. Both samples are greyish sandy clay, the latter with laminae of light-grey sand.

Figs. 12–15 — 80.55 m, 112.30 m, 124.45 m and 130.00 m, the palynological analyses indicate a brackish depositional environment for the beds represented by these samples. The sample at 124.45 m differs from the others in having very few palynomorphs and 57% of those found were the alga *Tasmanites* (see Plate 1, Fig. 3). These facts together with the general aspect of the palynofacies suggest a quiet shallow-water environment of deposition for this light-grey clayey sandstone sample.

Fig. 16 — 137.80 m, the palynofacies from this fine-grained light-grey sandstone was found to contain few palynomorphs. A freshwater environment is indicated by the presence of *Botryococcus*.

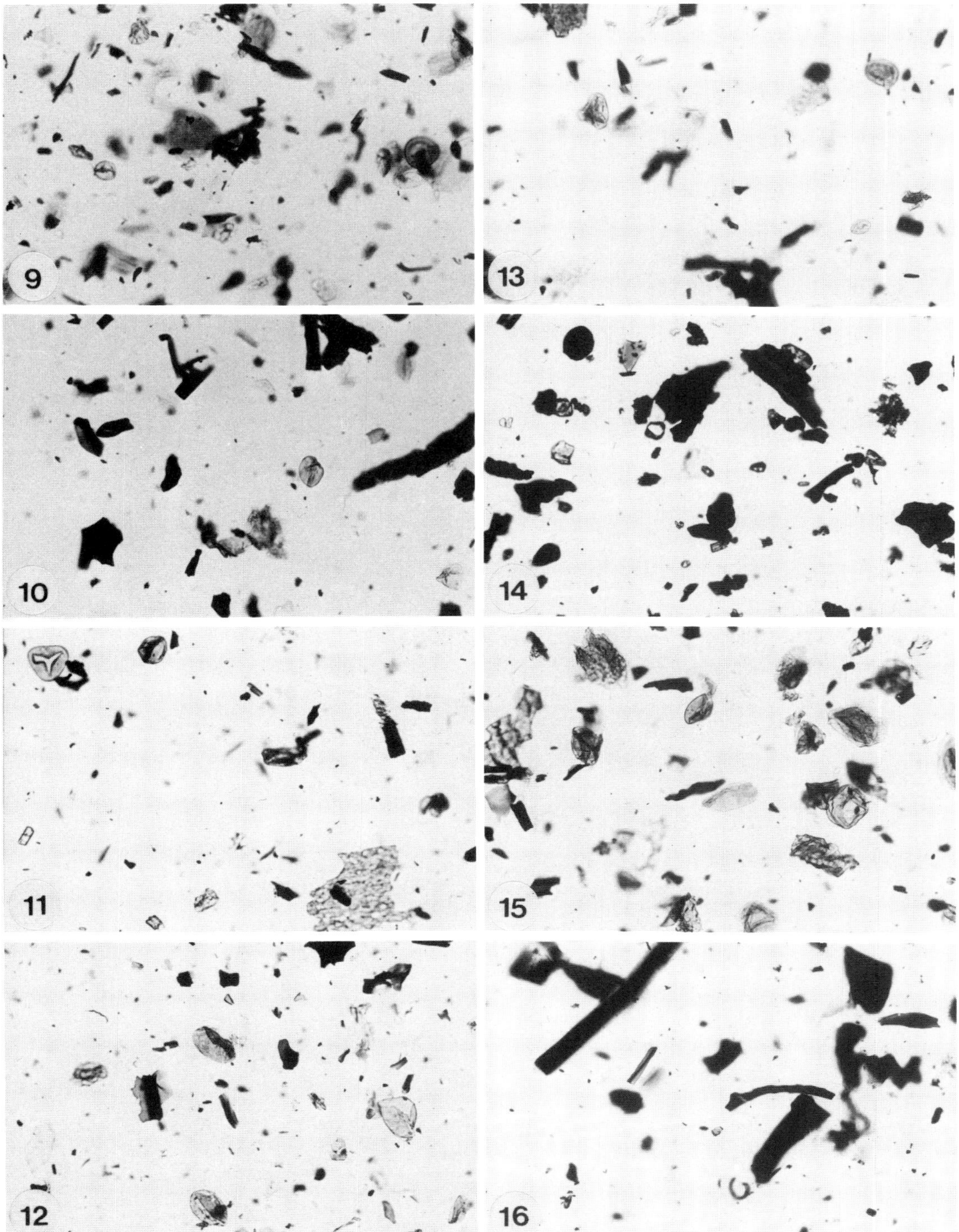
9
10
11
12
13
14
15
16

Plate 3—

Fig. 17 —*Cyathidites australis* Couper; SEM prep. 16.

Fig. 18 and 19 — *Cyathidites minor* Couper; proximal and distal views, SEM prep. 14.

Fig. 20 —*Baculatisporites comaumensis* (Cookson) Potonié; distal view, SEM prep. 24.

Fig. 21 —*Osmundacidites wellmanii* Couper; SEM prep. 23.

Fig. 22 —*Leptolepidites paverus* Levet-Carette; SEM prep. 26.

Fig. 23 —*Uvaesporites argentaeformis* (Bolkhovitina) Schulz; SEM prep. 25.

Fig. 24 —*Lycopodiumsporites reticulumsporites* (Rouse) Dettmann; distal surface, SEM prep. 11.

Fig. 25 —*Leptolepidites equatibossus* (Couper) Tralau; SEM prep. 20.

Fig. 26 —*Lycopodiacidites rugulatus* (Couper) Schulz; SEM prep. 41.

Fig. 27 —*Lycopodiumsporites semimuris* Danzé-Corsin & Laveine; SEM prep. 23.

Fig. 28 —*Lycopodiumsporites semimuris* Danzé-Corsin & Laveine; distal surface, SEM prep. 23.

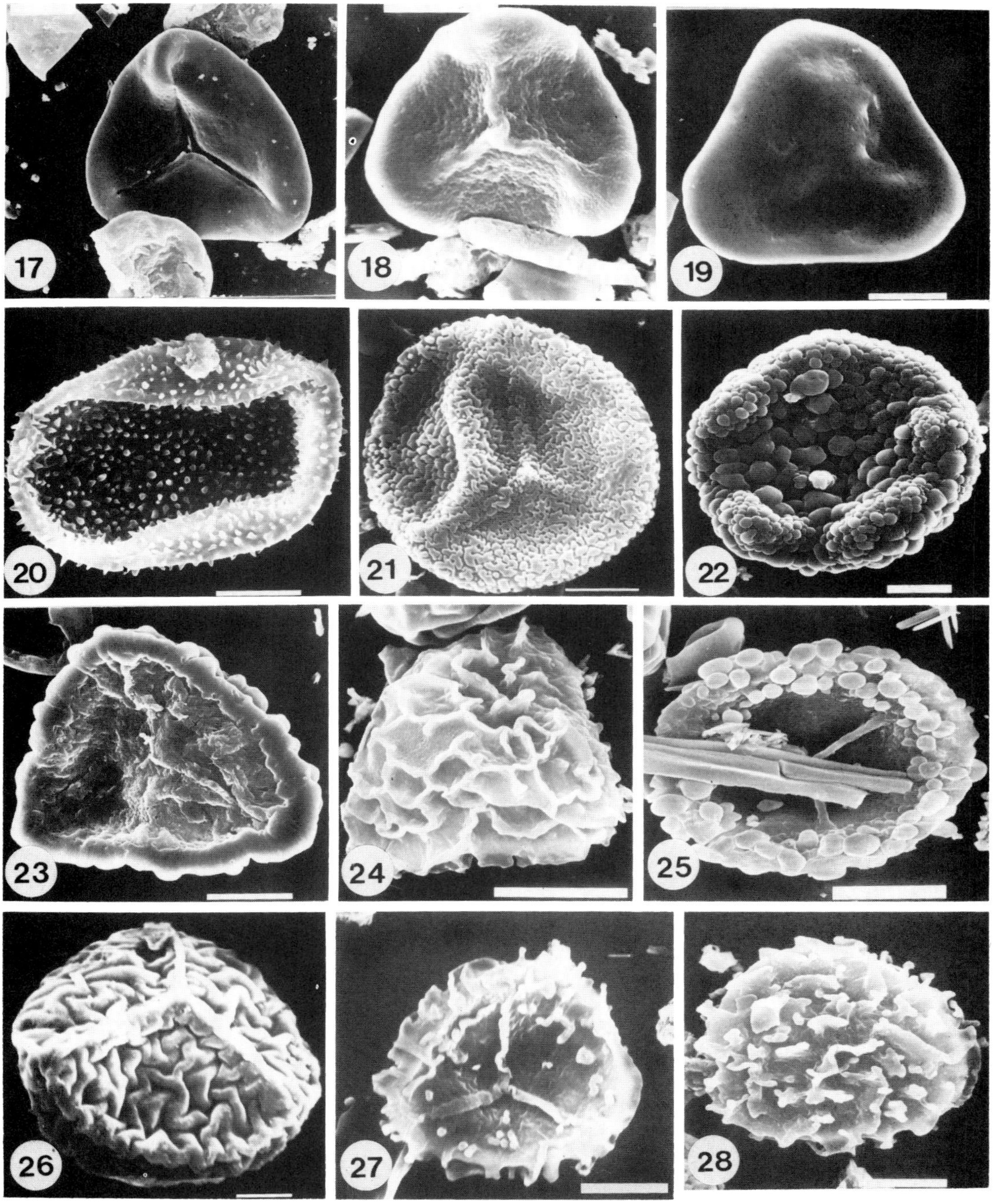
17
18
19
20
21
22
23
24
25
26
27
28

Plage 4 —

Fig. 29 — *Contignisporites problematicus* (Couper) Döring; SEM prep. 14.

Fig. 30 — *Gleicheniidites senonicus* Ross; SEM prep. 25.

Fig. 31 — *Gleicheniidites senonicus* Ross; distal surface, SEM prep. 14.

Fig. 32 — *Neoraistrickia gristhorpensis* (Couper) Tralau; distal surface, SEM prep. 20.

Fig. 33 — Cf. *Staplinisporites caminus* (Balme) Pocock; SEM prep. 19.

Fig. 34 — *Cerebropollenites mesozoicus* (Couper) Nilsson; SEM prep. 20.

Fig. 35 — *Polycingulatisporites triangularis* (Bolkhovitina) Playford & Dettmann; SEM prep. 21.

Fig. 36 — *Callialasporites trilobatus* (Balme) Dev; SEM prep. 28.

Fig. 37 — *Callialasporites minus* (Tralau) Guy; SEM prep. 19.

Fig. 38 — *Parvisaccites enigmatus* Couper; SEM prep. 24.

Fig. 39 — *Brachysaccus microsaccus* (Couper) Mädler; SEM prep. 16.

Fig. 40 — *Podocarpidites* sp.; SEM prep. 16.

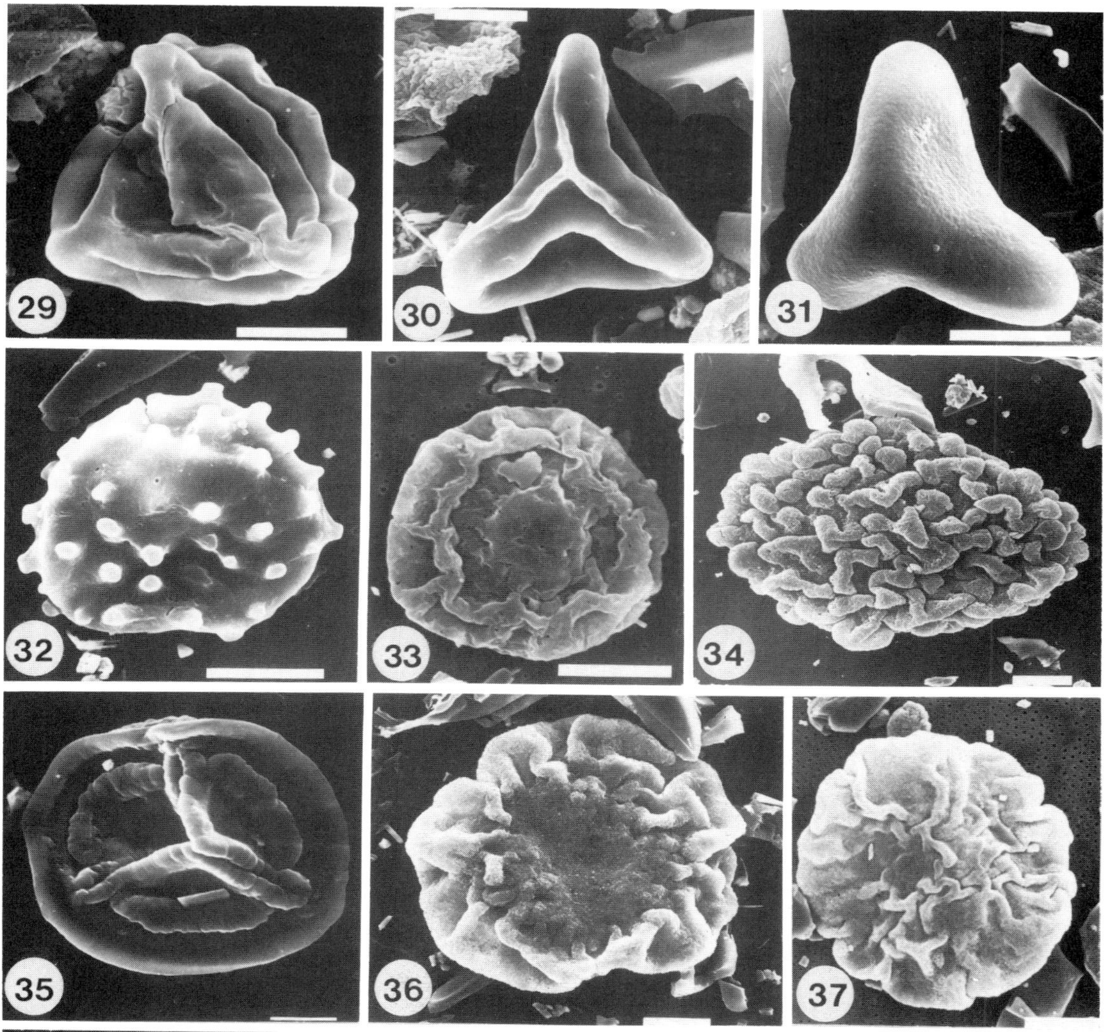
29
30
31
32
33
34
35
36
37
38
39
40

Plate 5 —

Fig. 42 — *Alisporites robustus* Nilsson; SEM prep. 24.

Fig. 43 — *Pityosporites scaurus* (Nilsson) Schulz; SEM prep. 24.

Fig. 44 — *Vitreisporites pallidus* (Reissinger) Nilsson; SEM prep. 16, scale bar represents 1 μm.

Fig. 45 — *Araucariacites australis* Cookson; SEM prep. 22.

Fig. 46 — *Ginkgocycadophytus* sp; SEM prep. 24.

Fig. 47 — *Nannoceratopsis gracilis* (Alberti) Evitt; SEM prep. 14.

Fig. 48 — *Classopollis classoides* (Pflug) Pocock & Jansonius; SEM prep. 25.

Fig. 49 — *Exesipollenites tumulus* Balme; SEM prep. 26.

Fig. 50 — *Nannoceratopsis gracilis* (Alberti) Evitt; SEM prep. 23.

All magnifications in Figs. 1–16 × 575. The scale bars in Figs. 17–50 represent 10 μm except where otherwise stated.

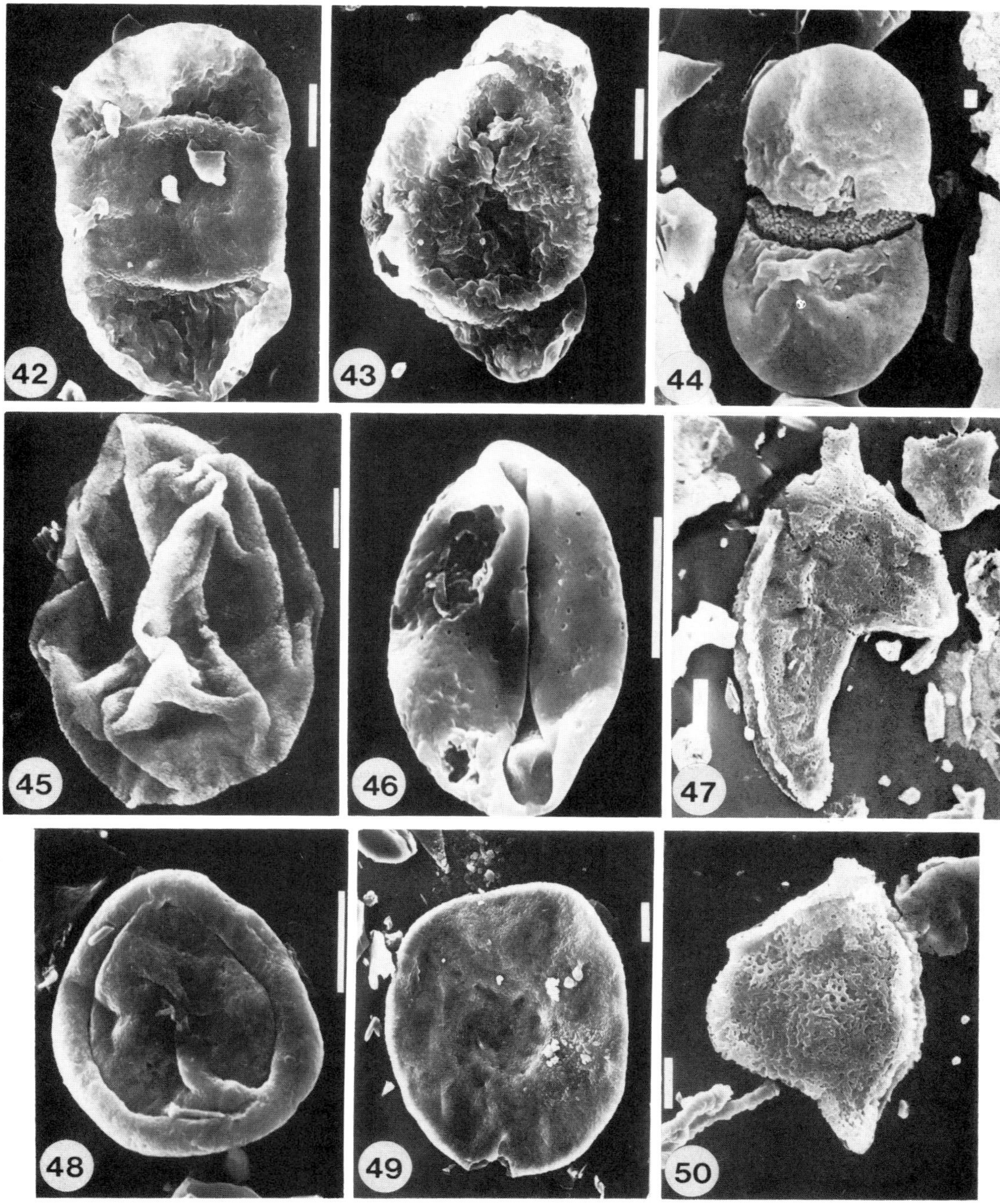
42
43
44
45
46
47
48
49
50

REFERENCES

Achilles, H., Kaiser, H. & Schweitzer, H.-J. (1984). Die räto-jurassischen Floren des Iran und Afghanistans. 7. Die Mikroflora der obertriadisch-jurassischen Ablagerungen des Alborg-Gebirges (Nord-Iran). *Palaeontographica B.* **194** 14–95.

Arjang, B. (1975). Die räto-jurassischen Floren des Iran und Afghanistans. 1. Die Mikroflora der räto-jurassischen Ablagerungen des Kermaner Beckens (Zentral-Iran). *Palaeontographica B* **152** 85–148.

Ashraf, A. R. (1977). Die räto-jurassischen Floren des Iran und Afghanistans. 3. Die Mikrofloren der rätischen bis unterkretazischen Ablagerungen Nordafghanistans. *Palaeontographica B* **161** 1–97.

Batten, D. J. (1978). Early Cretaceous to Middle Jurassic miospores and palynofacies of the northwest European Continental Shelf. In: Thusu, B. (ed.), *Continental Shelf Institute (Trondheim, Norway) Publication* **100** 97–101.

Batten, D. J. (1981). Palynofacies, organic maturation and source potential for petroleum. In: Brooks, J. (ed), *Organic maturation studies and fossil fuel exploration.* Academic Press, London, pp. 201–223.

Batten, D. J. & Morrison, L. (1981). Methods of palynological preparation for palaeoenvironmental, source potential and organic maturation studies. *Norwegian Petroleum Directorate Bulletin* **2** 35–53.

Černjavska, S. (1986). Lower and Middle Jurassic palynostratigraphy. *Geologica Balcania* **16** 21–32.

Couper, R. A. (1958). British Mesozoic microspores and pollen grains. *Palaeontographica B* **103** 75–179.

Del Rio, M. (1984). Palynology of Middle Jurassic black organic shale of Tacco di Laconi, central Sardinia, Italy. *Bollettino della Societa Paleontologica Italiano* **23** 325–342.

Filatoff, J. (1975). Jurassic palynology of the Perth Basin, western Australia. *Palaeontographica B* **104** 1–113.

Guy, D. J. E. (1971). Palynological investigations in the Middle Jurassic of the Vilhelmsfält boring, southern Sweden. *Publications from the Institutes of Mineralogy, Palaeontology and Quaternary Geology, University of Lund* **168** 1–104.

Guy-Ohlson, D. (1976). Additional palynomorphs from the Middle Jurassic of the Vilhelmsfält boring, southern Sweden. *Acta Universitatis Stockholmiensis, Contributions in Geology* **30** 81–91.

Guy-Ohlson, D. (1978). Jurassic biostratigraphy of three borings in NW Scania (a brief palynological report). *Sveriges Geologiska Undersökning, Rapporter och Meddelanden* **11** 1–41.

Guy-Ohlson, D. (1985). Palynology of Swedish Upper Jurassic sediments. *Palynology* **9** 243–244.

Guy-Ohlson, D. (1986). Jurassic palynology of the Vilhelmsfält Bore No. 1, Scania, Sweden. Toarcian–Aalenian. *Publication, Section of Palaeobotany, Swedish Museum of Natural History Stockholm,* 1–127.

Guy-Ohlson, D. & Malmquist, E. (1985). Lower Jurassic biostratigraphy of the Oppegård Bore No. 1, NW Scania, Sweden. *Sveriges Geologiska Undersökning, Rapporter och Meddelanden* **40** 1–27.

Guy-Ohlson, D. & Norling, E. (1988). Upper Jurassic litho- and biostratigraphy of NW Scania, Sweden. *Sveriges Geologiska Undersökning, Serie Ca* **72** 1–37.

Herngreen, G. F. W. & De Boer, K. F. (1974). Palynology of Rhaetian, Liassic and Dogger strata in the eastern Netherlands. *Geologie en Mijnbouw* **53** 343–368.

Higgs, K. & Beese, A. P. (1986). A Jurassic microflora from the Colbond Clay of Cloyne, County Cork. *Irish Journal of Earth Sciences* **7** 99–109.

Hoelstad, T. (1985). Palynology of the uppermost Lower to Middle Jurassic strata on Bornholm, Denmark. *Bulletin of the Geological Society of Denmark* **34** 112–132.

Ilyina, V. I. (1985). Jurassic palynology of Siberia. *Academy of Sciences of the USSR, Siberian Branch, Institute of Geology and Geophysics Transactions* **638** 1–237.

Ilynia, V. I. (1986). Subdivision and correlation of the marine and non-marine Jurassic sediments in Siberia based on palynological evidence. *Review of Palaeobotany and Palynology* **48** 357–364.

Lam, K. & Porter, R. (1977). The distribution of palynomorphs in the Jurassic rocks of the Brora Outlier, NE Scotland. *Journal of the Geological Society, London* **134** 45–55.

Levet-Carette, J. (1964). Etude de la microflore bajocienne d'un sondage effectue dans les sous-sol de Boulogne-sur-Mer (P.-de.C). *Annales de la Société Géologique du Nord* **84** 91–121.

Lund, J. J. & Pedersen, K. J. (1985). Palynology of the marine Jurassic formations in the Vardekloft ravine, Jameson Land, east Greenland. *Bulletin of the Geological Society of Denmark* **33** 371–399.

Mahin, K. (1968). Sporenpaläontologischer Nachweis von mittlerer Jura in der Bohrung von Höllviken, Sudschweden. *Geologiska Föreningens i Stockholms Förhandlingar* **90** 121–124.

Norling, E. (1972). Jurassic stratigraphy and Foraminifera of western Scania, southern Sweden. *Sveriges Geologiska Undersökning, Serie Ca* **47** 1–120.

Norling, E. (1978). Berggrund. In: *Beskrivning till Jordartskartan Höganäs NO/Helsingborg NV. Sveriges Geologiska Undersökning, Serie Ae* **25** 20–32.

Norling, E., (1981). Upper Jurassic and Lower Cretaceous geology of Sweden. *Geologiska Föreningens i Stockholms Förhandlingar* **103** 253–269.

Pocock, S. A. J. (1970). Palynology of the Jurassic sediments of western Canada. Part 1. Terrestrial species. *Palaeontographica B* **130** 12–136.

Riding, J. B. (1987a). Dinoflagellate cyst stratigraphy of the Nettleton Bottom Borehole (Jurassic: Hettangian to Kimmeridgian), Lincolnshire, England. *Proceedings of the Yorkshire Geological Society* **46** 231–266.

Riding, J. B. (1987b). Contribution to section on Jurassic palynomorphs. In: Lord, A. R., & Bown, P. R. (eds), *Mesozoic and Cenozoic stratigraphical micropalaeontology of the Dorset coast and Isle of Wight, southern England. Field guide for the 20th European Micropalaeontological Colloquium. British Micropalaeontological Society Guidebook 1.*

Schmidt, J.-P. (1987). Les microflores du Jurassique intérêtes stratigraphiques, paleogéographiques et paléoclimatiques étude palynologique du Jurassique en Alsace. *Doctoral thesis,* Université Louis Pasteur, Strasbourg 1, pp. 1–90.

Schulz, E. (1966). Uber einige neue Sporae dispersae aus dem älteren Mesophytikum Deutschland. *Beihefte zum Geologischen Jahrbuch* **55** 130–151.

Schweitzer, H.-J., Ashraff, A. R. & Weiss, M. (1987). Korrelation der Sporenzonen der Oberen Trias und des Jura in Mittleren Orient und in Suddeutschland. *Geologische Rundschau* **76** 923–943.

Srivastava, S. K. (1987). Jurassic spore-pollen assemblages from Normandy (France) and Germany. *Geobios* **20** 5–79.

Thusu, B. & Vigran, J. O. (1975). A review of the palynostratigraphy of the Jurassic system in Norway. *Continental Shelf Institute (Trondheim, Norway) Publication* **100** 1–12.

Tralau, H. (1968). Botanical investigations into the fossil flora of Eriksdal in Fyledalen, Scania. II: The Middle Jurassic microflora. *Sveriges Geologiska Undersökning, Serie C* **633** 1–185.

Tralau, H. & Artursson, K. (1971). New Middle Jurassic pollen and spore floras from southern Sweden and the Öresund. *Grana* **12** 57–63.

Volkheimer, W. (1968). Esporas y granos de polen del Jurasico de Neuquen (Republica Argentina) I. Descripciones sistematicas. *Ameghiniana* **5** 333–370.

Volkheimer, W., (1969). Esporas y granos de polen del Jurasico de Neuquen (Republica Argentina) II. Asociaciones microfloristicas, aspectos paleoecologicos y paleoclima. *Ameghiniana* **6** 127–145.

Wikman, H., Norling, E., Sivhed, U. & Karis, L. (1981). Berggrundskartan 3B Höganäs NO/3C Helsingborg NV. *Sveriges Geologiska Undersökning, Serie Af* **129.**

Woollam, R. & Riding, J. B. (1983). Dinoflagellate cyst zonation of the English Jurassic. *British Geological Survey Report* **83/2** 1–42.

4

Palaeoenvironments of the late Bathonian–mid-Callovian in the Inner Moray Firth

A. M. MacLennan and **N. H. Trewin**

ABSTRACT

Use of palynofacies, macropalaeontology and sedimentology in a comparison of three late Bathonian–mid-Callovian sections at Brora, Cadh'an Righ and in the Beatrice oilfield indicates that, at Brora and Cadh'an Righ, lagoonal conditions of the Inverbrora Member succeeded alluvial plain deposits of the Doll Member beneath the Brora Coal. Low-diversity dinoflagellate cyst assemblages dominated by *Ctenidodinium sellwoodii* (Sarjeant 1975) Stover & Evitt 1978 are found in the lagoonal shales in association with *Neomiodon*, *Isognomon*, juvenile *Viviparus?*, and fish scales, indicating intermittent marine connection and brackish-water conditions. In the Beatrice area the absence of macrofauna beneath the Brora Coal and the low organic recovery characterized by oxidized plant debris, reworked organic matter and non-marine algae attest to continued deposition in an alluvial plain environment. The widespread Brora Coal was deposited in the silted-up lagoonal area, although initial organic accumulation was from drifted plant material. Marine invasion of the area, probably from the northeast, brought about an influx of open marine fauna and dinoflagellate cysts and resulted in the deposition of a thin marine transgressive sandstone (the Roof Bed) at Brora and Cadh'an Righ. In the succeeding Brora Shale Member, diverse bivalve and dinoflagellate cyst assemblages associated with uniform palynofacies provide evidence for a more stable open marine environment. At Beatrice, the coal is overlain by shales containing several shallow marine sandstones (the 'B' Sand). The presence of coarsening-up sandstones with bioturbation, combined with pre-depositional oxidation and fragmentation of the organic debris and an erratic distribution of a less diverse dinocyst assemblage, suggest off-shore bar and shoreline environments.

A *discus* to early *macrocephalus* zonal age is proposed for the Inverbrora Member lagoonal shales based on the abundance of *Ctenidodinium sellwoodii* and the incoming of *Mendicodinium groenlandicum* (Pocock & Sarjeant 1972) Davey 1979 together with localized common occurrences of *Meiourogonyaulax* spp. and *Korystocysta gochtii* (Sarjeant 1976) Woollam 1983. At Brora the *C. sellwoodii* acme terminates above the coal, suggesting a *macrocephalus* age for the base of the Roof Bed. The subsequent common occurrence of *Ctenidodinium combazii* Dupin 1968 (immediately prior to its extinction) and proliferation of *C. ornatum* (Eisenack 1935) Deflandre 1938 and *C. continuum* Gocht 1970 are consistent with a *calloviense* age for the remainder of the Roof Bed. Brora Shale dinoflagellate cyst assemblages are typical of the *calloviense* and possibly *jason* Zones.

INTRODUCTION

The aim of this investigation is to relate palynofacies to macropalaeontology and sedimentology in an attempt to refine environmental interpretations. Three sections from the Inner Moray Firth (Fig. 1) were chosen for the project.

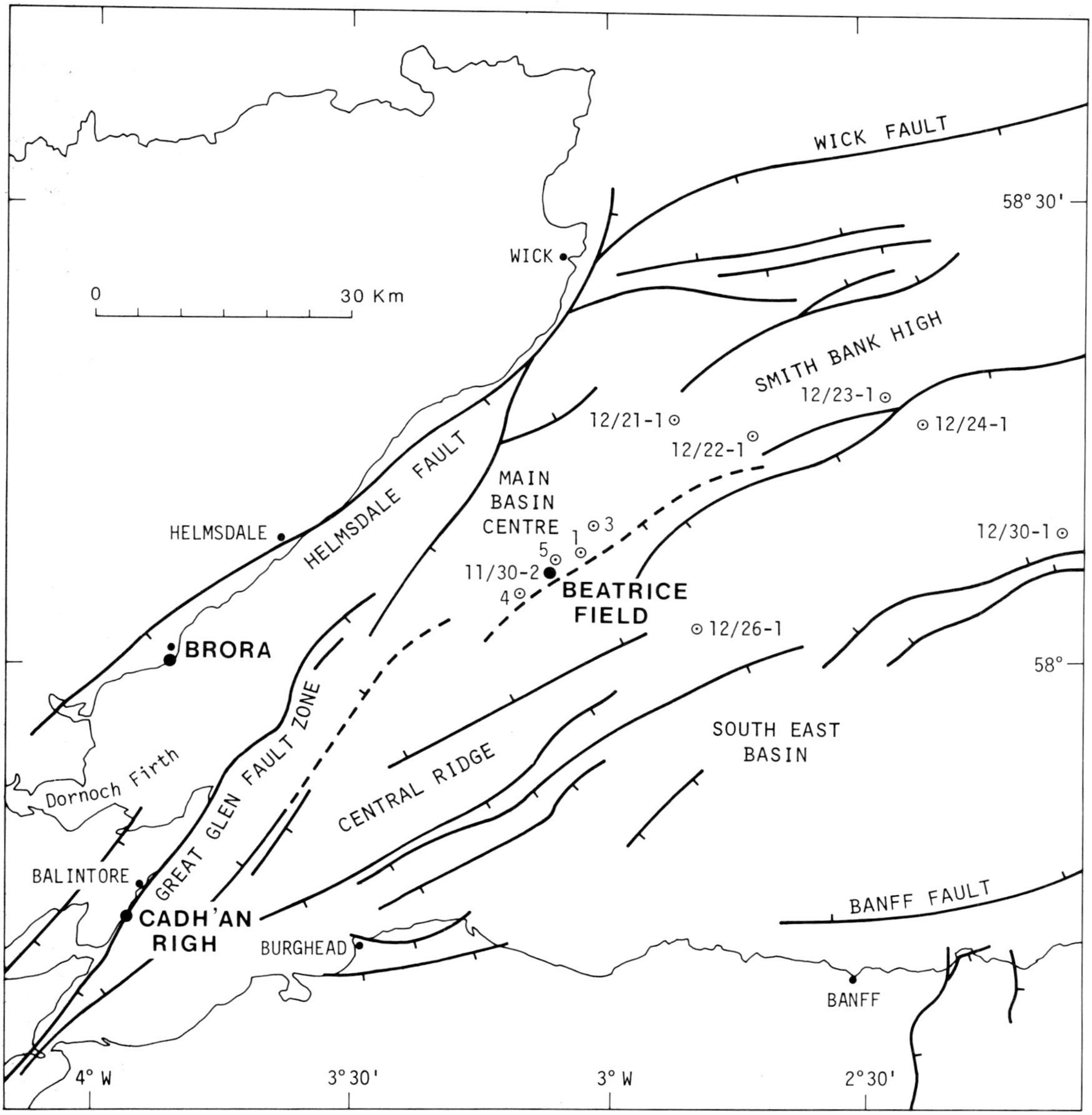

Fig. 1 — Locality map and main structural elements of the Inner Moray Firth. (Modified from Chesher & Lawson 1983.)

The onshore sections at Brora and at Cadh'an Righ near Balintore were sampled and compared with the cored section in well 11/30-2 of the Beatrice oilfield (Figs. 1 and 2). Fig. 2 is based on the authors' measured sections.

The three sections all include a major environmental transition generally referred to as the Callovian marine transgression (e.g. Sykes 1975a). They commence with Bathonian freshwater to marine-influenced lagoonal sediments capped by the extensive Brora Coal. At Brora and Cadh'an Righ the coal is overlain by thin transgressive sandstones which pass rapidly upwards into offshore marine shales. In Beatrice well 11/30-2, lagoonal sediments are virtually absent beneath the coal and erosion is not apparent immediately above it. The section above the coal contrasts with that at Brora in containing several shallow marine coarsening-up sandstone units.

HISTORY OF RESEARCH

The Jurassic of the Brora area has received considerable attention from early times, largely because of economic interest in the Brora Coal which was worked as early as 1598 and then

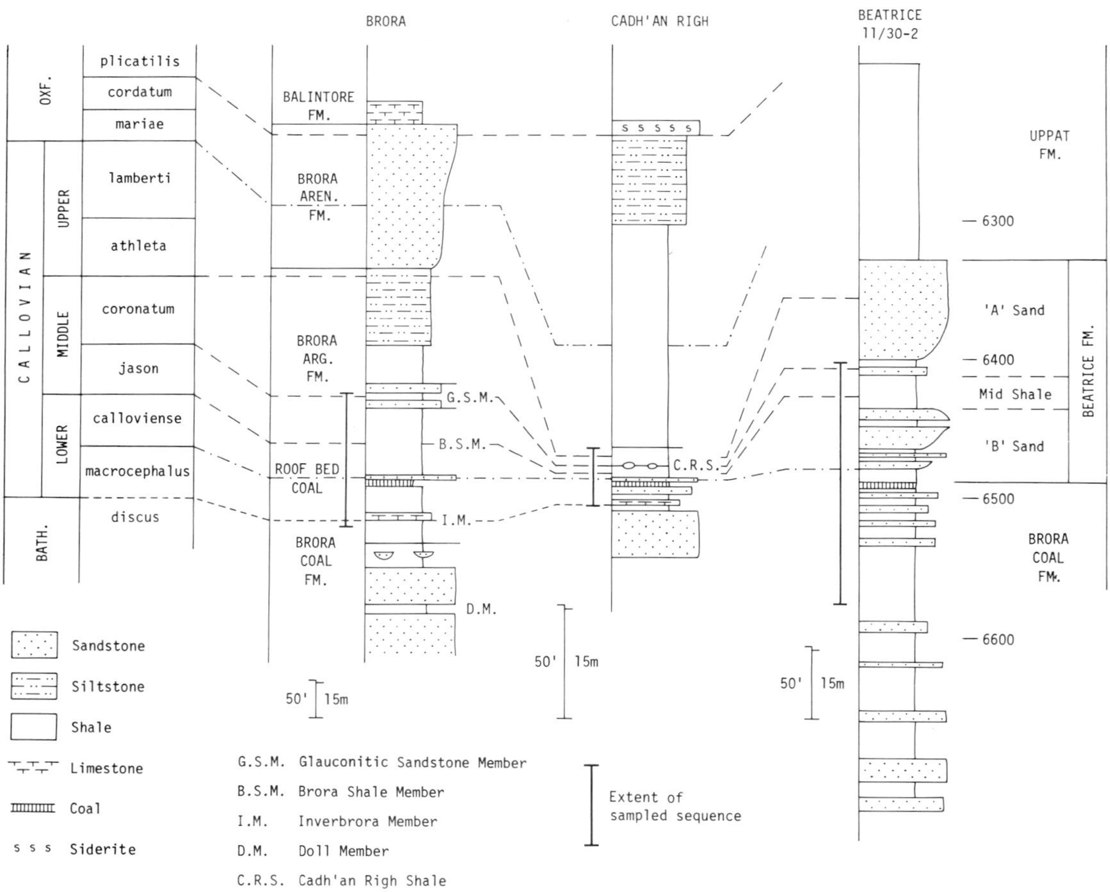

Fig. 2 — Stratigraphic sections, nomenclature and correlations for the three studied sections, based on the observations of Sykes (1975a), Judd (1873) and the present authors; note the difference in scales.

intermittently until 1974 (for summary of working see Flett (in: Lee 1925) and Sykes (1975b)). Murchison (1827) investigated the geology of the coalfield area, and Judd (1873) provided an early account of the succession. The *Memoir of the Geological Survey* on the area (Lee 1925) added much detail to knowledge of the succession. Plant fossils from the Brora Coal Formation were described by Stopes (1907) and more recently by Harris & Rest (1966).

The major modern contributions are those of Sykes (1975a, b) on the marine Callovian to Oxfordian succession and the establishment of ammonite zonal divisions, and the work of Hurst (1981) on the stratigraphy and sedimentology of the Brora Coal Formation at Brora. Neves & Selley (1975) provided an account of the entire Jurassic succession of the Brora–Helmsdale Outlier and, in a palynological investigation of the succession, Lam & Porter (1977) discovered dinocysts in association with the 'brackish' molluscan faunas of the Brora Coal Formation. The palynology of the coal was discussed by Hughes & Couper (1958).

The general geology of the Beatrice Field has been described by Linsley *et al.* (1980) and some further detail given by Johnson (1984). Andrews & Brown (1987) have illustrated the development of an unconformity apparently below the equivalent of the Brora Coal in well 12/26-1 (Fig. 1) which cuts out the equivalent of the Brora Coal Formation. In wells 12/23-1 and 12/30-1 the Jurassic section is considerably reduced with only thin Lias present beneath the unconformity in well 12/23-1; sandstone is found overlying the unconformity and the Brora Coal is absent.

METHODS

Samples were chosen to include any obvious macrofaunas present to enable direct comparison between lithological, macrofaunal and palynological data. A range of lithologies was sampled to provide links with sedimentology. Palynological samples were prepared using standard acid digestion techniques (Batten & Morrison 1983) and were sieved through a 10 µm mesh sieve which minimizes loss of the finer components but still allows 'cleaning' of the preparation.

PRESENTATION OF RESULTS

The data from each sampled section are considered separately under the following headings, the Brora Coal Formation, the Brora Argillaceous Formation (at Brora and Cadh'an Righ) and the Beatrice Formation (at Beatrice). The Brora Coal Formation is further subdivided into the interval below the coal and the coal itself.

Data collected from the three sections are presented in Figs. 3–11. In Figs. 3, 6 and 9 the macrofaunal data are plotted against lithology and the sample positions indicated on the accompanying scaled section. Figs. 4, 7 and 10 record the amount of organic matter and relative abundance of the main organic components recovered from each sample. Certain of the categories used require explanation. The term 'brown wood' is used for vitrinite laths and partially degraded vitrinite, whereas the category 'black wood' is almost entirely inertinite. The amorphous organic matter recovered is interpreted to be largely the product of land plant debris, based on the details reported by Batten (1983). Gelified vitrinite similar to that illustrated by Combaz (1980, Plate 3.6, Fig. 12) and Batten (1983, Fig. 7) is also incorporated in the amorphous category. In some samples from the Brora Argillaceous Formation and the Beatrice Formation, minor amounts of granular amorphous debris which may be of marine algal origin (Batten 1983) are recognized.

Most of the reworked material recognized comprises subrounded to well-rounded inertinitic fragments. Derived vitrinite is rarely recovered. Reworked Upper Palaeozoic miospores (*Emphanisporites?* sp.) were also noted in the Inverbrora Member shales at Brora.

The quantitative data in Figs. 5, 8 and 11 represent the most frequently occurring miospore genera and selected dinocyst taxa encountered in assessments of 300 palynomorphs per sample.

Where less than 30 miospores or dinocysts were recovered the prominent taxa are denoted as present (P). The state of preservation of palynomorphs is fair to poor in preparations from Brora and Cadh'an Righ but is sufficient to allow counts to be made on all samples. At Beatrice, extensive pre-depositional fragmentation and/or oxidation has resulted in a generally poor state of preservation which prevents specific identification and counting of palynomorphs in some samples. Finally, the results from all three sections are integrated and interpreted in their local and regional context.

BRORA

The Brora shore section is not well exposed, and shallow angles of dip make thickness measurements subject to considerable potential error. Minor faulting parallel to the shore also complicates the picture. The measured section (Fig. 3) utilizes a 9° average dip which can be measured at the top of the Roof Bed. The resultant thicknesses below the Roof Bed (Fig. 3) are much greater than previous estimates which place the shell beds approximately 3 m and 4 m beneath it (Hurst 1981). Borings in the area recorded by Judd (1873) place the shell beds about 8 m below but also indicate considerable local thickness variations.

Brora Coal Formation, Inverbrora Member

Below the Brora Coal

The Brora Coal Formation at Brora comprises the Doll Member and the Inverbrora Member. The lower Doll Member (not sampled) consists of sandstones and mudstones with sideritic bands and represents an alluvial plain environment (Hurst 1981). Towards the top of the Doll Member *Unio* and freshwater ostracods have been found (Hurst 1987, pers. comm.). The shales and mudstones of the Inverbrora Member underlying the Brora Coal (unexposed) are generally black, carbonaceous and frequently bituminous. They approach the appearance of oil-shale and have total organic carbon contents values of 11–31% (Sheridan 1987). Mica is abundant in fissile shales, and pyrite occurs as small nodules and as a replacement of the abundant plant debris which includes equisetalean and ginkgoalean remains (Stopes 1907).

Palynological assemblages from the dark shales are characterized by an abundance of amorphous organic matter, both brown and black woody debris and miospores (Fig. 4). As the preservation of extensive quantities of amorphous matter is linked with anoxic conditions (Batten 1983) either during deposition or in the early stages of burial, and in the absence of any apparent aerobic benthic fauna, it seems likely that the bottom waters of the lagoon were frequently anoxic and stagnant during the time of accumulation of the Inverbrora Member shales and mudstones. However, the persistent occurrence of dinoflagellate cysts within this interval provides confirmation of water circulation in an oxic photic zone and implies the existence of a stratified water column.

Dinocyst assemblages from the Inverbrora Member are characterized by the consistent association of the proximate forms *Ctenidodinium sellwoodii*, *Pareodinia ceratophora* Deflandre 1947, *Mendicodinium groenlandicum*, *Valensiella ovula* (Deflandre 1947) Eisenack 1963, *Escharisphaeridia pocockii* (Sarjeant 1968) Erkmen & Sarjeant 1980, *Chytroeisphaeridia chytroeides* (Sarjeant 1962) Downie & Sarjeant 1965 and *Sentusidinium sparsibarbatum* Erkmen & Sarjeant 1980. *Sentusidinium sparsibarbatum* and *Meiourogonyaulax* spp. are numerically most abundant in the lowermost assemblages (Fig. 5), whereas *Ctenidodinium sellwoodii* is dominant in the richer microplankton assemblages between the upper shell bed and the Brora Coal (Fig. 5). The dominance of low-diversity assemblages by a single taxon is characteristic of brackish-water conditions (Davey & Rogers 1975, Batten 1982, MacLennan 1985) and suggests deposition occurred in a moderately shallow sheltered lagoon. However, conditions were by no means constant during the development of the Inverbrora Member.

Some 13 m below the Brora Coal a more

BRORA

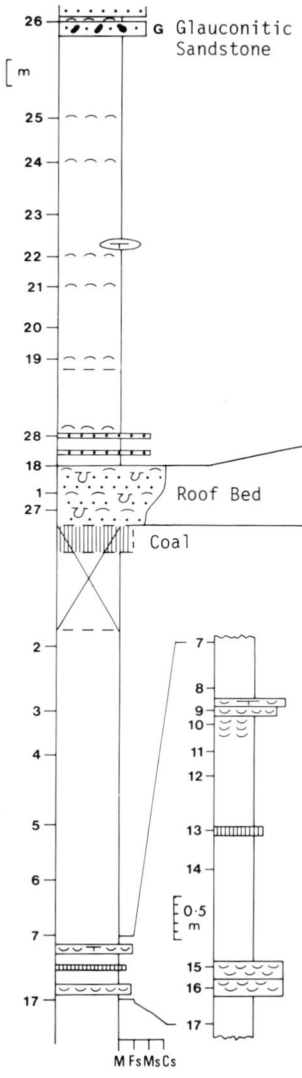

Sample No	General lithology	Ammonites	Belemnites	Marine bivalves	M.biv. sp. diversity	Angular biv. frags	Juvenile bivalves	Fish scales	Plant debris	Neomiodon	Isognomon	Juv. ?Viviparus	Ostracods	Bioturbation
26	Muddy glauc sst		⬤											⬤
25	Blk mdst		●	⬤	4	⬤		•						
24	Blk sh/mdst			⬤	5	⬤	●							
23	Blk sh/mdst		●			•			●					
22	Blk sh	●		⬤	4	⬤	•		●					
21	Blk silty mdst		●	⬤	3	⬤			•					
20	Blk sh							•	•					
19	Blk sh			●	3	●	●		●					
28	Glauc sandy mdst		●	●	2	●			•					⬤
18	Sst, brn, f-crs, calc			●	1+	●			●					⬤
1	Sst, grey, f-med			●	2+	•			●					⬤
27	Sst, grey, fine			●	3				●					⬤
2	Blk sh								●					
3	Blk sh								●					
4	Blk sh							•	●					
5	Blk sh, lams f sst							•	●					
6	Blk mdst													
7	Blk sh/mdst													
8	Blk sh								●					
9	Green sh, shelly							•		⬤				
10	Blk sh, shelly lam							•		⬤		●	●	
11	Blk sh								•	•		•		
12	Blk sh								●				•	
13	Coal, impure								⬤					
14	Blk-brn shl, silty								⬤					
15	Green shl, shelly							•		⬤	●			
16	Green sh, shelly								●	⬤	•			
17	Blk sh													

Key

⬤ Abundant

● Common

• Present

Numbers refer to diversity of marine bivalve species.

Fig. 3 — Log of Brora sequence to show sample positions, lithology of samples and macropalaeontological content. For key to section, see Fig. 6.

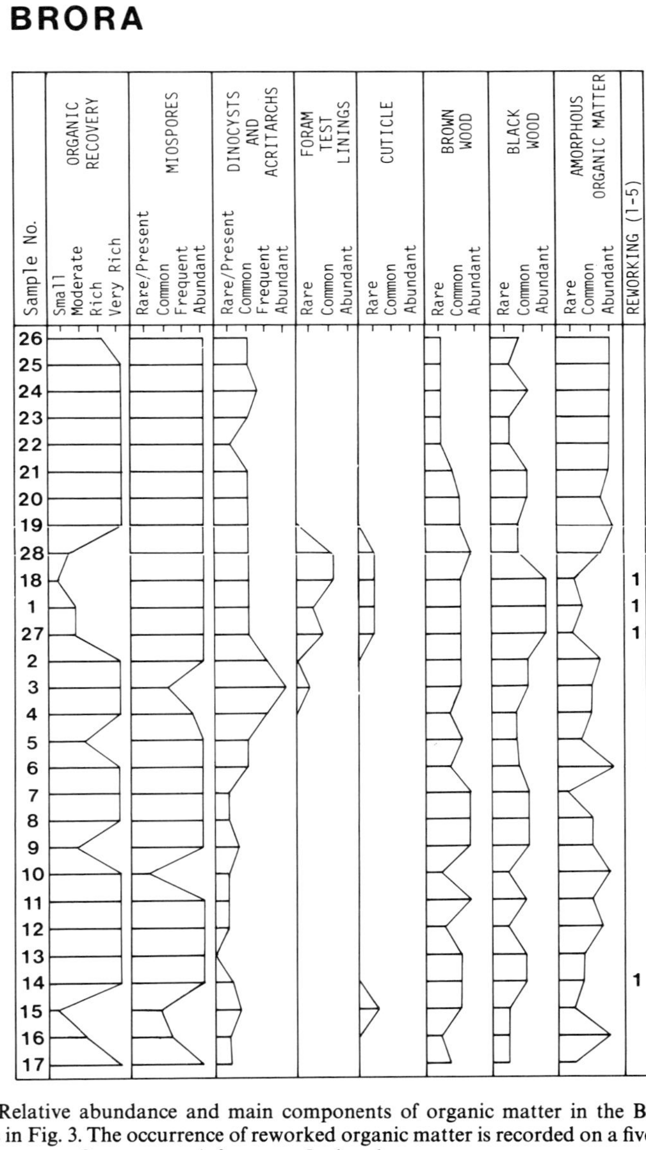

Fig. 4 — Relative abundance and main components of organic matter in the Brora section. Samples as in Fig. 3. The occurrence of reworked organic matter is recorded on a five-point scale: 1, trace; 2, present; 3, common; 4, frequent; 5, abundant.

varied lithological sequence includes two green shale beds with abundant bivalves, separated by black shales including a thin impure coal (Fig. 3). The lower shell bed (samples 15 and 16) is dominated by *Neomiodon* sp. with common *Isognomon murchisonii* (Forbes) (Fig. 3). Rounded shell debris and sparse organic matter are present at the top of the bed, indicating oxygenation and current activity, probably due to wave winnowing. Increased dinocyst diversity and abundance (Fig. 5) by comparison with the enclosing shales suggests an opening of the lagoon to greater marine influence. An abundance of amorphous organic matter and a relatively sparse dinocyst

BRORA — MIOSPORES

Sample No.	Bisaccates	*Cyathidites*	*Perinopollenites*	*Cerebropollenites*	*Callialasporites*	*Classopollis*	*Eucommiidites*	*Inaperturopollenites*	*Araucariacites*
26	24	10	17	18	2	26			
25	28	3	27	11	7	22			2
24	38	8	24	13	3	14			
23	25	9	23	18	10	14			
22	12	6	26	19	5	29	1	1	1
21	19	8	31	12	8	19			
20	12	3	34	20	12	15		2	1
19	8	9	55	7	2	10		3	2
28	39	5	26	9	15	3			3
18	46	10	13	7	8	4		5	
1	47	11	15	14	4	2		4	
27	51	6	15	12	7	3		4	2
2	23	5	23	37	2	8		2	
3	25	6	16	39	5	5	1		2
4	15	8	25	32	5	5	5		
5	18	12	29	28	3		4		3
6	22	14	24	28	1			2	2
7	24	15	34	15		3			3
8	13	11	48	15	2			4	
9	19	16	33	13	3	4		4	
10	9	30	28	6	1	17	2		4
11	19	21	36	7	1	1		2	3
12	19	18	40	2	1	6	4		
13	18	16	37	5	12	9			2
14	12	25	42	4	2	7			4
15	23	24	36	3	2	7			
16	20	23	39	4	2	2			5
17	21	25	38	3	3				

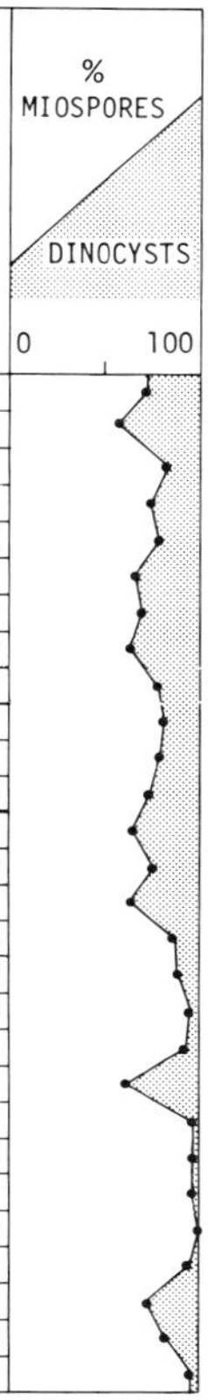

BRORA — DINOCYSTS

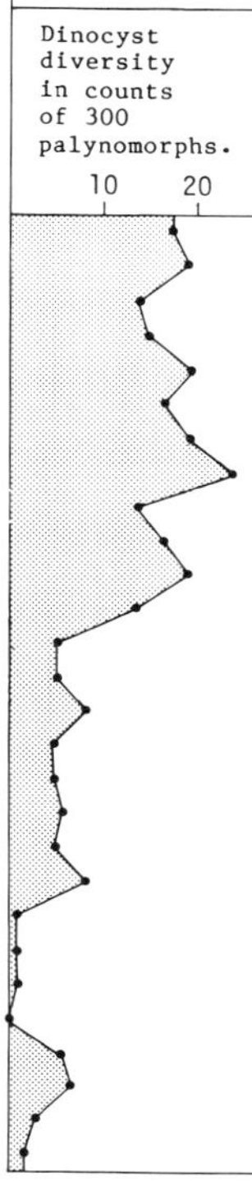

Valensiella spp.	*C. sellwoodii*	*E. pocockii*	*P. ceratophora*	*Meiourogonyaulax* spp.	*Chytroeisphaeridia* spp.	*M. groenlandicum*	*C. varispinosum*	*S. rioultii*	*S. sparsibarbatum*	*N. pellucida*	*K. gochtii*	Sample No.
15	1	10	13	17	7					1	9	26
18	1	1	4	5	19	5	1	7		1	19	25
4		2	13	13	17	8	13	4				24
	4	3	8	11	4	18	25	4		3		23
7	4	3		10	10	10	6	16		4	10	22
3	2	13	6	6	14	22	4	4		5	2	21
7	3	9	5	10	27	7	2	7			1	20
2	5	13	5	16	2	18	3	3		4	4	19
4	1	6	1	52		4	3	4	3	6		28
12	2	14		14			2	3		2	21	18
6	5	19	2	24	3			5	3	2	3	1
	36	5		23	4				4			27
	65	1		22							10	2
	62	18		9		6			3		2	3
1	72	4	1	1	2				1			4
	81	6		3		3			6			5
	69	28				10			3			6
	P	P			P	P			P		P	7
P	P	P				P			P		P	8
2	13	16		16		4			9		40	9
		P										10
	P											11
												12
												13
		33		20	7	7						14
		8		14	4	14			34			15
				25	4				71			16
				P					P			17

Fig. 5 — Distribution of selected miospore genera and dinocyst taxa recovered from the Brora section. Samples as in Fig. 3. P denotes the presence of prominent taxa in a small assemblage (less than 30 palynomorphs).

assemblage dominated by *Sentusidinium sparsibarbatum* characterizes the palynoflora from the base of the shell bed.

In shales between the shell beds is a 5 cm seam of impure coal, rich in miospores as well as brown and black woody debris. Dinocysts are absent in the impure coal and rare in the enclosing shales. A dominantly freshwater environment of deposition seems likely.

The upper shell bed (Fig. 3, samples 9 and 10) is composed of seams of complete *Neomiodon* sp. with ostracods and juvenile *Viviparus?* in black shale passing up into a green shale crowded with shell debris, and topped by a calcite-cemented shell bed in which the shells are fragmented and rounded. The palynofacies trend is similar to that in the lower shell bed; a moderate to large organic content at the base is dominated by amorphous material and contains only a few miospores and dinocysts. In the upper part of the bed, woody organic matter is associated with common diverse microplankton. *Escharisphaeridia pocockii* and *Meiourogonyaulax* spp. (predominantly forms referable to *M. caytonensis* (Sarjeant 1959) Sarjeant 1969 and *M. cristulata* (Sarjeant 1959) Sarjeant 1969) form prominent elements in both shell bed assemblages. The latter group also increases in abundance in the uppermost shales of the Inverbrora Member and is well represented in the Roof Bed and above. The shell bed dinocyst acmes coincide with intervals of increased water circulation and probable increased salinities in the basin and are typical of late Bathonian *discus* Zone and early Callovian *macrocephalus* Zone assemblages (Woollam & Riding 1983). Both shell beds, with their winnowed tops rich in dinocysts, record entry of marine waters, increased oxygenation and an increase in energy (relative to the enclosing shales), resulting in reworking and winnowing of the shell beds.

The incoming of *Mendicodinium groenlandicum* in the lower shell bed pre-dates the main pulse of the transgression (above the Brora Coal) and the previously recognized datum at the base of Riley & Fenton's (1982) *Dichadogonyaulax* (*Korystocysta*) *gochtii* Zone (*macrocephalus*–*calloviense* (pars) Zones). An abundance of *K. gochtii*, which typifies the above zone, is also noted here in the uppermost shell bed of the Inverbrora Member, i.e. below the previously assumed position of the Bathonian–Callovian boundary at the base of the Roof Bed (Sykes 1975a).

Between the upper shell bed and the coal a further increase in marine influence is noted with a gradual rise in the relative abundance of dinocysts (Fig. 5) and the incoming of the forms *Rhynchodiniopsis cladophora* (Deflandre 1938) Below 1981, *Aldorfia aldorfensis* (Gocht 1970) Stover & Evitt 1978 and *Ctenidodinium* sp. among others, marking an increase in diversity. A more even distribution of gymnospermous pollen and a concomitant decrease in numbers of *Cyathidites*- type fern spores is also noted in the shales below the coal and is attributed to greater mixing of land-derived components due to increased marine influence and improved circulation in the lagoon.

The coal

The coal has long been 'worked out' on the shore at Brora but is known to have been about a metre thick with no apparent rooting at the base (Judd 1873). Other minor seams are present beneath the main coal as noted on the shore section (Fig. 3). Judd (1873) recorded masses of crushed *Equisetum* stems in the coal and *Equisetum* roots in a pyritic dirt band within the coal. Harris & Rest (1966) showed the main components of the coal to include may kinds of fusainized conifer wood interpreted as charcoal resulting from forest fires, together with drifted masses of *Equisetum*. Harris & Rest considered the coal to have formed from accumulations of waterlogged wood and floating masses of *Equisetum* reed bed in a 'moderately deep' lagoon in which the water depth was too great to allow rooted growth of *Equisetum*.

Palynological preparation of a sample of coal from the old mine at Brora yielded a woody assemblage dominated by vitrinite laths and containing common miospores. Gymnospermous pollen dominate the miospore assemblage, with 36% being undifferentiated bisaccate grains, 23% *Cerebropollenites mesozoicus* (Couper 1958) Nilsson 1958 and 11% *Perinopollenites elatoides* Couper 1958. *Cyathidites* spp. and *Dictyophyllidites harrisii* Couper 1958 fern spores collectively represent 23% of the assemblage, possibly indicating the presence of a swampy vegetation. In their brief discussion on the palynology of the Brora Coal, Hughes & Couper (1958) reported the common occurrence of *Araucariacites australis* Cookson 1947 and *Eucommiidites troedssonii* Erdtman 1948 at Brora. In the coal sample studied here, *A. australis* forms 4% of the assemblage but *E. troedssonii* is rare (1% only).

Brora Argillaceous Formation, Brora Shale Member

The Roof Bed is the basal bed of the Brora Shale Member (Sykes 1975a), which consists mainly of black shales and mudstones. It is a transgressive marine quartzose sandstone which is porous at the base with quartz overgrowth cement and authigenic kaolinite but is non-porous at the top with a calcite cement. Pyrite is also abundant and occurs as cement. Near the base of the bed, sorting is moderate in fine sandstone but intense bioturbation resulted in the incorporation of mud into the matrix. Coarse sand and small pebbles are concentrated towards the top of the bed.

No exposures of the contact between the Roof Bed and coal remain in the area, but Sykes

(1975b) recorded 5–15 cm of planty shale between the two. There is, at present, no evidence of erosion beneath the Roof Bed other than the presence of reworked carbonaceous material in the Roof Bed which is presumed to have come from the Brora coal.

In boreholes in the coalfield the Roof Bed is reduced in thickness, such that only 30 cm is present (Sykes 1975b). However, this reduction is associated with a general thickening of sand-bearing strata; from 2.6 m on the shore to 6.4 m in boreholes 2 km to the northwest (Sykes 1975b).

Woody and derived coal debris is abundant in hand specimens in the Roof Bed. A rich and diverse shallow marine shelly fauna is present as recorded by Judd (1873) and Sykes (1975a, b). Bivalves include *Pleuromya uniformis* (J. Sowerby 1813), *Corbulomima obscura* (J. de C. Sowerby 1827), *Myophorella* sp., *Protocardia striatula* (J. de C. Sowerby 1829), *Entolium corneolum* (Young & Bird 1828) and others. Sykes (1975a) also recorded belemnites and the gastropod *Pietteia* sp. with shell apices orientated southwest on the top of the bed. Sykes (1975a) dated the Roof Bed as *calloviense* Zone on the basis of rare ammonites. However, palynological data (below) indicate that the *macrocephalus* Zone may also be represented.

Palynofacies assemblages from the Roof Bed also differ markedly from those in the underlying shales (Fig. 4). An abundance of woody debris (including blocky and lath-like inertinite fragments and minor reworked material) is associated with pale oxidized miospores, minor plant cuticle, and common dinoflagellate cysts and foraminiferal test linings. Although syndepositional turbulence and oxidation undoubtedly removed some of the dinocysts, the diversity of the population preserved is considerably increased by comparison with that in the underlying shales. Inceptions of *Chytroeisphaeridia hyalina* (Raynaud 1978) Lentin & Williams 1981, *Energlynia acollaris* (Dodekova 1975) Sarjeant 1978, *Gonyaulacysta pectinigera* (Gocht 1970) Fensome 1979, *G. jurassica* (Deflandre 1938) Norris & Sarjeant 1965, *G. eisenackii* (Deflandre 1938) Dodekova 1967, *Rigaudella aemula* (Deflandre 1938) Below 1982 and *Cleistosphaeridium varispinosum* (Sarjeant 1959) Woollam & Riding 1983 in the Roof Bed are consistent with the development of a more open marine environment and corroborate the world-wide diversification which Davies & Norris (1980) associated with the Callovian transgression. The restriction of *Ctenidodinium combazii* to the Roof Bed indicates environmental control on the distribution of this taxon, since it is only in the latest stages of its range that the open marine conditions necessary for its proliferation (Woollam & Riding 1983, Riding *et al.* 1985) developed in this area. The abundance of bisaccate grains in the sandstone (Fig. 5) is also environmentally controlled, their grain size and density being such that they may have been concentrated by flotation after sorting of the miospore grains.

Termination of the *Ctenidodinium sellwoodii* abundance zone at the base of the Roof Bed (Fig. 5, sample 27) is consistent with the top of the *C. combazii*–*C. sellwoodii* dinocyst zone of Woollam & Riding (1983) and indicates the presence of *macrocephalus* age sediments above the coal. The extinction of *C. combazii* no longer coincides with the top of this zone since Riding (1984) extended its range into the *calloviense* Zone. At Brora, *C. combazii* forms a relatively common element (9.5%) of the dinocyst assemblage from sample 1. The Roof Bed also contains first and common occurrences of *Ctenidodinium continuum* and *C. ornatum* (the zonal indices on which Woollam & Riding (1983) based their succeeding zone), indicating a probable *calloviense* age for the upper part of the bed.

Brora Shale

Glauconitic sand laminae occur near the base of the shales above the Roof Bed and are occasionally present higher in the shales. Sykes (1975a) reported the thickness of shales to be about 25 m based on borings in the coalfield but stated that only the 'lowest 15.6 m' can be seen on the beach. Our measurements give 17.5 m for the full thickness on the beach from the Roof Bed to the first bed of the Glauconitic Sandstone Member.

The lithology varies from fissile shale with

fine-grained mica to shale–mudstones with a smooth subconchoidal fracture. Bivalves are abundant in some samples and a varied fauna is present (Fig. 3). *Meleagrinella braamburiensis* (Phillips 1829) is abundant and is accompanied by *Nicaniella* (*Trautscholdia*) *phillis* (d'Orbigny 1850), *Myophorella* sp., *Protocardia striatula*, *Thracia depressa* (J. de C. Sowerby 1823), *Entolium corneolum*, *Oxytoma inequivalve* (J. Sowerby 1819) and seams of *Bositra buchii* (Roemer 1836) (sample 24). Sykes (1975b) discussed the molluscan fauna in detail on the basis of extensive collecting. Angular shell debris is abundant in some samples (Fig. 3) and is probably the result of predation by fish and/or arthropods on the molluscan population. Belemnites are common and occasional ammonites indicative of *calloviense* Zone (Sykes 1975a) are present. Local acmes in the distribution of the dinocysts *Cleistosphaeridium varispinosum*, *Chytroeisphaeridia hyalina* and members of the *Meiourogonyaulax caytonenisis* complex, together with the presence of *Rigaudella aemula*, *Mendicodinium groenlandicum* and *Gonyaulacysta jurassica* support a *calloviense* age. Sporadic occurrences of *Gonyaulacysta pectinigera* and *Aldorfia aldorfensis* are also consistent, as is the persistent occurrence of *Nannoceratopsis pellucida* Deflandre 1938. It is probable that the *jason* Zone is also represented (Sykes 1975a).

Fine-grained carbonaceous debris is recognized in hand specimens throughout the shales and is associated with a rich but remarkably homogeneous microscopic organic assemblage (Fig. 4). Miospores and amorphous organic matter of land plant and possible marine algal origin (recognition based on criteria given by Batten (1983)) predominate in palynological preparations, although poorly preserved dinocysts are consistently common. Such an abundance of amorphous organic matter implies that anaerobic conditions prevailed, but the presence of a rich benthic bivalve fauna precludes continuous anoxia of the depositional substrate.

Dinoflagellate cysts are relatively common, moderately diverse and more evenly distributed in the samples from the Brora Shale than in the Inverbrora Member (Figs. 4 and 5), implying that more open marine conditions prevailed. All the dinocyst taxa recorded in the Inverbrora Member of the Brora Coal Formation are present, suggesting that they have a broad salinity tolerance, but for the most part they are subordinate. Although the relative proportions of the different taxa fluctuate, the strong similarity between species composition in successive samples suggests a fairly stable environment. Miospore distribution is also more even (Fig. 5) because the greater effect of pre-depositional mixing of terrestrial debris in the marine environment reflects the regional terrestrial flora.

The base of the Glauconitic Sandstone Member consists of bioturbated dark green glauconitic muddy sandstone with phosphatic nodules. Belemnites (*Cylindroteuthis*) and woody debris are abundant and *Rhaxella* spicules are common. Intense bioturbation due to *Chondrites*, *Thalassinoides* and *Diplocraterion* is present.

CADH'AN RIGH

This section near Balintore (Fig. 1) provides the only locality where the Brora Coal is now exposed. The outcrops are steeply dipping and immediately adjacent to the Great Glen Fault. The main sedimentological and palaeontological features of the exposed succession together with sample positions are illustrated in Fig. 6.

Brora Coal Formation, Inverbrora Member

Below the Brora Coal

Only 2.4 m of strata are seen below the coal in the measured section, and softer lithologies, presumably shale, were not exposed at the time that the samples were collected for this study. Judd (1873) recorded a further 8 m + of strata including 'fire clay', argillaceous limestone and 'about 20 ft' (about 7 m) of white sandstone. The lowest exposure seen is a micritic mottled limestone which is concretionary in nature and has a host lithology of green shale. Ostracods are abundant in the limestone and shales, together with plant debris and fish scales (Fig. 6). The shales also

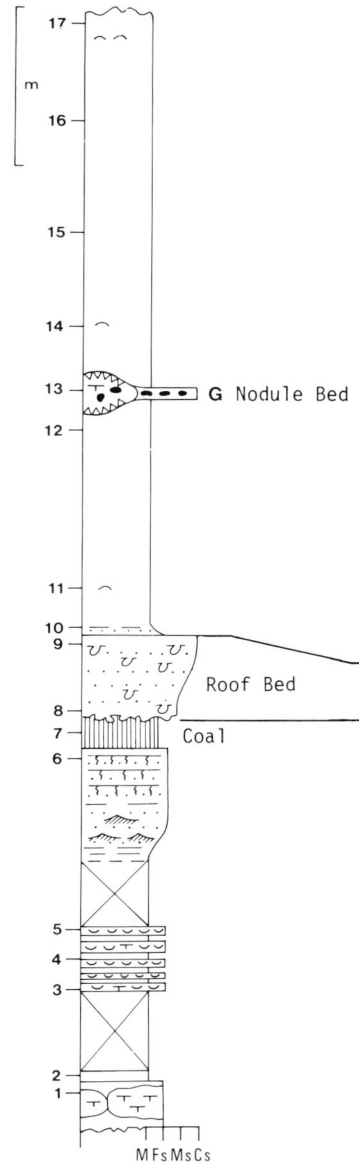

CADH'AN RIGH

Sample No.	General lithology	Ammonites	Belemnites	Marine bivalves	M. biv. sp. diversity	Angular biv. frags	Juvenile bivalves	Fish scales	Plant debris	Neomiodon	Isognomon	Juv. ? Viviparus	Ostracods	Bioturbation
17	Grey sh-mdst			●	**3**	●	●	•	•					
16	Blk sh		●					•	•					
15	Blk sh					•		•						
14	Blk sh					•	•		•					
13	Lst, phos nods, sandy		●											
12	Blk sh		●											
11	Blk sh		●	•	**1**									
10	Blk sh sandy, glauc		●											•
9	Sst med-crs, calc		●						•					●
8	Sst med-crs, porous			●	**2+**				•					●
7	Coal								●					
6	Sst, fine, white								•					
5	Mdst, sideritic shelly							•	•	●				
4	Green sh, + lst shelly						•		•	●	•	•	•	
3	Green sh, + lst shelly									●	•		•	
2	Green sh						•						●	
1	Lst micritic						•	•	•				●	

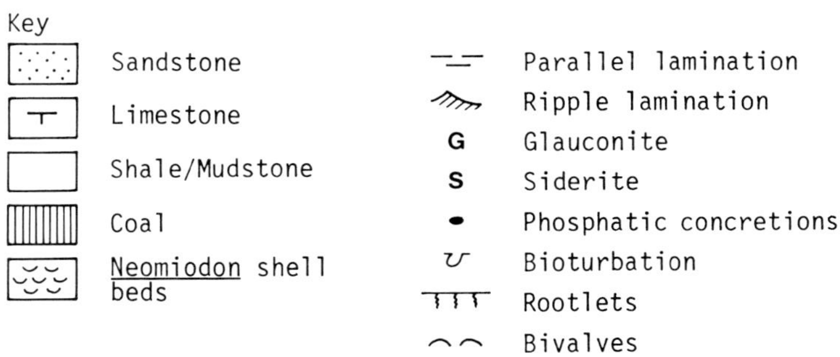

Fig. 6 — Log of Cadh'an Righ section with sample positions, lithology and macropalaeontological content.

yield abundant *Cyathidites*-type spores and vitrinite-dominated woody debris (Fig. 8). Such strong polarization of the floral and faunal components suggests local derivation, perhaps in or around ponds in a low-lying swampy area.

Above the limestones (Fig. 6) are thin-bedded shales and calcareous horizons with a fauna dominated by *Neomiodon* and with subordinate

Isognomon. Ostracods, *Euestheria* and juvenile gastropods (*Viviparus?*) are also present. Fibrous calcite is present in cone-in-cone layers that have disrupted the shell beds. The composition of both bivalve and dinocyst assemblages indicate that these shell beds at Cadh'an Righ are comparable in age and were deposited in an environment similar to those in the Brora section. As at Brora, this level is characterized by common *Meiourogonyaulax* spp. and marks the inception of *Ctenidodinium sellwoodii* and *Korystocysta gochtii*. However, the oganic content of the condensed sequence below the coal at Cadh'an Righ is considerably poorer than that from the comparable sequence at Brora. Those dinocysts, miospores and other land-derived organic debris that remain are in a poor state of preservation and show signs of extensive oxidation and microbial decay which pre-date formation and incorporation of finely disseminated pyrite in a reducing environment. Therefore, although *Ctenidodinium sellwoodii* is dominant in one and prominent in another of the three dinocyst assemblages from the upper part of the Inverbrora Member, it is difficult to draw conclusions from such scant evidence.

Immediately below the coal, 0.7 m of fine-grained pale sandstone is laminated at the base and shaly but becomes cleaner upwards and is ripple laminated (Fig. 6). Carbonaceous rootlet casts are present near the top of the bed. The microscopic organic content (sample 6) is dominated by a varied gymnosperm pollen assemblage (Fig. 8). Representatives of a number of spore genera, including *Cyathidites*, *Ischyosporites*, *Leptolepidites*, *Converrucosisporites* and *Osmundacidites* suggest proximity to a swampy source vegetation. The details of a channel fill, including coal, described by Sykes (1975b) could not be verified because of lack of lateral exposure at the time of sampling.

The coal

The coal is up to 20 cm thick but the top surface is highly irregular owing to borings and burrows filled with sand of the overying Roof Bed. It was well compacted at the time of burrowing, perhaps having the consistency of a firm peat, since there is only minor subsequent compaction of the sand-filled burrows. The larger subhorizontal burrow sections (probably *Thalassinoides*) are up to 100 mm in diameter and were probably excavated by arthropods; smaller borings resemble those of the bivalve *Lithophaga*. The base of the coal is sheared, owing to bedding plane slip caused by movement against the nearby Great Glen Fault, but does not appear to be underlain by a rooted soil horizon.

A palynological preparation of the coal differs somewhat from those of the Beatrice and Brora coals in that it is almost wholly composed of amorphous organic matter (Fig. 7), a large proportion of which is gelified vitrinitic material and possibly degraded *Botryococcus* algae. Miospores are present but not common. The assemblage is dominated by spores which are thought to represent a swampy fern-rich vegetation. Although these observations differ markedly from those for the Brora and Beatrice coals, they compare closely with Fisher's (1980) observations on non-rooted sapropelic coals from the Jurassic of Yorkshire. On the basis of their high vitrinite:inertinite ratio and the common occurrence of *Botryococcus*, Fisher (1980) suggested these non-rooted coals formed as detrital accumulations in a lagoonal setting.

Brora Argillaceous Formation, Cadh'an Righ Shale Member

The Roof Bed forms the base of the Cadh'an Righ Shale Member (Sykes 1975a) which is a condensed sequence, probably including a non-sequence at the nodule bed. The member ranges from *calloviense* Zone in the Roof Bed to *coronatum* Zone (Sykes 1975a, Fig. 1a). The Roof Bed is a brown, medium- to coarse-grained porous quartzose sandstone and the top is well cemented with calcite. Quartz overgrowths and patches of pyrite also cement the sandstone. Intense bioturbation is present and a variety of poorly preserved bivalves include *Pleuromya uniformis*, *Ctenostreon* sp. and pectinids. Belemnites (*Cylindroteuthis*) are abundant on the top surface of the bed and

CADH'AN RIGH

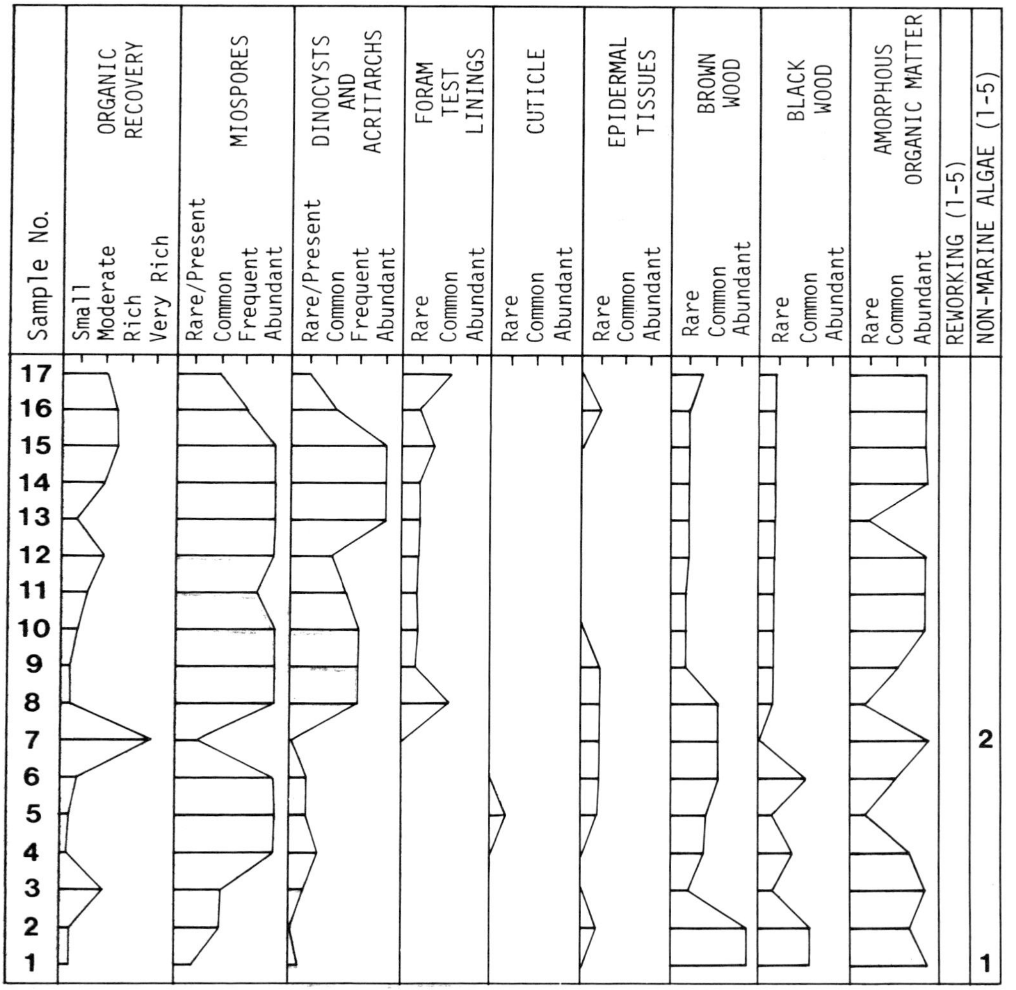

Fig. 7 — Relative abundance and main components of organic matter in the Cadh'an Righ section. Samples as in Fig. 6. The occurrence of reworking and non-marine algae is recorded on a five-point scale: 1, trace; 2, present; 3, common; 4, frequent; 5, abundant.

large coalified wood clasts, up to 20 cm in diameter, are present.

Although there is proportionately less woody debris and a larger number of palynomorphs in palynofacies assemblages from the Cadh'an Righ Roof Bed compared with Brora, there are close similarities between the two dinocyst and miospore populations. At Cadh'an Righ a similar dramatic increase in abundance and diversity of dinocysts is marked by the inception of *Chytroeisphaeridia hyalina*, *Cleistosphaeridium varispinosum*, *Gonyaulacysta pectinigera*, *Sentusidinium verrucosum* (Sarjeant 1968) Sarjeant & Stover 1978, *Aldorfia aldorfensis*, *Pareodinia prolongata* Sarjeant 1959, *Rigaudella aemula*, *Gonyaulacysta jurassica* and others.

Above the Roof Bed there is a rapid transition over 5 cm to black shale. This transition contains laminae of reworked sand from the Roof Bed together with glauconite. The shales are poorly fossiliferous with occasional belemnites, pectinid bivalves and angular bivalve debris. Sykes (1975a) also recorded ammonites and *Lingula*.

Between the Roof Bed and the concretionary bed (Fig. 6) the dinocyst assemblage character is similar to that seen in the Brora Shale, with *Valensiella* spp., *Meiourogonyaulax* spp., *Chytroeisphaeridia* spp. and *Pareodinia ceratophora* making up the common elements of the moderately diverse associations (Fig. 8). Another similarity to the Brora sequence is the subordinate representation of *Ctenidodinium sellwoodii* and triradiate spores, and the more even distribution of gymnospermous pollen above the Roof Bed (Fig. 8). Despite these similarities the actual amount of organic matter recovered from the Cadh'an Righ samples was small compared with Brora and the proportion of woody material considerably reduced (Fig. 7). This difference may reflect an increase in distance from a distributary source.

The concretionary bed (Fig. 6) is generally 5–30 cm thick, the greater thickness being due to concretionary swelling. In places the bed is reduced to 10 mm of laminated sandy shale. The concretions are calcitic and have cone-in-cone margins in places. Smaller phosphatic concretions are present within the bed together with detrital sand and glauconite. Belemnites are frequent and *Rhaxella* spicules are seen in thin sections together with fish debris. The peak in dinocyst diversity recorded in the concretionary horizon is not accompanied by an influx of different taxa as might be expected. Indeed, with the exception of *Energlynia acollaris* and *Ctenidodinium* sp., all the taxa are represented lower in the sequence. This horizon also marks a change in the balance of dinocysts preserved, from a diverse population with several common elements to one dominated by *Sentusidinium sparsibarbatum* in the upper part of the Cadh'an Righ Shale Member (*coronatum* Zone). Coupled with a gradual decrease in the relative number of dinocysts preserved (Fig.

CADH'AN RIGH — MIOSPORES

Sample No.	Bisaccates	*Cyathidites*	*Perinopollenites*	*Cerebropollenites*	*Callialasporites*	*Classopollis*	*Eucommiidites*	*Inaperturopollenites*	*Araucariacites*
17	30	4	14	7	17	24			2
16	30	10	15	18	10	15	1		1
15	28	9	19	11	19	12		2	
14	34	9	27	10	10	9			1
13	31	7	30	9	12	9			2
12	22	5	24	21	10	18			
11	26	5	20	20	10	18	1		1
10	30	9	20	13	5	23			
9	27	12	22	19	8	10			3
8	39	6	24	11	10	8			1
7	5	64	6		6	17	1		
6	23	12	24	30	5	3			2
5	29	36	12	6	11	2			
4	34	18	14	5	14	4		3	3
3	27	28	19	2	14	6			1
2	12	80			1	1			
1	35	31	5	4	8	12			

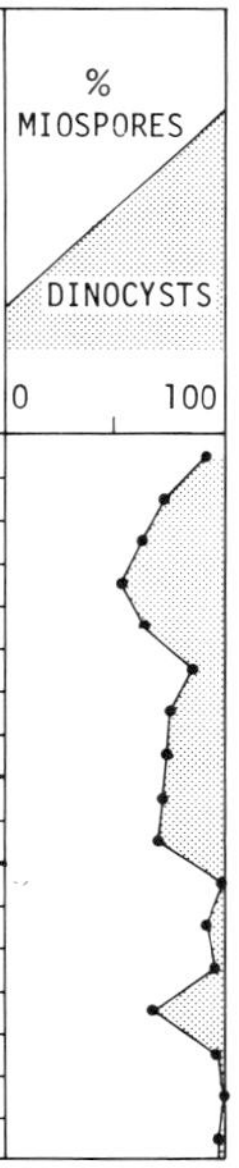

CADH'AN RIGH — DINOCYSTS

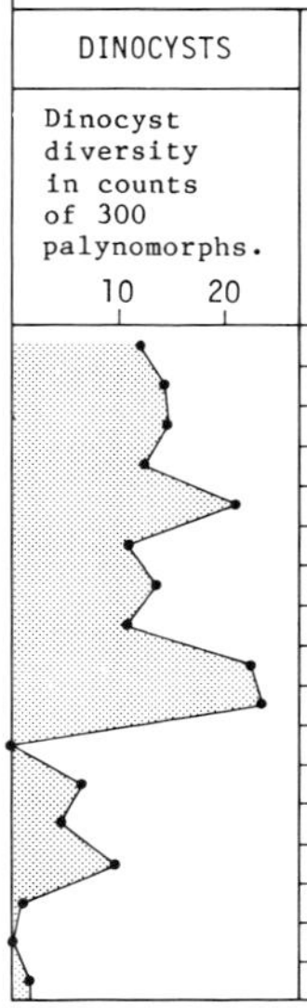

Valensiella spp.	*C. sellwoodii*	*E. pocockii*	*P. ceratophora*	*Meiourogonyaulax spp.*	*Chytroeisphaeridia spp.*	*M. groenlandicum*	*C. varispinosum*	*S. rioultii*	*S. sparsibarbatum*	*N. pellucida*	*K. gochtii*	Sample No.
4		8	4		19	15			19		4	17
6		3	5	1	9	6		18	35		2	16
2		10	1	3	17	2		10	35	1		15
	3	7	1	5	6	1			70			14
3	2	8		1	4			2	51		4	13
17	5	17	14	5	10			12			7	12
38		1	8	29	3		3			5	3	11
19		1	29	10	27		3	5	1			10
14	3	1	13	19	19		2					9
6	3	7	3	34	8		3	1		1	5	8
												7
	P	P		P								6
				P					P			5
	6	16		19							3	4
				P								3
												2
					P	P						1

Fig. 8 — Distribution of the main miospore genera and selected dinocyst taxa recovered from the Cadh'an Righ section. Samples as in Fig. 6. P denotes the presence of prominent taxa in a small assemblage (less than 30 palynomorphs).

8) this trend may have been a response to shallowing of the basin and the onset of a regression.

BEATRICE

The general geology of the Beatrice oilfield has been described by Linsley *et al.* (1980) who compared the section with that of the Brora area. A coal is present at the base of the Callovian which is correlated with the Brora Coal. Well 11/30-2 was cored from near the base of the Oxfordian to near the base of the Jurassic with excellent recovery. Cores were examined and logged for about 120 ft (36 m) both above and below the coal and sampled for palynology and lithological variation. The logged section (Fig. 9) covers the 'B' Sand and Mid Shale of the Beatrice Formation (Johnson 1984). Andrews & Brown (1987) did not recognize the Mid Shale as a separate unit and included it in the base of the 'A' Sand. Correlation with the Brora and Cadh'an Righ sections is illustrated in Fig. 2.

Brora Coal Formation

Below the coal

The dominant lithology is greenish-grey mudstone, highly kaolinitic but with subordinate illite and mixed-layer clays. The mudstone is variably sandy and massive and contains sideritic spheres up to 1 mm in diameter. The sandstone beds range from 10 cm to 1 m in thickness and are generally fine to medium grained and muddy and appear massive. Parallel lamination, ripple lamination and loading features are developed occasionally. Brown sideritic spheres are also present in the sandstones which contain abundant authigenic and detrital kaolinite and detrital illite. The sedimentology and mineralogy are similar to those of the top part of the Doll Member at Brora (Hurst 1981) and the depositional environment was probably a similar alluvial plain.

With the exception of the black shale immediately underlying the coal in well 11/30-2, organic recovery from sampled intervals below the coal is extremely low (Fig. 10). The common occurrence of inertinitic debris (much of which shows signs of reworking or redistribution) together with the presence of pale oxidized cuticle and/or epidermal tissues, and the initial paucity of brown wood and absence of dinocysts, is consistent with an alluvial plain environment. Moderate depositional energy and oxidizing conditions are also implied by the miospore assemblages; in many cases only fragmented 'ghost' outlines of the grains remain.

Raising of the water table associated with the marine incursion noted below the coal at Cadh'an Righ and Brora may account for the preservation of brown wood in the assemblages at 6521 ft and above (Fig. 10). However, the first evidence of raised salinities at Beatrice is not recorded until 6500 ft where isolated specimens of the dinocysts *Dissiliodinium* sp., *Escharisphaeridia pocockii*, *Chytroeisphaeridia chytroeides* and *Mendicodinium groenlandicum* were recovered (Fig. 11). Conditions were still oxidizing prior to deposition at this stage. This is indicated by the fact that the cuticular debris and palynomorphs are extremely pale and have lost all primary surface features. Indeed, the few dinocysts recovered may have been 'washed in' during a phase of increased marine influence, such as that associated with the shell beds below the coal at Brora and Cadh'an Righ. The macrofauna recorded below the coal at the latter two localities is absent in well 11/30-2 (Figs. 3, 6 and 9).

Immediately below the coal, 30 cm of organic-rich (Fig. 10) dark carbonaceous laminated shale, with plant debris and laminae of fine sandstone, overlie 20 cm of sandy mudstone with disrupted laminae. There is no evidence for rooting of plants or soil formation directly beneath the coal seam, but vague carbonaceous traces in the grey mudstones below might represent rootcasts. However, the composition and abundance of amorphous matter and the increase in the vitrinite:inertinite ratio suggest that the black shale reflects deposition in a reducing environment, probably at the extreme margin of the Inverbrora Member lagoon.

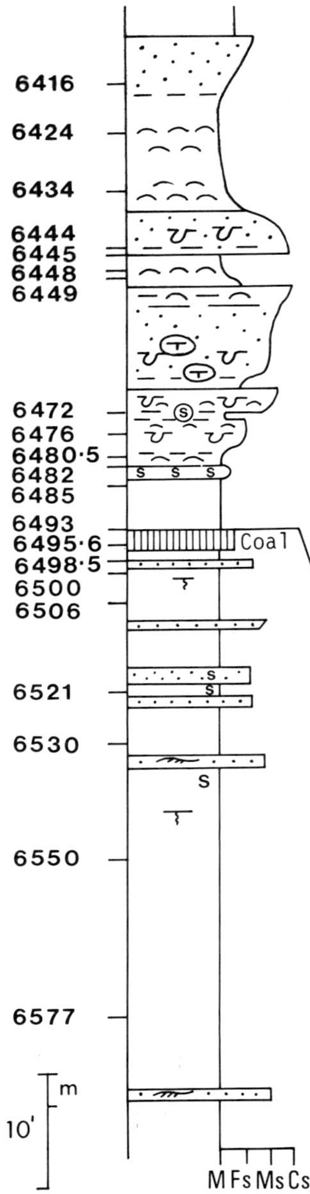

BEATRICE 11/30-2

Sample No.	General lithology	Ammonites	Belemnites	Marine bivalves	M.biv.sp.diversity	Angular biv.frags	Juvenile bivalves	Fish scales	Plant debris	Bioturbation
6416	Blk silty mdst			●	2+	●				●
6424	Blk sh		○	●	2	•				
6434	Blk sh		○	⬤	3+	●	•		•	
6444	Sst, brn, med-crs									●
6445	Sst, brn, crs			●					●	
6448	Blk sh		○	⬤	3+	●				
6449	Blk sh		○			•				
6472	Blk sh					•			●	
6476	Sst, med, calc					●				⬤
6480·5	Blk sh			○	2	•			⬤	
6482	Sideritic conc					•				
6485	Mdst grey-green								•	
6493	Blk sh, pyritic								⬤	
6495·6	Coal								⬤	
6498·5	Blk sh, pyritic								●	
6500	Green sandy mdst									
6506	Green sandy mdst								•	
6521	Green mdst								●	
6530	Green sandy mdst									
6550	Green sandy mdst									
6577	Green sandy mdst									

Fig. 9 — Log of Beatrice well 11/30-2 with sample positions, lithology of samples and macropalaeontological content. Sample numbers represent the core depth in feet as recorded on Department of Energy core samples. For keys and comparison, see Figs. 3 and 6.

The coal

The Beatrice coal is about 1.1 m thick, pyritic and blocky with a dull laminated band in the centre of the bed. In microscopic detail it shares features in common with the coal from the Brora mine and Cadh'an Righ. Amorphous matter, which appears to be the degradational product of cuticular and epidermal tissues, is common but laths and equant grains of vitrinite are dominant. The abundance of vitrinite (Fig. 10) and common

BEATRICE 11/30-2

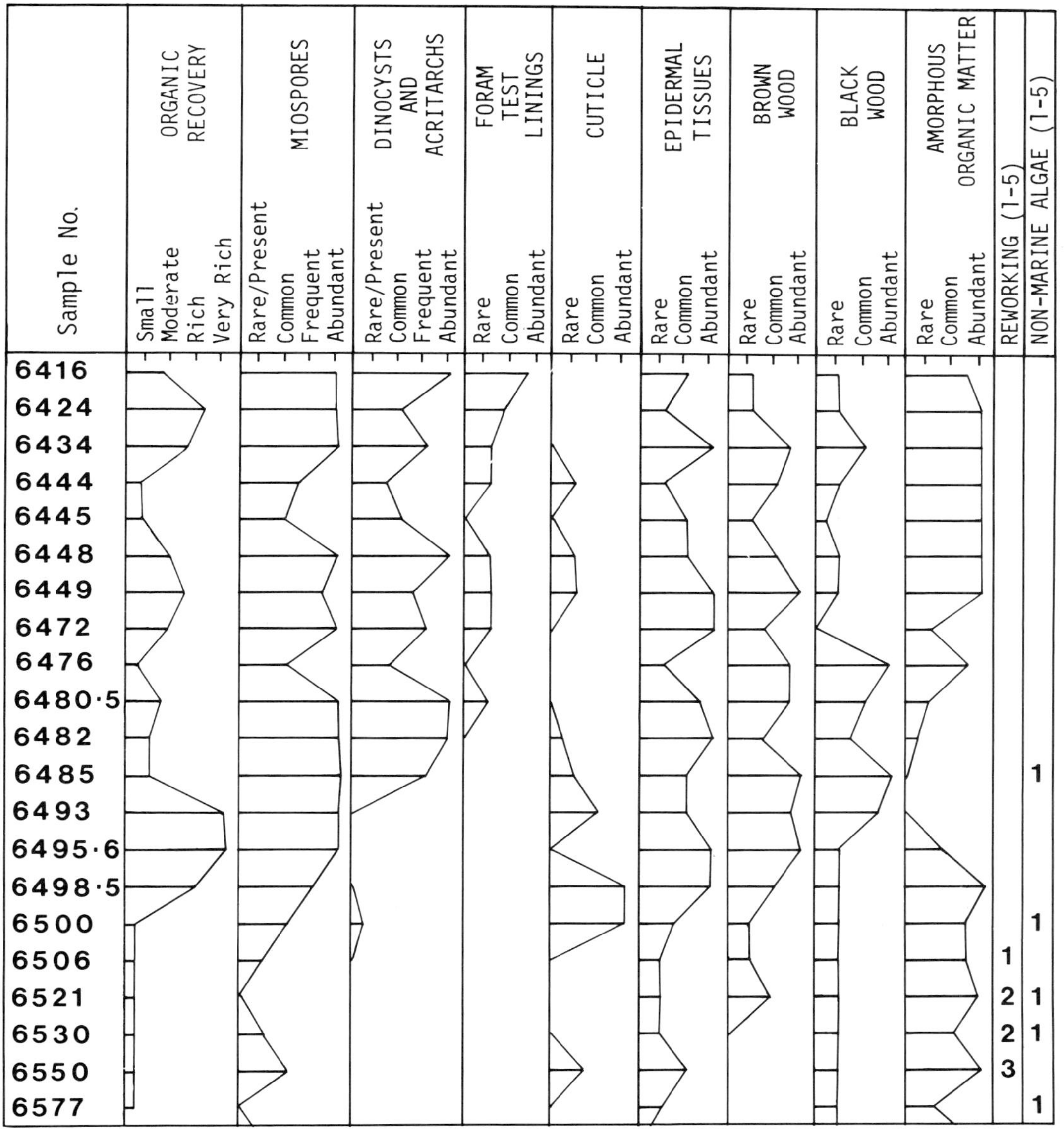

Fig. 10 — Relative abundance and main components of organic matter in Beatrice well 11/30-2. Sample depths as in Fig. 9. Representation of reworking and non-marine algae as in Fig. 7.

occurrence of the miospores *Perinopollenites elatoides*, *Cerebropollenites mesozoicus*, *Cyathidites australis* Couper 1953 and bisaccate pollen are features in common with the coal at Brora, whereas the coal from Cadh'an Righ is almost entirely composed of amorphous matter of land plant and algal origin.

Beatrice Formation

The strata above the coal comprise shales and sandstones, the latter making up the 'B' Sand of the Beatrice Reservoir (Johnson 1984). The sandstones, with one exception, coarsen upwards, are generally bioturbated and contain a varied shelly marine fauna (Fig. 9).

BEATRICE 11/30-2 — MIOSPORES

Sample No.	Bisaccates	*Cyathidites*	*Perinopollenites*	*Cerebropollenites*	*Callialasporites*	*Classopollis*	*Eucommiidites*	*Inaperturopollenites*	*Araucariacites*
6416	22	3	17	33	8	11		5	2
6424	28	13	9	17	5	26			2
6434	24	9	14	23	9	18		1	1
6444	9	10	24	23	8	23			3
6445									
6448	28	8	21	10	9	20			3
6449	17	17	34	8	1	21			1
6472	43	13	15	14	5	4			5
6476									
6480·5	41	15	9	17	6	5			4
6482	38	8	17	11	9	9			7
6485	24	8	11	24	21			7	3
6493	16	13	26	33	1		4	1	3
6495·6	25	20	23	16	3	1	1	1	2
6498·5	35	18	16	14	3	3	2		5
6500	50	9		3	14			4	7
6506	P	P						P	
6521									
6530	P	P	P	P					P
6550	P	P			P				P
6577									

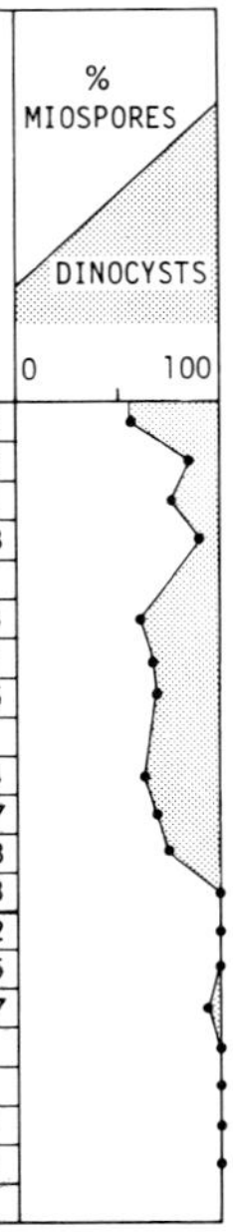

BEATRICE 11/30-2 — DINOCYSTS

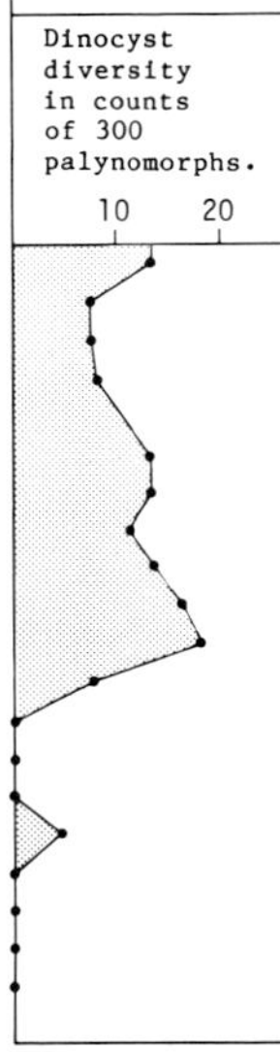

Valensiella spp.	*C. sellwoodii*	*E. pocockii*	*P. ceratophora*	*Meiourogonyaulax* spp.	*Chytroeisphaeridia* spp.	*M. groenlandicum*	*C. varispinosum*	*S. rioultii*	*S. sparsibarbatum*	*N. pellucida*	*K. gochtii*	Sample No.
28	8	1	7		3	1		7		3		6416
			8	13		5						6424
7			14	4		1	4	64	4			6434
22		22	6									6444
												6445
		25	9	6	9	11		3		4	7	6448
		12	3	18	22	5		13		6		6449
9	7	10	1	38	14	2		5				6472
												6476
23	6	14	3	11	10	8	4				7	6480·5
11	3	7	3	7		2	7	3		2		6482
37		15	4	6	14	6	12					6485
												6493
												6495·6
												6498·5
		P			P	P						6500
												6506
												6521
												6530
												6550
												6577

Fig. 11 — Distribution of the main miospore genera and selected dinocyst taxa recovered from Beatrice well 11/30-2. Sample depths as in Fig. 9. P denotes presence of prominent taxa in a small assemblage (less than 30 palynomorphs).

There is no equivalent of the 'Roof Bed' of Brora immediately above the coal; instead 9 ft (3 m) of black-brown, variably micaceous shale with pyrite and occasional macroscopic plant debris overlies the coal. Microscopic plant debris is abundant at the base of these shales (from 6493 ft) along with a rich palynological assemblage comprising miospores, cuticle, epidermal material, vitrinite and inertinite fragments (Fig. 10). The composition and preservation of these terrestrially derived components and the absence of marine derived organic debris imply that at Beatrice the clastic sedimentation which resumed after coal formation took place initially in a dominantly non-marine environment.

Gradual marine invasion took place within the period of accumulation of the shaley horizon; dinocysts are common in the sample from 6485 ft (Figs. 10 and 11). This first assemblage is similar in species diversity and composition (although the proportions of taxa differ) to those recorded below the coal from Brora and Cadh'an Righ with one or two stratigraphically and environmentally significant differences. *Ctenidodinium sellwoodii*, the dominant dinocyst species in the upper part of the Inverbrora Member at Brora and Cadh'an Righ, was not recorded from 6485 ft at Beatrice (Fig. 11) but *Cleistosphaeridium varispinosum* is present. The absence of the former species may be linked with the shallow littoral depositional conditions envisaged, whereas the inception of *C. varispinosum* is more or less concurrent with its appearance elsewhere in the basin. Common occurrences of *Valensiella* spp. (mainly forms referable to *V. ovula* and *V. vermiculata* Gocht (1970) are also present above the coal at Cadh'an Righ.

The first evidence of marine macrofauna above the coal is the presence of bioturbation (*Chondrites*) in a sandy concretionary sideritic bed at 6482 ft, followed by 'oysters' and pectinid bivalves in black shale at 6480 ft 2 in. The establishment of a macrofaunal community at 6482 ft coincides with palynological evidence for a more pronounced marine influence on sedimentation (Fig. 11). Dinoflagellate cysts show a peak in species diversity (Fig. 11) which marks the incoming of a number of taxa associated with fully marine conditions at Brora and Cadh'an Righ. Included in this influx are *Ctenidodinium continuum*, *Adnatosphaeridium caulleryi* (Deflandre 1938) Williams & Downie 1969, *Rigaudella aemula*, *Aldorfia aldorfensis* and others. Together with the presence of *Ctenidodinium combazii* and relatively sparse occurrence of *Ctenidodinium sellwoodii* at 6482 ft, this association suggests an early *calloviense* age equivalent to the Roof Bed at Brora and Cadh'an Righ. The bed contains kaolinite-replaced spheres that may be altered *Rhaxella* spicules.

Preservation of marine microfloral assemblages is generally poor and strongly facies dependent. The marine fauna increases in abundance upwards with molluscan faunas in both shales and sandstones. The shales contain abundant *Thracia depressa*, *Oxytoma inequivalve*, *Eocyprina?* sp. and *Myophorella* sp. The bivalves occur in seams, possibly as winnowed lags, and angular fragments are commonly scattered in the shales, as described above from the Brora Shale Member at Brora.

Following an initial influx of dinoflagellate cysts at 6482 ft and 6480.5 ft in association with abundant miospores, palynological preparations from the Beatrice Formation shales (Fig. 10) are dominated by amorphous organic matter, much of which appears to be of terrestrial origin. This is in keeping with the preparations from the Brora Argillaceous Formation at Brora and Cadh'an Righ (Figs. 4 and 7). However, at Beatrice and Cadh'an Righ an abundance of finely disseminated pyrite included in palyomorphs and amorphous masses indicates that the sediment in these areas was reducing within the zone of active diffusion of seawater, i.e. the oxic–anoxic interface may have been relatively close to the sediment surface, similar to that described by Morris (1980) for the Lower Oxford Clay in the English Midlands. The presence of benthic fauna at these localities precludes any possibility of prolonged development of a fully anoxic sedimentary substrate and reducing bottom waters. In addition to the orange–brown to dark-brown amorphous matter discussed above, there are small amounts of pale-yellow oxidized palynomorphs and plant fragments in most assemblages. This component is more important in sandy lithologies where, in addition, the amorphous masses are well sorted but fragmented, a reflection of the more turbulent, oxic conditions during transportation.

Dinocyst assemblage diversity is consistently lower and individual species abundance more erratic than in the equivalent intervals at Brora and Cadh'an Righ. However, in terms of the taxa represented, the Beatrice section is closely comparable with the other two localities. Therefore, although conditions were fully marine at Beatrice, it appears that the dinoflagellate population was variable and unstable. Possible factors which might induce such sensitivity include fluctuations in water depth and depositional energy in a nearshore environment, changes in basinal circulation and a variable nutrient supply.

Sandstone units which coarsen upwards (Fig. 9) are generally shaley and bioturbated at the base, with coarse- to medium-grained laminated sandstone at the top. The top of the sandstone at 6451 ft has concentrations of sorted and rounded bivalve debris, and the transition to shale is sharp with only minor reworking. Such coarsening-up units are interpreted as shallow marine offshore bars and can be regarded as 'regressive' in nature. They represent decreasing water depth and increased reworking in shallow water, probably due to wave action.

One sandstone unit 2.1 m thick (6438 – 6445 ft) has a coarse-grained sharp erosive base with marine bivalve debris and woody fragments, and it fines up into bioturbated sandstone followed by shale with fauna of bivalves and belemnites. This sandstone records increasing water depth and resembles in character the Brora Roof Bed, but there is no reason to conclude they are synchronous in view of the distance and fault separation between the localities. The sandstone could represent a minor transgressive phase or the fill of a tidal channel cutting an offshore bar but since it can be correlated on well logs from the Beatrice area over a distance of 20 km, the transgressive sheet sand hypothesis is preferred.

Immediately above the logged section (Fig. 9) is the 70 ft thick 'A' Sand (6330–6400 ft) (Fig. 2) which is a coarsening-up offshore marine bar sequence. Near the base of the 'A' Sand (6395 ft), phosphatic concretions are present, centred on bivalves. This horizon appears to correlate with the nodule bed at Cadh'an Righ and the Glauconitic Sandstone at Brora.

INTERPRETATION

The alluvial sediments with freshwater faunas of the Doll Member of the Brora Coal Formation (Hurst 1981) at Brora and Cadh'an Righ are overlain by the organic-rich shales with thin coals and shell beds of the Inverbrora Member. This member represents a varied complex of lagoonal facies with a 'brackish' bivalve fauna (recognized as such by Judd (1873)) associated with low-diversity dinocyst assemblages indicating periodic marine invasion of the lagoon and variable, but low, salinities. Shell beds topped by rounded shelly debris and incorporating dinocyst concentrations formed as wave-winnowed lags representing higher-energy conditions associated with marine invasions. Judd (1873) recorded thin bone beds with fish teeth and scales which might also represent winnowed lags.

Periodically the lagoon became stagnant, bottom waters were probably anoxic and highly organic-rich shales and thin coals of drifted plant material accumulated. Anoxic conditions allowed the accumulation of waterlogged vegetation and formation of amorphous organic matter by chemical and anaerobic bacterial decay. Minor pyrite in the shales is probably a reflection of available sulphate from sea water entering the lagoon. Sulphate might also have been available to the coal at the time of the major marine transgression which deposited the Roof Bed, which is itself rich in pyrite. Bacterial degradation of organic matter in the lagoon in the presence of H_2S may also account for the presence of pyrite in these shales as suggested by Hurst (1981).

At Brora, virtually continuous marine influence is present in shales between the shell beds and the Brora Coal but, in the higher-energy less-organic-rich sequence at Cadh'an Righ, very variable faunas and floras with low diversity and high abundance record local environments; the ostracod–*Cyathidites* association of sample 2, for example, might represent deposition in freshwater ponds in a marsh area. Macrofaunas are similar to those of the Brora sequence shell beds. The rippled sandstone with *Equisetum* rootlets and dinocysts below the coal at Cadh'an Righ records sandy marginal facies to the lagoon. In comparison with Brora, the Cadh'a Righ sediments were deposited in a shallower higher-energy lagoonal situation. If the shell beds can be correlated, the depositional rate was slower at Cadh'an Righ, probably reflecting a reduced subsidence rate relative to Brora.

In Beatrice well 11/30-2 the Brora Coal Formation closely resembles the Doll Member of Brora, and lagoonal facies comparable with the Inverbrora Member are virtually absent. Only very rare dinocysts were encountered below the Brora Coal which rests on alluvial plain facies sandy mudstones with thin sandstones. The lagoonal sediments are thus best developed near the western margin of the basin at Brora where the deeper anoxic lagoon phase persisted for longest, but the lagoon shore can only have lain a few kilometres to the west against the Scottish landmass (Fig. 12).

The Brora Coal has an extensive distribution from Cadh'an Righ to Brora and Beatrice and other wells in that area. At all localities studied, the coal lacks a well-defined seat-earth and initial accumulation was from drifted material comprising waterlogged 'conifer' wood, charcoal and masses of *Equisetum* (Harris & Rest 1966). Spore assemblages in coal samples from the three localities are variable; this might reflect local variation in vegetation patterns or could be a response to progressive changes in vegetation due to ecological succession as the coal swamp matured. More detailed sampling of the coal would be needed to test this hypothesis. At the time coal accumulation commenced, the landward lagoon margin may have consisted of a floating swamp type of environment with only

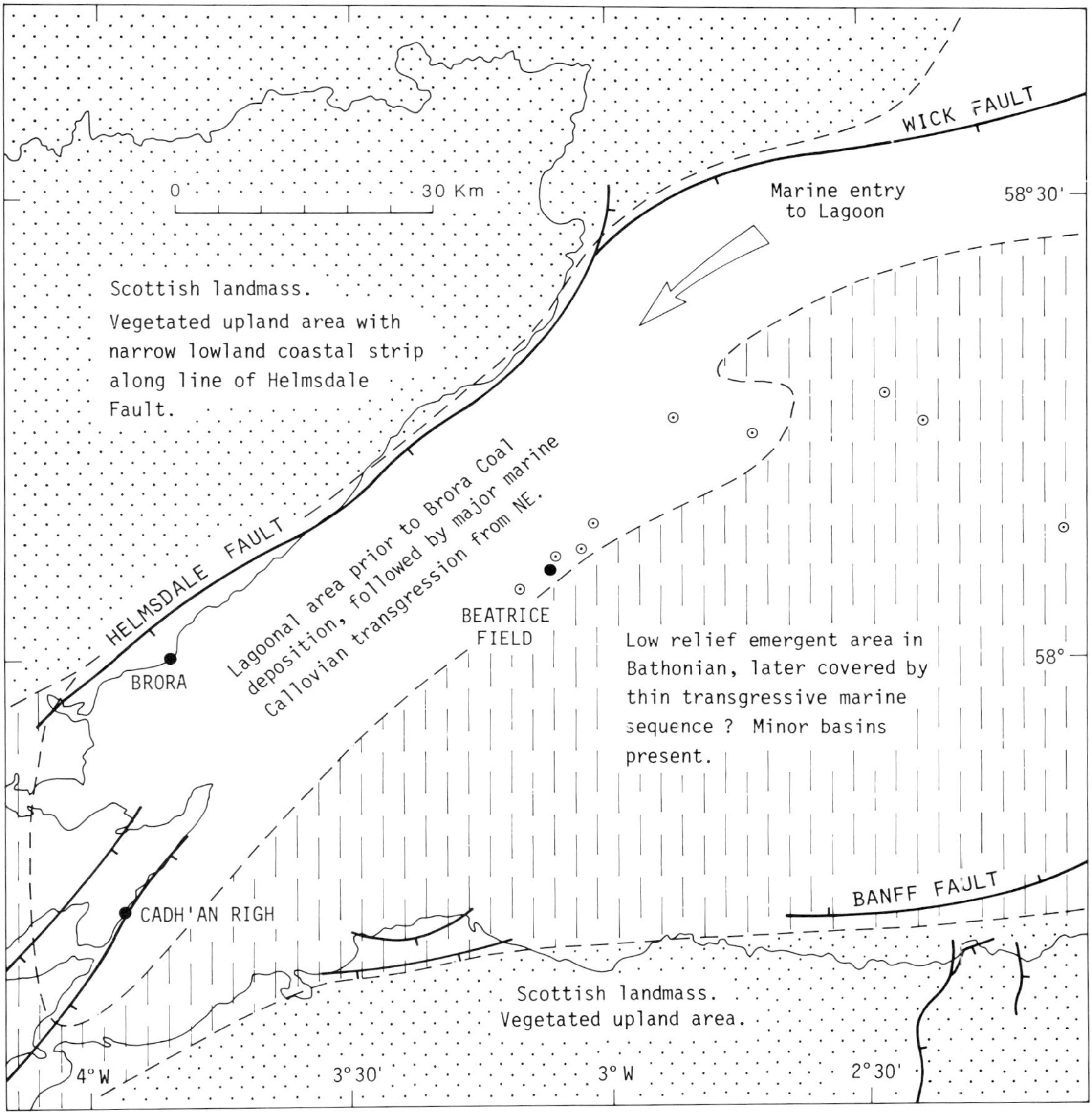

Fig. 12 — Generalized palaeogeographic setting in the Inner Moray Firth during the early Callovian, showing the probable extent of the lagoonal area. This coincides with the area of greatest thickness of Rhaetian–mid-Oxfordian inverval in the Inner Moray Firth, as shown by Andrews and Brown (1987, Fig. 3a).

sluggish streams entering the lagoon. Fringing marsh environments effectively isolated the lagoon from supply of land-derived detritus, and eventually the whole lagoon area was vegetated and became a low-lying swamp. McCabe (1984) discussed similar recent examples of evolving peat accumulations in the Okefenokee Swamp, Georgia. Lack of current activity in such environments might result in variable spore assemblages owing to lack of mixing and transport.

At Brora and Cadh'an Righ the Brora Coal is overlain by the transgressive marine sandstone of the Brora Roof Bed. An erosion surface is present below the Roof Bed with evidence of marine burrows and borings in the underlying coal (Cadh'an Righ) which at the time was probably a hard peat. The Roof Bed is extensively bioturbated and has a varied shallow marine bivalve fauna and diverse dinocyst assemblage. The Roof Bed is a marine shoreline sand which at Brora thins to the west into the coalfield, but thicker sandier facies are present immediately above the Roof Bed representing near-shore variation (Sykes 1975b).

Rapid deepening following Roof Bed deposition in the Brora and Cadh'an Righ areas resulted in shale deposition in open marine conditions close to the Scottish landmass. In the Beatrice well studied, the Brora Coal is immediately overlain by shales within which the transition to open marine conditions takes place, as evidenced by the introduction and rapid increase in diversity of dinocysts. The sandstones of the 'B' Sand of Beatrice (Johnson 1984) comprise small coarsening-up shallow marine bar sands and a single thin fining-up sequence reminiscent of the Roof Bed. The individual sandstones in this interval can be correlated with some thickness variations between individual Beatrice wells and were probably deposited in response to shoreline migration and depth changes affecting the whole Beatrice area. It seems probably that the 'B' sandstones are broadly equivalent to the Roof Bed of the Brora shore section (Fig. 3) and the Roof Bed and its associated sandstones in the Brora coalfield area, as summarized by Sykes (1975b).

The overlying shales (Brora Shale, Cadh'an Righ Shale and Mid Shale of Beatrice (Fig. 2)) were deposited in open marine conditions, generally below local wave base, although some winnowed shell lags are present at Brora and at Beatrice. At Brora and Beatrice, macrofauna is abundant; the bivalves *Bositra* and *Meleagrinella* may have lived attached to floating weed or may have been pendant from plants rooted to the sea floor (Duff 1978). The presence of *Myophorella* and *Nicaniella* (shallow burrowers) and *Thracia* (deep burrower) indicates that bottom surface conditions were generally oxygenated. The abundant angular shell debris present at Brora and Beatrice is probably the product of predation on the molluscan population, possibly by fish. High continuous levels of land-derived organic debris in the palynological preparations attest to the continued proximity of the Scottish landmass throughout deposition of the shales. The lack of variation reflects efficient mixing in the marine environment prior to deposition and the presence of a stable regional flora in the hinterland.

The thin shale at Cadh'an Righ contains an impoverished pelagic–nectonic macrofauna and may have accumulated in a more offshore situation, possibly on a sediment-starved basin rise (cf. Sykes 1975a) which may have been an extension of the Central Ridge (Fig. 1) or a local high over nearby Moine basement. Another possibility is that the locality lay at the head of the shallow embayment and was further from the open sea. Depositional rate was low, continuing the situation observed beneath the Roof Bed (see above).

The Brora Shales at Brora are overlain by the Glauconitic Sandstone Member which represents minor regression and shallower conditions. Hurst (1985, Fig. 7) demonstrated a decrease in kaolinite and increase in illite–smectite towards the top of the Brora Shale—a feature consistent with slow deposition in a marine environment. The Glauconitic Sandstone correlates with the condensed phosphatic nodule bed at Cadh'an Righ and a level approximately at the base of the 'A' Sand of Beatrice where phosphate is also present. Clastic sand deposition in the form of the coarsening-upward sequence of the 'A' Sand was concentrated at the eastern margin of the basin at this time but later, Oxfordian clastics were concentrated at the western margin to form the marine bar sandstones of the Brora Arenaceous Formation, whilst shale deposition took place at Beatrice (Fig. 2).

PALAEOGEOGRAPHY

Regional reconstructions are hampered by uncertainities about the amount of strike-slip movement associated with the Great Glen Fault. Using the reconstructon of Holgate (1969) with 29 km of dextral transcurrent movement since the Late Jurassic, Sykes (1975a) placed the Cadh'an Righ section some 15 km offshore from Brora. Speight & Mitchell (1979) deduced a post-Permo-Carboniferous 7–8 km dextral shift on the basis of a displaced dyke swarm in Argyll, whilst Bacon & Chesher (1975) considered the Mesozoic faulting to be essentially normal in the Inner Moray Firth. McQuillin *et al.* (1982) argued that the Inner Moray Firth Basin developed from Triassic times as a result of about 8 km dextral shift on the Great Glen Fault which resulted in an extension of about 5–6 km between the Wick and Banff boundary faults (Figs. 1 and 12).

Whichever model is adopted it is clear that the area over which the Brora Coal was deposited was considerable, occupying the Great Glen Sub-basin of the Inner Moray Firth and extending from Cadh'an Righ to Beatrice, some 50–60 km in a northeasterly direction and a similar distance easterly from Brora to well 12/26–1. To the northeast the coal is absent (e.g. in wells 12/23-1 and 12/30-1), suggesting that this area may have been emergent.

The large 'lagoonal' area postulated for the deposition of the Brora Coal must have been delimited by the Scottish landmass to the northwest and a series of emergent highs in the east and southeast such as the West Bank High and Smith Bank High (Figs. 1 and 12) (Andrews & Brown 1987). The lagoonal phase (here recognized as *discus* to *macrocephalus* in age) preceding formation of the Brora Coal is best developed at the western side of the basin close to the Scottish landmass in a position coincident with the thickest basin fill and greatest subsidence in the Rhaetian–mid-Oxfordian interval (Andrews & Brown 1987). It is probable that the sea entered the lagoonal area from the northeast and that the latter may have been an embayment some 100 km long by 30 km wide between the Scottish landmass and the Central Ridge (Fig. 12).

The main Callovian transgression brought open marine conditions to the area. Sandstones of shallow marine bar and shoreline origin were deposited on both sides of the basin but they vary greatly in thickness and stratigraphic position owing to differential subsidence, probably controlled by basin-margin faults.

CONCLUSIONS

Combined palynofacies, macropalaeontological and sedimentological data provide strong evidence for lagoonal conditions with intermittent marine influence in strata below the Brora Coal at Brora and Cadh'an Righ but lagoonal sediments are virtually absent at Beatrice. Highly variable dinocyst assemblages dominated by one or two taxa occur in the lagoonal shales and are associated with reworking of *Neomiodon–Isognomon* bivalve faunas due to marine invasion. The Brora Coal was deposited in an isolated freshwater lagoon, initially from drifted plant material. Subsequent marine invasion of the lagoon area, probably from the northeast, resulted in open marine conditions as indicated by macrofaunas, but palynological preparations reflect continuing close proximity of land supplying abundant miospores and land-derived plant debris. Miospore assemblages are more uniform in marine sediments due to greater transport and mixing prior to deposition. The 'B' Sand above the coal in Beatrice represents thin shallow marine coarsening-up bar sandstones and a single thin fining-up sandstone which resembles the transgressive Roof Bed of Brora and Cadh'an Righ. The lagoonal embayment, 100 km long by 30 km broad, lay between the Scottish landmass and the Central Ridge of the Moray Firth. Entry of marine waters was probably from the northeast.

The incoming of *Mendicodinium groenlandicum* and the common occurrence of *Korystocysta gochtii* at the level of the shell beds in the Inverbrora Member at Brora indicates a *discus* to early *macrocephalus* age for the uppermost part of the Brora Coal Formation. A *calloviense* age is proposed for the Roof Bed, based on the extinction of *Ctenidodinium combazii* and common occurrence of *Ctenidodinium ornatum* and *C. continuum*, concomitant with the inception of a host of other dinocyst taxa. At Brora, an abundance of *Ctenidodinium sellwoodii* may indicate that the *macrocephalus* Zone is also represented at the base of the Roof Bed.

ACKNOWLEDGEMENTS

We would like to thank the Department of Energy for allowing access and giving permission to sample cores from well 11/30-2. Stewart Brown and Andrew Hurst provided much useful discussion on the succession and the manuscript benefited from David Batten's constructive suggestions. Dr N. J. Morris kindly discussed bivalve identification with N.H.T. Britoil kindly supplied composite logs from released Beatrice wells. Barry Fulton provided valuable drafting services.

REFERENCES

Andrews, I. J. & Brown, S. (1987). Stratigraphic evolution of the Jurassic, Moray Firth. In: Brooks, J. & Glennie, K. W. (eds), *Proceedings of the 3rd Conference on Petroleum Geology of north west Europe. Barbican Centre, October 1986*. Graham & Trotman, London, pp. 785–795.

Bacon, M. & Chesher, J. A. (1975). Evidence against post-Hercynian transcurrent movement on the Great Glen Fault. *Scottish Journal of Geology* **11** 79–82.

Batten, D. J. (1982). Palynofacies and salinity in the Purbeck and Wealden of southern England. In: Banner, F. T. & Lord, A. R. (eds), *Aspects of Micropalaeontology*. Allen & Unwin, London, pp. 278–308.

Batten, D. J. (1983). Identification of amorphous sedimentary organic matter by transmitted light microscopy. In: Brooks, J. (ed.), *Petroleum geochemistry and exploration of Europe. Geological Society, London Special Publication* **12** 275–289.

Batten, D. J. & Morrison, L. (1983). Methods of palynological preparation for palaeoenvironmental source rock potential and organic maturation studies. *Norwegian Petroleum Directorate Bulletin* **2** 35–53.

Chesher, J. A. & Lawson, D. (1983). The geology of the Moray Firth. *Report, Institute of Geological Sciences, London* **83/5** 1–32.

Combaz, A. (1980). Les kerogenes vus au microscope. In: Durand, B. (ed.), *Kerogen, Insoluble organic matter from sedimentary rocks*. Editions Technip, Paris, pp. 55–111.

Davey, R. J. & Rogers, J. (1975). Palynomorph distribution in Recent offshore sediments along two traverses off south west Africa. *Marine Geology* **18** 213–225.

Davies, E. H. & Norris, G. (1980). Latitudinal variations in encystment modes and species diversity in Jurassic dinoflagellates. In: Strangway, D. W. (ed.), *The continental crust and its mineral deposits. Geological Association of Canada Special Paper* **20** 361–373.

Duff, K. L. (1978). Bivalvia from the English Lower Oxford Clay (Middle Jurassic). *Palaeontographical Society Monograph* **132** 1–137.

Fisher, M. J. (1980). Kerogen distribution and depositional environments in the Middle Jurassic of Yorkshire, UK. *Proceedings of the 4th International Palynological Conference, Lucknow*, Vol. 2, pp. 574–580.

Harris, T. M. & Rest, J. A. (1966). The flora of the Brora Coal. *Geological Magazine* **103** 101–109.

Holgate, N. (1969). Palaeozoic and Tertiary transcurrent movements on the Great Glen Fault. *Scottish Journal of Geology* **5** 97–139.

Hughes, N. F. & Couper, R. A. (1958). Palynology of the Brora Coal of the Scottish Middle Jurassic. *Nature* **181** 1482–1483.

Hurst, A. (1981). Mid Jurassic stratigraphy and facies at Brora, Sutherland. *Scottish Journal of Geology* **17** 169–177.

Hurst, A. (1985). The implications of clay mineralogy to palaeoclimate and provenance during the Jurassic in N.E. Scotland. *Scottish Journal of Geology* **21** 143–160.

Johnson, R. W. (1984). The development of the Beatrice Field. *Institute of Petroleum, Annual Conference 1984, Offshore Europe*, pp. 1–12.

Judd, J. W. (1873). The secondary rocks of Scotland. *Quarterly Journal of the Geological Society, London* **29** 97–195.

Lam, K. & Porter, R. (1977). The distribution of palynomorphs in the Jurassic rocks of the Brora Outlier, N.E. Scotland. *Journal of the Geological Society, London* **134** 45–55.

Lee, G. W. (1925). Mesozoic rocks of East Sutherland and Ross. In: Read, H. H., Ross, G. & Phemister, J. (eds), *The geology of the country around Golspie, Sutherlandshire. Memoir of the Geological Survey*, pp. 65–115.

Linsley, P. N., Potter, H. C., McNab, G. & Racher D. (1980). The Beatrice Oil Field, Inner Moray Firth, UK North Sea. In: Halbouty M. T. (ed.), *Giant oil and gas fields of the decade, 1968–1978. American Association of Petroleum Geologists, Memoir* **30** 117–129.

MacLennan, A. M. (1985). Biostratigraphic and palynofacies analysis of selected Lower Cretaceous sections from Portugal. *Ph.D. Thesis*. University of Aberdeen, pp. 1–314.

McCabe, P. J. (1984). Depositional environments of coal and coal-bearing strata. In: Rahmani R. A. & Flores R. M. (eds), *Sedimentology of coal and coal-bearing sequences. Special Publication of the International Association Sedimentologists* **7** 13–42.

McQuillin, R., Donato, J. A. & Tulstrup, J. (1982). Development of basins in the Inner Moray Firth and the North Sea by crustal extension and dextral displacement of the Great Glen Fault. *Earth and Planetary Science Letters* **60** 127–139.

Morris, K. A. (1980). Comparison of major sequences of organic-rich mud deposition in the British Jurassic. *Journal of the Geological Society, London* **137** 157–70.

Murchison, R. I. (1827). On the coal-field of Brora in Sutherlandshire. *Transactions of the Geological Society, London* **2** 293–326.

Neves, R. & Selley, R. C. (1975). A review of the Jurassic rocks of north-east Scotland. In: Finstad, K. G. & Selley, R. C. (eds), *Proceedings of the Jurassic Northern North Sea*

Symposium, Stavanger, September 1975. Norsk Petroleumsforening **JNNSS/5** 1–29.

Riding, J. B. (1984). Dinoflagellate cyst range-top biostratigraphy of the uppermost Triassic to lowermost Cretaceous of northwest Europe. *Palynology* **8** 195–210.

Riding, J. B., Penn, I. E. & Woollam, R. (1985). Dinoflagellate cysts from the type area of the Bathonian stage (middle Jurassic; southwest England). *Review of Palaeobotany and Palynology* **45** 149–169.

Riley, L. A. & Fenton, J. P. G. (1982). A dinocyst zonation for the Callovian to Middle Oxfordian succession (Jurassic) of northwest Europe. *Palynology* **6** 193–201.

Sheridan, N. C. (1987). The organic geochemistry of selected onshore and offshore samples from the Middle Jurassic, Inner Moray Firth. *M.Sc. Thesis.* University of Aberdeen, pp. 1–52.

Speight, J. M. & Mitchell, J. G. (1979). The Permo-Carboniferous dyke-swarm of northern Argyll and its bearing on dextral displacement of the Great Glen Fault. *Journal of the Geological Society, London* **136** 3–11.

Stopes, M. C. (1907). The flora of the Inferior Oolite of Brora (Sutherland). *Quarterly Journal of the Geological Society, London* **63** 375–382.

Sykes, R. M. (1975a). The stratigraphy of the Callovian and Oxfordian stages (Middle–Upper Jurassic) in northern Scotland. *Scottish Journal of Geology* **11** 51–78.

Sykes, R. M. (1975b). Facies and faunal analysis of the Callovian and Oxfordian Stages (Middle–Upper Jurassic) in northern Scotland and east Greenland. *Ph.D. Thesis.* University of Oxford.

Woollam, R. & Riding, J. B. (1983). Dinoflagellate cyst zonation of the English Jurassic. *Report, Institute of Geological Sciences, London* **83/2** 1–42.

5

Visual and chemical kerogen analyses of the Lower Kimmeridge Clay, Westbury, England

Henrik Nøhr-Hansen

ABSTRACT

Visual and chemical kerogen analyses are described and compared from 50 samples from a 40 m section of the Jurassic Kimmeridge Clay exposed in a clay pit in Westbury, southern England. The kerogen indicates a marine depositional environment, with thermally immature amorphous kerogen, type II dominating. Three kerogen facies and two subfacies have been distinguished on the basis of visual kerogen analysis; these are discussed in terms of anoxic–oxic depositional conditions. Comparison between the chemical pyrolysis data and visual kerogen data shows that the total organic carbon (TOC) content, hydrocarbon potential S_2 and relative amorphous kerogen content fluctuate in a similar way throughout the sequence; high TOC and S_2 values occur in samples with high amorphous kerogen content, whereas the TOC and S_2 values are clearly lower in samples dominated by other kerogen assemblages. The results suggest that it is possible to predict the relative quantity and quality of kerogen from the composition of a palynological slide.

INTRODUCTION

The type locality for the Kimmeridge Clay is on the Dorset coast of southern England, but in this area the lower part of the formation is rather poorly exposed. In contrast, a sequence of Lower Kimmeridge Clay is extremely well exposed in the clay pit belonging to the Blue Circle Portland Cement Company at Westbury, Wiltshire (Grid Reference ST 882528; Fig. 1). The 40 m of section have been described lithologically and divided biostratigraphically into three ammonite zones (Birkelund *et al.* 1983) and two dinoflagellate subzones (Nøhr-Hansen 1986). The sediments consist predominantly of clay, with subordinate bituminous clay, oil shale, septarian nodules and limestone. They range in age from the late *cymodoce* Zone to the early *eudoxus* Zone of the lower Kimmeridgian (*sensu anglico*).

During the last 20 years, much work has been done on the kerogen composition of source rocks in order to elucidate the potential of kerogen particles for oil and gas generation. Staplin (1969), Burgess (1974), Combaz (1980), Batten (1981, 1983), Venkatachala (1981) and Tyson (1984, 1987) have studied kerogen visually

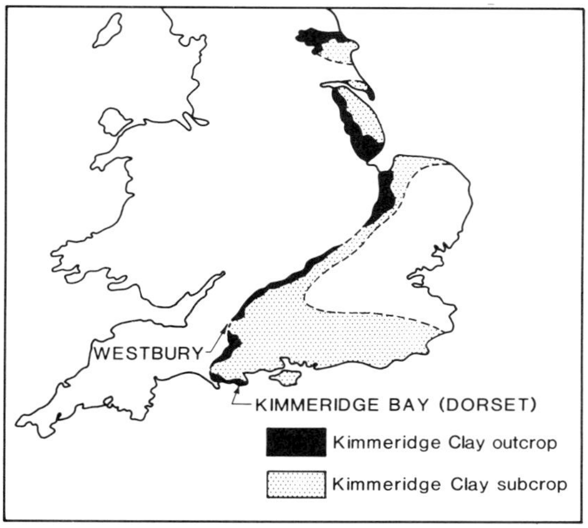

Fig. 1 — Map showing Kimmeridge Clay outcrop and subcrop distribution in England and location of Westbury.

under the optical microscope in order to describe kerogen facies and levels of maturation. Visual kerogen analysis is very useful for determining the oil–gas potential of a new area. Chemical kerogen analysis carried out by Rock Eval pyrolysis (Espitalié *et al.* 1977a, b, Peters 1986) is a second, but more expensive, screening method. Both types of analysis are used in this detailed kerogen study which aims to describe the variation in kerogen composition and to correlate the results produced by the two methods.

The word 'kerogen' is used in the sense of Brooks (1981, p. 2), namely 'The disseminated organic matter of rocks that is insoluble in non-oxidising mineral acids, bases and organic solvents'.

METHODS

5 g of each of the 55 sediment samples collected were prepared using standard palynological techniques. Carbonates and silicates were removed by hydrochloric and hydrofluoric acids. The organic residue was sieved using a 20 µm mesh nylon screen and then strew mounted in the permanent medium 'eukitt' (produced by O. Kindler) on a glass slide for transmission optical microscopy. Each sample represents less than 5 cm of vertical thickness in the investigated sequence.

57 samples, of which the 50 are from the same horizons as those prepared for palynological examination, were chemically analysed using Leco and Rock Eval instruments. Sample numbers given in the text and in Table 1 correlate with the bed numbers in Figs. 2 and 3 and in Birkelund *et al.* (1983, Fig. 2). Sample M 21 155 indicates that it is 1.55 m above the base of bed 21 in the *mutabilis* Zone.

VISUAL KEROGEN ANALYSES

From each sample, 300 kerogen particles were randomly chosen by a point counter in a slide of sieved unoxidized material and identified as yellow to orange–brown amorphous kerogen, marine microplankton (dinoflagellate cysts, acritarchs, etc.), saccate pollen, non-saccate spores and pollen, 'multicellular' micro-organisms (e.g. scolecodonts), cuticles, resins, brownish supporting tissue from land plants, black supporting tissue from land plants with tracheids or other recognizable cellular structures, and black opaque and structureless coal. The different types of kerogen are presumed to have different hydrocarbon potential. An attempt has been made to classify them by formulating three kerogen assemblages which are expected to have similar hydrocarbon potential.

Kerogen assemblage A consists of yellow to orange–brown amorphous kerogen. In the analysed samples, this occurs in lumps or occasionally as fine particles in a greyish mineral-like gel. Kerogen assemblage B is composed of marine microplankton, saccate pollen, non-saccate spores and pollen, multicellular plankton, cuticles and resins. In this study it is dominated by non-saccate spores and pollen, but saccate pollen, dinoflagellate cysts and foraminiferal linings are common, whereas cuticles, tasmanitids, resins and multicellular organisms only con-

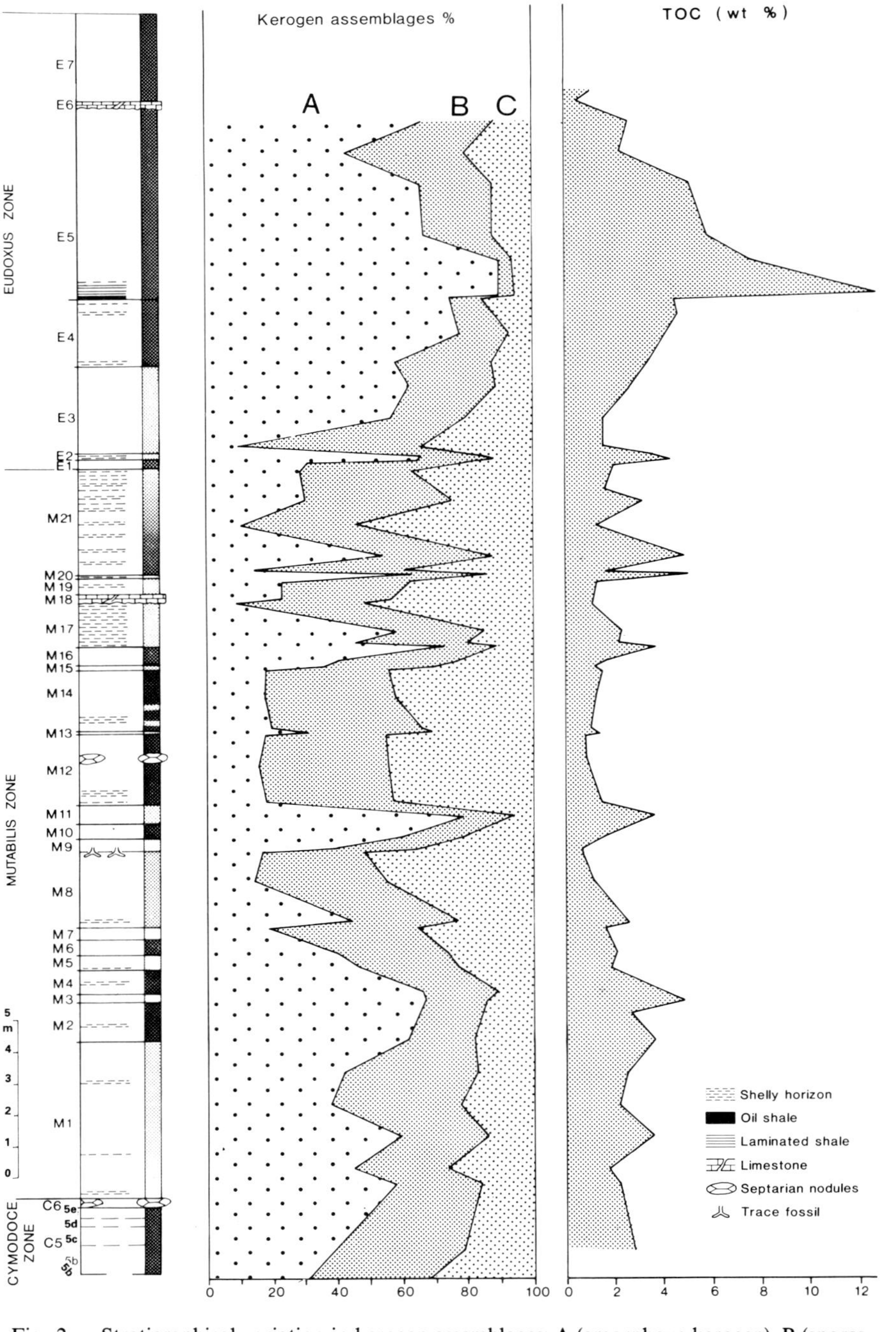

Fig. 2 — Stratigraphical variation in kerogen assemblages; A (amorphous kerogen), B (spores, pollen, dinocysts, acritarchs, cuticles, and foraminiferal linings) and C (woody particles and coal) compared with TOC distribution in the Westbury section. Bed numbers correlate with those used by Birkelund *et al.* (1983). C represents the *cymodoce* Zone, M the *mutabilis* zone and E the *eudoxus* zone). The shading corresponds to weathering colour at outcrop.

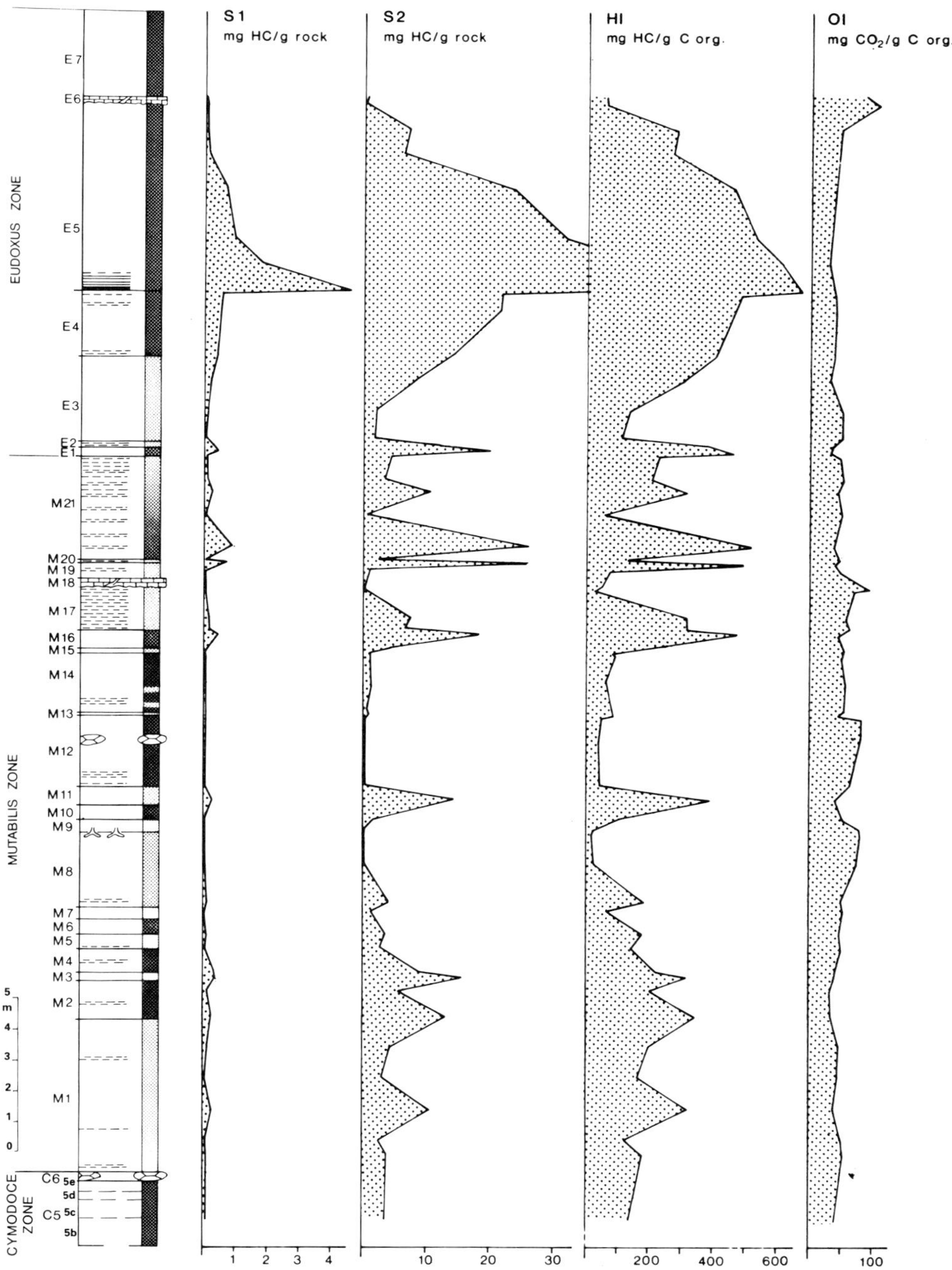

Fig. 3 — Stratigraphical distribution of the chemical parameters S_1 (the quantity of hydrocarbons generated in the samples), S_2 (the quantity of hydrocarbons that the sample could generate if thermal maturation was to continue), HI and OI in the Westbury section. For the bed numbers and legend, see section on methods and Fig. 2).

stitute a minor part (less than 2%). Kerogen assemblage C consists of angular coal particles of unknown affinity and slightly less abundant brown and black tracheid particles.

The distribution of the three kerogen assemblages throughout the sequence (Fig. 2) indicates that kerogen assemblage A is dominant (more than 50%) in the *eudoxus* Zone and makes up a major component of the kerogen in the lower, and in intervals of the upper, parts of the *mutabilis* Zone (Fig. 2). Kerogen assemblage B dominates in only one sample but is the most abundant type in the middle and upper part of the *mutabilis* Zone (Fig. 2). Kerogen assemblage C dominates in only three samples but, in a similar manner to assemblage B, it is the most abundant type in the middle and upper parts of the *mutabilis* Zone (Fig. 2).

Three kerogen facies have been distinguished as illustrated in the ternary diagram of kerogen assemblage composition (Fig. 4). Kerogen facies 1 contains more than 50% of kerogen assemblage A (Plate 1, Fig. 1), kerogen facies 2 contains more than 25% and less than

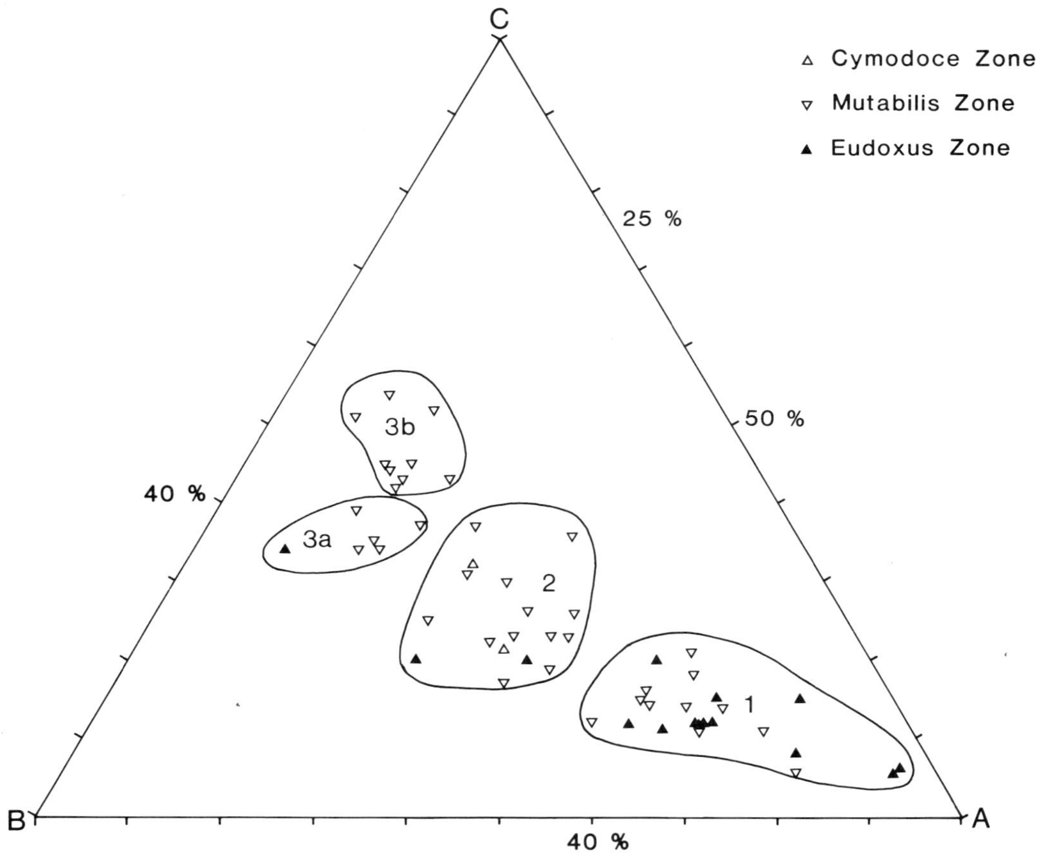

Fig. 4 — Ternary kerogen assemblage composition and stratigraphical position of the analysed samples: A, kerogen assemblage A (amorphous kerogen); B, kerogen assemblage B (spores, pollen, dinocysts, acritarchs, cuticles, and foraminiferal linings); C, keregon assemblage C (woody particles and coal); 1, kerogen facies 1 (more than 50% kerogen assemblage A); 2, kerogen facies 2 (more than 25% and less than 50% kerogen assemblage A); 3a, kerogen subfacies 3a (less than 25% kerogen assemblage A and more than 40% kerogen assemblage B); 3b, kerogen subfacies 3b (less than 25% kerogen assemblage A and more than 40% kerogen assemblages C).

50% of kerogen assemblage A (Plate 1, Fig. 2), whereas kerogen facies 3 contains less than 25% of kerogen assemblage A. Furthermore, kerogen facies 3 can be divided into two subfacies: subfacies 3a contains 40% or more of kerogen assemblage B (Plate 1, Fig. 3) and subfacies 3b contains 40% or more of kerogen assemblage C (Plate 1, Fig. 4).

CHEMICAL KEROGEN ANALYSIS

The data from the chemical analyses are presented graphically in Figs. 2 and 3 and numerically in Table 1. On the basis of the total organic carbon (TOC) values the sediments are divided into clay (less than 2 wt% TOC), bituminous clay (2–10 wt% TOC), and oil shale (more than 10 wt% TOC). Clay dominates in the middle and upper part of the *mutabilis* Zone, bituminous clay dominates in the *eudoxus* Zone and in the lower part of the *mutabilis* Zone, whereas oil shale only occurs in the *eudoxus* Zone (Fig. 2). This oil shale can be correlated with the earliest Kimmeridgian oil shale which Cox & Gallois (1981) report from the *eudoxus* Zone in the Kimmeridgian type area on the Dorset coast.

Values for the amount of hydrocarbons already generated (S_1) are in general low (Fig. 3 and Table 1), as expected from the thermal alteration index (TAI) $1-1+$ (spore colour) and the thermal stage of the organic matter, $T_{max} < 430°C$. According to Espitalié *et al.* (1977a, p. 440), the 'temperature T corresponds to the maximum hydrocarbon-type products produced by the cracking during pyrolysis' and 'is characteristic of the evolution level of the organic matter'; in other words, T_{max} is a thermal maturity parameter. However, the S_1 peak in the oil shale sample indicates that some of the organic matter has reached an oil-generating level. The occurrence of oil in immature surface samples is not uncommon according to Brooks *et al.* (1987, p. 19) who mention that 'oil shales with potential large pyrolytic yields of hydrocarbons usually originate in shallow, low temperature environments'.

(S_2) values for the maximum amount of hydrocarbons which can be generated if the sediments are submitted to pyrolysis are high in the *eudoxus* Zone and in both the upper and the lower parts of the *mutabilis* Zone (Fig. 3). The hydrogen index HI($= S_2$/TOC) follows the same trend. The oxygen index OI($= S_3$/TOC, where S_3 is the CO_2 produced by the pyrolysis of the organic matter) is fairly constant throughout the sequence (Fig. 3). However, it is notable that the OI values decrease in intervals with high generation potential S_2 and high HI values. The OI values are high compared with most good potential source rocks; this could be caused by surface oxidation.

In the modified van Krevelen diagram (Fig. 5) the HI versus OI values of kerogen from the *eudoxus* Zone plot relatively close to the type II trend. This indicates a good oil-prone source rock, whereas the kerogen from the main part of the *mutabilis* Zone is a mixture of types II and III, indicating reasonably good potential for oil and gas. A few samples, especially those from the middle part of the *mutabilis* Zone, plot close to the type III trend; these probably have only gas potential.

COMPARISON OF THE VISUAL AND CHEMICAL ANALYSES

Examination of the data obtained from the visual and chemical kerogen analyses reveals a strong relationship between the optical characteristics of the kerogen assemblage (Figs. 2 and 4) and the bulk geochemical parameters derived from pyrolysis measurements (Figs. 2, 3 and 5–8). The degree of correlation that is possible between these two independent methods makes them particularly useful for the identification of source rock potential and for the interpretation of depositional environments.

A significant proportion of the Westbury sequence yields TOC values that are above the average for mudrocks. Elevated TOC values have commonly been used as a criterion for identifying good petroleum source rocks. Unfor-

Table 1 — Data from chemical analyses: sample number (see section on methods), TC (total carbon), TOC (total organic carbon), T_{max} (thermal maturity parameter) (see section on chemical kerogen analysis), S_1 (quantity of hydrocarbon generated), S_2 (quantity of hydrocarbon that the samples could generate if thermal maturation was to continue), S_3 (CO_2 produced by the pyrolysis of the organic matter, HI ($=S_2$/TOC) (hydrogen index) of OI ($=S_3$/TOC) (oxygen index).

Sample No.	*Depth* (mm)	TC (wt%)	TOC (wt%)	T_{max} (°C)	S_1 (mg HC/g rock)	S_2 (mg HC/g rock)	S_3 (mg HC/g rock)	HI (mg HC/g C_{org})	OI (mg CO_2/g C_{org})
E 7 0, 10	0	6.21	1.19	429	0.03	0.72	1.00	61	84
E 6 0, 20	200	10.10	0.49	413	0.11	0.30	0.54	60	110
E 5 5, 40	900	5.15	2.44	424	0.15	7.12	1.22	292	50
E 5 4, 40	1900	4.75	2.25	426	0.12	6.22	1.05	276	47
E 5 3, 45	2850	6.64	5.08	418	0.67	23.56	2.00	464	39
E 5 1, 80	4500	8.60	5.92	418	0.98	31.84	2.11	538	36
E 5 1, 00	5300	9.96	7.88	415	1.80	48.14	2.61	611	33
E 5 0, 00	6300	15.40	12.88	411	4.52	87.03	4.60	676	36
E 4 2, 15	6400	6.34	4.51	419	0.60	21.79	1.91	483	42
E 4 1, 60	7450	6.60	4.66	421	0.51	21.58	1.93	463	41
E 4 0, 05	8400	5.19	3.50	424	0.40	14.03	1.38	401	39
E 3 2, 10	9150	4.22	2.74	424	0.21	8.29	0.99	303	36
E 3 1, 10	10150	3.12	1.58	429	0.11	2.17	0.82	137	52
E 3 0, 20	11050	3.14	1.54	430	0.05	1.81	0.79	118	51
E 2 0, 05	11400	4.73	3.68	422	0.28	13.96	1.44	379	39
E 1 0, 30	11450	6.03	4.34	423	0.42	19.85	1.44	457	33
E 1 0, 10	11650	4.48	2.09	426	0.10	4.81	1.02	230	49
M 21 2, 70	12450	5.24	1.73	426	0.10	3.56	0.94	206	54
M 21 2, 30	12850	5.70	3.23	423	0.26	10.47	1.43	324	44
M 21 1, 55	13600	3.29	1.28	423	0.04	0.77	0.66	60	51
M 21 0, 55	14600	6.26	4.99	420	0.83	25.88	1.88	519	38
M 21 0, 55	15100	3.62	1.68	429	0.09	2.24	0.81	134	48
M 20 0, 05	15200	7.66	5.08	418	0.66	25.19	2.07	496	41
M 19 0, 30	15450	2.98	1.33	426	0.05	1.05	0.67	79	50
M 18 0, 00	16050	8.32	0.45	414	0.01	0.23	0.41	51	92

M 17 1, 35	16100	4.79	1.07	411	0.03	0.26	0.75	24	70
M 17 0, 40	17050	4.90	2.35	426	0.17	7.46	1.29	318	55
M 17 0, 00	17450	5.52	2.10	424	0.18	6.67	1.33	318	63
M 16 0, 55	17500	6.11	3.85	421	0.44	18.28	1.90	475	49
M 16 0, 05	18000	4.70	1.79	423	0.15	3.39	0.95	190	53
M 15 0, 05	18150	3.97	1.37	430	0.07	1.19	0.76	87	56
M 14 1, 90	18250	4.94	1.48	428	0.07	1.46	0.76	99	52
M 14 1, 10	19050	3.83	1.27	423	0.05	0.81	0.72	64	57
M 14 0, 05	20100	3.19	1.15	424	0.07	0.96	0.67	84	58
M 13 0, 05	20200	2.94	1.39	425	0.07	1.16	0.62	84	45
M 12 2, 30	20250	2.49	0.86	414	0.03	0.43	0.70	50	81
M 12 1, 50	21050	2.94	0.86	413	0.03	0.37	0.69	43	80
M 12 0, 05	22500	3.00	1.23	416	0.04	0.54	0.78	44	63
M 11 0, 20	22950	5.26	3.64	422	0.26	14.39	1.54	395	42
M 10 0, 05	23600	3.48	1.71	429	0.05	1.85	0.90	108	52
M 9 0, 05	24000	2.07	0.71		0.02	0.21	0.56	29	78
M 8 2, 35	24100	2.25	0.70		0.01	0.14	0.56	20	80
M 8 1, 40	25050	2.68	1.02	415	0.02	0.26	0.75	25	74
M 8 0, 20	26250	3.85	2.30	428	0.11	4.25	1.15	185	50
M 7 0, 40	26450	2.67	1.64	427	0.04	1.14	0.88	69	54
M 6 0, 05	27250	3.59	2.04	429	0.09	3.60	0.97	176	48
M 5 0, 05	27750	3.58	1.90	429	0.05	2.71	0.97	143	51
M 4 0, 05	28500	4.16	3.90	413	0.31	8.39	1.61	215	41
M 3 0, 05	28750	5.22	4.94	416	0.39	15.55	1.92	315	39
M 2 0, 85	29200	2.95	2.76	429	0.13	5.52	0.87	200	32
M 2 0, 00	30050	3.88	3.70	425	0.24	12.65	1.18	342	32
M 1 4, 00	31050	3.82	2.25	429	0.11	4.46	1.00	198	45
M 1 3, 00	32050	3.18	2.03	430	0.08	3.33	0.91	164	45
M 1 2, 00	33050	3.69	3.44	422	0.23	10.95	1.34	318	39
M 1 1, 00	34050	2.61	1.86	427	0.09	2.25	0.93	121	50
M 1 0, 50	34550	3.01	2.09	427	0.11	3.77	1.06	180	51
C 5b 0, 80	35950	2.50	2.39	427	0.10	3.54	0.94	148	39

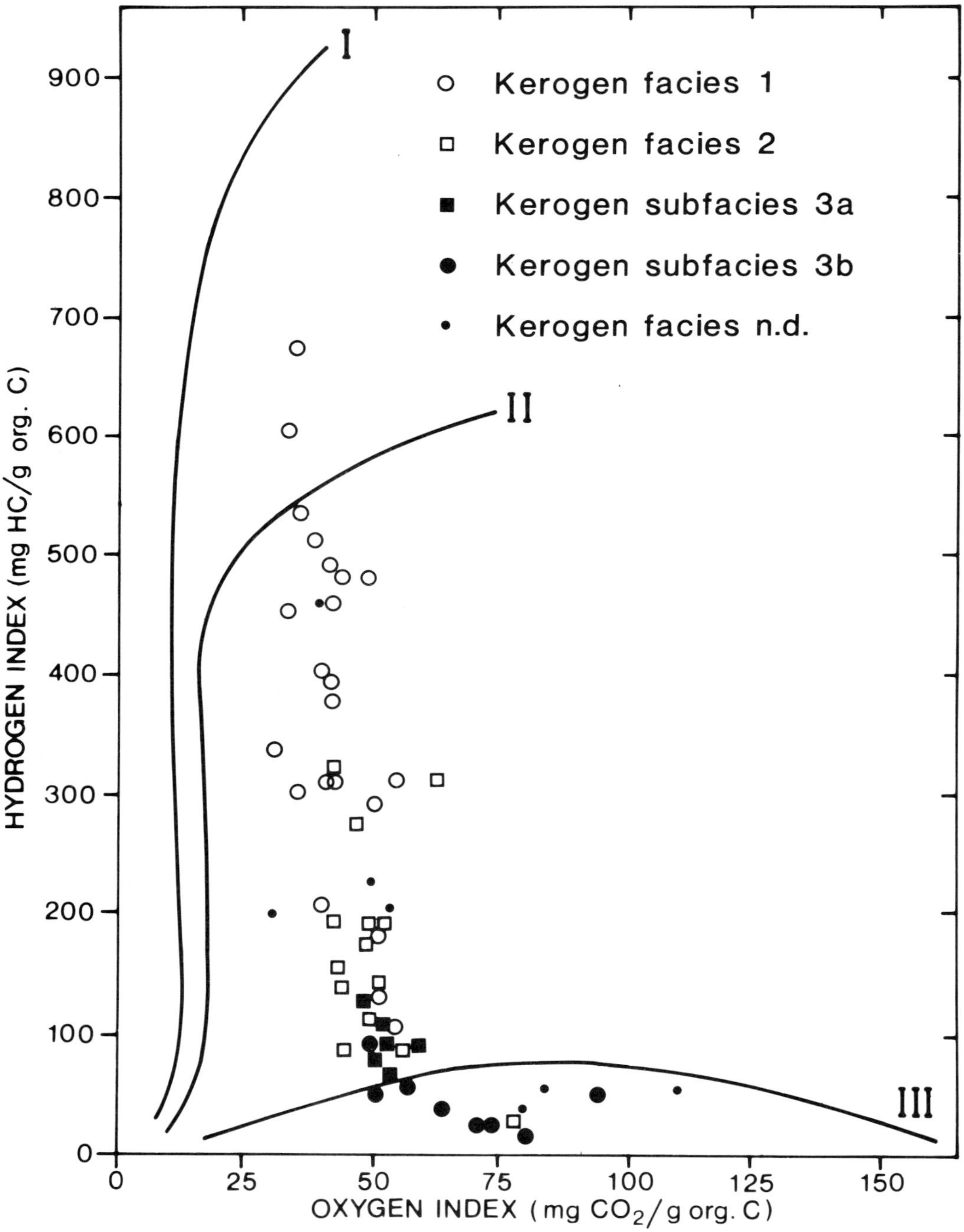

Fig. 5 — Modified van Krevelen diagram illustrating the relationship between the kerogen facies (see Fig. 4) and the kerogen type: n.d., not determined.

tunately this parameter does not provide any information on the remaining generative potential of the sediment (determined largely by the level of maturation) or the kind of hydrocarbons that might be produced (determined by the type of kerogen). However, TAI and T_{max} values from the Westbury sequence indicate that the oil window has not yet been reached and that the generative potential of the sediments is consequently still largely intact. The marine depositional environment of these dark-coloured clays (Birkelund *et al.* 1983, Nøhr-Hansen 1986) would lead one to expect marine-dominated or at least mixed kerogen compositions. This has been generally substantiated by the visual kerogen analyses.

Intervals dominated by amorphous-rich kerogens (kerogen facies 1) are supposed to have been formed in periods with anoxic bottom conditions (e.g. Batten 1981, 1983, Tyson 1987). The supposition is supported by the relatively high TOC values (greater than 2.5%) for most of the kerogen facies 1 samples (Figs. 2 and 6).

The dominance of spores, pollen and woody material in kerogen facies 3 might indicate an increase in terrestrial input. The low TOC content (less than 2 wt%) for this facies (Figs. 2 and 6) seems more likely, however, to indicate oxic bottom conditions. Such conditions are further supported by the occurrence of numerous shelly horizons and a limestone layer in the middle to upper part of the *mutabilis* Zone where kerogen facies 3 dominates (Figs. 2 and 6).

Kerogen facies 2 dominates the lower part of the *mutabilis* Zone (Fig. 2). The relatively low TOC values (Fig. 6, more than 1.5 wt% and less than 2.5 wt%) for this facies also seem to indicate periods when bottom conditions were oxic, although the shelly horizons are less numerous than in kerogen facies 3.

The S_1 values (Fig. 3) indicate that very small amounts of hydrocarbons have already been generated from the organic matter because of low thermal influence. The S_2 values (Fig. 3) show that many of the samples from the sequence have a good hydrocarbon potential, but the yields are highly variable. Comparison between the S_2 values and the kerogen assemblage A curve (Fig. 2) reveals identical fluctuations. One deviation from this trend is observed in sample E 3 110 where a relatively high content of kerogen assemblage A is not followed by an increase in the S_2 value. The reason for this particular inconsistency is clearly related to the fact that the major part of the amorphous kerogen in this sample occurs as fine amorphous particles in a grey mineral-like gel which is evidently devoid of any hydrocarbon potential. The general trend is that high S_2 values occur in samples with high kerogen assemblage A content, whereas the S_2 values are clearly reduced with increasing kerogen assemblage B and C content. The good correlation between S_2 and kerogen assemblage A shows that oil-prone organic matter is mainly composed of the lumpy yellow to orange–brown amorphous kerogen.

The quality of total hydrocarbon generation potential is predicted by the $S_1 + S_2$ versus TOC plot (Fig. 7) which illustrates that kerogen facies 1 has a generation potential above average. Kerogen facies 2 has a fair generation potential, whereas that for kerogen facies 3 is poor. It is especially poor for subfacies 3b, owing to the dominance of woody material.

The modified van Krevelen diagram (Fig. 5) classifies the organic matter from Westbury as type II, type II–III and type III. Correlation of the hydrogen index and percentage of kerogen assemblage A in each sample (Fig. 8) obviously shows that kerogen type II (mainly oil prone: HI >300 mg HC/g C_{org}) chiefly occurs in kerogen facies 1. It also demonstrates that kerogen type III (mostly gas prone: HI <75 mg HC/gC_{org}) is dominated by subfacies 3b whereas kerogen type II–III (oil and gas prone: HI 75–300 mg HC/g C_{org}) mainly occurs in kerogen facies 2 and subfacies 3a. The trend in Fig. 8 is in general similar to the HI versus liptinite content plot of Mukhopadhyay *et al.* (1985, Fig. 1).

Comparison of the relative kerogen assemblage composition with the absolute TOC values (Fig. 2) for each sample illustrates that kerogen assemblages B and C constitute nearly identical percentages of the TOC value throughout the

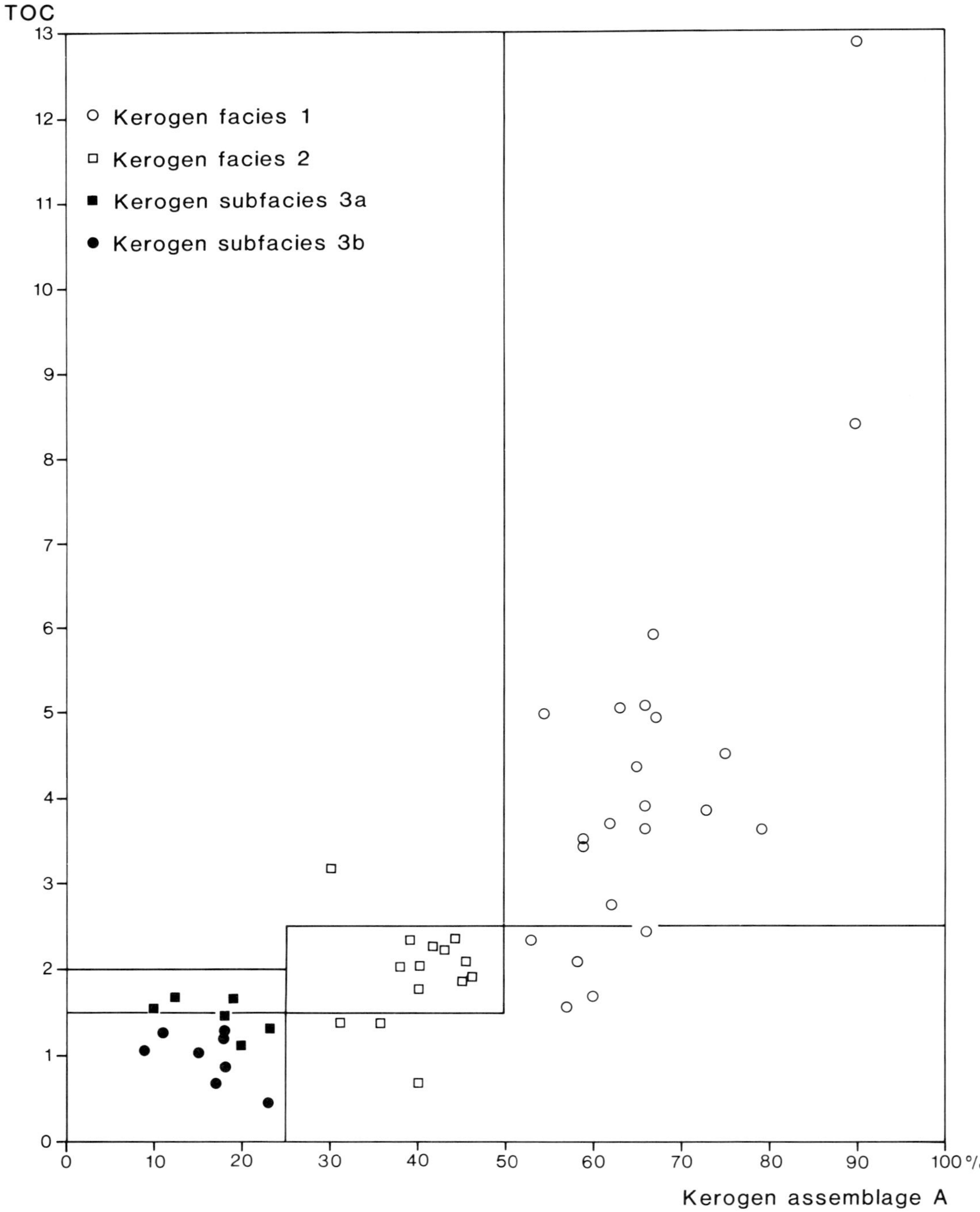

Fig. 6 — Kerogen assemblage A (amorphous kerogen) content related to TOC values and kerogen facies (see Fig. 4).

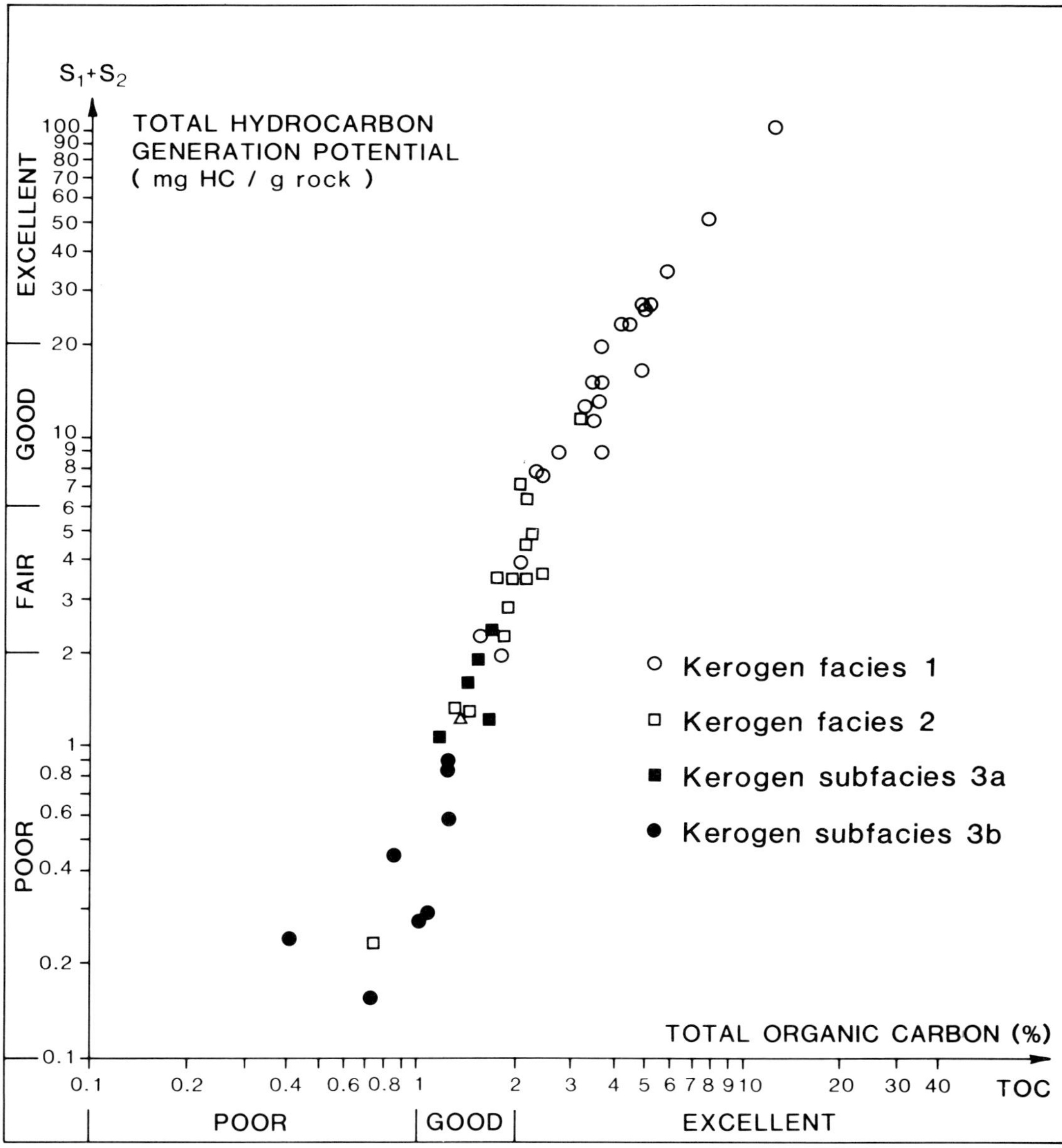

Fig. 7 — Total hydrocarbon generation potential according to TOC and kerogen facies (see Fig. 4).

profile. Increasing values of the TOC are followed by decreasing values for the kerogen assemblages B and C, and an increase in the kerogen assemblages B and C is followed by a decrease in TOC values. The kerogen assemblage A content, however, increases mainly in samples with more than 2.5 wt% TOC. From this it is supposed that kerogen assemblages B and C accumulated in conditions that were conducive to preservation throughout the interval but were

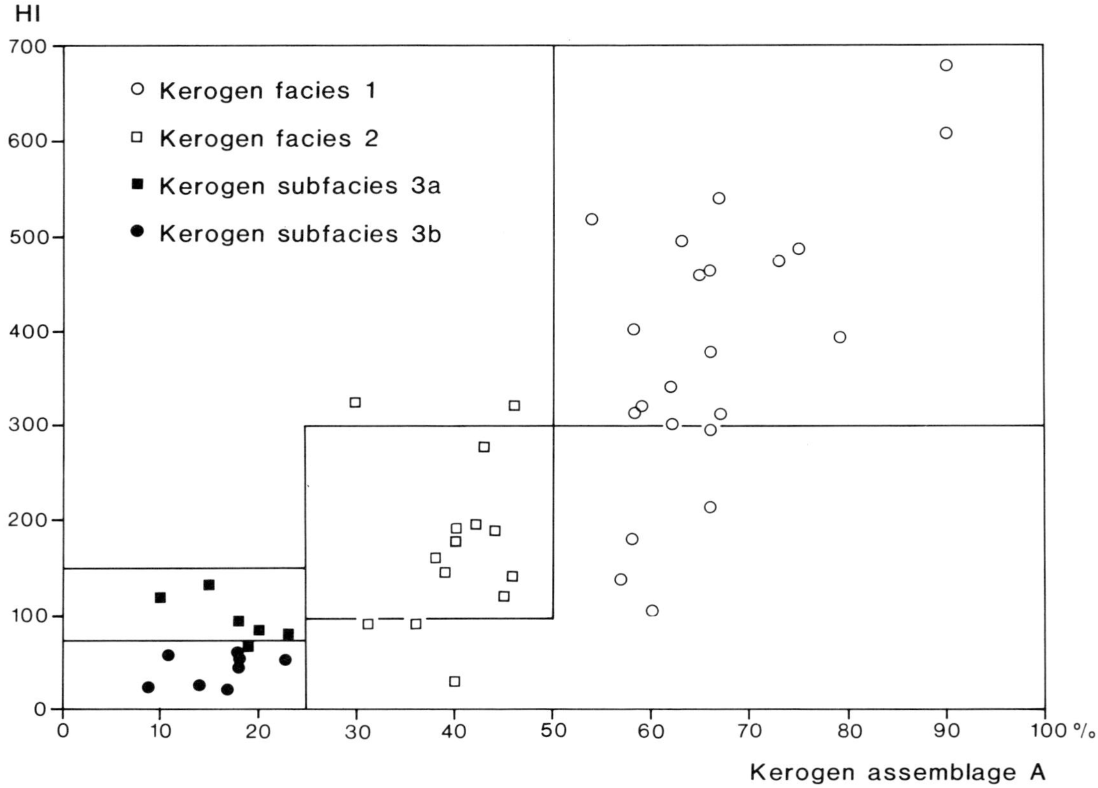

Fig. 8 — Kerogen assemblage A (amorphous kerogen) content related to HI and kerogen facies (see Fig. 4).

never abundant because of modest levels of supply, whereas preservation of kerogen assemblage A depended on periods with anoxic or near anoxic bottom conditions. The kerogen composition in the Kimmeridge Clay is therefore probably controlled by the oxidizing or reducing nature of the depositional environment and only to a lesser extent on variations in the organic input (see also chapter 6 by Tyson in this volume). These observations agree with the earlier models of Tyson *et al.* (1979), Morris (1979, 1980) and Ebukanson & Kinghorn (1985) which explained the variations in the organic content of the Kimmeridge Clay and other Jurassic mudrocks in terms of vertical fluctuations in the anoxic–oxic boundary with respect to the sediment water interface.

CONCLUSION

The correlation between visual and chemical kerogen analyses demonstrates that, if palynological data are dealt with in a systematic way, they can be very useful in predicting relative source potential and depositional condition in bituminous clay. This study of an immature Kimmeridge Clay sequence has shown that the relative amount of amorphous kerogen (kerogen assemblage A), expressed as kerogen facies 1–3, correlates in detail with the fluctuation of the chemical parameters TOC and S_2 and with it also HI and OI.

Nearly all the samples dominated by kerogen assemblage A (more than 50%, kerogen facies 1) contain more than 2.5 wt% TOC and have a good to excellent hydrocarbon generation potential. Kerogen facies 2 (more than 25% and less then 50%, kerogen assemblage A) is dominated by samples with 1.5–2.5% TOC and has a fair hydrocarbon generation potential. Kerogen facies 3 (less than 25%, kerogen assemblage A) has been divided into two subfacies of which subfacies 3a (dominated by spores, pollen and dinocysts, kerogen assemblage B) generally has TOC values between 1.5 wt% and 2.0 wt% and a slightly higher hydrocarbon generation potential than subfacies 3b (dominated by woody material and coal, kerogen assemblage C) with TOC values below 1.5 wt% and very poor potential.

The distribution of the three kerogen facies is explained by vertical fluctuation in the anoxic–oxic boundary at the sediment–water interface.

ACKNOWLEDGMENTS

I am grateful to Messrs M. A. Easton and E. D. Knights, General Works Managers, and their colleagues at the Blue Circle Portland Cement Company's Westbury Works for allowing me to visit their pit and for their hospitality during two field visits. I also thank F. G. Christiansen, S. Piasecki and the two referees for discussions and suggestions on how to improve the manuscript. J. S. Peel improved the English, J. Lautrup prepared the photographs, B. Sikker Hansen and J. Halskov produced the figures, and M. Larsen and B. Larsen typed the manuscript. The Geological Institute, University of Copenhagen, is thanked for fieldwork grants.

Plate 1 — The scale bars represent 100 μm

Fig. 1 — Sample E 5 100, kerogen facies 1, dominated by kerogen assemblage A (more than 50% amorphous kerogen).

Fig. 2 — Sample M 21 230, kerogen facies 2 (more than 25% and less than 50% kerogen assemblage A), representing the intermediate group in which none of the three kerogen assemblages A, B and C dominates.

Fig. 3 — Sample E 3 20, kerogen subfacies 3a (less than 25% kerogen assemblage A and more than 40% of kerogen assemblage B) (spores, pollen, dinocysts, cuticles).

Fig. 4 — Sample M 21 155, kerogen subfacies 3b (less than 25% kerogen assemblage A and more than 40% of kerogen assemblage C) (woody particles and coal).

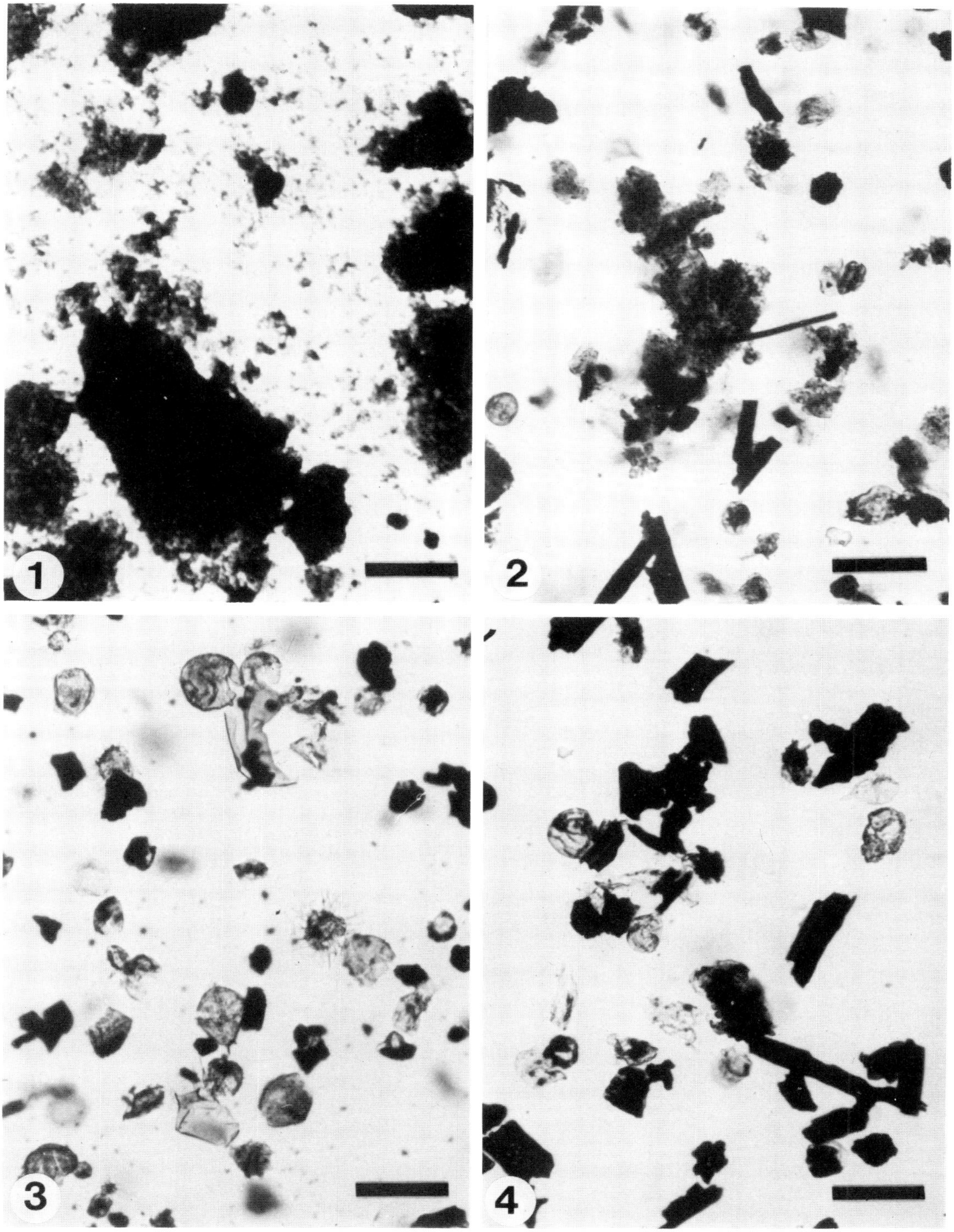
1
2
3
4

REFERENCES

Batten, D. J. (1981). Palynofacies, organic maturation and source potential for petroleum. In: Brooks, J. (ed.), *Organic maturation studies and fossil fuel exploration.* Academic Press, London, pp. 201–223.

Batten, D. J. (1983). Identification of amorphous sedimentary organic matter by transmitted light microscopy. In: Brooks, J. (ed.), *Petroleum geochemistry and exploration of Europe. Special Publication, Geology Society, London* **12** 275–287.

Birkelund, T., Callomon, J. H., Clausen, C. K., Nøhr-Hansen, H. & Salinas, I. (1983). The Lower Kimmeridge Clay at Westbury, Wiltshire, England. *Proceedings of the Geologists' Association, London* **94** 289–309.

Brooks, J. (1981). Organic maturation of sedimentary organic matter and petroleum exploration: a review. In Brooks, J. (ed.), *Organic maturation studies and fossil fuel exploration.* Academic Press, London, pp. 1–37.

Brooks, J., Cornford, C. & Archer, R. (1987). The role of hydrocarbon source rocks in petroleum exploration. In: Brooks, J. & Fleet, A. J. (eds), *Marine petroleum source rocks. Special Publication, Geological Society, London* **26** 17–46.

Burgess, J. D. (1974). Microscopic examination of kerogen (dispersed organic matter) in petroleum exploration. *Geological Society of America, Special Paper* **153** 19–30.

Combaz, A. (1980). Les kérogènes vus au microscope. In: Durand, B. (ed.), *Kerogen: insoluble organic matter from sedimentary rocks.* Éditions Technip, Paris, pp. 55–111.

Cox, B. M. & Gallois, W. (1981). The stratigraphy of the Kimmeridge Clay of the Dorset type area and its correlation with some other Kimmeridgian sequences. *Report Institute of Geological Sciences, London* **80/4**, 1–44.

Ebukanson, E. J. & Kinghorn, R. R. F. (1985). Kerogen facies in the major Jurassic mudrock formations of southern England and the implication on the depositional environments on their precursors. *Journal of Petroleum Geology* **8** 435–462.

Espitalié, J., Laporte, J. L., Madec, M., Marquis, F., Leplat, P., Paulet, J. & Boutefeu, A. (1977a). Méthode rapide de caractérisation des roches mères de leur potentiel pétrolier et de leur degré d'évolution. *Revue de l'Institut Français du Pétrole* **32** 23–42.

Espitalié, J., Madec, M. & Tissot, B. (1977b). Source rock characterization method for petroleum exploration. *Proceedings of the 9th Annual Offshore Technology Conference, Houston, 1977*, pp. 439–444.

Morris, K. A. (1979). A classification of Jurassic marine shale sequences: an example from the Toarcian (Lower Jurassic) of Great Britain. *Palaeogeography, Palaeoclimatology, Palaeoecology* **26** 117–126.

Morris, K. A. (1980). Comparison of major sequences of organic-rich mud deposition in the British Jurassic. *Journal of the Geological Society, London* **137** 157–170.

Mukhopadhyay, P. K., Hagemann, H. W. & Gormly, J. R. (1985). Characterization of kerogens as seen under the aspect of maturation and hydrocarbon generation. *Erdöl und Kohle–Erdgas–Petrochemie vereinigt mit Brennstoff-Chemie* **38** 7–18.

Nøhr-Hansen, H. (1986).Dinocyst stratigraphy of the Lower Kimmeridge Clay, Westbury, England. *Bulletin of the Geological Society of Denmark* **35** 31–51.

Peters, K. E. (1986). Guidelines for evaluation of petroleum source rocks using programmed pyrolysis. *American Association of Petroleum Geologists Bulletin* **70** 318–329.

Staplin, F. L. (1969). Sedimentary organic matter, organic metamorphism, and oil and gas occurrence. *Bulletin of Canadian Petroleum Geology* **17** 47–66.

Tyson, R. V. (1984). Palynofacies investigation of Callovian (Middle Jurassic) sediments from DSDP Site 534, Blake-Bahama Basin, western central Atlantic. *Marine and Petroleum Geology* **1** 3–13.

Tyson, R. V. (1987). The genesis and palynofacies characteristics of marine petroleum source rocks. In: Brooks, J. & Fleet, A. J. (eds), *Marine petroleum source rocks. Special Publication, Geological Society, London* **26** 47–67.

Tyson, R. V., Wilson, R. C. L. & Downie, C. (1979). A stratified water column environmental model for the type Kimmeridge Clay. *Nature* **277** 377–380.

Venkatachala, B. S. (1981). Differentiation of amorphous organic matter types in sediments. In: Brooks, J. (ed.), *Organic maturation studies and fossil fuel exploration.* Academic Press, London, pp. 177–200.

6

Late Jurassic palynofacies trends, Piper and Kimmeridge Clay Formations, UK onshore and northern North Sea

R. V. Tyson

ABSTRACT

Quantitative analyses of kerogen and palynomorph variations have been made on core and outcrop material of a variety of Late Jurassic facies from the Piper and Kimmeridge Clay Formations. The most important controls on the nature of the kerogen assemblage are identified as the grain size of the sediment, proximity of the depositional site to fluvio-deltaic sources of terrestrial organic matter, and the prevalent redox conditions. Percentage particle abundances of amorphous, phytoclast and exinitic organic matter can be used to plot ternary amorphous–phytoclast–exinitic diagrams; these allow recognition of at least 10 distinct shale palynofacies. Combination of percentage particle abundances with total organic carbon data are used to examine stratigraphic and palaeogeographic trends in the distribution of terrestrial and marine organic matter. Cross-plots of phytoclast categories permit the discrimination of true basinal facies from shelf and proximal basin (slope?) facies; more distal facies are characterized by a relative enrichment in the percentage particle abundance of granular 'black wood' particles (regardless of redox conditions). Quantitative analysis of the palynomorph components shows that distal facies are comparatively enriched in bisaccate pollen and chorate dinocysts. Distal anoxic shale facies are characterized by a low dinocyst content and relative enrichment in prasinophycean algae.

INTRODUCTION

This paper presents an analysis of palynofacies trends in Late Jurassic sediment sequences from the UK onshore (Dorset, Sutherland) and offshore (northern North Sea). The conclusions are based on the results of an extensive integrated palynofacies and sedimentological study which involved the logging of over 431 m of onshore section and 584 m of core, quantitative analysis of 512 palynological preparations, 215 total organic carbon (TOC) determinations and 100 pyrolysis assays. All geochemical analyses were provided by Robertson Research. The samples examined are immature to marginally mature with respect to their organic maturation. Full sample details, statistical, stratigraphic, and sedimentological data can be found in the work by Tyson (1985).

The objectives of the study were to investigate the interrelationship of:

(1) palynofacies to both lithology and lithofacies,
(2) palynofacies and geochemical parameters, and
(3) total kerogen and palynomorph parameters.

The Late Jurassic sequences examined encompass a diversity of depositional environments ranging from shallow to deep, aerobic to anoxic, distal to highly proximal, and cyclic to uniform (Table 1; see also Figs. 1, 4 and 6). Palynofacies variations are examined both within and between the individual sections. Stratigraphic trends are described for the lower Kimmeridgian of the Brora–Helmsdale Outlier and the late Oxfordian to mid-Volgian of well 16/17-4 from the 'T Block' area of the northern North Sea.

METHODOLOGY

All palynological processing was carried out by Robertson Research using standard preparation techniques similar to those described by Barss & Williams (1973). Quantitative palynological analysis of the overall kerogen composition was carried out on sieved unoxidized material.

Kerogen counts were made by making a series of three or more traverses across the slide. The standard number of counts per slide was 500. The 10 descriptive categories used for counting are described in Table 2; in the text the Wu + Wd and I categories are informally referred to as 'brown wood' and 'black wood', respectively. The slides were also briefly examined under blue-light fluorescence in order to provide a qualitative assessment of the character of the amorphous organic matter (AOM) and the condition of the exinitic components. Those slides that had at least 10% palynomorphs were re-examined in order to quantify the character of the palynomorph assemblage.

The data generated by the kerogen counting procedure are relative percentage particle abundances (%PAs). These should not be equated with percentage volumes of different kinds of

Table 1 — Summary of location, age and generalized facies characteristics of the Late Jurassic sediments examined in this study. For consistency with North Sea sections the table uses Boreal stratigraphic terminology (short 'Kimmeridgian' ending in *autissiodorensis* Zone, which is equivalent to the lower Kimmeridgian of the text).

Section	Well 15/17-4-7	Type Kimmeridge Clay Formation	Brora	Well 16/17-4, 16/17-1	16/29-2	Kimmeridge Clay Formation
Location	Piper	Dorset	Sutherland	'T Block'	Maureen	Piper
Age	Upper Oxfordian	Lower Volgian	Kimmeridgian	Upper Oxf–mid Volgian	Volgian?	Volgian
Character						
Proximal	?		×	×		
Distal		×			×	×
Shallow shelf	×					
Deep shelf		×	×	?		
Basinal			×	×	×	×
Aerobic	×	×				
Dysaerobic		×	×			
Anoxic		×	×	×	×	×
Hemipelagic		×	×	×	×	×
Turbiditic			×	×		
Cyclic		×				

Table 2 — Characteristics of informal kerogen categories utilized in this study. Equivalent coal macerals are approximate only (see Stach *et al.* 1982).

Symbol	*Colour*	*Description*	*?Equivalent maceral*
Wu	Orange–brown	Elongate; angular; sharp outline; may be splintered; parallel-sided or fusiform thickenings (banding); ± bordered or scalariform pitting. Mostly tracheid (xylem) tissue of higher plants	Vitrinite or semifusinite
Wd	Orange–brown	Elongate to equidimensional; rounded; irregular; sharp to more diffuse outline; usally lacking regular internal structure. Biodegraded macrophyte tissues	Vitrinite or semifusinite
I	Black or opaque	Equidimensional > elongate; sharp outline; usually no internal structure but more elongate particles may show pits; generally smaller than Wu or Wd particles. Charcoal or oxidized woody tissue	Inertinite: semifusinite, fusinite, inertodetrinite
C	Pale yellow	Irregular, thin, sheetlike; sharp outline; translucent; cellular; one cell layer thick. Waxy epidermal (cuticle) tissue of higher plants. (Usually < 1%PA in analysed samples)	Cutinite
Tp	Yellow, orange or light brown	Palynomorphs, subdivided into: miospores (M) acritarchs, dinocysts and prasinophycean algae (P), undifferentiated sphaeromorphs (U)	Exinite: sporinite, alginite
E	Yellow or orange	Irregular, thin particles; sharp outline; translucent; non-cellular; sheet like to membraneous	Exinite
AOM	Yellow, orange, brown or grey	Irregular particles; rounded to angular; gradational margins; heterogeneous; speckled to clotted; varied common micro-inclusions	Bituminite, collinite?[a] liptodetrinite, alginite B, amorphinite
F	Dark > light brown	Planispiral or trochospiral > serial and biserial. No internal structure; pyrite framboids common. Tectin "chitinous" foraminiferal linings	None

[a]Unless specified, all text references to AOM concern non-collinitic material.

organic matter. The different kerogen particle types can have inherent differences in size and volume depending on their biological source, mechanical strength and degree of fragmentation. Owing to their varying physical properties, sorting according to hydrodynamic equivalence often results in assemblages where particle size varies considerably with particle type. Determination of the relative volumes of kerogen components is clearly essential for the meaningful comparison of optical and geochemical data. However, the influence of hydrodynamic equivalence on the composition of natural kerogen assemblages may mean that %PA values are more meaningful for facies studies. While AOM and phytoclast particles may be roughly equal in volumetric terms, palynomorphs are generally both smaller and hollow. The %PA (and percentage area) data consequently tend to exaggerate the apparent abundance of the exinitic components.

For meaningful interpretation of the data it is vital that percentages are not confused with absolute amounts. A high percentage of woody material can reflect either a high absolute content of this material in the sediment or an oxidizing situation where only the more refractory woody materials have survived. It is therefore necessary to have some form of calibration in order to determine variations in the gross amount of kerogen present. Weight percentage TOC values are the most convenient means of doing this, even though the TOC content and the total kerogen content are not, by definition, the same.

Just as the different sizes of the particles have not been corrected for, neither have their different densities. Particles of the same volume can have different masses of material. Furthermore, different kerogen types contain different proportions of carbon; compared with other kerogen types, lower volumes of carbonaceous 'black wood' are required to give any particular TOC value. Weight percentage TOC is also a relative concentration term expressing the ratio of organic carbon to mineralic matter; it is therefore strongly affected by clastic and carbonate dilution effects, organic matter supply and preservation.

I have found it useful to depict variations in the %PA composition of kerogen assemblages graphically by plotting ternary amorphous–phytoclast–exinitic (A–P–E) organic matter diagrams. Ternary plots of the main phytoclast components are also given. Absolute abundance trends are illustrated by means of a combination of TOC values with %PA values. These data are used to derive two parameters: PhytOC (TOC × %PA of phytoclast materials), and AmexOC (TOC × %PA of fluorescent AOM and exinitic components). The term 'amex' was first coined by Senftle (Larter 1985, Senftle *et al.* 1987) to denote the areal percentage of the liptinite (or exinite) and fluoramorphinite (fluorescent AOM) macerals. General aspects of palynofacies were reviewed by Pocock (1982); the palynofacies of organic-rich sediments were discussed by Tyson (1987).

INDIVIDUAL SECTION DESCRIPTIONS

The facies characteristics of each of the Late Jurassic sections are briefly described to provide the sedimentological context for the palynofacies observations. This is followed by a summary of the specific palynofacies features of each sequence.

Piper Formation, Piper Oil Field (Block 15/17, northern North Sea)

General Description

Background information on the Late Jurassic of the Piper area can be found in Williams *et al.* (1975), Maher (1980, 1981), Turner *et al.* (1984), Harker *et al.* (1987) and Boote & Gustav (1987). This study utilized four wells from the Piper area (Fig. 1). Palynofacies investigations were based on comparatively good core coverage from the I Shale and G and H Sandstone divisions of Maher. The I Shale is late Oxfordian in age, and the overlying part of the formation is intra-late Oxfordian–early Kimmeridgian (L. A. Riley, pers. comm.). The transgressive 'I Shale' forms the base of the redefined Piper Formation (Harker *et al.* 1987, Boote & Gustav 1987).

The I Shale ranges from 2 m to 9 m in thickness and varies from a sandy and silty mudstone in the two eastern wells to a hard siltstone in the two western wells. It has a highly bioturbated fabric and, apart from occasional indications of indistinct ripple bedding, all traces of sedimentary structures have been destroyed. In well 15/17-4 the unit is developed in the mudstone facies and contains common to abundant bioclastic debris (including *Trautscholdia* and *Dentalium*). Rare ammonite fragments were also recorded. Macrofossils and bioclastic debris

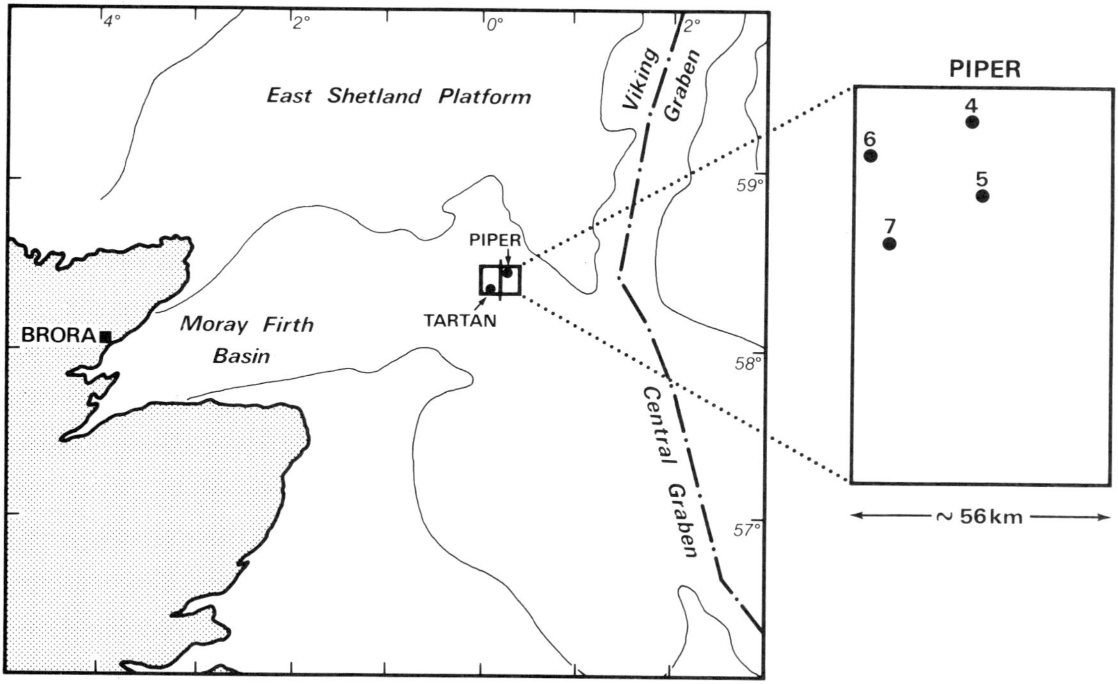

Fig. 1 — Location of the Piper Oilfield (Block 15/17) and the four Piper wells discussed in the text.

were not observed in the other wells. Occasional *Chondrites* and *Teichichnus* burrows are present in the siltstone facies.

I interpret the upward-coarsening sequence of the I Shale to lower F Sandstone as a shelf mud–lower submarine bar–upper submarine bar sequence. This is in broad agreement with Maher (1980) who interpreted the H and G Sandstones as 'classical offshore bars', but the lower F Sandstone as an 'upper shoreface or beach deposit'. The I Shale represents deposition under aerobic conditions in the upper part of the lower mixed layer (i.e. between fairweather and storm wave base, but closer to the former than the latter).

No cored material from the Kimmeridge Clay Formation was available at the time of the study, but Robertson Research provided access to six oxidized palynological preparations from an undisclosed well in this area. The laminated dark-grey to black shales of the Kimmeridge Clay Formation were deposited following a hiatus and period of reworking at the end of the deposition of the Piper Formation (Maher 1980). They are Volgian–Ryazanian in age (see also Turner *et al.* 1984, Harker *et al.* 1987).

Palynofacies Characteristics

The palynofacies characteristics of the I Shale, the lower, more argillaceous and carbonaceous parts of the H–G Sandstone, and the local Kimmeridge Clay Formation are summarized in Table 3. The Piper Formation samples are characterized by a dominance of the phytoclast components, low contents of AOM and comparatively high amounts of exinitic debris and palynomorphs. The range of kerogen compositions is shown in Fig. 2 (as noted previously, the apparent volumetric significance of the exinitic debris is exaggerated by the %PA data). Pyrolysis analyses reveal a Type IV kerogen composition that indicates the phytoclast material has suffered extensive oxidation; much of the 'brown wood' is comparatively dark and

Table 3 — Palynofacies summary for I Shale and H–G Sandstone units of the Piper Formation (mean kerogen and palynomorph characteristics). Includes palynomorph characteristics of the local Kimmeridge Clay Formation. See Table 2 for explanation of categories; AOM is for the non-collinitic variety only; *n* is the sample size; Tp is the total of palynomorphs.

	Mean %PA Kerogen composition									
	n	Wu	Wd	I	Tp	AOM	E	Wu:Wd	TOC	*n*
I Shale	13	13	42	7	10	9	17	0.37	2.6	11
H + G Sandstone	21	10	49	8	7	4	14	0.21	0.8	3
All Piper	53	12	45	8	8	7	15		2.6	26

	Mean palynomorph composition		
	I Shale	*H + G Sandstone*	*Kimmeridge Clay Formation*
% chorates (of dinocysts)	47	27	63
% prasinophyceae (of plankton)	—	—	21
% bisaccates (of spores + bisaccates)	20	36	84
% 'heavy' spores[a] (of all spores)	71	70	40
Number of samples counted	9	8	6

[a]'Heavy' spores are comparatively thick-walled ornamented spore morphotypes.

may correspond to semifusinite rather than to vitrinite. Some of the carbonaceous material may represent reworked Middle Jurassic 'coal' from the underlying Sgiath Formation.

Lithological contrasts between the mudstone–siltstone and sandstone samples are not especially apparent from the %PA kerogen data. The principal difference is seen in the absolute values. Available data suggest that the mudstone–siltstone generally has a mean TOC content two to three times greater than the sandstone. The eastern mudstone facies of the I Shale also has about twice as much TOC (average, 4.0 wt%) as the western siltstone facies (although again without any apparent consistent difference in the nature of the gross kerogen assemblage). For both the I Shale and the H–G Sandstone, the %PA of phytoclast material is significantly higher (by 20–50%) in well 15/17-6. The I Shale in this well also shows TOC values 40 wt% higher than those in well 15/17-7 (both in siltstone facies). These observations together suggest that at least some of the phytoclast material may enter the area from the north east. The Wu:Wd ratios (see Table 2) of I Shale samples are all greater than 0.35 (equal to or less than 0.52), except in well 15/17-6 where values are usually less than 0.20.

The most striking contrasts between the I Shale and H–G Sandstone are seen in the palynomorph data (Table 3). While lithology influences only the absolute abundance of the major kerogen categories (and in a rather non-discriminatory fashion), palynomorphs appear far more sensitive to differential sorting. One of the most marked differences is that seen in the ratio of chorate (long-spined) to proximate and proximo-chorate (non- or short-spined) dinocysts. This is generally less than 1 in the sandstones and equal to or more than 1 in the I Shale. Chorate cysts are usually observed to be preferentially concentrated in distal low-energy shelf environments (Tyson 1987). Their proportion is here higher in the eastern wells in both

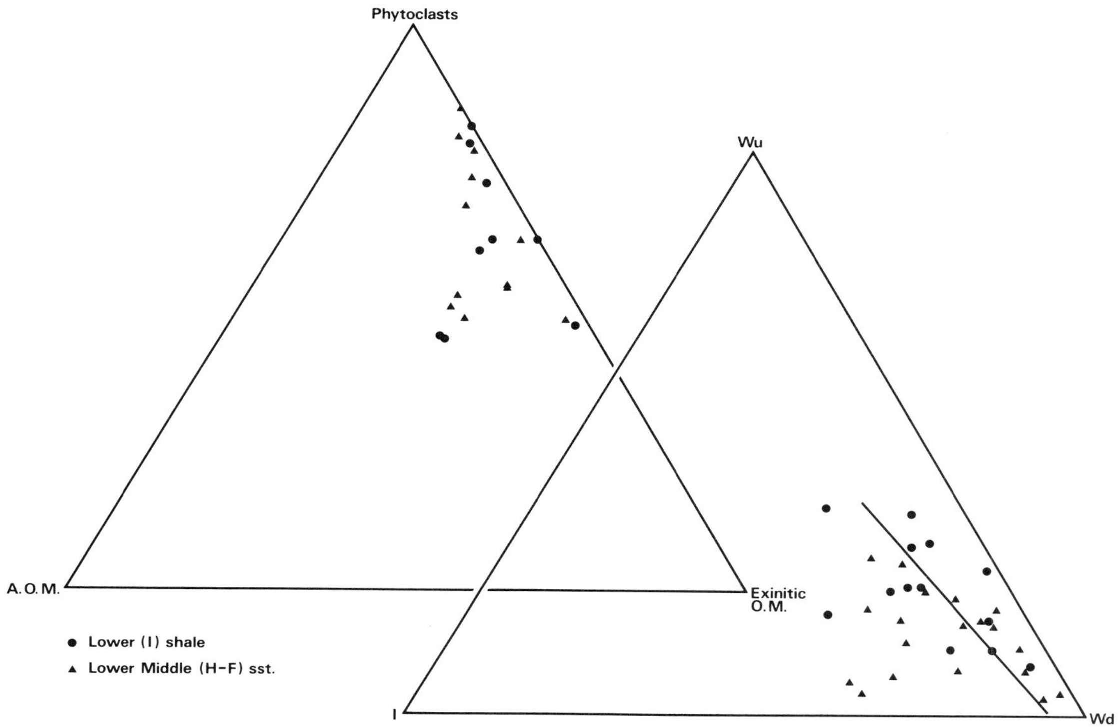

Fig. 2 — A–P–E diagram and ternary phytoclast plot for kerogen assemblages from the I Shale and H–G Sandstone units of the Piper Formation (subdivisions of Maher (1980)). Ternary phytoclast plot shows regression line A derived from Fig. 15.

the shale and the sandstone units; this trend and grain size variations both support an eastward decrease in the energy of the depositional environment.

The percentage of total plankton is highest in the argillaceous lithologies, except in well 15/17-6 where the plankton is apparently diluted by a higher input of terrestrial material (as noted above). Most palynofacies studies indicate that bisaccate pollen are relatively more abundant in distal low-energy environments (Tyson 1987). On a well-by-well basis, the ratio of bisaccate pollen to spores is here surprisingly two to three times higher in the H–G Sandstones than in the I Shale. However, the consistent dominance of spores over bisaccate pollen, and 'heavier' over 'lighter' spore morphotypes suggests an imbalance in supply, or insufficient distance or duration of transport for significant sorting to have occurred. The Piper Formation was deposited on a comparatively extensive marine shelf; true marginal facies have not been identified, although deltaic influences are thought to exist in the adjacent Tartan area (Turner *et al.* 1984). Boote & Gustav (1987) have recently reinterpreted the Piper sands as the products of shoreface accretion in a wave-dominated delta complex.

Kerogen preparations of the Kimmeridge Clay Formation are dominated by AOM, the proportion of phytoclast material having declined considerably from that in the Piper Formation. This major change in palynofacies is also reflected in the palynomorph composition (Table 3). The bisaccate pollen are far more abundant than spores, and the percentage of 'heavier' spore morphotypes is signficantly lower than in the Piper Formation. The overall percentage of plankton is similar to the mean for the Piper samples, but it includes about 20% prasinophycean algae (virtually absent in the Piper Formation), and a dominance of chorate dinocysts in the plankton. These observations indicate a marked distal shift in facies and the establishment of anoxic (or at least dysaerobic) bottom water conditions. No TOC or other geochemical data were available.

Wheatleyensis to *pectinatus* Zones, type Kimmeridge Clay, Dorset

General Description

Background information on the Kimmeridge Clay of the Dorset type section can be found in the papers by Tyson *et al.* (1979), Cox & Gallois (1981), Farrimond *et al.* (1984), Ebukanson & Kinghorn (1985), Williams (1986) and Tyson (1985). The late *wheatleyensis* to early *pectinatus* Zones of the type Kimmeridge Clay are distinguished by the frequency of highly organic-rich interbeds (including true oil shales with more than 25 wt% TOC), organic-poor dolostone beds and distinctive coccolith-rich lithologies. The 75 m interval characterized by this lithological association is a cyclic sequence consisting of 70% mudstones, 12% kerogenous shales, 8% kerogenous coccolithic laminites, 4.5% dolostones, 3% oil shales and 2.5% coccolith limestones. Oil shale interbeds in the lower *hudlestoni* Zone contain the highest TOC values for the Kimmeridge Clay (up to 57 wt%), but they are mostly less than 17 wt%. Modal values in the dominant mudstone facies lie between 2 wt% and 8 wt% (unpublished data compilation). In the lower *pectinatus* Zone the dominant TOC values are between 2 wt% and 6 wt%; peak values are generally lower than in the *hudlestoni* Zone (25 wt% or less) owing to dilution by coccoliths.

This interval of the type Kimmeridge Clay was deposited in the central part of a subsiding shelf basin at water depths close to the extreme limit of storm influence. Palaeogeographic reconstruction is limited by the rarity of definitive marginal facies, but the nearest landmass was probably at least 100 km distant. The shelf watermass appears to have varied between two states, one having a seasonal stratification cycle like that of modern shelf seas, and the other being characterized by permanent to semi-permanent stratification (marine meromixis). These states alternated in an apparent response to astronomically influenced climatic forcing, resulting in cycles with an estimated duration of about 40 000 years (Dunn 1974, House 1985). Highly organic-rich kerogenous shales, oil shales, and laminated (variably organic-rich) coccolithic marls and limestones were deposited under dysaerobic to anoxic conditions during the meromictic episodes.

Less organic-rich, light- to dark-coloured massive to shaley mudstones were deposited during the 'aerobic' part of the cycles. Many of the darker mudstones have impoverished faunas and high levels of TOC (up to 10 wt%). These were probably deposited under fluctuating conditions and include organic matter that was deposited during short-lived oxygen-deficient episodes and only partially modified or consumed during subsequent bioturbation. Laminated sedimentary fabrics are apparently more sensitive to modification during transient improvement in oxygenation than are most geochemical characteristics of the sediment (Tyson 1987). The sediment geochemistry is therefore often 'out of equilibrium' with the sedimentary fabric.

Palynofacies Characteristics

Only 60 palynofacies samples were examined in detail, and many of these were from the atypical oil shale facies. What is striking about all the Dorset samples analysed is the very low content of phytoclasts (%PA less than 5% in the oil shales and rarely up to 17% in the mudstones (Fig. 3)).This observation is entirely consistent with those previously made by Tyson *et al.* (1979), Williams & Douglas (1980, 1983), Farrimond *et al.* (1984), Ebukanson & Kinghorn (1985) and Williams (1986). As noted earlier by Williams & Douglas (1983), observed variations in TOC content do not correspond to any marked

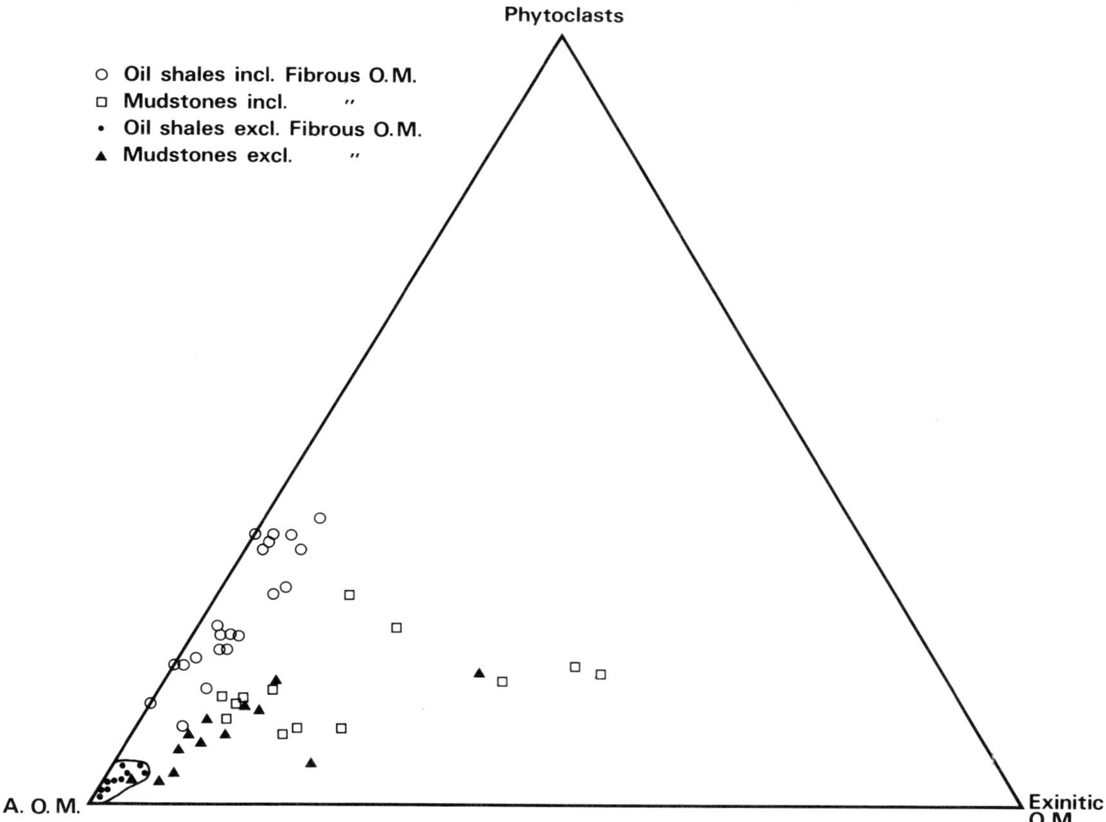

Fig. 3 — A–P–E diagram for oil shales and organic-rich mudstones from the upper *hudlestoni* and lower *pectinatus* Zones of the Upper Kimmeridge Clay, Dorset. The data are shown with and without the problematic pseudo-amorphous fibrous organic matter (Type 3 particles) included in the phytoclast category; see also Nøhr-Hansen (chapter 5 in this volume).

changes in the nature of the kerogen assemblage. Terrestrial inputs are consistently low, and TOC is almost entirely controlled by AOM content.

Pyrolysis data show that the oil shales have a Type II to Type I composition. Optical examination reveals a kerogen assemblage that appears to consist almost entirely of AOM. However, detailed study shows that the 'amorphous' material is made up of four distinct particle types. In addition to 'normal' AOM, there are three other varieties. Variety 1 particles are platey, brown, homogeneous and pseudo-amorphous with angular to irregular (but distinct) margins, common coccolith imprints, with no internal structure, and occasional tears. Variety 2 particles are platey or flake-like, thin, yellowish and 'lamellar' (sometimes almost membraneous in appearance) without an internal structure (possibly a thinner version of variety 1?). Variety 3 particles are brown and pseudo-amorphous with a tendency for elongated shapes and a distinct length-parallel fibrous internal structure.

These all exhibit strong fluorescence under blue-light excitation; the fibrous structure of variety 3 particles is even more apparent when observed under fluorescence. These particle types are particularly characteristic of the oil shales and rich kerogenous shales (%PA average 90), while normal AOM usually dominates in the other lithologies (%PA 35–75). Variety 1 particles are conspicuously richer in the Best Blackstone ('Kimmeridge Coal') oil shale horizon where TOC values are between 40 wt% and 52 wt%. These particle types were not conspicuous in any of the other Late Jurassic sediments studied.

Williams (1986) concludes that the oil shale facies consists dominantly of lamellar alginite. This clearly implies a correspondence between lamellar alginite and my particle varieties 1 and 2. Dominance of 'lamalginite' does not appear to be typical for most marine oil shales (Hutton 1987). The variety 3 particles may represent bacterially degraded debris of poorly lignified marine macrophytes, a source which may explain the high iodine contents that have been recorded from the oil shales (Cosgrove 1970). None of the 'amorphous' material represents terrestrially-derived doppleritic gels as has been suggested by Ioannides *et al.* (1976).

Terrestrial phytoclasts are rare (5% or less) in the oil shales. 'Black wood' particles are typically two to seven times more numerous than 'brown wood'. This agrees with observations by Ebukanson & Kinghorn (1985) and Williams (1986) who both record inertinite as being more abundant than vitrinite. The relative percentage of 'black wood' appears to be more variable in the normal mudstones, either approximately equal to or two to three times less than the 'brown wood'. This difference between the two facies must result from differential transport or supply rather than preservational factors. The mudstone kerogens are dominated by normal AOM although, in the lighter-coloured mudstones and as TOC decreases, it becomes progressively more degraded. It is increasingly more dull and heterogeneous under fluorescence, more greyish-brown in colour, matt rather than lustrous, less cohesive, of lower relief and less resistant to palynological oxidation treatment and sieving. Such variations result in the mudstone kerogens ranging in composition from Type II to Type III.

Palynomorphs are rare in the oil shales (even after oxidation treatment to remove some of the resistant 'amorphous' components). The relative abundance of sporomorphs in the Blackstone oil shale reported by Ioannides *et al.* (1976) is probably a reflection of a very low dinocyst content. Palynomorphs generally represent about 7–9% (40% or less) of the kerogen assemblage in the mudstones. Sporomorphs (i.e. terrestrial palynomorphs) dominate the assemblage, with bisaccate pollen generally more abundant than spores.

The total microplankton content of the mudstones ranges up to 40% of the palynomorphs (Riley 1974). Proximate dinocysts are the dominant plankton observed in the mudstones, but chorate dinocysts appear to be the single most abundant dinocyst group in the laminated coccolith limestones. Prasinophycean algal phycomata were always found to represent less than 3% of the total palynomorphs but 0–16% (generally less than 10%) of the microplankton. Higher values seem to be most typical of the more organic-rich mudstones. Williams (1986) also notes the low abundance of tasmanitids, which are apparently significantly overestimated by Mebradu (1978).

Early Kimmeridgian, Brora–Helmsdale Outlier, northeast Scotland

General Description

The classic description of the Kimmeridgian to earliest Portlandian 'boulder beds' sequence is that given by Bailey & Weir (1932). Pickering (1984) has given the only recent published account to follow Surlyk's (1978) trend-setting study of the Late Jurassic fault scarp fan facies associations in east Greenland. The most detailed studies of the Brora–Helmsdale sequence are all unpublished (Linsley 1972, Tyson 1985, MacDonald 1985). The following description of the lower Kimmeridgian interval is based on the published accounts, supplemented by a summary of new observations on the Lothbeg outcrops (Fig. 4) (Tyson 1985).

With the exception of the Lothbeg area, the succession consists of shales with subordinate (25% or less) intercalated proximal turbiditic sandstones and clast-rich debris flows ('boulder beds'). The shales are typically dark grey to black, laminated, organic rich and without *in situ*

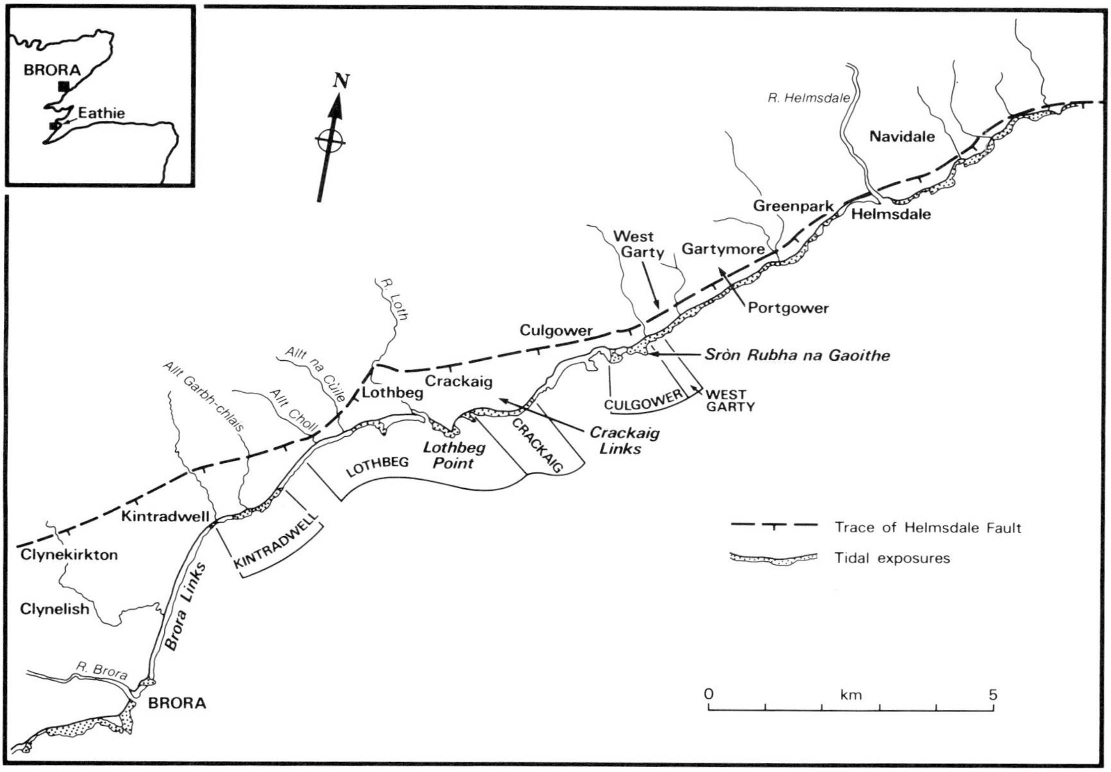

Fig. 4 — Location map of the Brora–Helmsdale Outlier showing the position of the outcrop areas discussed in the text.

benthos. They were deposited below wave base in a dysaerobic to anoxic basinal environment developed on the downthrow side of an active submarine fault scarp. A variety of proximal turbiditic and debris flow sediments occur within these shales at Kintradwell and northwards from Lothbeg Point.

Bioclastic debris (including corals and large brachiopods) is a major constituent of many of the sandstones from the *eudoxus* Zones onwards. Proximal high-density clast-rich turbidites and debris flow units containing Old Red Sandstone megaclasts are also most typical of the post-*mutabilis* interval. This partly reflects the progressive northward proximal facies shift as the outlier narrows and the outcrop moves increasingly closer to the line of the Helmsdale Fault. The 'boulder bed' facies dies out away from the fault and is absent from the coeval sequence in the adjacent Beatrice Field (Linsley *et al.* 1980). Only the Kintradwell and Lothbeg sequences warrant more detailed discussion in the present context.

The oldest Kimmeridgian sediments exposed in the Brora–Helmsdale Outlier are the so-called 'Kintradwell Boulder Beds'. Available ammonite data suggest a probable mid-*cymodoce* Zone age equivalent to the 'uralensis' subzone of Morris (1968) (see Linsley 1972, pp. 55, 77). The sequence consists predominantly of shale with less than 20% sandstone and only occasional 'boulder beds'.

The Kintradwell outcrops are followed by a gap in exposure but appear to be conformably overlain by the 'Allt na Cùile Sandstones' of the Lothbeg area. Detailed field work has shown that the 'Allt na Cùile Sandstone' of earlier workers actually consists of two distinct lithofacies (probably of slightly differing ages). In its type area in the western part of the Lothbeg region the Allt na Cùile Sandstone consists of low-angle current-bedded, fine- to very-fine-grained yellow friable sandstones with intercalated massive horizons, pebbly lenses and impressive intraformational breccias. Horizons of decalcified casts of shelly benthos are present, as are intervals with intense *Skolithos* ichnofacies bioturbation.

I favour an intermittently fault-scarp-modified submarine fan delta interpretation for these sandstones. They appear similar to the Raukelv Formation submarine fan delta sands of east Greenland (Surlyk *et al.* 1973, Surlyk 1975), although less extensive, finer grained and probably more distal. Available ammonite evidence is less than definitive but indicates a probable late *cymodoce* Zone age equivalent to the 'askepta' Subzone of Morris (1968) (see Linsley 1972, p. 55; see also Brookfield 1976, p. 184). The maximum age of the unit is uncertain but it is probably intra-*cymodoce.*

Continuous exposure of the Allt na Cùile Sandstone is lost to the east of its type area, but the available exposures indicate a rather abrupt, probably fault-controlled, change to more shaley facies with lenses of light-grey to pinkish massive structureless sandstones.

The eastern facies exposed in the vicinity of Lothbeg Point is distinctly different from that in the type area. It seems to represent a large, possibly composite body with an overall lenticular geometry. These sandstones are light grey, fine grained, moderately sorted and largely structureless. They have common siliceous fracture cementation ('veining'). Interesting sedimentological features are confined to the lateral and basal margins of the sandstones; these include carbonaceous laminae, thin carbonaceous shale lenses, microfaulting and very rare low-angle cross-bedding. Occasional casts of fine bioclastic material are sometimes present.

The bulk of the massive sandstones are believed to represent channelized liquified–grain flow deposition (although this is based primarily on the absence of other diagnostic features). They are very similar to some of the grain flow and high-density turbidity current deposits of the Hareelv Formation of east Greenland (Sykes 1975, p. 191, Surlyk *et al.* 1973, Surlyk 1987). The age of the eastern Allt na Cùile facies cannot be defined precisely; field relationships suggest that it is predominantly early *mutabilis* Zone (possibly latest *cymodoce* Zone at its base). The clear contrast in facies strongly supports a difference in age between the eastern and western Allt na Cùile Sandstone.

The revised interpretation is summarized below. Increased late *cymodoce* tectonism produced a regressive facies change that led to the distal parts of a prograding submarine fan delta extending out across the line of the Helmsdale Fault (western facies of the Allt na Cùile Sandstone). Episodic faulting periodically formed a fault scarp and led to the deposition of localized rock fall avalanche and debris flow breccia deposits. Shallowing (and/or greater current activity) occasionally permitted intense bioturbation in the late *cymodoce* sandstone and shale facies. The subsequent early *mutabilis* sea level rise led to persistent anaerobic bottom water conditions. Shallow water sands were then redeposited as channelized grain flows (eastern facies of the Allt na Cùile Sandstone). The scale of sand redeposition progressively declined through the *mutabilis* and *eudoxus* Zones, probably in response to further sea level rise and the progressive exhaustion of the loose sand supply on the upthrow side of the Helmsdale Fault.

Palynofacies Characteristics

Kerogen characteristics of the early Kimmeridgian sediments are summarized in Table 4. The shales are all organic-rich and have an average TOC content of 7.5 wt%. The highest value recorded was 17.5 wt%, but only 13% of the shale samples analysed gave a TOC content in excess of 10 wt%. Even the highly carbonaceous sands from the Lothbeg area contain only approximately half as much TOC as the average shale. Pyrolysis data indicate that the kerogen has a Type III–IV to Type III–II composition,

Table 4 — Kerogen summary for lower Kimmeridgian lithologies of the Brora–Helmsdale Outlier (includes TOC and hydrogen index values). See Table 2 for explanation of categories; *n* is the number of samples; T_w is the total phytoclasts; HI is the hydrogen index.

	Mean %PA Kerogen composition							
	n	T_w	I	Tp	AOM	Wu:Wd	TOC (*n*)	HI (*n*)
Shales	175	58	10	8	29	1:3–4	8 (62)	148 (24)
Argillaceous sst	8	65	12	10	18			
Argillaceous boulder beds	16	62	10	9	22			
Argillaceous clasts	5	69	15	8	19			
Carbonaceous sands	21	88	21	3	2	1:2	4 (7)	48 (2)
Kaolinitic clays	3	81	19	3	13		0.3 (4)	0 (3)

with hydrogen indices approximately three times higher in the shales than in the carbonaceous sandstones. The shales have a mean %PA composition containing about 70% phytoclast material; this is even higher in the carbonaceous sands, but the absolute phytoclast content is actually about 20% higher in the shales. Most of the phytoclast material is rather dark, but it has only been referred to the 'black wood' category when it is completely opaque. Some of the darker 'brown wood' material may equate with the coal maceral semifusinite rather than vitrinite.

The Wu:Wd ratio is usually two to four times higher in the sandstones than in the shales, indicating that the influx of sand is associated with the supply of fresh phytoclasts that are little modified after deposition. 'Black wood' values are up to 30% higher in the sandstones; this may reflect *in situ* post-depositional oxidation as well as supply factors.

Palynomorphs reach up to 10% of the kerogen assemblage in Kintradwell samples, but most of the shales show values between 3% and 8%. The palynomorph assemblage is strongly dominated by sporomorphs; the mean bisaccate pollen:spore ratio is about 1:2, and simple 'light' morphotypes dominate the spores. My observations on unoxidized preparations (likely to give an underestimate) indicate that microplankton constitutes less than 10% of the total palynomorphs. On the basis of oxidized palynostratigraphic preparations, but much fewer samples, both Lam & Porter (1977) and Barron (in chapter 8 of this volume) indicate values from 20–25% up to 30%. The ratio of chorate to other dinocysts is generally less than 1.0 Prasinophycean algae form less than 5% of the observed plankton in my data, but Lam & Porter (1977, p. 54) indicate they are 'common' from the *mutabilis* Zone onwards. Comparatively little contrast is seen in the palynomorph assemblages from different lithologies, but the total plankton content appears to be significantly higher in the argillaceous debris flows.

Stratigraphic Trends in Palynofacies

Stratigraphic variations are apparent in the geochemical and %PA kerogen data at the ammonite zone level, but individual sections do not usually show any obvious trends. This probably reflects the rather proximal facies setting with respect to terrestrial supply. Stratigraphic variations in the kerogen characteristics of the early Kimmeridgian sequence can be seen in Fig. 5 and Table 5. The data are given for each of the main outcrop areas studied (Fig. 4).

The A–P–E plot for the Kintradwell (*cymodoce*) sediments (Fig. 5a) shows a relatively high %PA of exinitic organic matter (which appears to be a fairly good indicator of shelf influence). The mean %PA for palynomorphs is at least twice as much as that in the younger sediments (Table 5), and the microplankton also shows maximum values. Section A2, 37 m of shales from

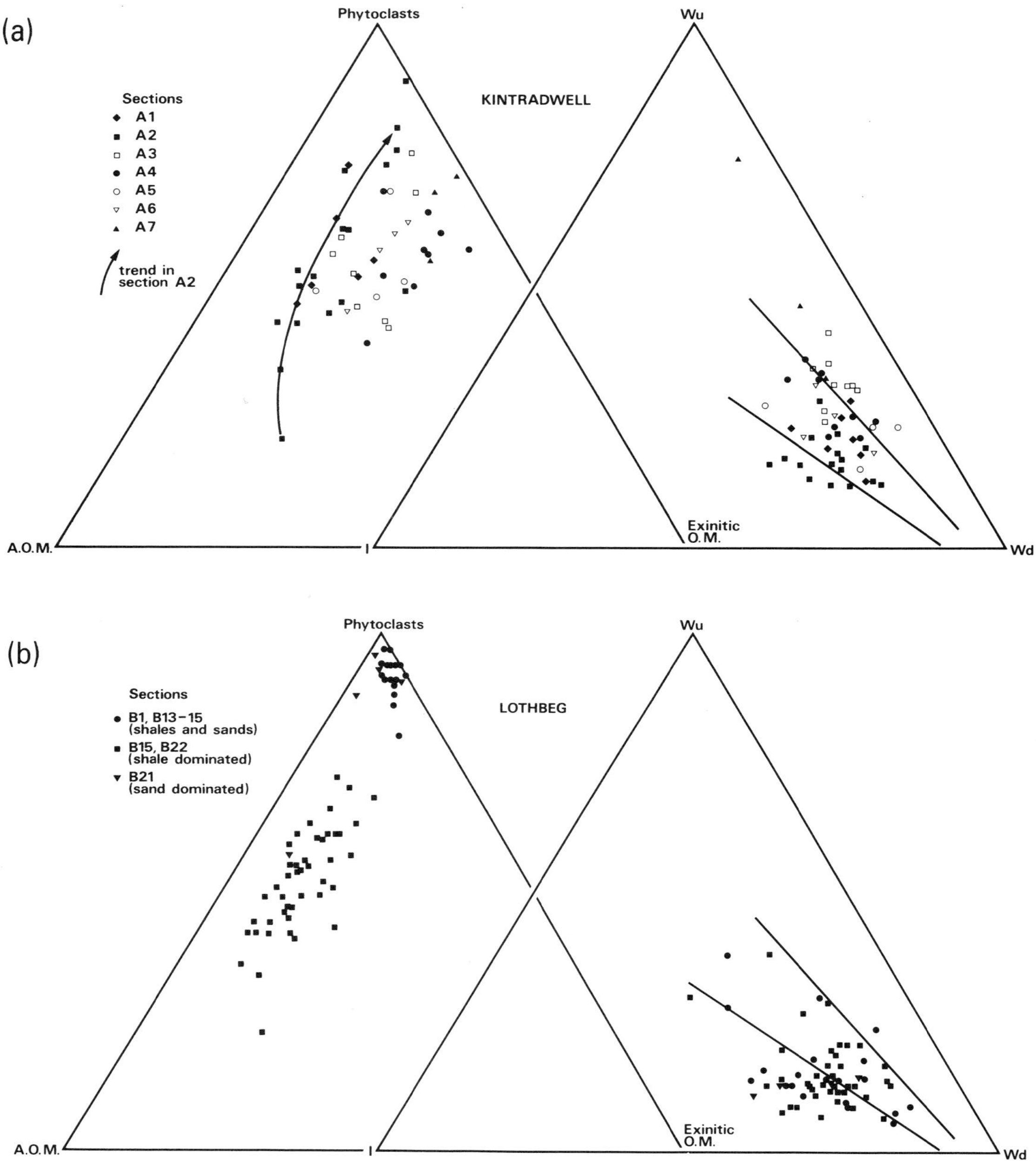

Fig. 5 —; for caption, see p. 150.

the west Kintradwell foreshore, exhibits a fairly consistent upward increase in the %PA of phytoclasts (trend shown by the arrow in Fig. 5(a)). Lam & Porter (1977) have previously recorded an upward increase in the abundance of microplankton in the *cymodoce* sediments of Kintradwell (probably from the same section). Together these observations suggest a regressive paly-

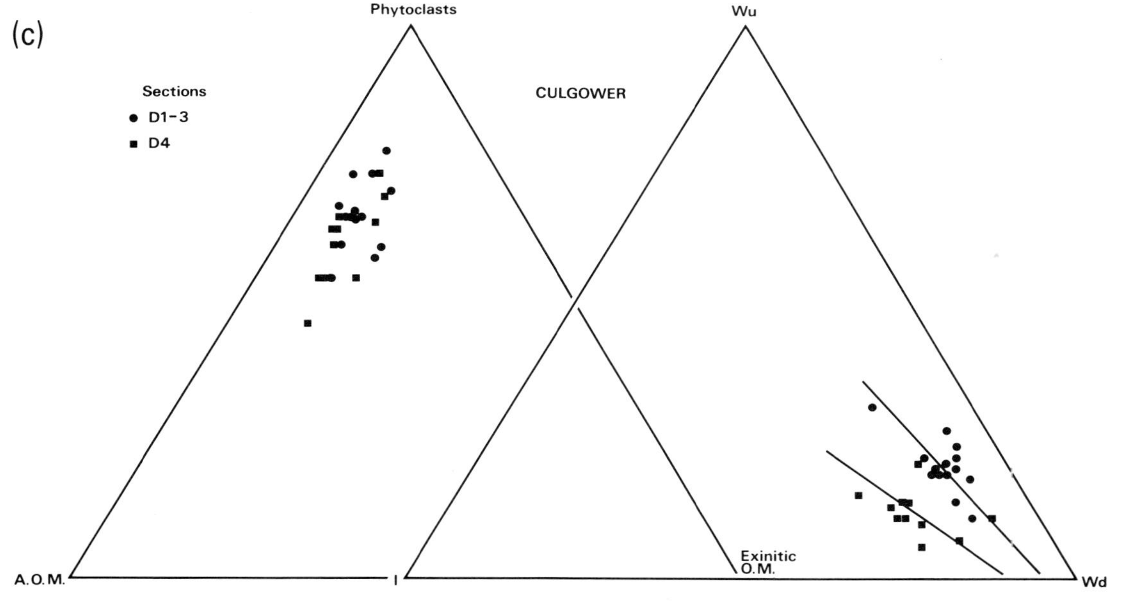

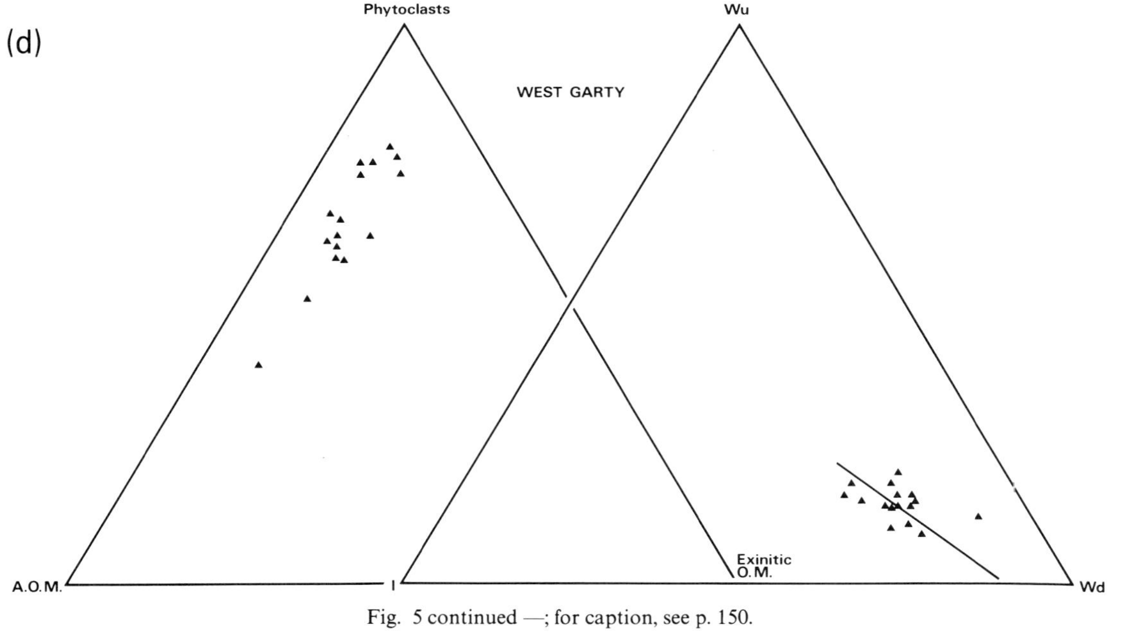

Fig. 5 continued —; for caption, see p. 150.

nofacies change with increasing shelf influence (shallowing, deltaic progradation, increased redeposition of shallow shelf material). Such a trend is in agreement with the independent sedimentological evidence for a late *cymodoce* regressive facies change (from shales to the western facies of the Allt na Cùile Sandstone).

Data from the stratigraphically higher

(e)

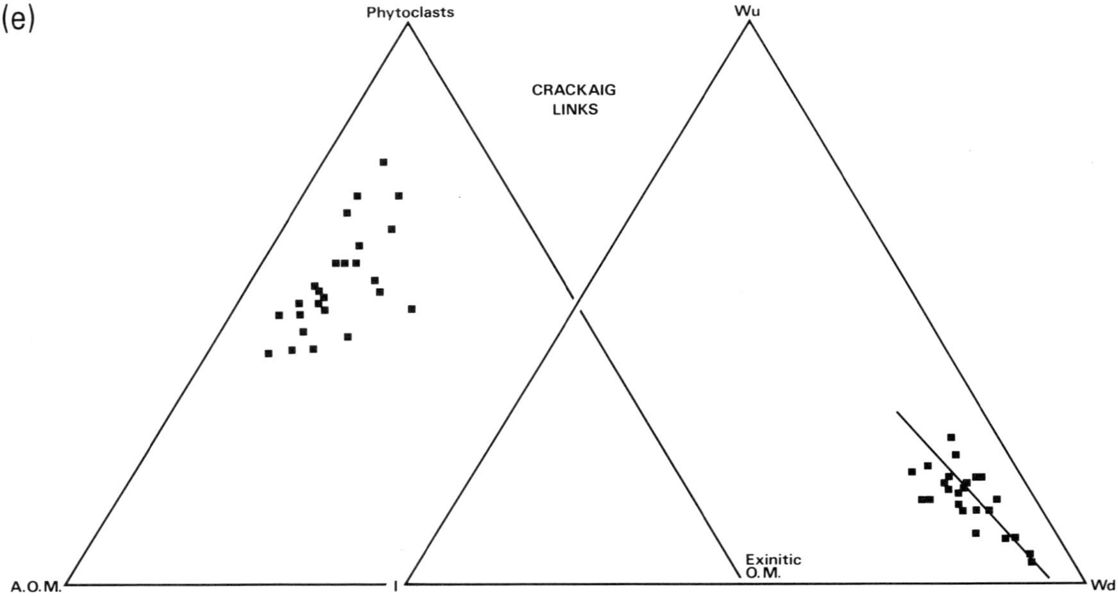

Fig. 5 — A–P–E diagrams and ternary phytoclast plots for the six main outcrop areas of lower Kimmeridgian sediments in the Brora–Helmsdale Outlier. Ternary phytoclast plots show regression lines derived from Fig. 15.

Table 5 — Stratigraphic trends in the mean shale kerogen parameters for the main outcrop areas of the lower Kimmeridgian sediments in the Brora–Helmsdale Outlier (see also Figs. 5(a)–5(f). The abbreviations are as in Table 4; the AmexOC and PhytOC values are from Figs. 10 and 11.

Area	*n*	T_w	I	Tp	AOM	Wu:Wd	TOC	(*n*)	AmexOC	PhytOC	HI (*n*)
A	58	57	⩽20	14	20	0.26–1.3	6.5	(24)	2.6	3.2	123 (10)
B	60	72	≈20	4	19	0.2 –0.25	6.2	(12)	2.9	3.4	143 (5)
E	21	65	22–18	6	25	0.2 –0.25	6.5	(6)	2.1	3.9	274 (2)
D	28	63	18–6	7	26	0.2 –0.25	8.1	(10)	2.9	5.1	227 (3)
C	27	54	11–7	7	34	0.2 –0.25	9.3	(6)	3.8	5.3	183 (2)

Area A, Kintradwell (mid-*cymodoce*).
Area B, Lothbeg (*mutabilis*).
Area C, Crackaig Links (*eudoxus–autissiodorensis*).
Area D, Culgower (mainly *eudoxus*).
Area E, West Garty (mainly *eudoxus*).

(A1 → A7) sections of the Kintradwell area tend to plot progressively farther to the right in Fig. 5(a). This is a consequence of a decreasing %PA of AOM (due to dilution by phytoclasts, and/or lower AOM preservation in more oxic sediments—note the presence of bioturbation in late *cymodoce* sediments). This shift is also accompanied by a 50% fall in the mean content of 'black wood' (as a percentage of the phytoclasts). The Wu:Wd ratio also progressively rises from 0.26

to 1.3, indicating an increase, and eventual dominance, of fresh angular material in the 'brown wood' component—another regressive signal. In the other sections the Wu:Wd values normally lie between 0.2 and 0.25.

The A–P–E diagram for the *mutabilis* Zone sediments of the Lothbeg area (Fig. 5(b)) reveals significantly lower %PA values for exinitic organic matter. The bulk of the shale samples show a more basinal pattern (with 20–50%PA of AOM), but there is also a concentration of data near the phytoclast pole. The latter group of data comes from carbonaceous sands and from shales interbedded within sand sequences. This pattern emphasizes the importance of lithofacies and lithological controls. The more typical shales have similar TOC, PhytOC and AmexOC values to those from Kintradwell, but the mean hydrogen indices are apparently slightly higher.

The A–P–E plots for data from the *eudoxus* Zone sediments of Culgower and West Garty (Figs. 5(c) and 5(d)) are very similar to each other and indicate a basinal pattern. However, the Culgower sediments have significantly higher TOC, PhytOC and AmexOC values, and a lower proportion of 'black wood' (Table 5). The latter trends are continued in the *eudoxus–autissiodorensis* Zone sediments of Crackaig Links, which show a significant increase in the %PA of AOM (Fig. 5(e)), and a corresponding AmexOC maximum (Table 5; see also Fig. 9).

Qualitative assessment of relative fluorescence intensity of AOM under blue-light excitation (performed in random sequence) indicates higher lipid preservation in the post-*cymodoce* sediments, agreeing well with the higher (but still only moderate) hydrogen indices in these sections. This trend suggests an increase in the intensity of oxygen deficiency in the lower *mutabilis* Zone. This agrees with the widespread transgressive facies changes observed at this time throughout Europe, the coeval geochemical trends seen at Westbury in southern England (see chapter 5 by Nøhr-Hansen in this volume), and the post-*cymodoce* reduction in foraminiferal faunas reported from Brora by Gregory (in chapter 7 of this volume). This probably represents a change from dysaerobic–anaerobic to anaerobic–anoxic conditions.

Late Jurassic fault scarp fan sequence, Toni–Thelma Oil Field

General Description

No published description exists for these sequences apart from a generalized summary by Kessler & Moorhouse (1984). However, very similar facies have been described from the nearby Brae Oil Field (Stow *et al.* 1982, Stow 1984, Turner *et al.* 1987, Harris & Fowler 1987). The present study involved the examination of cores from Phillips wells 16/17-4A and 16/17-1 (Fig. 6). The major part of the work was based on 270 m of almost continuous core from well 16/17-4A.

Well 16/17-4

The cored interval consists of two similar fining-upward sequences that are each composed of three distinct lithofacies (Fig. 6).

Facies A (Units 1 and 4) consists of a thinly bedded alternation of brown–grey or white fine-grained sandstone and black to dark-grey laminated shale. The proportion of shale in individual cores does not exceed 50%. Individual shale beds are generally less than 5 cm thick (as also noted by Kessler & Moorhouse (1984). Individual sandstone beds (uninterrupted by shale partings) are generally less than 10–15 cm thick, and thin, irregular to lenticular sandstone 'stringers' (2 cm or less) are common throughout the shales. The sandstones appear to be mostly planar laminated, but the lenticularity of the thinner 'stringers' is very suggestive of current ripples (like those described by Stow *et al.* (1982). Microfaulting is common in the shale-rich intervals, and low-relief loading and irregular sandstone injection structures are also observed. The upper parts of Units 1 and 4 contain small amounts of bivalve and belemnite debris. No macrobenthos or bioturbation is present in the shales.

Facies B (units 2 and 5) is similar but characterized by the presence of graded or structureless pebbly sandstones (between 14% and

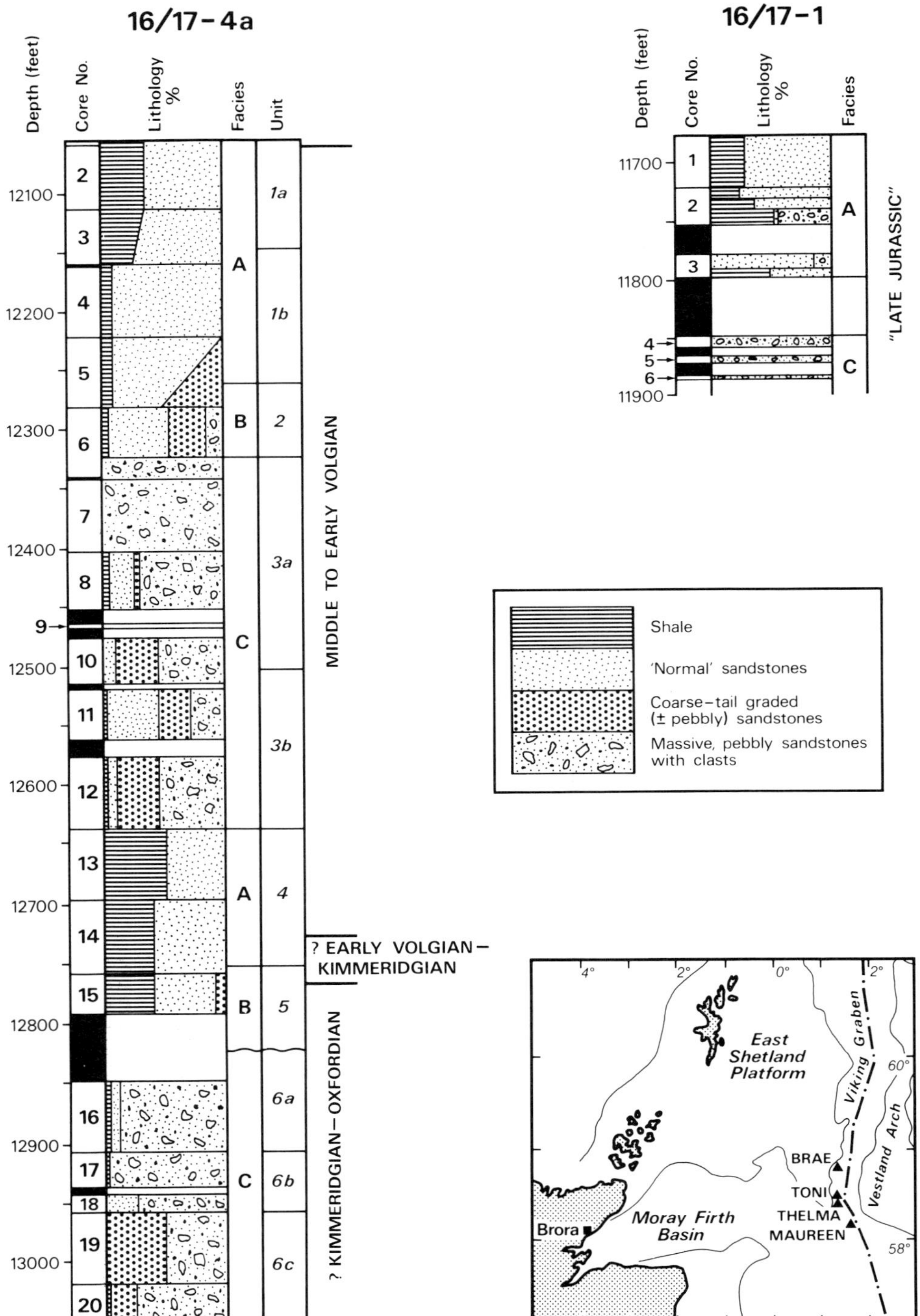

Fig. 6 — Location of Toni–Thelma and Maureen Oilfields, and summary logs for cored intervals from wells 16/17-4A and 16/17-1. The age of the well 16/17-1 sequence was given only as 'intra-Late Jurassic'.

56% of the thickness). These occur as coarse-tail-graded (CTG) units that show downward deterioration in matrix sorting, and an increase in the size and abundance of quartzitic pebbles (2 cm or less in diameter). The upper 20–50% of the CTG units is generally pebble free; the basal parts sometimes contain occasional clasts of wood, shale, claystone or sandstone. The mean thicknesses of the CTG units is 40 cm in Unit 2, and 26 cm in Unit 5.

Facies C (Units 3 and 6) is characterized by its very low shale content (less than 5%) and an abundance, and variety, of angular sandstone clasts (ranging up to at least 46 cm in diameter). The facies is dominated by matrix-supported, poor to very poorly sorted, generally unorganized medium-grained pebbly clast-rich sandstones. Short intervals of clast-supported sandstone occur locally throughout the facies. The distribution of recognizable CTG units (like those in facies B, but with clasts), has been used to subdivide the units (Fig. 4). The mean thickness of the CTG units lies between 70 cm and 80 cm; their number is either equal to, or greater than, that in facies B.

Within each CBA sequence the fining-upward trend is reflected by the distribution of sandstone grain size and sorting, the overall lithological ratios, the distribution of clasts and pebbles, the mean thickness of the CTG sandstone units, and the mean and maximum values of both clast and pebble size. Both fining-upward sequences show similar ratios of the three facies.

Sedimentological features are very similar to those previously described from Brae sequences (see references cited above). The latter studies support sedimentation within a submarine fault scarp fan setting ('faulted slope apron' of Stow (1985)). The hemipelagic shale and fine-grained turbidites of facies A are believed to represent interfan, fan abandonment or basinward subenvironments. The CTG units which characterize facies B probably represent deposits of channelized grain flows, liquified flows and high-density turbidity flows. Such processes were also active during the deposition of the more proximal facies C, but the coarser clast-bearing intervals represent debris flow (and possibly rock fall avalanche) deposits.

Well 16/17-1

The studied interval consists of only one fining-upward sequence (Fig. 6). The sediments are generally very similar to those in well 16/17-4A except no facies B equivalent is developed in the cored section. The upper three cores are very similar to facies A, but calcite cementation and bioclastic debris are much more common, and the sandstones tend to be more argillaceous. Individual sandstone beds range up to 6 cm in thickness, and uninterrupted shale units are up to 4 cm thick (but generally about 1 cm). The lower three cores are equivalent to facies C. They exhibit a downward increase in maximum clast size and clast density; the sandstone matrix is argillaceous and often dark grey-green in colour. Sedimentological interpretations are as given for well 16/17-4A.

Palynofacies Characteristics

Well 16/17-4A

The black laminated shales are all very organic rich with TOC values in the range 3.0–30.0 wt% and an overall mean of 9.6 wt%. The kerogen assemblages are generally strongly dominated by the phytoclast component. The %PA of AOM is usually only about 20%, but it consistently shows strong fluorescence under ultraviolet light and is clearly oil prone. The nature of the kerogen assemblages from all the Phillips wells are shown in Fig. 7. Significant stratigraphic variations are evident in the data and the mean values of the main geochemical and kerogen parameters have therefore been calculated for each lithological unit or sub-unit (Table 6). Such calculations indicate that only some of the stratigraphic variations correspond to lithofacies changes. The lithological symmetry and similarity between Units 1–3 and Units 4–6 are not reflected in any comparable way in the kerogen data.

The subdivision of Unit 1 is based entirely on geochemical and kerogen data; both indicate a significant event at the 1a–1b boundary that cannot be detected by ordinary sedimentological means. The marked increase in hydrogen indices, the decrease in the %PA of phytoclasts, and the

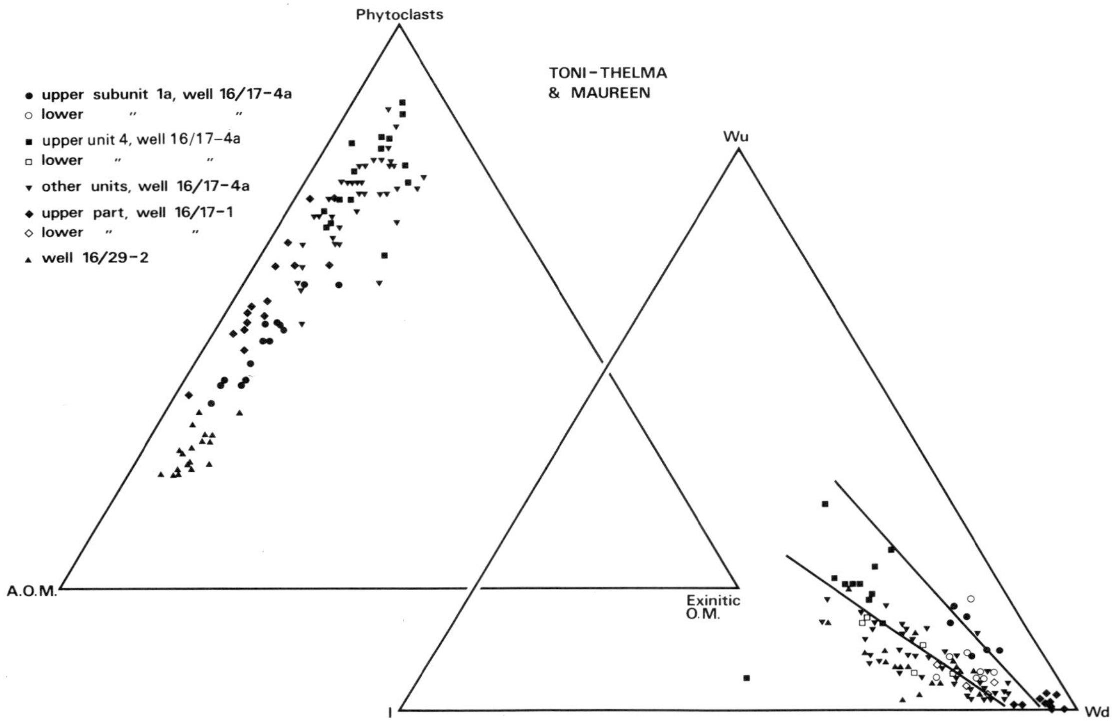

Fig. 7 — A–P–E diagram and ternary phytoclast plot for wells 16/17-4A, 16/17-1 and 16/29-2. Ternary phytoclast plot shows regression lines derived from Fig. 15.

Table 6 — Summary of mean kerogen parameters for shales from Phillips wells 16/17-4A, 16/17-1 and 16/29-2X (see also Fig. 7). Abbreviations are as in Table 4; the arrows indicate an upward increase in values through the unit; the AmexOC and PhytOC values are from Figs. 10 and 11.

Well	*n*	T_w	I	Tp	AOM	Wu:Wd	TOC	(*n*)	AmexOC	PhytOC	HI (*n*)
16/17-4A											
Subunit 1a	14	37	4	1.5	50	0.14↑	10	(13)	6.1	4.2	606 (6)
Subunit 1b	13	55	8	1.5	28	0.04↑	11	(6)	4.3	6.5	330 (4)
Unit 2	3	60	7	1.9	21	0.09	9	(3)			282 (2)
Subunit 3a	5	58	17	1.2	17	0.09	13	(2)			380 (1)
									3.0	9.0	
Subunit 3b	5	58	14	1.2	19	0.29	17	(3)			239 (2)
Unit 4	16	58	16	1.1	18	0.34↑	6	(13)	1.4	4.1	208 (5)
Unit 5	8	58	14	1.7	20	0.20	11	(4)	2.7	8.2	253 (2)
Unit 6	4	54	17	2.8	20	0.20					
16/17-1	16	50	4	0.9	31	0.03↑	6	(6)	2.8	3.7	380 (3)
16/29-2	18	20	5	2.7	68	0.15	10	(13)	6.8	2.4	306 (4)

corresponding relative increase in AOM are believed to reflect the basinwide mid-Volgian 'hot shale' event (Rawson & Riley 1982, Tyson 1985). Although identical in lithofacies, Unit 4 is conspicuously less organic rich than Unit 1 (and all other units). This clearly illustrates the influence of large-scale secular variations as well as general facies controls. However, the graben environment appears to have been continuously anaerobic to anoxic from the late Oxfordian onward.

A number of samples were taken from atypical argillaceous lithologies (shale or claystone clasts, argillaceous matrices to debris flows, etc.). The kerogen parameters of these samples were each separately compared with the mean values for the bedded shales in the unit or subunit from which the samples were taken. These atypical (more allochthonous) sediments show significant differences to the normal shales; they are two to four times less organic rich, have about three times as many palynomorphs, and a third to half the phytoclast content. This emphasizes the need for sedimentological awareness during sampling.

The palynomorph content is consistently low and generally less than 2%. Such values are too low for quantitative analysis to be practical. As the AOM 'content' is usually only about 20%, these low abundances should not just be due to 'masking'. Robertson Research gave me access to a suite of 10 oxidized palynostratigraphic preparations that confirmed a low original palynomorph density. The palynomorph assemblages are strongly dominated by sporomorphs, and microplankton is virtually absent except in the upper part of facies A (a similar trend to that of bioclastic debris). Chorate dinocysts are usually the most abundant component of the microplankton. The general rarity of the microplankton lends support to the sedimentological argument for the existence of only a very narrow marine shelf at the eastern edge of the Shetland Platform.

Well 16/17-1

The general %PA composition of the kerogen assemblages from the upper three cores is similar to those in well 16/17-4A, but the values for AOM tend to be higher, and those for 'black wood' and palynomorphs somewhat lower (Table 6 and Fig. 7). In lithofacies terms the sediments are clearly equivalent to either Unit 1 or Unit 4 of facies A in well 16/17-4A (no biostratigraphic correlation was possible). Although the TOC values are most like those of Unit 4, the comparatively high hydrogen indices, high AOM, low 'black wood' and low Wu:Wd ratios suggest an equivalence to Unit 1. This is also supported by various phytoclast cross-plots that indicate an overall higher affinity with Subunit 1b than with Subunit 1a (see Figs. 7, 14, 15 and 16). These plots also indicate that the lower interval (core 3 and very base of core 2) differs from the upper one in a similar fashion to that observed in Subunit 1a of well 16/17-4A.

Kimmeridge Clay Formation, Maureen Oil Field

General Description

A single 9 m core of the Kimmeridge Clay Formation was examined from Phillips well 16/29-2X from the Maureen area (Fig. 6). The core is made up of a monotonous sequence of dark-grey, organic-rich, micaceous, sometimes pyritic shale. The shale is featureless except for occasional concentrations of pyrite and fish scales. It is unfossiliferous and clearly laminated in thin section. These characteristics point to distal, hemipelagic sedimentation in a stable anaerobic–anoxic basin. The age of the sediments was given as 'Kimmeridgian' (probably corresponding to the Volgian?).

Palynofacies Characteristics

The shales have TOC contents greater than 8 wt%, hydrogen indices generally in excess of 300 and AOM-dominated kerogen assemblages with only about 25% phytoclasts (Table 6 and Fig. 7). The %PA of palynomorphs is very low, but this is undoubtedly partly due to masking by AOM. Examination under blue-light fluorescence reveals an abundance of strongly fluorescent alginitic–bituminitic debris and common to relatively abundant prasinophycean algae. The

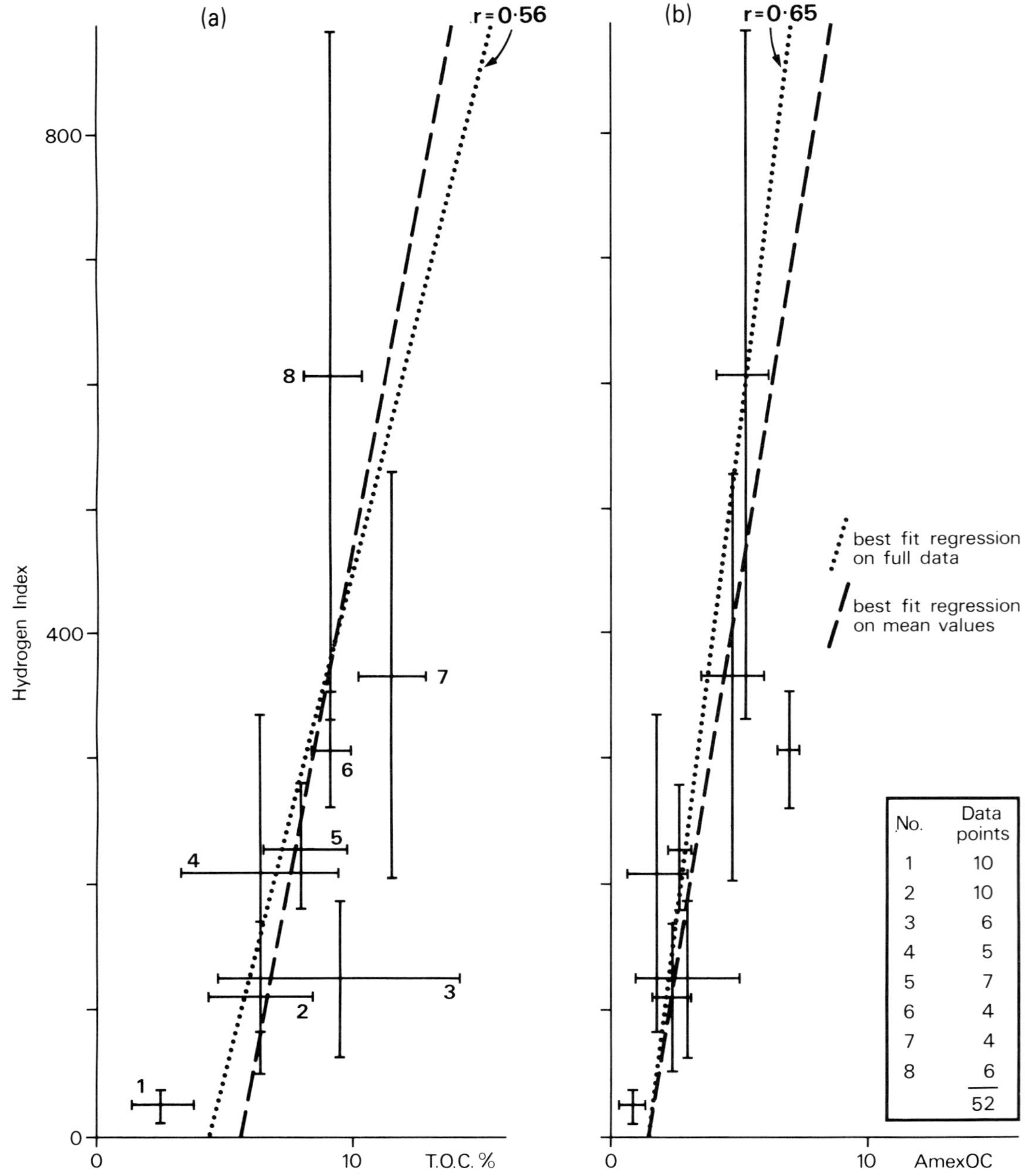

Fig. 8 — Plots of (a) TOC versus hydrogen index and (b) AmexOC versus hydrogen index: 1, Piper; 2, Kintradwell; 3, Lothbeg; 4, Unit 4 (well 16/17-4a); 5, other Brora; 6, well 16/29-2; 7, Subunit 1b (well 16/17-4a); 8, Subunit 1a (well 16/17-4a); centre of crosses represent the mean of values; vertical and horizonal bars show standard deviations. Regression lines are shown for both mean values and total data.

palynofacies assemblage is clearly comparatively distal.

RELATIONSHIPS BETWEEN GEOCHEMICAL AND KEROGEN DATA

Fig. 8(a) shows the tendency for the hydrogen index to increase with increasing TOC. Hydrogen indices greater than 300, which indicate a good potential oil source (Peters 1986), are primarily confined to samples with TOC values in excess of 8 wt%. Fig. 8(b) shows the better relationship between hydrogen index and AmexOC values; hydrogen indices greater than 300 are only characteristic of samples with AmexOC values over 3.0. Data from Nøhr-Hansen (in chapter 5 of this volume) also agree well with this trend.

The residual scatter shown by Fig. 8(b) reflects the fact that the %PA data used to calculate the AmexOC values do not accurately represent the relative volume of the 'AMEX' kerogens, or their preservation state. For data from well 16/17-4A the correlation between hydrogen index and AmexOC values is considerably higher for samples with hydrogen index values below 400 ($r = 0.79$). This suggests that above this level the geochemical characteristics are controlled not so much by the abundance of AOM but by its nature (e.g. alginite content and preservation state). Some of the (non-cutinite) phytoclast materials may also be more hydrogen-rich than others (depending on their original nature, state of degradation and degree of bacterial modification).

The plot of PhytOC versus AmexOC (Fig. 9) shows the distinctive nature of the shales from well 16/29-2 and Subunit 1a of well 16/17-4A. They are characterized by TOC values over 8.0 wt% and AmexOC values greater than 5.0 (Fig. 8). Fig. 9 shows their relative excess of AmexOC over PhytOC. The lower Kimmeridgian shales described by Nøhr-Hansen (in chapter 5 of this volume) show an even more marked excess of AmexOC (predominantly 0.2–4.5) over PhytOC (0.2–0.8). These observations suggest that the origin of the AOM-enriched organic-rich shales is not causally related to the input of terrestrial organic matter. The less organic-rich, more 'woody' shales, with AmexOC values less than 4.0, show an opposite trend with an irregular tendency for AmexOC to increase with PhytOC.

COMPARATIVE PALYNOFACIES

Meaningful comparisons between the individual sections are best made by using ranked PhytOC and AmexOC data and A–P–E diagrams. Cross-plots of the different phytoclast categories are also informative.

PhytOC ranking

As most terrestrial organic matter reaches the sea via rivers, the absolute amount of woody material in a sediment provides an indication of proximity to fluvial input. The studied sections have been ranked according to their mean PhytOC content in order to assess their relative proximity to major fluvio-deltaic sources of terrestrial organic matter (Fig. 10). The fault scarp fan facies are clearly associated with a prolific supply of woody material into the marine basin. The values were not even diminished by probable high sedimentation rate. By comparison, both Dorset and the Piper area clearly represent shelf areas significantly removed from phytoclast sources. The Piper area may have been comparatively shallow and sand dominated, but it was not 'proximal'; the 'wave-dominated delta' of Boote & Gustav (1987) was apparently not a major source of phytoclasts. The equally low PhytOC values given for the Dorset oil shale facies appear to be reasonably representative for the onshore Kimmeridge Clay; data from Nøhr-Hansen (in chapter 5 of this volume) also give a mean PhytOC value of 0.7.

Scotchman (1987) interprets kerogen patterns in the late *eudoxus* Zone of the onshore Kimmeridge Clay on the basis of 13 samples. He states that the shallower flanks of the basin had a 'strong terrestrial flux' and a 'higher content of

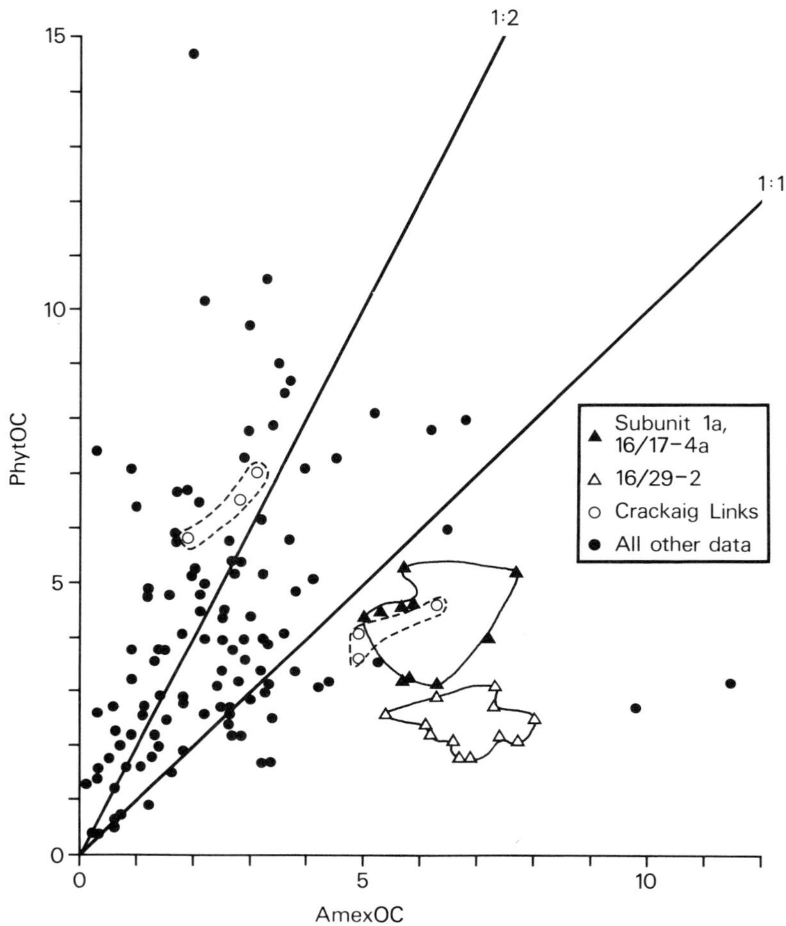

Fig. 9 — Plot of PhytOC values versus AmexOC values.

Type III humic kerogen'. Calculations based on his Table 2 (p. 257) indicate a significant range of PhytOC values 0.7–4.3; mean, 1.7). This shows the input of terrestrial organic matter was not 'relatively constant', and the two cited examples of basin flank situations actually have among the lowest PhytOC values (0.8). The highest PhytOC value actually occurs in the 'deeper water and more anaerobic' facies of Marton in Yorkshire (which does also show the highest AmexOC value of 5.2). These examples show that the calculation of PhytOC values can significantly clarify palynofacies interpretations and palaeogeographic reconstructions.

AmexOC ranking

The AmexOC ranking is not the reverse of the PhytOC ranking (Fig. 11). The low AmexOC content of Piper shales reflects deposition under aerobic conditions; all the other data are from laminated black shales. If the main control on the preservation of sedimented marine organic matter is the degree of bioturbation, the non-bioturbated character of these shales apparently precludes variations in preservation as a viable explanation for the significant differences. What else could explain these differences? Temporal fluctuations in palaeo primary productivity are one possibility.

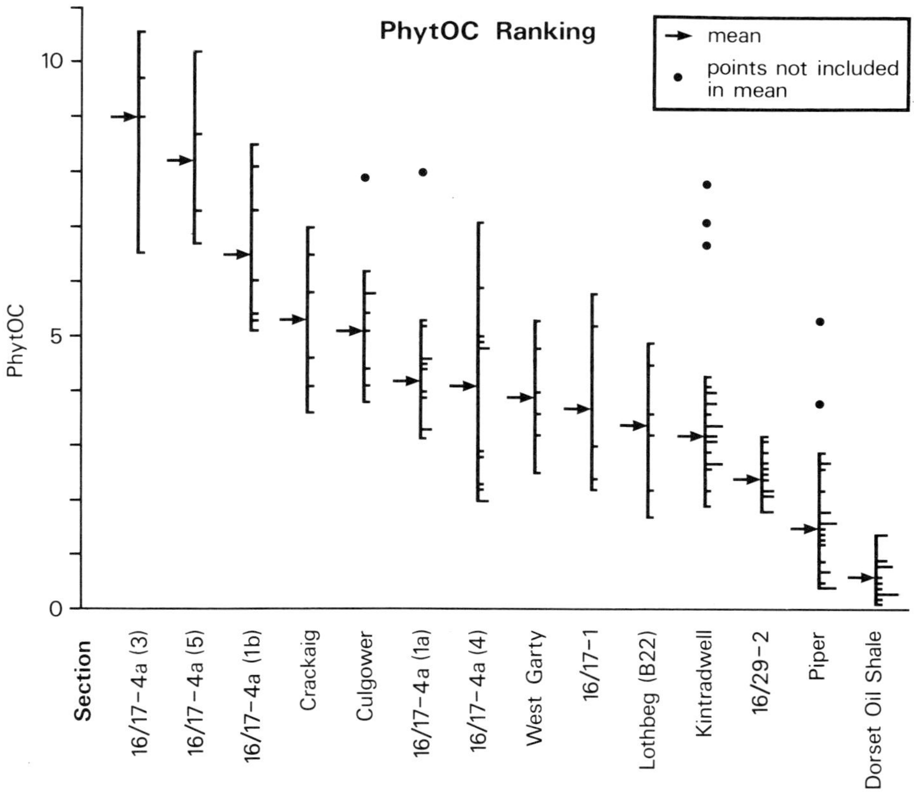

Fig. 10 — Ranking of examined shales according to mean PhytOC values: vertical bars show range of values; the length of the horizonal lines indicates number of samples corresponding to any particular value.

A more basic explanation is the most probable. Weight percentage TOC is not a truly absolute parameter and it is significantly affected by variations in sedimentation rate (i.e. the 'clastic dilution factor'). Unfortunately, I do not have the appropriate information to calculate the organic carbon accumulation rates in order to assess this effect (e.g. Ehrmann & Thiede 1985, Summerhayes 1987). However, the highest AmexOC values characterize the most distal facies (removed from sediment and terrestrial organic matter supply). Such sediments are likely to have accumulated at significantly lower rates than those adjacent to active fault scarp fans and may, by virtue of this alone, exhibit higher TOC values as well as higher %PA contents of AOM. Eustatic (and tectonic) trends may have also influenced sedimentation rates, producing periodic starved-basin conditions.

Summary A–P–E diagram

Fig. 14 is a summary and interpretation of all the %PA data from all the sediments examined during this study. The diagram shows a surprising degree of resolution in its ability to distinguish between the different shale facies. The diagram also has some interesting implications for determining phytoclast (and sediment) dispersal patterns. The phytoclast material obviously originates from the phytoclast pole at the apex of the diagram, but then proceeds offshore by several routes (arrows in Fig. 12b).

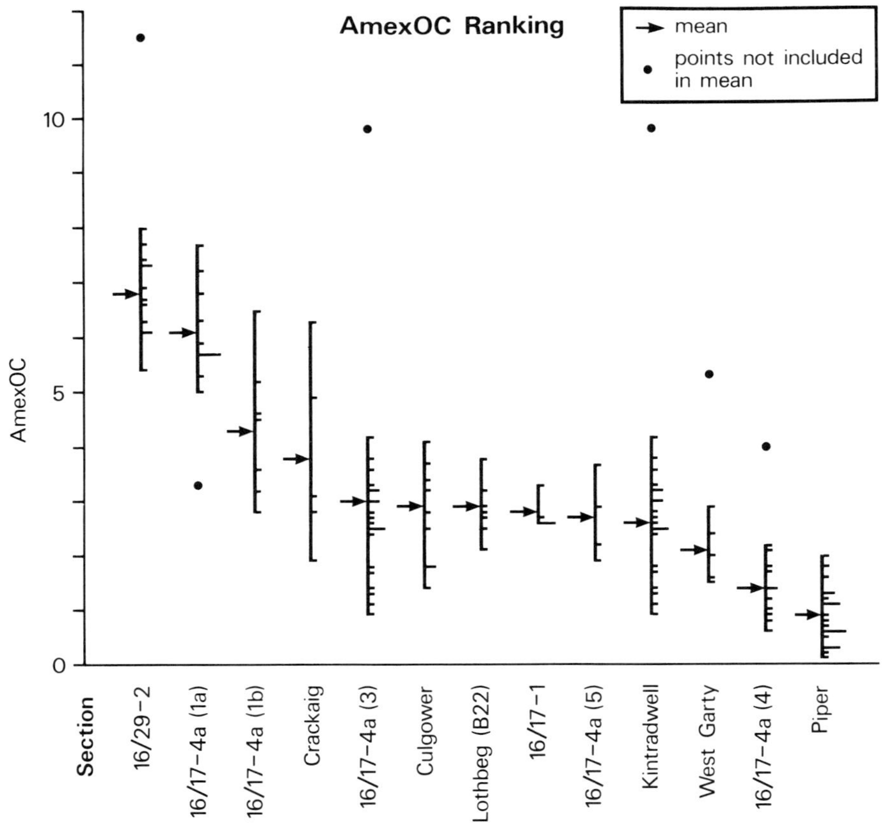

Fig. 11 — Ranking of examined shales according to mean AmexOC values.

Route A is a very direct path straight from the source into the anoxic marine basin, represented by a line parallel to the left side of the diagram. PhytOC values decrease distally towards the AOM pole. This path is typical of the fault scarp fan facies.

Route B phytoclasts move away from their source out across a shallow marine shelf. PhytOC values decrease progressively but %PA values remain high because the aerobic conditions prevent significant preservation of AOM. Much of the phytoclast material may become partially oxidized. Increasing marine influence and particle sorting during transport lead to a significant increase in the %PA of exinitic organic matter. This path initially parallels the right hand side of the diagram.

Route C involves the redirection of phytoclasts into the basin by tapping it at an early stage of its passage along route B. It is still a comparatively short path into the basin; partial sorting has taken place on the shelf and more exinitic material has become included. This sharp shelf to basin (slope)? transition is represented by the *cymodoce* shales of Brora.

Route D represents a continuation of route B with further reduction in the PhytOC values and progressive sorting of the phytoclasts and palynomorphs. The route is shown as swinging around towards the AOM pole to indicate a trend to increasingly reducing conditions.

Route E is a much more poorly defined shelf to basin pathway. If the problematic (variety 3) fibrous material is excluded, most of the data

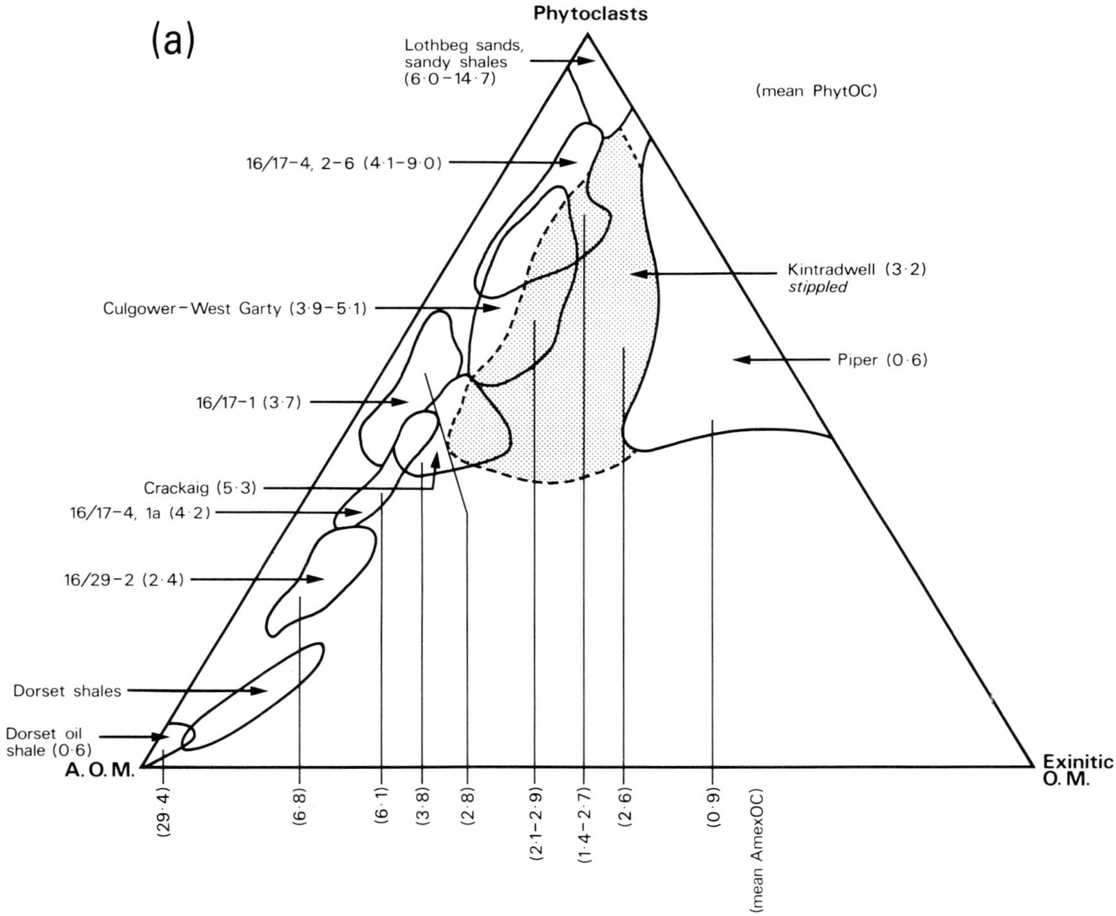

Fig. 12 — For explanation, see p. 162.

from the type Kimmeridge Clay plots in a fairly consistent pattern about a line from the AOM pole to the middle of the right side of the diagram (see also chapter 5 by Nøhr-Hansen in this volume). This line also happens to bisect the bulge in the field defined by the Piper data. Inner-shelf-derived turbidites (Tyson 1984) also interestingly plot about this line. It appears to represent the basinward redirection of material from route B at a point intermediate between routes C and D. It may reflect an increase in the relative importance of turbiditic or storm-generated redeposition over normal bottom currents.

Phytoclast cross-plots

%PA Wu + Wd versus %PA I (see Table 2)

The cross-plot of 'brown wood' and 'black wood' shows that both the range and the maximum value for 'black wood' increase as the total 'brown wood' category increases (Fig. 13). Maximum values for both are particularly associated with sandstone samples, or shale samples from thin interbeds within sandstones. The only samples which exhibit a dominance of 'black wood' over 'brown wood' are those from the type Kimmeridge Clay of Dorset (but note the very

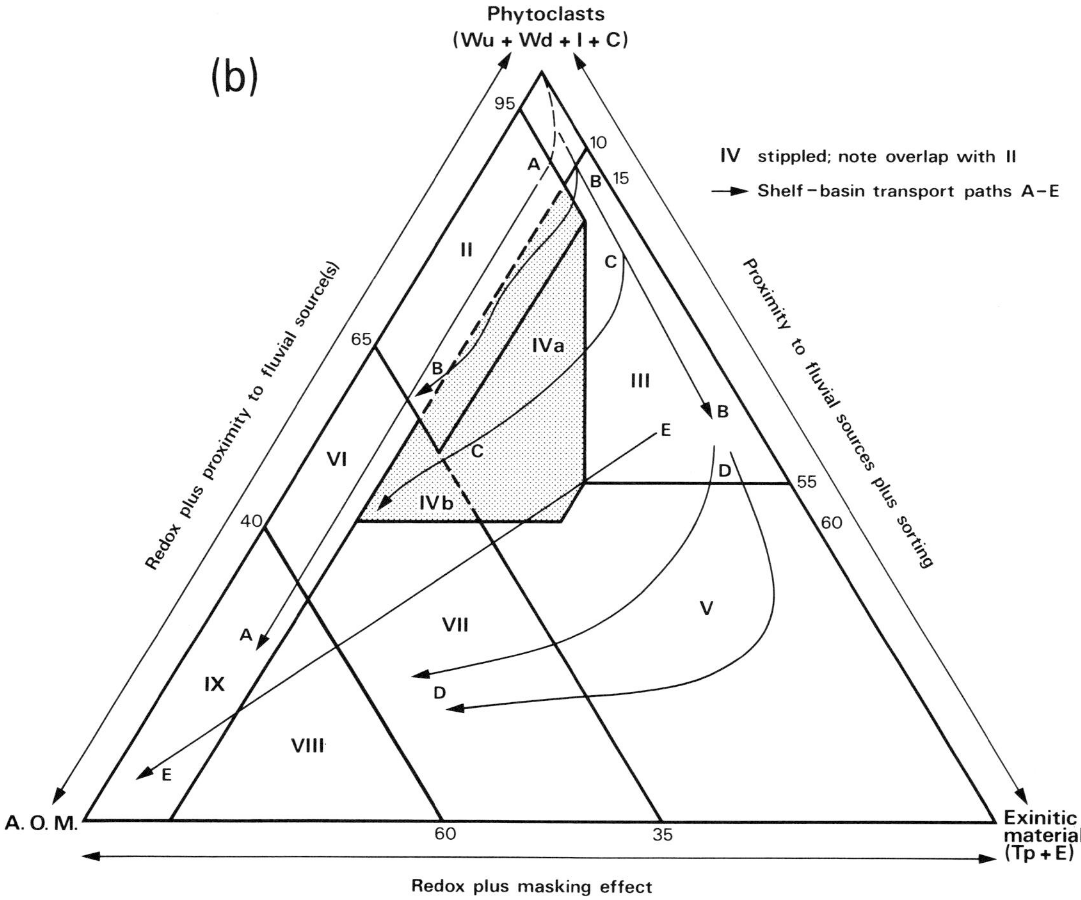

Fig. 12 — A–P–E Summary diagrams: (a) characteristic fields for some of the various shales examined, plus their mean AmexOC and PhytOC values; (b) schematic diagram showing derived palynofacies fields and hypothetical shelf to basin transport paths; see Table 7.

low abundances). Data from some of the studied sections tends to form fairly distinctive clusters which suggests this cross-plot may have limited potential for correlation purposes when biostratigraphic resolution is impracticable. For example, the shales from well 16/17-1 appear to be transitional between those of Subunits 1a and 1b of well 16/17-4A, but closer to the latter than the former.

This cross-plot should not be expected to produce very definitive results because the phytoclast categories are expressed as a percentage of the total kerogen assemblage. Percentage values are really an attribute of the population rather than any individual component, and the total population includes diverse elements many of which are unlikely to have any inherent relationship to each other. It is much more meaningful to recalculate the data as percentages of the phytoclast population, whose different particle types can reasonably be expected to show some degree of logical interrelationship through either provenance or sorting.

Wd (% of phytoclasts) versus Wu (% of phytoclasts)

The cross-plot of rounded (degraded) to angular (fresh) 'brown wood' particles consistently shows

Table 7 — Characteristics of the palynofacies fields defined by the A–P–E summary diagram (Fig. 12(b)).

	Field and environment	*Comments*	*Spores:bisaccates*	*Microplankton*	*Phytoclasts*	*PhytOC*	*AmexOC*	*Kerogen type*
I	Highly proximal shelf or basin	High phytoclast supply dilutes all other components	Usually high	Very low	Wd $\gg$ Wu $>$ I	Very high	Low $\leqslant 1.0$	III, gas prone
II	Marginal dysaerobic–anoxic basin	AOM diluted by high phytoclast input, but AOM preservation moderate to good. AmexOC values dependent on basin redox	High	Very low	Wd $\gg$ Wu $>$ I	High to very high	Moderate	III, gas prone
III	Sand-dominated aerobic shelf 'proximal shelf')	Generally low AOM preservation, PhytOC values dependent on actual proximity to fluvial source. Oxidation and reworking common	High	Common to abundant. Dinocysts dominant	Wd $\gg$ Wu $>$ I	Low (to moderate)	Low	III or IV, gas prone
IV	Shelf to basin transition	Passage from shelf to basin in time (e.g. increased subsidence and water depth) or space (e.g. basin slope). PhytOC values depend on proximity to fluvial source and degree of redeposition. AmexOC values depend on basin redox: IVa, dysaerobic–anaerobic; IVb, anaerobic–anoxic	Moderate to high	Very low to low	Wd $\gg$ Wu $>$ I	Moderate, 2.0–3.0	Moderate	III or II, mainly gas
V	Mud-dominated aerobic shelf ('distal shelf')	Low to moderate AOM (usually degraded) Palynomorphs abundant. Includes light-coloured mudstones from the Kimmeridge, Speeton and Upper Oxford Clays	Usually low	Common to abundant. Dinocysts dominant	I $\geqslant$ Wu + Wd	Low, $\leqslant 1.0$	Low	III $>$ IV, gas prone
VI	Proximal anaerobic–anoxic basin	High AOM preservation due to reducing basin conditions. PhytOC values are moderate to high due to turbiditic input and/or general proximity to source	Variable, low to to moderate	Low to common. Dinocysts dominant	Wd $\gg$ I $>$ $<$ Wu	Moderate to high	Moderate to high	II, oil prone (includes 'waxy sapropel' of Barnard *et al.*)
VII	Distal dysaerobic–aerobic shelf	Moderate to good AOM preservation, low to moderate palynomorphs. Includes dark-coloured mudstones from the Kimmeridge, Speeton and Lower Oxford Clays	Low	Moderate to common. Dinocysts dominant	Wd $\gg$ Wu $>$ I	Moderate to low	Moderate to high	II, oil prone
VIII	Distal dysaerobic–anoxic 'shelf'	AOM-dominated assemblages, excellent AOM preservation. Low to moderate palynomorphs (partly due to masking). Typical of organic-rich shales deposited under meromictic shelf sea conditions	Low	Low to moderate. Dinocysts dominant, % prasinophyceae increasing	I $\geqslant$ (Wu $>$ Wd)	Low to very low	High	II $\gg$ I, oil prone
IX	Distal anaerobic–anoxic basin	AOM-dominated assemblages. Low abundance of palynomorphs partly due to masking. Frequently alginite rich. Deep basin or meromictic shelf sea deposits	Low	Generally low. Prasinophyceae often dominant	I $>$ Wu Wd $>$ $<$I + Wu	Low to very low, $\leqslant 1.0$	High to very high	II $\geqslant$ I, oil prone (AOM alginite-rich)

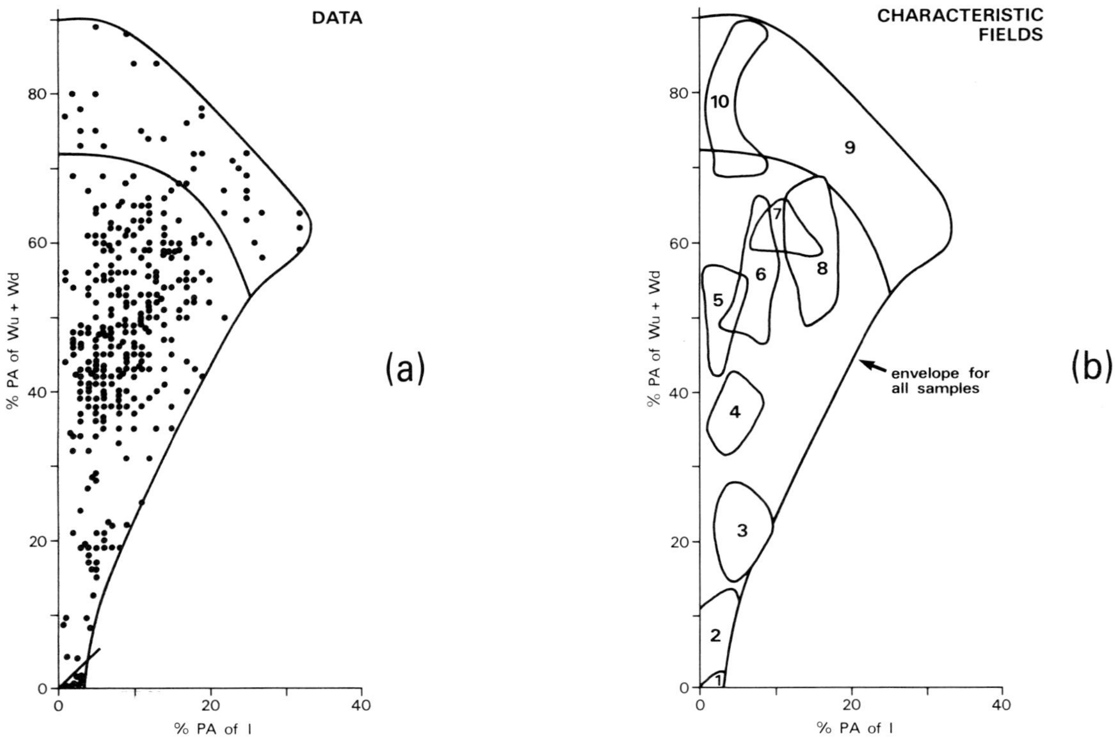

Fig. 13 — Plot of percentage particle abundances of 'brown wood' (Wu + Wd) versus 'black wood' (I) showing (a) all data and (b) diagnostic fields for some of the units studied: 1, 100% Dorset oil shale samples; 2, 100% of type Kimmeridge Clay samples; 3, 100% of well 16/29-2 samples; 4, 80% of Subunit 1a (well 16/17-4a); 5, 67% of well 16/17-1; 6, 85% of Subunit 1b (well 16/17-4a); 7, 80% of Unit 3 (well 16/17-4a); 8, 87% of Unit 4 (well 16/17-4a); 9, 64% of all sandstone samples analysed.

a negative correlation (Fig. 14). Two converging trends are apparent. Trend A appears to be indicative of greater shelf influence and includes all Piper samples except well 15/17-7 sandstones, and all the stratigraphically higher sections from Kintradwell (Brora) previously interpreted as more regressive and/or transitional between shelf and basin. It also includes samples from the upper parts of the sequences from well 16/17-1 and Subunit 1a from well 16/17-4A; these have previously been shown to be distinctive because they contain significant bioclastic material and microplankton.

Trend A also includes those *eudoxus–autissiodorensis* shales from Brora for which a case for more basinal, or at least more strongly reducing, conditions has been made. However, these shales do come from sections which show the first significant post-*cymodoce* reappearance of bioclast-rich sandstones. Similarly bioclast-rich sandstones also characterize the *eudoxus* Zone shales (Culgower, D1–D3) that plot on this trend. These samples may, therefore, reflect shelf influence by virtue of redeposition rather than *in situ* conditions (although not all sections with shelly turbidites plot on this trend). Trend A is more sharply defined than trend B (correlation coefficient r is 0.95 compared with 0.83; see Fig. 14).

Trend B appears to be indicative of more basinal conditions and includes the shales from well 16/29-2, most of those from well 16/17-4A (including the lower part of Subunit 1a), all the *mutabilis* Zone shales, and the lowermost

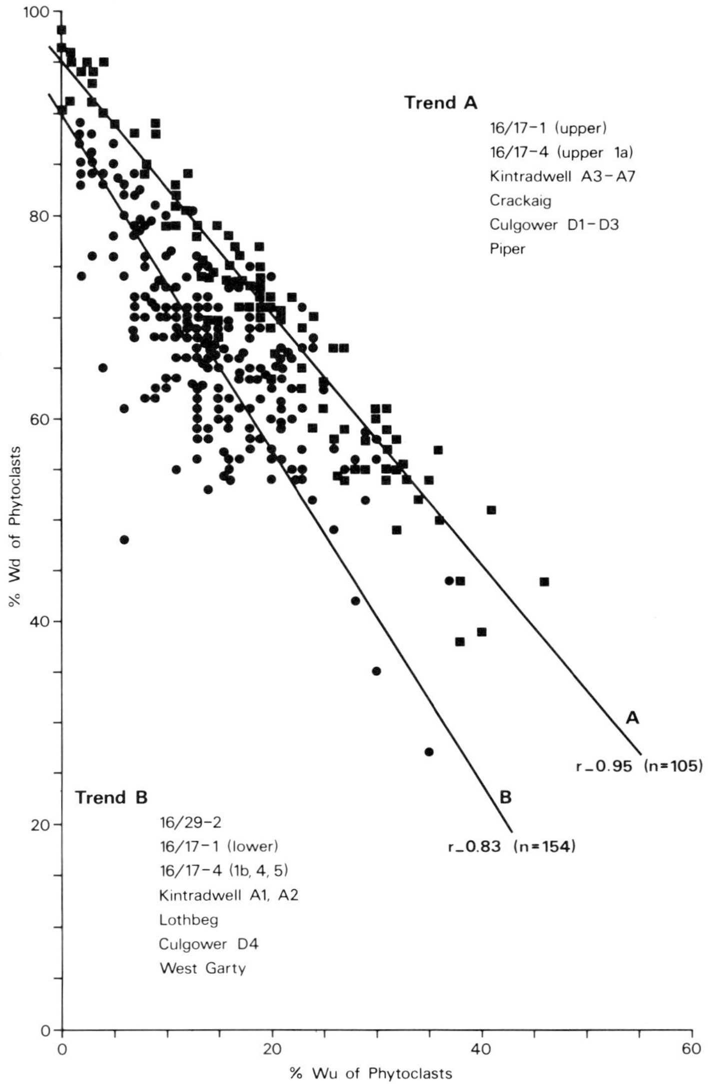

Fig. 14 — Plot of Wd versus Wu 'brown wood' categories as percentages of the total phytoclast population. Plot shows regression lines for trend A (shelf and shelf-influenced basin margin) and trend B (true basinal facies); *r* is the correlation coefficient.

cymodoce sections from Brora. The *eudoxus* Zone shales to the north of (and including) the Sròn Rubha na Gaoithe promontory (Culgower, Culgower D4, West Garty; see Fig. 4) also plot on this trend. The unanticipated sensitivity of this cross-plot promises to be useful in facies interpretation and correlation work. The degree of scatter may be substantial for individual sections, and some may, like Subunit 1a of well 16/17-4A, show vertical (facies) transitions from one trend to the other. The regression lines for both trends are shown on the ternary phytoclast plots accompanying the A–P–E diagrams. Much of the difference between the two trends probably reflects the differing behaviours of 'black wood' particles in shelf and basin settings (see below).

I (% of phytoclasts) versus Wu (% of phytoclasts)

Two significant parameters can be determined from this cross-plot: the sign or sense of the correlation, and whether the data falls above or below the 45° (1:1) line (Fig. 15). These two parameters are related; positive correlations characterize the portion above the 1:1 line, and negative correlations those below. The slope of the correlation varies from section to section. Data from individual sections rarely cross the 1:1 line, which appears to represent a naturally significant divide. The groups identified in this

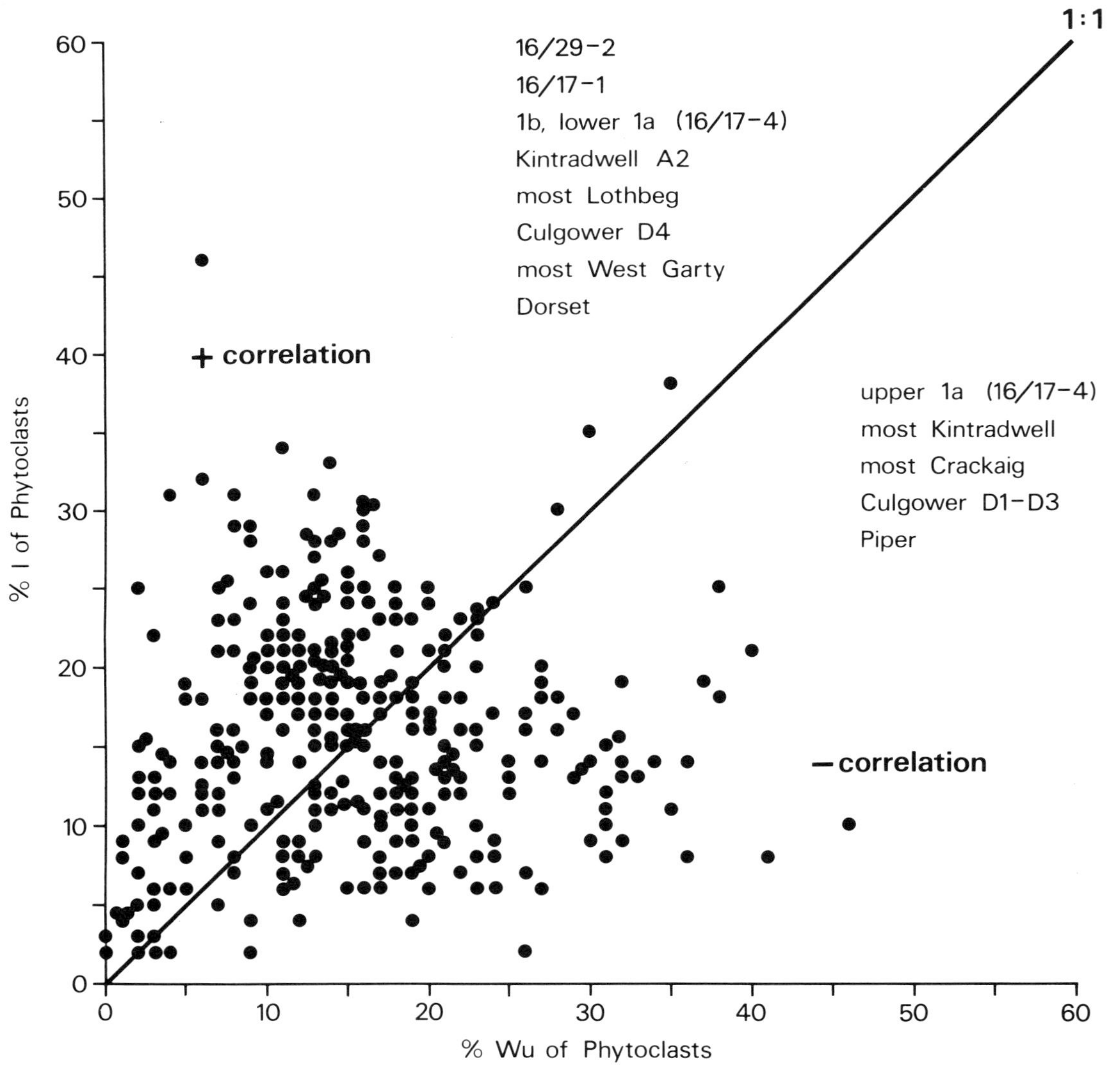

Fig. 15 — Plot of Wu ('brown wood') versus I ('black wood') categories as percentage of the total phytoclast population. Sections plotting below the 1:1 line correspond with trend A in Fig. 14, and those above with trend B.

manner correspond to those differentiated by trends A and B on the previous cross-plot. Trend A corresponds with a negative correlation between Wu and I, and an excess of Wu over I; trend B equates with a positive correlation between Wu and I and an excess of I over Wu.

The abundance and proportion of 'fresh' undegraded 'brown wood' material may be expected to be highest adjacent to source and therefore represent a proximal palynofacies indicator. A number of other workers (including Barnard & Cooper (1981), Barnard *et al.* (1981), Cooper & Barnard (1984) and Fisher & Miles (1983)) have previously stated that the inertinite:vitrinite ratio is highest in 'near-shore' situations.

My data indicate that the %PA proportion of 'black wood' in the phytoclast population is always relatively higher in more basinal sediments (although it is also frequently finer). As PhytOC values decrease distally, the absolute amount of the 'black wood' must decrease into the basin. The regression lines for the total data set (Fig. 14) indicate that the rate of the distal decline in the proportion of 'black wood' is two to three times higher in the basinal sediments than in the shelf-influenced ones. This presumably reflects the differing hydrodynamic equivalence of these particles and their consequent fractionation during transport. The tendency for I and Wu to be negatively correlated in shelf (or shelf-influenced) sediments presumably reflects dilution of the 'black wood' by a constant input of fresh 'brown wood' material.

The phytoclast population of shallow shelf sediments (like Piper) is likely to suffer secondary formation of 'black wood' (degrado-fusinite) material owing to oxidation during marine reworking (particularly in sandy sediments). As the opportunities for such oxidation are clearly higher on the shelf (as the basin is predominantly reducing), and the ultimate source of all phytoclasts is land, it is rather inevitable that the absolute amount of this material will be highest in near-shore settings. The case for a hydrodynamic concentration of inertinite in coastal sediments has not been proven. Most optical studies generally support a relative distal increase in this material (Summerhayes 1987).

Some of the reported trends appear to be based on pyrolysis data, even though pyrolysis assays are bulk analyses which provide only an averaged characterization of the sample. Pyrolysis data are influenced by the nature of the original organic constituents, their degradation states, and the level of maturation and reworking. While the objectivity and utility of the pyrolysis method cannot be denied, I find it a very questionable and undesirable practice to interpret pyrolysis data in terms of the supposed original maceral composition (particularly when such inferences are then used in geological modelling). This tends to lose sight of the fact that the real strength of pyrolysis methods is that they are a rapid, cost-effective and objective means of determining the generative potential of sediments (Peters 1986). Only the direct optical observation of what is actually in the sediment can meaningfully determine the nature of the kerogen assemblage, and thereby the geological factors which control its nature and distribution. Geochemical methods should be used in parallel with optical investigations but not as an alternative.

DISCUSSION OF LATE JURASSIC PALYNOFACIES

The observations on Late Jurassic palynofacies trends are largely compatible with previous studies. The Oxfordian to lower Kimmeridgian interval is characterized by phytoclast-dominated oxidized (inertinite-rich) kerogen assemblages (Claret *et al.* 1981, Bissada 1983, Fisher & Miles 1983, Whitaker 1984). This reflects the extensive shallow ('Piper') shelf seas that existed prior to the Kimmeridgian–Volgian phase of subsidence. Deep-water oxygen deficiency in the major grabens was apparently initiated by the late Oxfordian transgression and spread as the Late Jurassic sea level rose and rates of subsidence increased. Bottom waters became progressively more oxygen deficient during the *mutabilis* Zone to early Volgian interval. The general deepening during this period resulted in lower levels of oxdiation and reworking, and a consequent increase in the vitrinite:inertinite ratio (Fisher & Miles 1983, Pearson & Watkins 1983). This trend is not as clearly indicated by the 'brown wood':'black wood' ratio. This is because the 'black wood' (I) category was defined as being opaque, leaving a 'brown wood' category that incorporated particles with a wide range of translucency.

The middle Volgian regression resulted in widespread and significant changes that appear to partly reflect an increase in the intensity of anoxia, and partly a decrease in sedimentation rates. Together these resulted in the sedimentation of the rich source rocks of the 'hot shale' facies characteristic of the upper parts of the North Sea Kimmeridge Clay Formation and its lateral equivalents. This event appears to be reflected in the Subunit 1a–1b boundary in well 16/17-4A. A regression would normally be expected to result in an increased input of terrigenous organic matter into the basin. This does not seem to have occurred, but a more extensive data base is required to determine representative regional trends in PhytOC values. Decreased (or more localized) basinward transport of phytoclasts is likely to have correlated with the lowered rates of clastic sedimentation within the Volgian basins. This may partly reflect a dynamic sedimentary response to intensified watermass stratification.

Barnard *et al.* (1981) and Cooper & Barnard (1984) have argued that the very organic-rich nature of the Kimmeridge Clay Formation was the result of high primary productivity. They considered that this increase in primary productivity was based on nutrients provided by (or at least correlated with) a large input of terrestrial organic matter. This interpretation has also been suggested for other sequences but it has generally been found wanting (e.g. Waples 1983, p. 975). In addition, the AmexOC versus PhytOC cross-plot (Fig. 9) indicates that the especially organic-rich facies shows a clear and significant excess of marine organic matter.

The stabily stratified nature of the deeper graben basins implies that both the bottom waters and the sediments would have been major nutrient sinks. This would have made sustained high productivity rather unlikely. The typical AmexOC values also show that these rocks are not as amazingly good source rocks as the crude TOC data imply (and the values cited here are significantly higher than the generally reported North Sea means). The characteristic values do not require extravagant rates of palaeo productivity (especially when sedimentation rate effects are taken into account). It is possible that, at the peak(s) of anoxia, nutrient-rich bottom waters may have extended sufficiently high in the water column (in or near the photic zone) for their nutrients to have been at least periodically available to the microplankton. This could have produced episodes of basin eutrophication that may have lead to proliferation of alginitic plankton (such as cyanobacteria and some sulphur bacteria). Partial support for such processes comes from the carbon isotope signal near the top of the 'hot shale' facies (Buchardt & Cederberg 1986).

By far the most organic-rich sediments are the oil shales of the onshore Kimmeridge Clay Formation (particularly those of the Dorset type section). These have not only the highest TOC and AmexOC values but also the lowest PhytOC values. They clearly do not offer any support for terrestrially enhanced levels of palaeo productivity. Their organic richness may be partly emphasized by slow rates of sedimentation, but they are unique in having very high abundances of lamalginite. The latter indicates a mode of origin whose details differ significantly from that of the northern North Sea Kimmeridge Clay Formation.

CONCLUSIONS

I hope that this paper has demonstrated the value and potential of quantitative methods of palynofacies analysis. Such methods provide data that can be used to enhance facies recognition, palaeoenvironmental diagnosis, provenance studies, source rock characterization and correlation work.

The A–P–E diagram can be used to recognize at least 10 palaeoenvironmentally significant palynofacies and to identify several different sediment transport paths. It can also be used to define potential petroleum source rocks (fields IVb and VI–IX in Fig. 12(b) and Table 7). It cannot be guaranteed that the palynofacies fields defined by this study will have the same palaeoenvironmental significance for other data sets. The Late Jurassic basinal facies plot distinctively because they are all consistently dysaerobic to anoxic and therefore characterized by high AOM contents.

Quantitative analysis of the phytoclast components can also prove highly informative. Differences in the way in which these materials are transported allow distal basin facies to be distinguished from shelf and shelf-influenced proximal basin facies (Figs. 14 and 15). This differentiation is supported by independent sedimentological criteria. The phytoclast parameters are less affected by the redox status of the basins, although prolonged exposure to oxidizing conditions is likely to lead to conversion of 'brown wood' to 'black wood' material.

Some reliable form of absolute calibration is essential for the meaningful analysis of palynofacies trends. The combination of %PA data with TOC values has been shown to be extremely valuable. Such analysis allows a much better appreciation of real stratigraphic and regional variations in terrestrial carbon flux and marine organic matter preservation (Figs. 10 and 11). True proximal–distal relationships (with respect to terrestrial sources) can be established. The derived AmexOC parameter is also a much more reliable indicator of source rock character than are simple TOC values. This is borne out by the much better correlation between AmexOC values and hydrogen indices (Larter 1985). It is evident that it is not just the mere abundance of AOM which is important, but its character (maceral composition) and preservation state. These can be optically assessed by the source and intensity of the fluoresence observed under incident blue light; such assessment must form an integral part of any palynofacies investigation. Wherever possible, the data should also be combined with organic carbon accumulation rates.

A major strength of the quantitative palynofacies approach is its ability to identify significant differences in what are, from the sedimentological viewpoint, virtually indistinguishable mudrocks. This characteristic provides enormous potential for both improved facies recognition and subbiostratigraphic correlation studies. It also permits much higher levels of data recovery from monotonous mudrock sequences (assessment of variations in redeposition, terrestrial inputs, redox conditions and source rock potential; recognition of transgressive–regressive trends).

The importance of quantitative analyses of palynomorph assemblages has not been particularly emphasized in this paper. This is because most of this investigation was based on unoxidized kerogen preparations, and many of the examined facies exhibited low palynomorph contents even after oxidation treatment. However, it is clear that palynomorphs provide much more sensitive indices of sorting effects than do bulk kerogen or phytoclast parameters (especially the proportions of bisaccate pollen and chorate dinocysts).

The dinocysts: prasinophycean algae ratio appears diagnostic. A high relative abundance of prasinophycean algae is especially characteristic of stabily stratified basins where the *in situ* production of fossilizing dinocysts is at a minimum (Tyson 1984, 1987). This group of microplankton is therefore particularly characteristic of the best quality basinal 'black shale' petroleum source rocks (Table 7).

ACKNOWLEDGEMENTS

The bulk of this research was carried out at the Open University, Milton Keynes, while in receipt of a Natural Environmental Research Council CASE studentship with Robertson Research International. The cooperation of these bodies, and of Occidental and Phillips Petroleum, is gratefully acknowledged. Janet Baker (University College London) is thanked for her careful preparation of the figures. Colin Summerhayes and Erik Thomsen are thanked for their encouragement.

REFERENCES

Bailey, E. B. & Weir J. (1932). Submarine faulting in Kimmeridgian times: east Sutherland. *Transactions of the Royal Society of Edinburgh* **47** 431–467.

Barnard, P. C., Collins, A. G. & Cooper, B. S. (1981). Identification and distribution of kerogen facies in a source-rock horizon: examples from the North Sea. In: Brooks, J. (ed.), *Organic maturation studies and fossil exploration.* Academic Press, London, pp. 271–282.

Barnard, P. C. & Cooper, B. S. (1981). Oils and source rocks of the North Sea area. In: Illing L. V. & Hobson G. D. (eds), *Petroleum geology of the continental shelf of north-west Europe.* Heyden, London, pp. 169–175.

Barss, M. S. & Williams, G. L. (1973). Palynology and nannofossil processing techniques. *Geological Survey of Canada, Paper* **73—26** 1–25.

Bissada, K. K. (1983). Petroleum generation in Mesozoic sediments of the Moray Firth Basin, British North Sea. In: Bjorøy, M. *et al.* (eds), *Advances in organic geochemistry 1981.* Wiley, Chichester, pp. 7–15.

Boote, D. R. D. & Gustav, S. H. (1987). Evolving depositional systems within an active rift, Witch Ground Graben. In: Brooks, J. & Glennie, K. W. (eds), *Petroleum geology of north west Europe*, Vol. 2. Graham & Trotman, London, pp. 819–834.

Brookfield, M. E. (1976) The age of the Allt na Cùile Sandstones (Upper Jurassic, Sutherland). *Scottish Journal of Geology* **12** 181–186.

Buchardt, B. & Cederberg, T. (1986). Carbon-isotope composition of organic matter from the Jurassic sediments of the Danish Central Graben: implications for palaeoenvironmental interpretations and oil/source-rock correlations. In: *Abstracts of the 3rd Conference on Petroleum Geology of NW Europe, London, 26–29 October 1986.*

Claret, J., Jardine, S. & Robert, P. (1981). La diversité des roches mères petrolières: aspects géologiques et implications économiques à partir de quatre exemples. *Bulletin des Centres de Recherches Exploration – Production, Elf-Aquitaine* **5** 383–417.

Cooper, B. S. & Barnard, P. C. (1984). Source rocks and oils of the central and northern North Sea. In: Demaison, G. & Murris, R. J. (eds), *Petroleum geochemistry and basin evaluation. Memoir of the American Association of Petroleum Geologists* **35** 303–314.

Cosgrove, M. E. (1970). Iodine in the bituminous Kimmeridgian shales of the Dorset coast, England. *Geochemica Cosmochimica Acta* **34** 830–836.

Cox, B. M. & Gallois, R. W. (1981). The stratigraphy of the Kimmeridge Clay of the Dorset type area and its correlation with some other Kimmeridgian sequences. *Report, Institute of Geological Sciences, London* **80/4**, 1–44.

Dunn, C. E. (1974). Identification of sedimentary cycles through Fourier analysis of geochemical data. *Chemical Geology* **13** 217–232.

Ebukanson, E. J. & Kinghorn, R. R. F. (1985). Kerogen facies in the major Jurassic mudrock formations of southern England and the implications on the depositional environments of their precursors. *Journal of Petroleum Geology* **8** 435–462.

Ehrmann, W. U. & Thiede, J. (1985). History of Mesozoic and Cenozoic sediment fluxes to the North Atlantic Ocean. *Contributions to Sedimentology* **15** 1–109.

Farrimond, P., Comet, P., Eglinton, G., Evershed, R. P., Hall M. A., Park, D. W. & Wardroper, A. M. K. (1984). Organic geochemical study of the Upper Kimmeridge Clay of the Dorset type area. *Marine and Petroleum Geology* **1** 340–354.

Fisher, M. J. & Miles, J. A. (1983). Kerogen types, organic maturation and hydrocarbon occurrences in the Moray Firth and South Viking Graben, North Sea Basin. In: Brooks, J. (ed.), *Petroleum geochemistry and exploration of Europe. Special Publication, Geological Society, London* **12** 195–201.

Harker, S. D., Gustav, S. H. & Riley, L. A. (1987). Triassic to Cenomanian stratigraphy of the Witch Ground Graben. In: Brooks, J. & Glennie, K. W. (eds), *Petroleum geology of north west Europe*, Vol. 2. Graham & Trotman, London, pp. 809–818.

Harris, J. P. & Fowler, R. M. (1987). Enhanced prospectivity of the Mid–Late Jurassic sediments of the South Viking Graben, northern North Sea. In: Brooks, J. & Glennie, K. W. (eds), *Petroleum geology of north west Europe*, Vol. 2. Graham & Trotman, London, pp. 879–898.

House, M. R. (1985). A new approach to an absolute timescale from measurements of orbital cycles and sedimentary micro-rhythms. *Nature* **315** 721–725.

Hutton, A. C. (1987). Petrographic classification of oil shales. *International Journal of Coal Geology* **8** 203–231.

Ioannides, N. S., Stavrinos, G. N. & Downie C. (1976). Kimmeridgian microplankton from Clavell's Hard, Dorset, England. *Micropalaeontology* **22** 443–478.

Kessler, L. G. & Moorhouse, K. (1984). Depositional processes and fluid mechanics of Upper Jurassic conglomerate accumulations, British North Sea. In: Koster, E. H. & Steel, R. J. (eds), *Sedimentology of gravels and conglomerates. Memoir of the Canadian Society of Petroleum Geologists* **10** 383–397.

Lam, K. & Porter, R. (1977). The distribution of palynomorphs in the Jurassic rocks of the Brora Outlier, NE Scotland. *Journal of the Geological Society, London* **134** 45–55.

Larter, S. R. (1985). Integrated kerogen typing in the recognition and quantitative assessment of petroleum source rocks. In: Thomas, B. M. *et al.* (eds), *Petroleum geochemistry in exploration of the Norwegian Shelf.* Graham & Trotman, London. pp. 269–286.

Linsley, P. N. (1972). The Stratigraphy and sedimentology of the Kimmeridgian deposits of Sutherland. *Ph.D. Thesis.* University of London, pp. 1–214.

Linsley, P. N., Potter, H. C., McNab, G. & Racher, D. (1980). The Beatrice Field, Inner Moray Firth, U.K. North Sea. In: Halbouty, M. T. (ed.), *Giant oil and gas fields of the decade 1968–1978. Memoir of the American Association of Petroleum Geologists* **30** 117–129.

MacDonald, A. C. (1985). Kimmeridgian and Volgian fault margin sediments in the northern North Sea. *Ph.D. Thesis.* University of Strathclyde, pp. 1–255.

Maher, C. E. (1980). Piper oilfield. In: Halbouty, M. T. (ed.), *Giant oil and gas fields of the decade 1968–1978. Memoir of the American Association of Petroleum Geologists* **30** 131–172.

Maher, C. E. (1981). The Piper oilfield. In: Illing, L. V. & Hobson, G. D. (eds), *Petroleum geology of the continental shelf of north-west Europe.* Heyden, London, pp. 358–370.

Mebradu, S. (1978). Stratigraphic palynology of the Upper Jurassic of the Dorset coast, England. Part 1: statistical variables. *Palinologia, Special Issue* **1** 327–333.

Morris, N. J. (1968). Stratigraphical and palaeontological researches in the Upper Jurassic rocks. *Ph.D. Thesis.* University of Oxford, pp. 1–234.

Pearson, M. J. & Watkins, D. (1983). Organofacies and early maturation effects in Upper Jurassic sediments from the Inner Moray Firth Basin, North Sea. In: Brooks, J. (ed.) *Petroleum geochemistry and exploration of Europe. Special Publication, Geological Society, London* **12** 147–160.

Peters, K. E. (1986). Guidelines for evaluating petroleum source rocks using programmed pyrolysis. *Bulletin of the American Association of Petroleum Geologists* **70** 318–329.

Pickering, K. T. (1984). The Upper Jurassic 'Boulder Beds' and related deposits: a fault-controlled submarine slope, NE Scotland. *Journal of the Geological Society, London* **141** 357–374.

Pocock, S. A. J. (1982). Identification and recording of particulate sedimentary organic matter. In: Staplin, F. L., Dow, W. G., Milner, C. W. D., O'Connor, D. I., Pocock, S. A. J., van Gijzel, P., Welte, D. H. & Yukler, M. A. (eds), *How to assess maturation and paleotemperatures. Society of Economic Paleontologists and Mineralogists, Short Course Notes* **7** 13–131.

Rawson, P. F. & Riley, L. A. (1982). Latest Jurassic–Early Cretaceous events and the 'Late Cimmerian Unconformity' in North Sea Area. *Bulletin of the American Association of Petroleum Geologists* **66** 2628–2648.

Riley, L. A. (1974). A palynological investigation of Upper Jurassic–basal Cretaceous sediments from England France and Iberia. *Ph.D. Thesis.* Open University, Milton Keynes, pp. 1–289.

Scotchman, I. C. (1987). Relationship between clay diagenesis and organic maturation in the Kimmeridge Clay Formation, onshore U.K. In: Brooks, J. & Glennie, K. (eds), *Petroleum geology of north west Europe,* Vol. 1. Graham & Trotman, London, pp. 251–262.

Senftle, J. T., Brown, J. H. & Larter, S. R. (1987). Refinement of organic petrographic methods for kerogen characterization. *International Journal of Coal Geology* **7** 105–117.

Stach, E., Mackowsky, M. T., Teichmuller, M., Taylor, G. H., Chandra, D. & Teichmuller, R. (1982). *Stach's textbook of coal petrology,* 3rd edition. Gebruder Borntraeger, Berlin, pp. 1–535.

Stow, D. A. V. (1984). Upper Jurassic overlapping-fans slope-apron system: Brae Oilfield, North Sea. *Geo-Marine Letters* **3** 217–222.

Stow, D. A. V. (1985). Deep-sea clastics: where are we and where are we going? In: Brenchley, P. J. & Williams, B. P. J. (eds), *Sedimentology; recent developments and applied aspects.* Blackwell Scientific, Oxford, pp. 67–93.

Stow, D. A. V., Bishop, C. D. & Mills, S. J. (1982). Sedimentology of the Brae Oilfield, North Sea: fan models and controls. *Journal of Petroleum Geology* **5** 129–148.

Summerhayes, C. P. (1987). Organic-rich Cretaceous sediments from the North Atlantic. In: Brooks, J. & Fleet, A. J. (eds), *Marine petroleum source rocks. Special Publication, Geological Society, London* **26** 301–316.

Surlyk, F. (1975). Block faulting and associated marine sedimentation at the Jurassic–Cretaceous boundary, East Greenland. In: Finstad, K. G. & Selley, R. C. (eds), *Proceedings of the Jurassic Northern North Sea Symposium, Stavanger, September 1975.* Norwegian Petroleum Society. **JNNSS/7** 1–31.

Surlyk, F. (1978). Submarine fan sedimentation along fault scarps on tilted blocks (Jurassic–Cretaceous boundary, east Greenland). *Grønlands Geologiske Undersøgelse* **128** 1–108.

Surlyk, F. (1987). Slope and deep shelf gully sandstones, Upper Jurassic, east Greenland. *Bulletin of the American Association of Petroleum Geologists* **71** 464–475.

Surlyk, F., Callomon, J. H., Bromley, R. G. & Birkelund, T. (1973). Stratigraphy of the Jurassic–Lower Cretaceous sediments of Jameson Land and Scoresby Land, east Greenland. *Meddelelser om Grønland* **193** 1–76.

Sykes, R. M. (1975). Facies and faunal analysis of the Callovian and Oxfordian Stages (Middle–Upper Jurassic) in northern Scotland and east Greenland. *Ph.D. Thesis.* University of Oxford, pp. 1–312.

Turner, C. C., Cohen, J. M., Connell, E. R. & Cooper, D. M. (1987). A depositional model for the South Brae Oilfield. In: Brooks, J. & Glennie, K. W. (eds), *Petroleum geology of north west Europe.* Graham & Trotman, London, pp. 853–864.

Turner, C. C., Richards, P. C., Swallow, J. L., & Grimshaw, S. P. (1984). Upper Jurassic stratigraphy and sedimentary facies in the central Outer Moray Firth Basin. *Marine and Petroleum Geology* **1** 105–117.

Tyson, R. V. (1984). Palynofacies investigation of Callovian (Middle Jurassic) sediments from D.S.D.P. Site 534, Blake-Bahama Basin, western central Atlantic. *Marine and Petroleum Geology* **1** 3–13.

Tyson, R. V. (1985). Palynofacies and sedimentology of some Late Jurassic sediments from the British Isles and northern North Sea. *Ph.D. Thesis.* The Open University, Milton Keynes, pp. 1–623.

Tyson, R. V. (1987). The genesis and palynofacies characteristics of marine petroleum source rocks. In: Brooks, J. & Fleet, A. J. (eds), *Marine petroleum source rocks. Special Publication, Geological Society, London* **26** 47–67.

Tyson, R. V., Wilson, R. C. L. & Downie, C. (1979). A stratified water column environmental model for the type Kimmeridge Clay. *Nature* **277** 377–380.

Waples, D. W. (1983). Reappraisal of anoxia and organic richness with emphasis on Cretaceous of North Atlantic. *Bulletin of the American Association of Petroleum Geologists* **67** 963–978.

Whitaker, M. F. (1984). The usage of palynology in definition of Troll field geology. *Proceedings of the 6th Offshore Northern Seas Conference, Stavanger, 1984*, Paper G6.

Williams, J. J., Conner, D. C., & Peterson, K. E. (1975). Piper: a fault block structure with Upper Jurassic beach/bar reservoir sands. *Bulletin of the American Association of Petroleum Geologists* **59** 1585–1601.

Williams, P. F. V. (1986). Petroleum geochemistry of the Kimmeridge Clay of onshore southern and eastern England. *Marine and Petroleum Geology* **3** 258–281.

Williams, P. F. V. & Douglas, A. G. (1980). A preliminary organic geochemical investigation of the Kimmeridgian oil shales. In: Douglas, A. G. & Maxwell, J. R. (eds), *Advances in organic geochemistry 1979. Physics and Chemistry of the Earth* **12** 531–546.

Williams, P. F. V. & Douglas, A. G. (1983). The effects of lithologic variation on organic geochemistry in the Kimmeridge Clay of Britain. In: Bjorøy, M. *et al.* (eds), *Advances in organic geochemistry 1981.* Wiley Chichester, pp. 16–27.

7

Palaeoenvironmental interpretation and distribution of the lower Kimmeridgian foraminifera from the Helmsdale–Brora Outlier, northeast Scotland

Francis John Gregory

ABSTRACT

The foraminiferal assemblages recovered from the lower Kimmeridgian (*cymodoce* Zone to *eudoxus* Zone) of the Helmsdale–Brora Outlier are characterized by both low abundance and low diversity. 24 foraminiferal species are illustrated and briefly described. Poor preservation, the dominance of calcareous foraminifera associated with anoxic black shales, and sediments associated with syndepositional tectonics indicate that the assemblages have been transported. Originally, in the lower Kimmeridgian, two environments were present, separated by submarine faulting, with the foraminifera inhabiting the upthrown shallow-water block. The foraminifera were subsequently transported, owing to either seismic activity or climatic episodes, into the downfaulted anoxic basin during the lower Kimmeridgian. The *cymodoce* Zone is characterized by decreasing nodosariid and agglutinated foraminifera and increasing spirillinid dominance, which may be due to shallowing. The *mutabilis* and *eudoxus* Zones are characterized by a general absence of foraminifera.

INTRODUCTION

The coastal exposure from Golspie to the Ord of Caithness of east Sutherland is composed of an outlier of Mesozoic strata. These range from the Late Triassic to the Late Jurassic (Kimmeridgian), of which the latter can possibly be extended to the Portlandian according to palynological evidence presented by Riley (1980). The outlier is bounded to the west by the Helmsdale Fault. This was active during the Kimmeridgian, having profound effects upon sedimentation, including the deposition of the spectacular Boulder Beds (Bailey & Weir 1932). Old Red Sandstone, metamorphosed Moine Schists and the Helmsdale Granite crop out to the west of the Helmsdale Fault.

This study was initiated as a micropalaeontological reconnaissance of the lower Kimmeridgian of the outlier, from Kintradwell (Grid Reference NH 073933) to Sron Rubha na Gaoithe

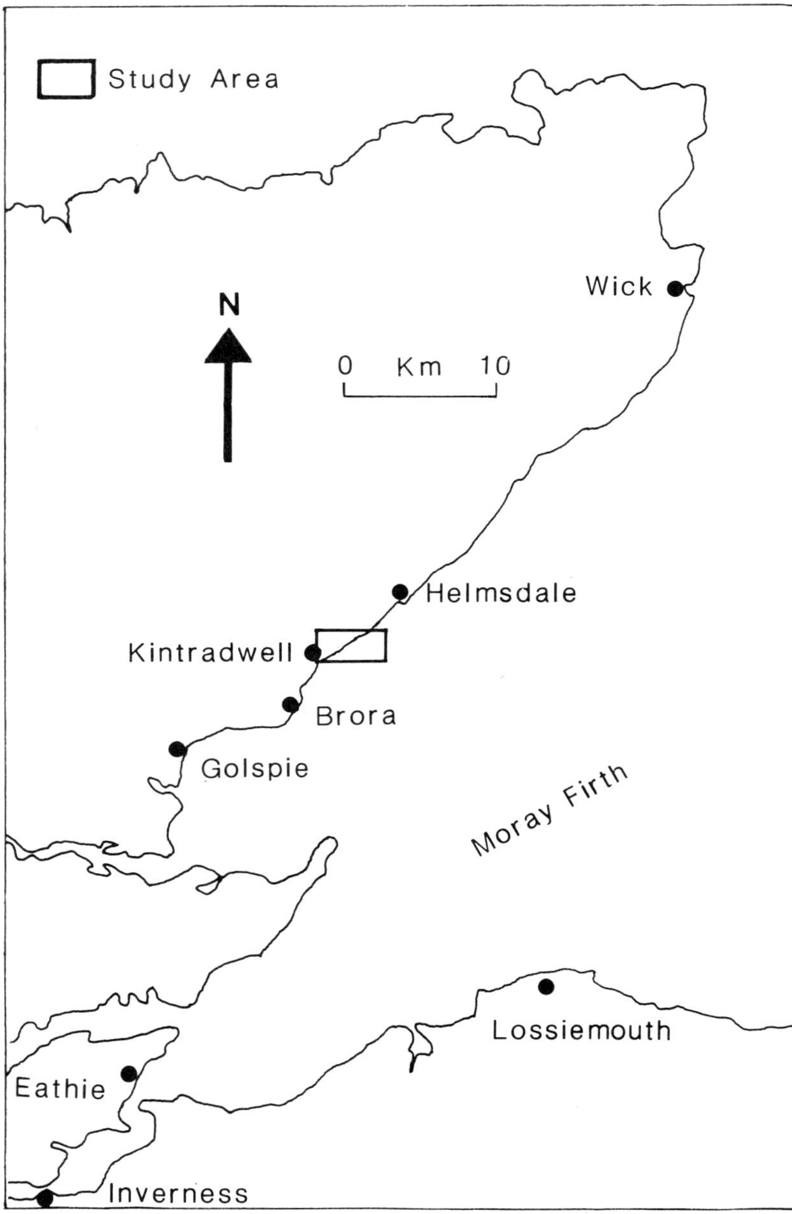

Fig. 1 — Map showing the location of the study area in northeastern Scotland.

(Grid Reference NH 125995) (Figs. 1 and 2). The exposures were logged in detail and 40 samples were taken from the shale facies developed in the *cymodoce* Zone (Kintradwell Boulder Beds), the *mutabilis* Zone (Lothbeg Point Shales) and the *eudoxus* Zone (the lower part of the Helmsdale Boulder Beds) (Fig. 3). Only the foraminifera recovered are presented in this paper.

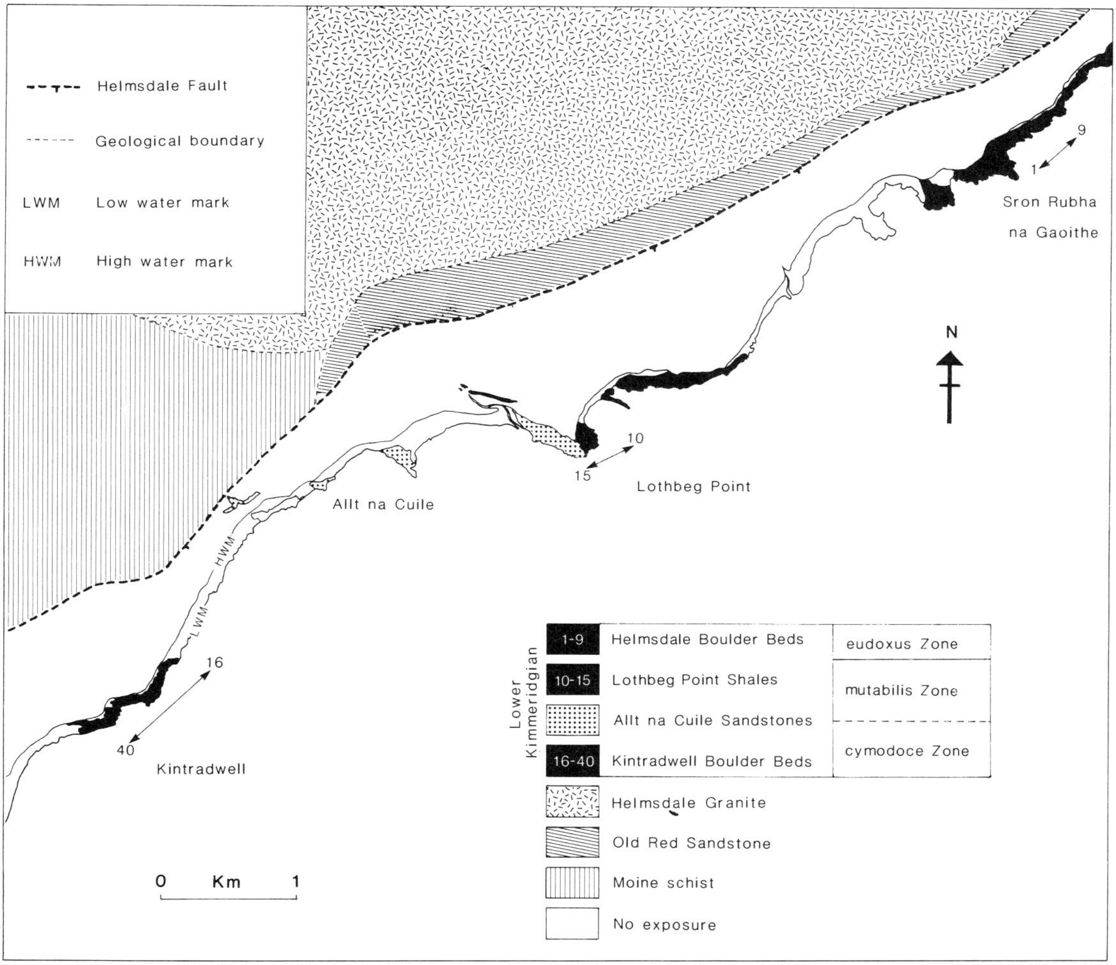

Fig. 2 — Locality map of the lower Kimmeridgian deposits and sampling points (based on the present author's logging) and showing the regional geology. (After Pickering 1984.)

STRATIGRAPHY

The stratigraphy used here follows the work of Neves & Selley (1975), Lam & Porter (1977) and Pickering (1984). This in turn is based upon earlier surveys carried out by Murchison (1829), Judd (1875), Lee (1925) and Lee & Pringle (1932). Lithological units defined from the above are, in ascending stratigraphical order (Fig. 3)

(1) the Kintradwell Boulder Beds,
(2) the Allt na Cuile Sandstones, which are in part laterally equivalent to the preceding unit (Neves & Selley 1975, Lam & Porter 1977),
(3) the Lothbeg Point Shales, and
(4) the Helmsdale Boulder Beds.

Since the coastal exposure is not continuous, the logs produced (Fig. 3) represent composite sections. It also must be noted that the laterally discontinuous nature of some of the lithofacies means that the sections as shown here in

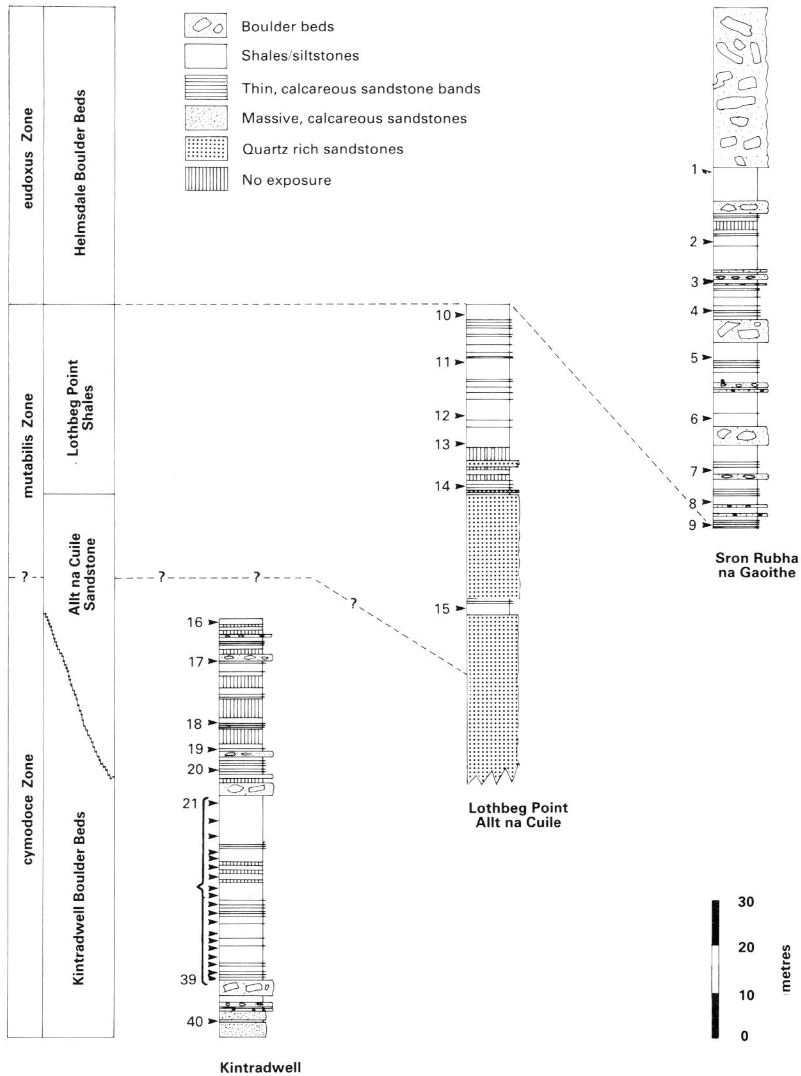

Fig. 3 — Stratigraphic sections, sampling points and the ammonite zonation of the lower Kimmeridgian.

Fig. 3 may not compare with other published data.

LITHOLOGY

The principal components of the lithological units are described here and will later be incorporated into the palaeoenvironmental model. These are boulder beds intercalated with shales, siltstones, calcareous sandstones and quartz-rich sandstones.

The boulder beds produce the most prominent erosive feature along the foreshore exposures. They occur as two stratigraphically separate units: the lower unit is placed in the *cymodoce* zone (Kintradwell Boulder Beds) while the upper extends from the *eudoxus* Zone to the *pallasioides?* Zone (Helmsdale Boulder Beds). Their origin and mode of deposition has provoked much speculation. Bailey & Weir (1932) were the first to attribute them to rapid deposition following periodic Kimmeridgian activity along the submarine Helmsdale Fault, a hypothesis that has been developed further by many workers, including Linsley (1972), Neves & Selley (1975) and Pickering (1984). These beds are notable for containing large numbers of randomly orientated angular blocks. These show little or no evidence of erosion and are mainly composed of well-bedded, yellowish-grey, well-sorted, quartz-rich sandstones. Bailey & Weir (1932) pointed out a similarity between these clasts and the Old Red Sandstone (Caithness Flagstone) exposed to the west of the Helmsdale Fault. From this it is evident that these blocks were derived from an Old Red Sandstone palaeo-surface. The matrix varies from mud to sand grade and often includes shell debris, echinoid spines and plant material, indicating derivation from near-shore marine and terrigenous sources. The bed form is characteristically irregular and laterally discontinuous. The thickness of individual beds varies from a few centimetres to several metres. Internally, the beds show major slumping and jumbling of the clasts, indicating a sliding emplacement down the Helmsdale Fault scarp (Crowell 1961, Pickering 1984). They interdigitate with and deform the underlying sediments, which are composed of black shales, interbedded irregularly with soft siltstones and thin hard fine-grained calcareous sandstones.

The shales occur as thin (0.5–2 mm) featureless laminae and are rich in organic carbon (Fig. 4), palynomorphs (Lam & Porter 1977) and plant debris (Van der Burgh & Van Konijnenburg-van Cittert 1984). Siltstones generally attain greater thickness than the shales (0.5 mm–10 cm). Sedimentary features include small-scale ripples and parallel lamination, both of which can be traced by the presence of organic debris. Bed form is irregular and is usually associated with scouring of the underlying sediments.

Thinly bedded hard fine-grained calcareous sandstones are interbedded in groups with the shales and siltstones (see Fig. 3). Bed thickness is between 1 mm and 10 cm, although it occasionally reaches 2 m. The few sedimentary features that are developed mainly consist of grain size fining upwards and parallel lamination. Poorly defined cross-bedding occurs in the thicker beds. The lower surfaces are sharp and erosive, while the upper surfaces can vary from sharp to transitional. Bed form is irregular and laterally discontinuous.

The Allt na Cuile Sandstone is a major unit (Fig. 3), exposed from Lothbeg Point (Grid Reference NH 095961) to the north of Kintradwell (Grid Reference NH 089937) (Fig. 2) and is composed of evenly bedded well-sorted, white to orange, coarse- to medium-grained, quartz-rich sandstones. Occasional autobrecciation can be seen, while some conglomerates occur in the lower parts.

The age of this unit has not been exactly defined because of the scarcity of fossils. The upper part of the Allt na Cuile Sandstone is overlain by the Lothbeg Point Shales which contain ammonites indicative of the *mutabilis* Zone. Brookfield (1976) tentatively identified the sparse bivalve and ammonite fauna preserved in these sandstones and indicated an age range from basal *cymodoce* Zone to the *mutabilis* Zone. Sedimentological data presented by Linsley (1972), Neves & Selley (1975) and Pickering (1984) suggest that these sandstones represent chasm deposits that bypassed the Helmsdale Fault scarp system (Fig. 5). If so this unit is, in part, a lateral equivalent of the Kintradwell Boulder Beds. However, direct field evidence of this is lacking.

METHODS

The samples were typically well indurated and resisted normal paraffin preparation. So they were boiled for several hours in a sodium bicarbonate solution. Every $\frac{1}{2}$ hr any residue present was removed to save damage by excess boiling. The residue was then washed and dried and all fractions above 63 μm were examined for micro-

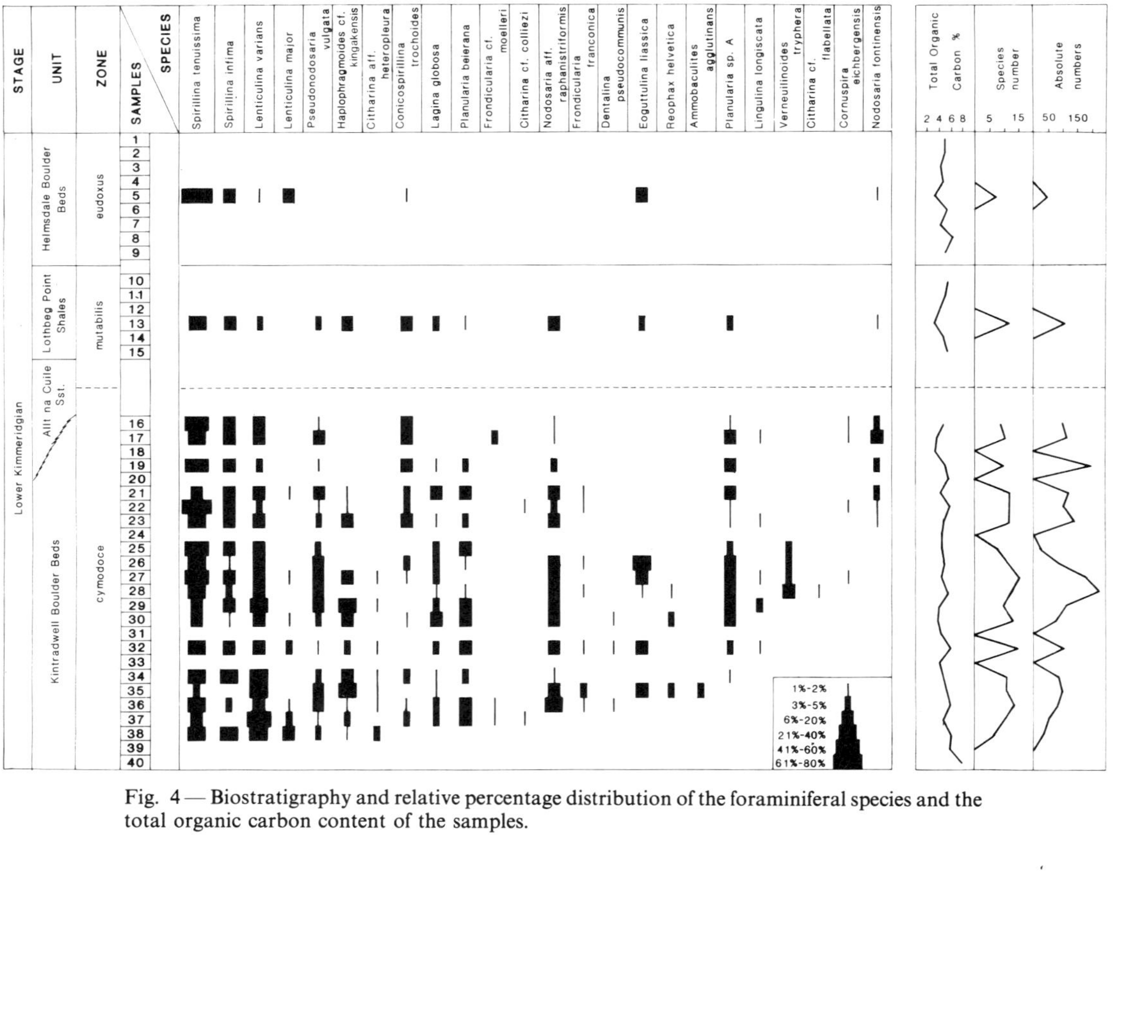

Fig. 4 — Biostratigraphy and relative percentage distribution of the foraminiferal species and the total organic carbon content of the samples.

fossils. The procedure for determining the organic carbon content follows the titrative methods of Walkley (1947) and Tinsley (1950).

FORAMINIFERAL ASSEMBLAGES

Of the 40 samples collected, 20 were found to be barren, especially in the *mutabilis* and *eudoxus* Zones. The assemblages recovered were typified by low abundance and low diversity (Fig. 4). Only 24 species were recovered, with most samples containing less than 14 species. Another unusual characteristic is the dominance of only a few calcareous species, especially *Spirillina tenuissima* Gümbel. This taxon accounts for 35% of all the foraminifera recorded. *Lenticulina varians* (Bornemann), *Pseudonodosaria vulgata*

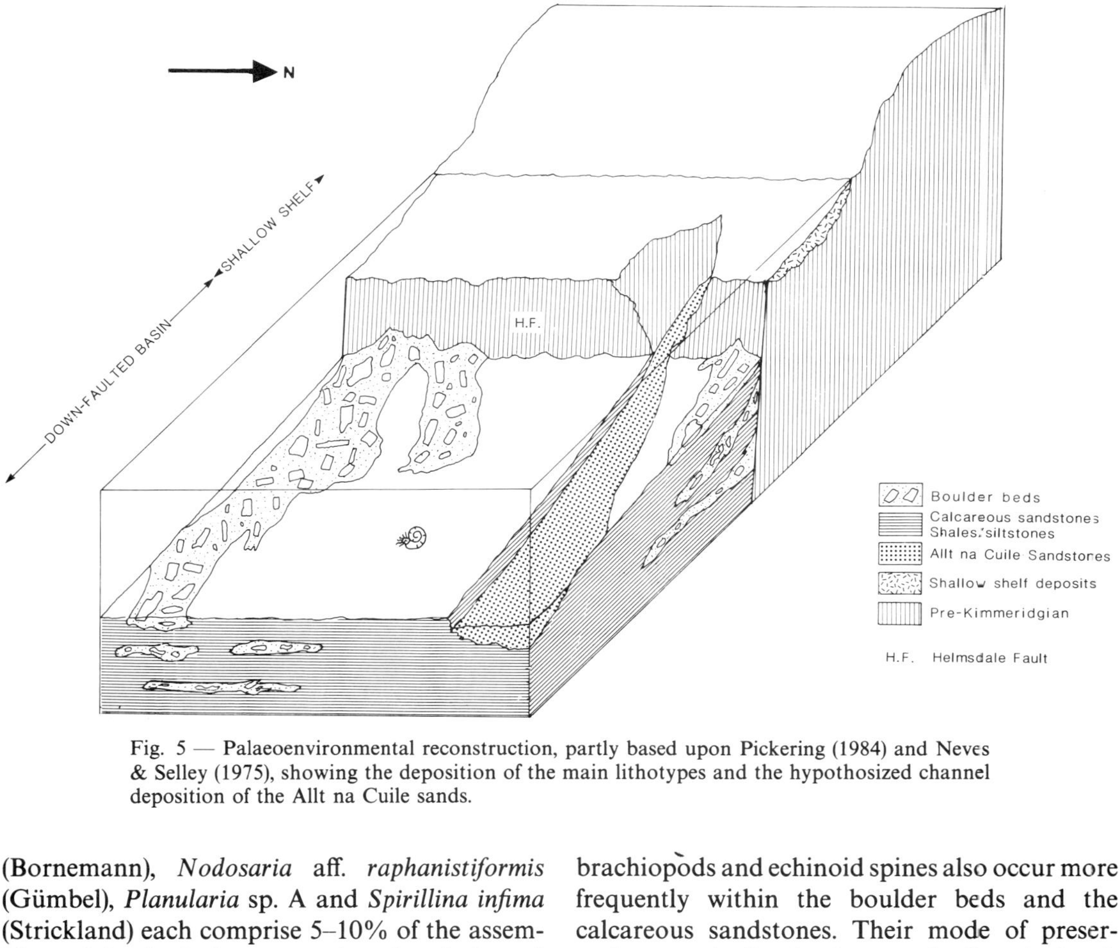

Fig. 5 — Palaeoenvironmental reconstruction, partly based upon Pickering (1984) and Neves & Selley (1975), showing the deposition of the main lithotypes and the hypothesized channel deposition of the Allt na Cuile sands.

(Bornemann), *Nodosaria* aff. *raphanistiformis* (Gümbel), *Planularia* sp. A and *Spirillina infima* (Strickland) each comprise 5–10% of the assemblages (Fig. 4). Only one species of agglutinated foraminifer, *Haplophragmoides* cf. *kingakensis* Tappan, occurs in any abundance, mainly in the lower part of the *cymodoce* Zone (Fig. 4). All the foraminifera were found to be poorly preserved, often broken and crushed.

Each sample possesses a high organic carbon content (Fig. 4). This, together with fine lamination and a lack of bioturbation indicates anoxic conditions. This is problematical, however, because of the dominance of calcareous foraminifera that normally imply well-oxygenated conditions. This has been shown by work on Recent benthic foraminiferal faunas (e.g. Boltovskoy & Wright 1976, Murray 1973). Associated with the foraminifera is a very limited *in situ* macrofauna of epifaunal bivalves. Corals, bivalves, brachiopods and echinoid spines also occur more frequently within the boulder beds and the calcareous sandstones. Their mode of preservation, typically broken and disarticulated, indicates some transportation. Ammonites are quite common in the *cymodoce* Zone as pelagic components of the macrofauna, tending to become rarer upwards through the *mutabilis* and *eudoxus* Zones.

FORAMINIFERA AND PALAEOENVIRONMENTS

The following palaeoenvironmental model presented for the foraminiferal assemblages (see Fig. 5) draws upon the sedimentological models of Neves & Selley (1975) and Pickering (1984). A high abundance of spirillinids and nodosariids with simple agglutinated foraminifera (Figs. 4

and 6) may be taken to indicate that the original environment was a shallow-water oxygenated shelf. This conclusion is based on comparable palaeoecological work of Boltovskoy & Wright (1976) and Murray (1973) on Recent microfaunas and the interpretations of Gordon (1970), Johnson (1977) and Stam (1986) on Jurassic microfaunas. The shelf is believed to have been created as an upthrown block during the Kimmeridgian by the reactivation of the Helmsdale Fault. For a more detailed discussion of the Helmsdale Fault in relation to the regional tectonics of the Moray Firth, see McQuillin *et al.* (1982) and Bird *et al.* (1986).

Foraminifera were then removed sporadically from this environment, with transport initiated either by the periodic seismic activity along the Helmsdale Fault or by climatic or tidal factors. Removal occurred by gravity-induced turbidite flows (Pickering 1984) down the fault scarp into the downfaulted basin, thus accounting for the mixed shallow-water foraminiferal assemblages associated with deeper-water sediments found here. The boulder beds, silts and calcareous sands were deposited by the same mechanisms, but the sands forming the Allt na Cuile Sandstones bypassed this fault scarp and represent chasm desposits (e.g. Neves & Selley 1975, Pickering 1984) (see Fig. 5).

The basinal hemipelagic shales were anoxic, indicated by their laminated nature and high organic carbon content, associated with a very limited *in situ* benthonic macrofauna of epifaunal bivalves and pelagic ammonites. These factors point to a generally restricted environment not conducive to foraminiferal colonization. This type of restricted marine environment was particularly prevalent in the Kimmeridgian as summarized by Morris (1980). Agglutinated species have been recorded in restricted environments, (Løfaldli & Nagy 1983), but high carbon content may produce extreme substrate pH (e.g. Kureshy 1969, Boltovskoy & Wright 1976). This can affect the general physiology and the ability of foraminifera to secrete a calcareous test (Bradshaw 1961).

Since these assemblages are derived, palaeoecological interpretations must be based on comparisons with work carried out by Johnson (1977) and Stam (1986) on similar Jurassic faunas. Contraction and expansion of niche availability resulting from sea-level changes may be a cause of fluctuations in the abundance of spirillinids and nodosariids in this region (Fig. 6). Thus the former are envisaged to have occurred in larger numbers during shallowing events, while an increase in nodosariids may reflect a return to deeper-water conditions on the shelf.

Overall in these lower Kimmeridgian deposits there is an increase of spirillinids with time, which is associated with a corresponding decrease in nodosariids and agglutinated foraminifera (Fig. 6). This may indicate a general localized shallowing throughout the Kimmeridgian.

The totally barren samples are harder to explain. It is possible that they represent periods of extreme sea-level drop, reducing the availability of niches on the upthrown block. Another possible factor is general oceanic anoxia affecting both environments. This is supported by the general rarity of ammonites in the *mutabilis* and *eudoxus* Zones.

The foraminiferal assemblages recovered from these onshore deposits are similar to those found in the Inner Moray Firth Basin (P. Copestake 1987 pers. comm.). These offshore assemblages are, however, even lower in diversity and abundance. This can be accounted for by the greater distance from the original near-shore source environment.

BIOSTRATIGRAPHY AND COMPARISON WITH OTHER KIMMERIDGIAN ASSEMBLAGES

There is little published work on foraminifera of the lower Kimmeridgian of Britain. Part of the present author's research project on the foraminifera of Scotland has been concerned with the lower Kimmeridgian deposits of Staffin Bay, Skye and Eathie Haven, northeast Scotland. The microfaunas of Staffin Bay are characterized by a dominance of agglutinated foraminifera, with nodosariids competing successfully. The fauna

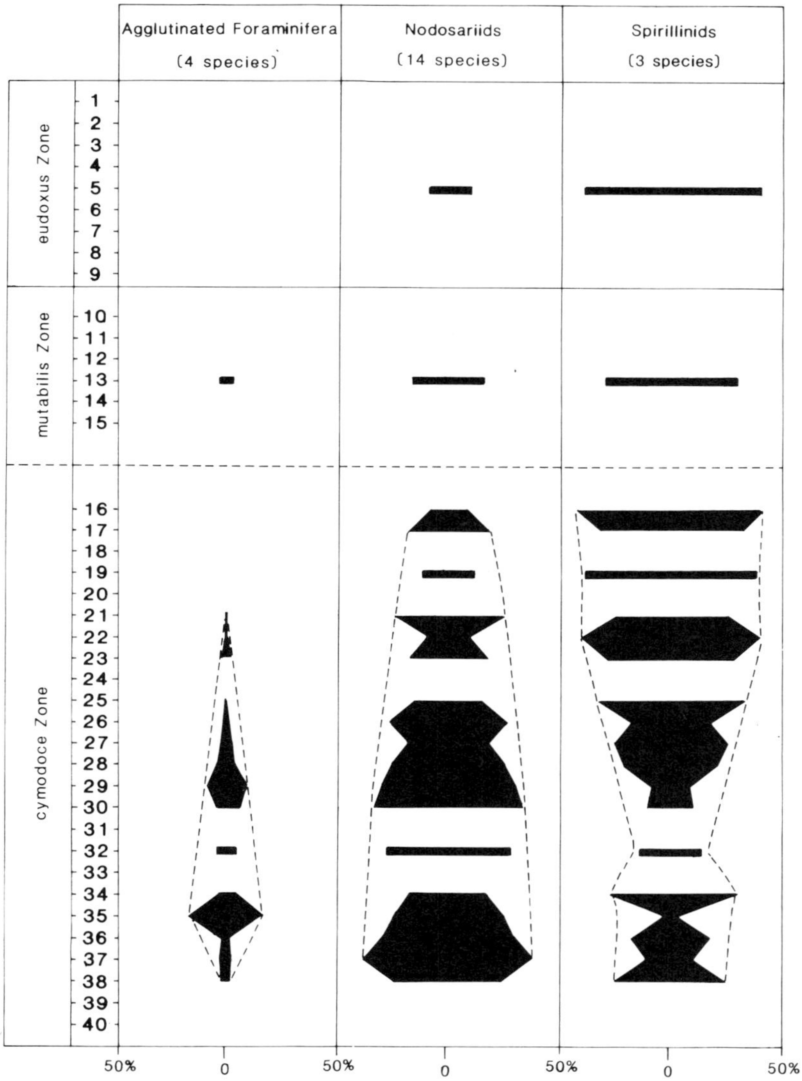

Fig. 6 — Percentage distribution of the major foraminiferal groupings, showing general trends and relative abundance.

recovered from Eathie Haven is similar to that found at Helmsdale–Brora, but the abundance and diversity are even lower. This represents deeper-water deposition, equivalent to facies developed in the Jurassic of the North Sea.

Lloyd (1959, 1962) published accounts on the type Kimmeridgian at Dorset, but only on the agglutinated, miliolid and rotalid faunas. However, Shipp's (unpublished data, 1981) studies indicate a dominance of nodosariids here. Barnard *et al.* (1981) also report an abundance of the nodosariids associated with agglutinated forms. Medd (1979) studied the lower Kimmeridgian of the Worlaby Borehole, north Humberside, which also shows a similar pattern of distribution.

Therefore it appears that nodosariids and

agglutinated foraminifera generally predominate in most of these deposits. So the abundance of the spirillinid fauna of Helmsdale–Brora is indicative of the unusual depositional environment present there.

Regarding the biostratigraphy of these deposits, it is evident that most of the foraminifera are long-ranging species, being characteristic of the Middle and Upper Jurassic. The upper extent, or final appearences of many forms have been extended here to the top of the *cymodoce* Zone and these include *Frondicularia franconica* (Gümbel), *Reophax helvetica* (Haeusler), *Lingulina longiscata* (Terquem) and *Verneuilinoides tryphera* Loeblich & Tappan.

One form, *Planularia* sp. A, appears to be restricted to the lower Kimmeridgian (*baylei* Zone to *mutabilis* Zone); this is from this work for Scotland and from Young's (1988) data for central and southern England. Further work is being carried out on this form to elucidate its true stratigraphical range (Gregory 1989).

CONCLUSIONS

During the Kimmeridgian, two distinct environments were created by submarine faulting:

(1) a shallow-water shelf (since removed by erosion) and
(2) a deeper-water half-graben-type basin, whose sediments are now exposed along the east Sutherland coastline.

The two were separated by the Helmsdale Fault scarp. Periodic seismic reactivation of the fault, associated with climatic events or sea level changes, is suggested to have transported material from the shelf down into the basin.

Foraminiferid faunas characterized by low abundances and low diversity were therefore transported from their original habitat to arrive in the basin as mixed, poorly preserved assemblages. The overall pattern in the *cymodoce* Zone is one of decreasing nodosariid and agglutinated forms with increasing spirillinids. This may indicate contracting niche availability associated with shallowing. The *mutabilis* Zone and *eudoxus* Zone are notable for the absence of foraminifera, which may be reflective of a more restrictive environment associated with general oceanic anoxia.

SYSTEMATIC DESCRIPTIONS

The classification scheme follows Loeblich & Tappan (1964), which they subsequently revised (Loeblich & Tappan 1974, 1984), although Haynes' (1981) simplified scheme is used for the Nodosariacea. For brevity, only brief descriptions are given.

Suborder Textulariina Delage & Hérouard 1896
Superfamily Lituolacea de Blainville 1827
Family Haplophragmoididae Maync 1952

Genus *Haplophragmoides* Cushman 1910
Haplophragmoides cf. *kingakensis* Tappan 1955
(Plate 1, Fig. 1)

Description. Test medium to large, planispiral, involute, biumbilicate; eight to ten chambers in the final whorl; sutures distinct, depressed, straight; test wall coarsely agglutinated, little cement; aperture obscured.
Remarks. Poor preservation of the tests prevents definite identification. This species occurs mainly in the lower part of the *cymodoce* zone. For comparison, see 1955 *Haplophragmoides kingakensis* Tappan; p.43, Plate 10, Figs. 1–6.
Material. 105 specimens.

Family Lituolidae de Blainville 1827

Genus *Ammobaculites* Cushman 1910
Ammobaculites agglutinans (d'Orbigny) 1846
(Plate 1, Fig. 2)

1846 *Spirolina agglutinans* d'Orbigny; p. 137, Plate 7, Figs. 10–12.
1937 *Ammobaculites agglutinans* (d'Orbigny);

Bartenstein & Brand, p. 186, Plate 4, Fig. 14, Plate 5, Fig. 7, Plate 6, Figs. 40a–b, Plate 8, Figs. 38a–c, Plate 10, Figs. 45a–b, Plate 11a, Figs. 19a–b, Plate 12a, Fig. 22, Plate 13, Fig. 23, Plate 14b, Fig. 19.

1981 *Ammobaculites agglutinans* (d'Orbigny); Barnard, Cordey & Shipp, p. 388, Plate 1, Fig. 2.

Description. Test small to medium, initial planispiral coil of three to four chambers followed by uncoiled uniserial section of three to four chambers; chambers subrectangular, final chamber inflated; sutures indistinct; aperture obscured, often broken.
Remarks. This agglutinated species can be distinguished from *Ammobaculites coprolithiformis* (Schwager) in having a smaller initial coil. *A. agglutinans* is a rare form, confined to the lower part of the *cymodoce* Zone.
Material. Three specimens.

Family Hormosinidae Haeckel 1894
Subfamily Hormosininae Haeckel 1894

Genus *Reophax* de Montfort 1808
Reophax helvetica (Haeusler) 1881
(Plate 1, Fig. 3)

1881 *Dentalina helvetica* Haeusler; p. 34, Plate 2, Fig. 45.
1883 *Reophax helvetica* (Haeusler); Haeusler, p. 27, Plate 2, Figs. 8–10.
1981 *Reophax helvetica* (Haeusler); Barnard & Shipp, p. 8, Plate 1, Fig. 2.

Description. Test large, uniserial; seven to ten chambers, proloculus spherical; sutures distinct to indistinct, depressed; aperture obscured.
Remarks. *Reophax helvetica* is a rare form, only recorded from the lowermost part of the *cymodoce* Zone.
Material. Five specimens.

Superfamily Verneuilinacea Cushman 1911
Family Verneuilinidae Cushman 1911

Genus *Verneuilinoides* Loeblich & Tappan 1949
Verneuilinoides tryphera Loeblich & Tappan 1950
(Plate 1, Fig. 4)

1950 *Verneuilinoides tryphera* Loeblich & Tappan; p. 42, Plate 11, Figs. 16a–b.
1983 *Verneuilinoides tryphera* Loeblich & Tappan; Medd, Plate 1, Fig. 5.

Description. Test small to medium, triserial; chambers initially subrectangular, becoming increasingly globular; sutures distinct, depressed; test wall finely agglutinated with considerable cement; aperture distinct, large, comma shaped with an arched lip.
Remarks. A distinctive form showing little variation. *Verneuilinoides tryphera* is fairly common in the lower part of the *cymodoce* Zone.
Material. 27 specimens.

Suborder Miliolina Delage & Hérouard 1896
Superfamily Cornuspiracea Schultze 1854
Family Cornuspiridae Schultze 1854
Subfamily Cornuspirinae Schultze 1854

Genus *Cornuspira* Schultze 1854
Cornuspira eichbergensis Kübler & Zwingli 1870
(Plate 1, Fig. 5)

1870 *Cornuspira eichbergensis* Kübler & Zwingli; pp. 17, 24, 38, Plate 2, Fig. 2, Plate 3, Fig. 2, Plate 4, Fig. 2.
1981 *Cornuspira eichbergensis* Kübler & Zwingli; Barnard, Cordey & Shipp, p. 395, Plate 1, Fig. 16.

Description. Test large, can be free or attached, an unseptated evolute planispiral tube, biconcave; proloculus large, spherical, followed by eight to ten whorls; whorl section rounded, twisted in places to give constrictions; spiral suture distinct; test wall calcareous, porcellaneous, imperforate; aperture simple, terminal opening.
Remarks. A rare large form which can be distinguished from larger specimens of *Spirillina tenuissima* Gümbel by its porcellaneous, imperforate test.
Material. Four specimens.

Suborder Lagenina Delage & Hérouard 1896
Superfamily Nodosariacea Ehrenberg 1838
Family Nodosariidae Ehrenberg 1838

Genus *Nodosaria* Lamarck 1812
Nodosaria fontinensis Terquem 1870
(Plate 1, Fig. 6)

1870 *Nodosaria fontinensis* Terquem; p. 251, Plate 26, Figs. 1–5.
1982 *Nodosaria fontinensis* Terquem; Ruget, p. 73, Plate 5, Fig. 17.

Description. Test large, uniserial; two to four spherical chambers, proloculus large, spherical; sutures distinct, straight, depressed; ornament consists of up to 12 coarse low broad ribs that taper proximally and flare distally, ribs join to form an arched ridge surrounding each chamber; aperture terminal, central, produced on a short neck.
Remarks. *Nodosaria fontinensis* is fairly common in the upper part of the *cymodoce* Zone. It also occurs in the *mutabilis* Zone and the *eudoxus* Zone.
Material. 26 specimens.

Nodosaria aff. *raphanistriformis* (Gümbel) 1862
(Plate 1, Fig. 7)

Description. Test uniserial, three to eight chambers, large spherical proloculus, occasionally larger than preceding chambers; sutures distinct, depressed, sometimes with a proximal constriction; ornament of up to 26 closely spaced cord-like ribs that taper distally and proximally; aperture simple, flush to slightly protruding.
Remarks. This common species bears closest affinity to *Nodosaria raphanistriformis*. There is, however, some difference in the ribbing which is more numerous and not as well defined. For comparison, see 1862 *Dentalina raphanistiformis* Gümbel; p. 219, Plate 3, Figs. 12 a–b.
Material. 168 specimens.

Genus *Planularia* Defrance 1826

Planularia beierana (Gümbel) 1862
(Plate 1, Fig. 8)

1862 *Marginulina beierana* Gümbel; p. 221, Plate 3, Figs. 20 a–b.
1959 *Planularia beierana* (Gümbel); Cifelli, p. 299, Plate 3, Figs. 1–10, Text-Fig. 3.
1983 *Planularia beierana* (Gümbel); Nagy, Løfaldi & Bomstad, p. 12, Plate 1, Figs. 23 and 24.

Description. Test compressed, lenticular, initial three or four chambers in a loose coil, remaining chambers in a gently curved uncoiled series; proloculus small, suboval; sutures indistinct, flush to depressed, sinuous; no ornament; aperture radiate, produced on a short neck.
Remarks. There is some variation of the test, especially in the degree of coiling and arrangement of the initial chambers. *Planularia beierana* is common throughout the sections.
Material. 92 specimens.

Planularia sp. A
(Plate 1, Fig. 9)

Description. Test medium to large, loose planispiral, lenticular coil; proximal periphery rounded, distal keeled; six to eight chambers, final chamber envelops the proloculus; proloculus small; sutures depressed, arcuate; ornament distinct, five ribs cover the final chamber, converging on the apertural neck; aperture peripheral, on a short neck, with a thick notched collar.
Remarks. This is a distinctive form which cannot be assigned to any known species. As such a fuller description is being prepared (Gregory 1989).
Material. 109 specimens.

Genus *Citharina* d'Orbigny 1839

Citharina cf. *colliezi* (Terquem) 1866
(Plate 1, Fig. 10)

Description. Test large, uniserial, curved; dorsal margin straight with a keel, ventral margin curved; eight to nine chambers; proloculus small, oval, sutures depressed, oblique, arcuate; ornament of numerous curved ribs, oblique to the dorsal margin, do not cross sutures; aperture peripheral, radiate.

Remarks. All the recovered specimens are broken. It closely resembles *Citharina colliezi* in possessing a straight keeled margin. For comparison, see 1866 *Marginulina colliezi* Terquem; p. 433, Plate 17, Figs. 10 a–c.

Material. Three specimens.

Citharina cf. *flabellata* (Gümbel) 1862
(Plate 1, Fig. 11)

Description. Test uniserial, robust, subtriangular, compressed; peripheries angular; large proloculus followed by four or five chambers; sutures flush; ornament of ten coarse cord-like ribs, occurring distally; peripheral ribs continue to aperture; aperture peripheral on a short neck, with thick notched collar.

Remarks. The single specimen recovered is similar to *Citharina serracostata* (Gümbel) except the ribs are continuous, narrow and bifurcate. It resembles *C. flabellata*, the only difference being the distally discontinuous ribbing. For comparison, see 1862 *Marginulina heteropleura* Terquem; p. 117, Plate 7, Figs. 22 and 23.

Material. 12 specimens.

Citharina aff. *heteropleura* (Terquem) 1868
(Plate 1, Fig. 12)

Description. Test medium to large, subtriangular, uniserial, rounded periphery; five to six chambers; proloculus large, followed by two or three chambers in a very loose coil; sutures depressed, oblique, curved; ornament of numerous coarse cord-like ribs, becoming discontinuous distally; aperture peripheral, with several ribs forming a thick notched collar.

Remarks. This form cannot be definitely identified as *Citharina heteropleura* since the ribs are coarser and not continuous. For comparisons, see *Marginulina heteropleura* Terquem (1868; p. 117, Plate 7, Figs. 22 and 23).

Material. 12 specimens.

Genus *Frondicularia* DeFrance 1826

Frondicularia franconica Gümbel 1862
(Plate 1, Fig. 13)

1862 *Frondicularia franconica* Gümbel; p. 219, Plate 3, Figs. 13 a–c.
1983 *Frondicularia franconica* Gümbel; Kalia & Chowdhury, p. 234, Plate 3, Figs. 12–14.

Description. Test elongate, uniserial; periphery rounded, unkeeled; up to seven chambers, inverted V to arcuate in shape; proloculus large, oval to spherical; sutures depressed, chevron shaped; surface smooth; aperture terminal on a short neck.

Remarks. This is a very variable species in form, a result of irregular chamber growth. It occurs sporadically throughout the *cymodoce* Zone.

Material. 12 specimens.

Frondicularia cf. *moelleri* Uhlig 1883
(Plate 1, Fig. 14)

Description. Test large, quadrate, compressed, uniserial; large spherical proloculus followed by five or six chevron-shaped chambers; ornament of several coarse smooth continuous ribs; aperture large, terminal on a short neck, with a thick notched collar.

Remarks. All the specimens are degraded, thus hindering positive identification. This form is distinct from *Frondicularia nikitini* Uhlig which has finer discontinuous arcuate ribbing. Barnard *et al.* (1981) consider that a transitional form exists between *F. moelleri* and *F. nikitini*. For comparison, see 1883 *Frondicularia moelleri* Uhlig; p. 758, Plate 9, Figs. 12–15.

Material. Five specimens.

Genus *Lenticulina* Lamarck 1804

Lenticulina major (Bornemann) 1854
(Plate 1, Fig. 15)

1854 *Cristellaria major* Bornemann; p. 40, Plate 4, Fig. 13
1960 *Lenticulina* (*Astacolus*) *major* (Bornemann); Lutze, p. 453, Text-Figs. 12 a–g.
1981 *Lenticulina major* (Bornemann); Barnard, Cordey & Shipp, p. 412, Plate 2, Figs. 17 and 18, Text-Fig. 19.

Description. Test very large, lenticular, loose planispiral coil; large proloculus enclosed by later chambers; sutures distinct to indistinct, arcuate, raised to flush; radiate peripheral aperture.
Remarks. This large species shows some variation in the degree of uncoiling of the final chambers. It occurs throughout the section but is rare.
Material. 14 specimens.

Lenticulina varians (Bornemann) 1854
(Plate 1, Fig. 16)

1854 *Cristellaria varians* Bornemann; p. 41, Plate 4, Figs. 32–34.
1954 *Lenticulina varians* (Bornemann); Bielecka & Pozaryski, pp. 32, 164, Table 4, Fig. 11.
1983 *Lenticulina varians* (Bornemann); Nagy, Løfaldli & Bomstad, p. 120, Plate 1, Figs. 20 and 21.

Description. Test small, lenticuline, flattened to globular, involute; numerous chambers, up to eight in the final whorl; proloculus small, oval; sutures distinct to indistinct, arcuate, oblique; aperture peripheral, radiate on a short neck.
Remarks. There is considerable variation within this species in size and degree of coiling of the final chambers, ranging from tight to loose. Loosely coiled forms occur more frequently. *Lenticulina varians* (Bornemann) is very common throughout the sections.
Material. 280 specimens.

Genus *Pseudonodosaria* Boomgaart 1949

Pseudonodosaria vulgata (Borenemann) 1854
(Plate 1, Fig. 17)

1854 *Glandulina vulgata* Bornemann; p. 31, Plate 2, Figs. 1 a–b and 2.
1956 *Pseudonodosaria vulgata* (Bornemann); Seibold & Seibold, p. 139, Text-Figs. 6 m–n and 6 r–t.
1981 *Pseudonodosaria vulgata* (Bornemann); Barnard & Shipp, p. 11, Plate 2, Figs. 21 and 22.

Description. Test medium to large, compressed uniserial; small proloculus, sometimes aplicate, followed by four or five subrectangular chambers; aperture central, radiate, flush or on a raised neck.
Remarks. Owing to morphological variation, *Pseudonodosaria vulgata* has been differentiated into many species. Norling (1972) grouped these variants together as their highly variable shape is a product of fluctuating growth rate. *P. vulgata* is common throughout the sections.
Material. 120 specimens.

Genus *Dentalina* d'Orbigny 1826

Dentalina pseudocommunis Franke 1936
(Plate 1, Fig. 18)

1936 *Dentalina pseudocommunis* Franke; p. 30, Plate 2, Figs. 20. a–b.
1981 *Dentalina pseudocommunis* Franke; Barnard & Shipp, p. 10, Plate 2, Figs. 16 and 17.

Description. Test arcuate, slender, uniserial; five to eight chambers, final chamber inflated, proloculus large, aplicate, sutures initially indistinct, depressed to flush, oblique; aperture radiate, peripheral.
Remarks. This is a rare form and all the specimens are broken.
Material. Three specimens.

Genus *Lagena* Walker & Jacob 1798

Lagena globosa (Montagu) 1803
(Plate 1, Fig. 19)

1803 *Vermiculum globosum* Montagu; p. 253.
1937 *Lagena globosa* (Montagu); Bartenstein

& Brand, p. 165, Plate 4, Fig. 68, Plate 5, Fig. 48, Plate 11A, Fig. 8, Plate 12A, Fig. 8, Plate 12B, Fig. 10a, Plate 15, Fig. 26.
1967 *Lagena globosa* (Montagu); Gordon, p. 456, Plate 3, Fig. 19.

Description. Test large, unilocular, globular, smooth to slightly hispid surface, aperture terminal, produced on a short neck.
Remarks. *Lagena globosa* is fairly common throughout the sections.
Material. 52 specimens.

Family Lingulinidae Loeblich & Tappan 1961
Genus *Lingulina* d'Orbigny 1826

Lingulina longiscata (Terquem) 1870
(Plate 1, Fig. 20.)

1870 *Frondicularia longiscata* Terquem; p. 318, Plate 22, Figs. 23 and 24.
1957 *Linqulina longiscata* (Terquem); Adams, p. 222, Text-Figs. 15, 16, 23 and 24.
1983 *Lingulina longiscata* (Terquem); Medd, p. 233, Plate 3, Fig. c.

Description. Test large, compressed uniserial, large spherical proloculus is followed by five or six arcuate to chevron-shaped chambers; sutures distinct, associated occasionally with basal constrictions; ornament of numerous fine, converging ribs, not continuous over sutures, aperture terminal, elongate slit, flush or on a short neck.
Remarks. *Lingulina longiscata* occurs sporadically throughout the sections. It is similar to *Frondicularia nodosaria* (Terquem) but can be distinguished by its elongate slit aperture.
Material. Nine specimens.

Family Polymorphinidae d'Orbigny 1839
Subfamily Polymorphininae d'Orbigny 1839
Genus *Eoguttulina* Cushman & Ozawa 1930
emend. Lloyd 1962

Eoguttulina liassica (Strickland) 1846
(Plate 1, Fig. 21)

1846 *Polymorphina liassica* Strickland; p. 31, Text-Fig. b.
1930 *Eoguttulina liassica* (Strickland); Cushman & Ozawa, p. 17, Figs. 2 a–c.
1981 *Eoguttulina liassica* (Strickland); Barnard, Cordey & Shipp, p. 426, Plate 3, Fig. 20, Text-Fig. 29a

Description. Test elongate to globular; proloculus small, angular to oblate, followed by three or four inflated chambers, spirally arranged, each less than 90° apart, final chamber inflated; sutures depressed to flush; aperture terminal, radiate.
Remarks. The morphological variation of *Eoguttulina liassica* corresponds with the morphotypes described by Lloyd (1962). Its broader test and less-well-defined sutures distinguish it from *E. polygona* (Terquem). It occurs throughout the sections studied, but is commonest in the lower part of the *cymodoce* Zone.
Material. 62 specimens.

Suborder Spirillinina Hohenegger & Piller 1975
Superfamily Spirillinacea Reuss & Fritsch 1861
Family Spirillinidae Reuss & Fritsch 1861
Genus *Spirillina* Ehrenberg 1843

Spirillina infima (Strickland) 1846
emend. Barnard 1952
(Plate 1, Fig. 22)

1846 *Orbis infimus* Strickland; p. 30, Text-Fig. a.
1952 *Spirillina infima* (Strickland); Barnard, p. 906, Figs. 1–3.
1986 *Spirillina infima* (Strickland); Stam, p. 125, Plate 6, Fig. 10.

Description. Test small to medium, non-septated planispiral tube, biconcave, evolute; four to six rounded whorls, last whorl greatly inflated and slightly involute; spiral suture initially indistinct; test coarsely perforate; aperture simple, open end of tube.
Remarks. Barnard (1952) emended this species as confusion existed concerning its generic position. It can be distinguished from *S. tenuissima* by its small size, and a greatly inflated final whorl. It

occurs frequently throughout all the sections.
Material. 220 specimens.

Spirillina tenuissima Gümbel 1862
(Plate 1, Fig. 23)

1862 *Spirillina tenuissima* Gümbel; p. 214, Plate 4, Figs. 12 a-b.
1986 *Spirillina tenuissima* Gümbel; Stam, p. 126, Plate 6, Fig. 9.

Description. Test medium to large, planispiral evolute to slightly involute, biconcave; proloculus small, followed by up to ten unseptated flattened whorls; spiral sutures, distinct, depressed; coarsely perforate test; aperture simple opening.
Remarks. *Spirillina tenuissima* can be distinguished from *S. infima* by its larger size, greater number of whorls, more regularly arranged perforations and its final uninflated chamber. It is by far the commonest form recorded, occurring throughout the sections.
Material. 722 specimens.

Genus *Conicospirillina* Cushman 1927

Conicospirillina trochoides (Berthelin) 1879
(Plate 1, Fig. 24)

1879 *Spirillina trochoides* Berthelin; p. 37, Plate 1, Figs. 20–22.
1954 *Conicospirillina trochoides* (Berthelin); Bielecka & Pozaryski, pp. 68, 197, Plate 11, Fig. 55.
1981 *Conicospirillina trochoides* (Berthelin); Copestake & Johnson, p. 100, Plate 6.1.5, Figs. 4 and 5.

Plate 1 —

Fig. 1 — *Haplophragmoides* cf. *kingakensis* Tappan 1955. Length, 0.31 mm.
Fig. 2 — *Ammobaculites agglutinans* (d'Orbigny) 1846. Length, 0.26 mm.
Fig. 3 — *Reophax helvetica* (Haeusler) 1881. Length, 0.52 mm.
Fig. 4 — *Verneuilinoides tryphera* Loeblich & Tappan 1950. Length, 0.26 mm.
Fig. 5 — *Cornuspira eichbergensis* Kubler & Zwingli 1870. Diameter, 0.48 mm.
Fig. 6 — *Nodosaria fontinensis* Terquem 1870. Length, 0.31 mm
Fig. 7 — *Nodosaria* aff. *raphinistiformis* (Gümbel) 1862. Length, 0.29 mm.
Fig. 8 — *Planularia beierana* (Gümbel) 1862. Length, 0.27 mm.
Fig. 9 — *Planularia* sp. A. Length, 0.31 mm.
Fig. 10 — *Citharina* cf. *colliezi* (Terquem) 1866. Length, 0.6 mm.
Fig. 11 — *Citharina* cf. *flabellata* (Gümbel) 1862. Length, 0.37 mm.
Fig. 12 — *Citharina* aff. *heteropleura* (Terquem) 1868. Length, 0.34 mm.
Fig. 13 — *Frondicularia* cf. *moelleri* Uhlig 1883. Length, 0.33 mm.
Fig. 14 — *Frondicularia franconica* Gümbel 1862. Length, 0.24 mm.
Fig. 15 — *Lenticulina major* (Bornemann) 1854. Length, 0.49 mm.
Fig. 16 — *Lenticulina varians* (Bornemann) 1854. Length, 0.25 mm.
Fig. 17 — *Pseudonodosaria vulgata* (Bornemann) 1854. Length, 0.32 mm.
Fig. 18 — *Dentalina pseudonodosaria* Franke 1936. Length, 0.37 mm.
Fig. 19 — *Lagena globosa* (Montagu) 1803. Diameter, 0.2 mm.
Fig. 20 — *Lingulina longiscata* (Terquem) 1870. Length, 0.31 mm.
Fig. 21 — *Eoguttulina liassica* (Strickland) 1842. Length, 0.24 mm.
Fig. 22 — *Spirillina infima* (Strickland) 1846. Diameter, 0.23 mm.
Fig. 23 — *Spirillina tenuissima* (Gümbel) 1862. Diameter, 0.26 mm.
Fig. 24 — *Conicospirillina trochoides* (Berthelin) 1879. Diameter, 0.22 mm.

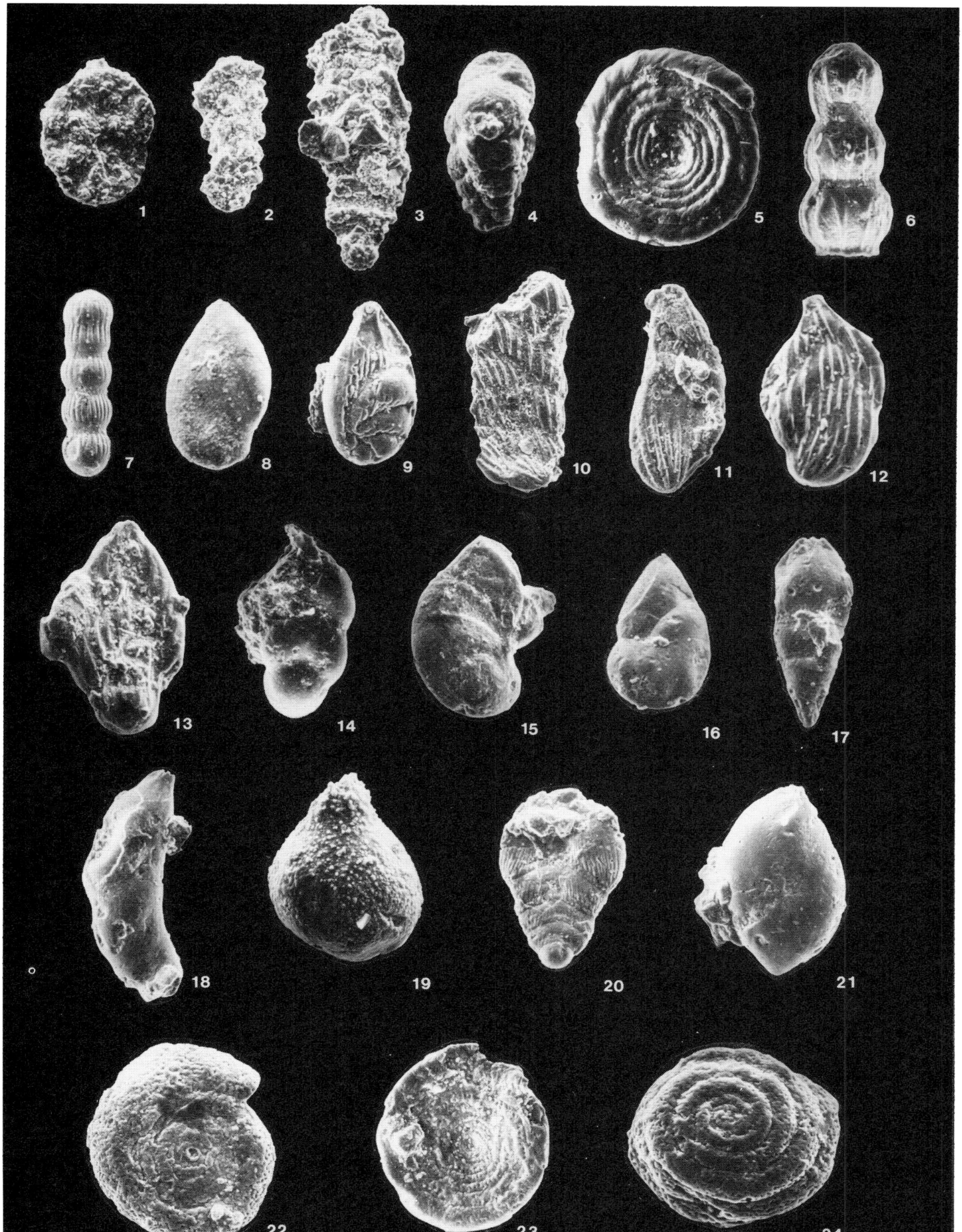
1
2
3
4
5
6
7
8
9
10
11
12
13
14
15
16
17
18
19
20
21
22
23
24

Description. Planispiral low conical domed test; small proloculus followed by six or seven unseptated rounded whorls, with an indistinct angular keel; ornament coarsely perforate, aperture simple, open end of final whorl.
Remarks. The material recovered differs from the description by Copestake & Johnson (1981) in having several rows of large circular pores on each whorl instead of a single row.
Material. 117 specimens.

ACKNOWLEDGEMENTS

Britoil plc kindly provided funding for the research and Dr P. Copestake (Britoil) was very helpful with comments and advice. Dr M. D. Brasier (Department of Geology, University of Hull) read and offered criticism of the text. The Department of Geology at Hull provided facilities for research.

REFERENCES

Adams, C. G. (1957). A study of the morphology and variation of some Upper Lias foraminifera. *Micropaleontology* **3** 205–226.
Bailey, E. B. & Weir, J. (1932). Submarine faulting in Kimmeridgian times, east Sutherland. *Transactions of the Royal Society of Edinburgh* **47** 431–467.
Barnard, T. (1952). Notes on *Spirillina infima* (Strickland) foraminifera. *Annals and Magazine of Natural History Series* **12** 905–909.
Barnard, T., Cordey, W. G. & Shipp, D. J. (1981). Foraminifera from the Oxford Clay (Callovian–Oxfordian of England). *Revista Espanõla de Micropaleontolgica* **13** 383–462.
Barnard, T. & Shipp, D. J. (1981). Kimmeridgian foraminifera from the Boulonnais. *Revue de Micropaléontologie* **24** 3–26.
Bartenstein, H. & Brand, E. (1937). Mikropalaontologische Untersuchungen zur Stratigraphie des nordwest deutschen Lias und Doggers. *Senckenbergische naturforschende Gesellschaft* **439** 1–244.
Berthelin, G. (1879). Foraminifères du Lias moyen de la Vendée. *Revue et Magasine de Zoologie Pure et Appliquée* **7** 24–41.
Bielecka, W. & Pozaryski, W. (1954). Micropalaeontological stratigraphy of the Upper Malm in Central Poland. *Instytut Geologiczny Prace* **12** 143–206.
Bird, T. J., Bell, A., Gibbs, A. D. & Nicholson, J. (1986). Aspects of strike-slip tectonics in the Inner Moray Firth Basin, offshore Scotland. *Norsk Geologisk Tidsskrift* **67** 353–369.
Boltovskoy, E. & Wright, R. (1976). *Recent Foraminifera.* W. Junk, The Hague, pp. 1–515.
Bornemann, J. G. (1854). *Über die Lias Formation im der Umgegend und von Gottingen und ihre organischen Einschlusse.* A. W. Schade, Berlin, pp. 1–45.
Bradshaw, J. S. (1961). Laboratory experiments on the ecology of foraminifera. *Contributions from the Cushman Foundation for Foraminiferal Research* **12** 87–106.
Brookfield, M. E. (1976). The age of the Allt na Cuile Sandstones (Upper Jurassic, Scotland). *Scottish Journal of Geology* **12** 181–186.
Copestake, P. & Johnson, B. (1981). The Hettangian to Toarcian. In: Jenkins, D. G. & Murray, J. W. (eds), *Stratigraphical atlas of fossil foraminifera.* Ellis Horwood, Chichester, pp. 81–105.
Cifelli, R. (1959). Bathonian Foraminifera of England. *Bulletin of the Museum of Comparative Zoology at Harvard College* **121** 265–368.
Crowell, J. C. (1961). Depositional structures from the Jurassic Boulder Beds, east Sutherland. *Transactions of the Edinburgh Geological Society* **18** 202–219.
Cushman, J. A. (1930). Note sur quelques foraminifères Jurassiques de Auberville (Calvados). *Bulletin de la Société linnéenne de Normandie Séries 8* **2** 132–135.
Cushman, J. A. & Ozawa, Y. (1930). An outline of a revision of the Polymorphinidae, recent and fossil. *Proceedings of the United States National Museum* **77** 1–195.
d'Orbigny, A. D. (1846). Tableau méthodique de la class des Cèphalodes. *Annales des sciences naturelles, Foraminifères* 245–315
Franke, A. (1936). Die foraminiferen des deutschen Lias. *Abhandlungen der Preussischen geologischen Landesantalt* **169** 1–140.
Gordon, W. A. (1967). Foraminifera from the Callovian (Middle Jurassic) of Brora, Scotland. *Micropaleontology* **13** 445–464.
Gordon, W. A. (1970). Biogeography of Jurassic Foraminifera. *Bulletin of the Geological Society of America* **81** 1689–1704.
Gregory, F. J. (1989). *Planularia kintradwellensis* n.sp., a foraminiferal biostratigraphical marker species for the Lower Kimmeridgian. To be published.
Gümbel, C. W. (1862). Die Streitberger Schwammlager und ihre Foraminiferen Einschlüsse. *Jahrescheft des Vereins fur vaterlandische Naturkunde in Wüttemberg* **18** 192–238.
Haeusler, R. (1881). Untersuchungen über die mikroskopischen Strukurverhaltnisse der Aargauer Jurakalke mit besonderer Berucksichtigung ihrer Foraminiferen fauna. *Dissertationen.* Zurich, pp. 1–47.
Haeusler, R. (1883). Notes on some Upper Jurassic Astrorhizidae and Lituolidae. *Quarterly Journal of the Geological Society, London* **39** 25–28.
Haynes, J. R. (1981). *Foraminifera.* Macmillan, London, pp. 1–433.
Johnson, B. (1977). Ecological ranges of selected Toarcian and Domerian foraminiferal species from Wales. *Maritime Sediments Special Publication No.* 1, *Part B: Paleoecology and Biostratigraphy* pp. 545–556.
Judd, J. W. (1875). The secondary rocks of Scotland, first paper. *Quarterly Journal of the Geological Society, London* **34**, 660–739.
Kalia, P. & Chowdhury, S. (1983). Foraminiferal biostratigraphy, biogeography and environment of the Callovian

sequence Rajasthan, NW India. *Micropaleontology* **29** 223–254.

Kübler, J. & Zwingli, H. (1870). *Die Foraminiferen des Schweizer Jura.* Winterthur, pp. 1–49.

Kureshy, A. A. (1969). Ecological studies of foraminifera of the Wash (England) and the relationship between their distribution and sedimentation. *Revue de Micropaléontologie* **11** 222–232.

Lam, K. & Porter, R. (1977). The distribution of palynomorphs in the Jurassic rocks of the Brora Outlier, NE Scotland. *Journal of the Geological Society, London* **134** 45–55.

Lee, G. W. (1925). Mesozoic rocks of east Sutherland and Ross. In: Read, H. H. & Phemister, J. (eds), *The geology of the country around Golspie, Sutherlandshire. Memoirs of the Geological Survey of Scotland.*

Lee, G. W. & Pringle, J. (1932). A synopsis of the Mesozoic rocks of Scotland. *Transactions of the Geological Society of Glasgow* **19** 158–224.

Linsley, P. N. (1972). The stratigraphy and sedimentology of the Kimmeridgian deposits of Sutherland, Scotland. *Ph.D. Thesis.* University of London, pp. 1–245.

Lloyd, A. J. (1959). Arenaceous Foraminifera faunas from the type Kimmeridgian. *Palaeontology* **1** 298–320.

Lloyd, A. J. (1962). Polymorphinid, Miliolid and rotaliform Foraminifera from the type Kimmeridgian. *Micropaleontology* **8** 369–383.

Leoblich, A. R. & Tappan, H. (1950). North American Jurassic Foraminifera 1. The type Redwater Shale (Oxfordian) of south Dakota. *Journal of Paleontology* **24** 39–60.

Loeblich, A. R. & Tappan, H. (1964). *Sarcodina, Treatise on invertebrate paleontology, Part C, Protista 1, 2.* Geological Society of America, University of Kansas Press.

Loeblich, A. R. & Tappan, H. (1974). Recent advances in the classification of the Foraminifera. In: Hedley, R. H. & Adams, C. G. (eds) *Foraminifera* **1** 1–53.

Loeblich, A. R. & Tappan, H. (1984). Suprageneric classification of the Foraminiferida (Protozoa). *Micropaleontology* **30** 1–70.

Løfaldi, M. & Nagy, J. (1983). Agglutinating Foraminifera in Jurassic and Cretaceous Dark Shales in southern Spitsbergen. In: Verdenius, J. G., Van Hinte, J. E. & Fortuin, A. R. (eds) *Proceedings of the 1st Workshop on Arenaceous Foraminifera*, pp. 91–107.

Lutze, G. F. (1960). Zur Stratigraphie und Paläontologie des Callovian und Oxfordian in Nord West Deutschland. *Geologische Jahrebuchen* **77** 391–532.

McQuillin, R., Donato, J. A. & Tulstrup, J. (1982). Development of basins in the Inner Moray Firth and North Sea by crustal extension and dextral displacement of the Great Glen Fault. *Earth and Planetary Science Letters* **60** 127–139.

Medd, A. W. (1979). The Mesozoic stratigraphy of two boreholes near Worlaby, Humberside. *Bulletin of the Geological Survey of Great Britain* **58** 1–24.

Medd, A. W. (1983). Foraminifera from the Lower Oxford Clay (Callovian stage) of the Norman's Cross Pit, near Peterborough, Cambridgeshire. *Revista Espanõla de Micropaleontologie* **15** 221–240.

Montagu, G. (1803). *Testacea Britannica.* Romsey, pp. 1–606.

Morris, K. A. (1980). A comparison of major sequences of organic-rich mud deposition in the British Jurassic. *Journal of the Geological Society, London* **137** 157–170.

Murchison, I. R. (1829). On the coal field of Brora in Sutherlandshire, and some other stratified deposits in the north of Scotland. *Transactions of the Geological Society, London, Second Series* **2** 293–326.

Murray, J. W. (1973). *Distribution and ecology of living benthic foraminiferids.* Heinemann, London pp. 1–274.

Nagy, J., Løfaldi, M. & Bomstad, K. (1983). Marginal marine microfaunas of the Jurassic (Bajocian) Yons Nab Beds of Yorkshire. In: Verdenius, J. G., Van Hinte, J. E. & Fortuin, A. E. (eds) *Proceedings of the 1st Workshop on Arenaceous Foraminifera*, pp. 111–128.

Neves, R. & Selley, R. C. (1975). A review of the Jurassic rocks of north-east Scotland. In: Finstad, K. G. & Selley, R. C. (eds), *Proceedings of the Jurassic Northern North Sea Symposium*, pp. 1–29.

Norling, E. (1972). Jurassic stratigraphy and foraminifera of western Scania, southern Sweden. *Sveriges Geologiska Undersökning* **47** 1–12.

Pickering, K. T. (1984). The Upper Jurassic 'Boulder Beds' and related deposits: a fault-controlled submarine slope, NE Scotland. *Journal of the Geological Society, London* **141** 357–374.

Riley, L. A. (1980). Palynological evidence of an early Portlandian age for the uppermost Helmsdale Boulder Beds, Sutherland. *Scottish Journal of Geology* **16** 29–31.

Ruget, C. (1982). Foraminifères du Lias moyen et superieur d'Obon. *Geobios* **15** 53–91.

Seibold, E. & Seibold, I. (1953). Foraminiferenfauna und kalkgehalt eines Profils im gebanken unteren Malm Schwages. *Neues Jahrbuch für Geologie und Paläontologie, Abhandlungen* **98** 28–86.

Seibold, E. & Seibold, I. (1956). Revision der Foraminiferen Bearbeitung C. Schwagers (1865) aus den Impressaschichten (Unterer Malm) Suddeutschland. *Neues Jahrbuch für Geologie und Paläontologie, Abhandlungen* **103** 91–154.

Stam, B. (1986). Quantitative analysis of Middle and Late Jurassic Foraminifera from Portugal and its implications for the Grand Banks of New Foundland. *Utrecht Micropaleontological Bulletins* **34** 1–216.

Strickland, H. E. (1846). On two species of microscopic shells found in the Lias. *Quarterly Journal of the Geological Society, London* **2** 31.

Tappan, H. (1955). Foraminifera from the Arctic Slope of Alaska; Part 2, Jurassic Foraminifera. *Professional Paper of the United States Geological Survey* **236(B)** 21–90.

Terquem, O. (1866). Sixième mémoire sur les Foraminifères du Lias des Départements de l'Indre. *Mémoire de l'Académie Impériale de Metz* **47** 459–532.

Terquem, O. (1868). Premier mémoire sur les foraminifères du Système Oolithique étude du Fullers-Earth de la Moselle. *Bulletin Société d'Historie Naturelle de Moselle.*

Terquem, O. (1870). Troisème mémoire sur les Foraminifères de Système Oolithique comprenant les genres *Frondicularia, Flabellina, Nodosaria* etc. de la Zone *Ammonites parkinsoni* de Fontoy. *Mémoires de l'Académie Impériale de Metz* **51** 299–380.

Tinsley, J. (1950). The determination of organic carbon in soils by dichromate mixtures. *Transactions of the 4th International Congress of Soil Science*, pp. 161–164.

Uhlig, V. (1883). Über Foraminiferen aus dem rjasan'schen

ornatenthone. *Jahrebuch der K.K. geologischen Reinsanstalt* **33** 735–774.

Van Der Burgh, J. & Van Konijnenburg-van Cittert, J. H. A. (1984). A drifted flora from the Kimmeridgian (Upper Jurassic) of Lothbeg Point, Sutherland, Scotland. *Review of Palaeobotany and Palynology* **43** 359–396.

Walkey, A. (1947). A critical examination of a rapid method for determining organic carbon content in soils. Effects of variations in digestion conditions and of inorganic soil constituents. *Soil Science* **63** 251–264.

Young, R. L. A. (1988). The Palaeoecology of Jurassic Foraminifera from selected sequences in Great Britain with particular emphasis on the Suborder Robertinina. *Ph.D. Thesis.* University of Hull, pp. 1–487.

8

Dinoflagellate cyst biostratigraphy and palaeoenvironments of the Upper Jurassic (Kimmeridgian to basal Portlandian) of the Helmsdale region, east Sutherland, Scotland

Hugh F. Barron

ABSTRACT

Palynomorph assemblages from the Upper Jurassic fault-controlled submarine slope deposits of east Sutherland have yielded miospores and microplankton in varying abundances. The *Adnatosphaeridium caulleryi* group, *Cribroperidinium* spp. and *Cyclonephelium hystrix* are abundant throughout the section. Shorter-ranging dinocyst taxa include *Cribroperidinium longicorne*, *Dichadogonyaulax? pannea*, *Egmontodinium expiratum*, *Gochteodinia mutabilis*, *Kleithriasphaeridium porosispinum*, *Oligosphaeridium pulcherrimum sensu* Ioannides *et al.* 1977 and *Senoniasphaera jurassica*. The base of the *Pavlovia rotunda* and the top of the *Virgatopavlovia fittoni* ammonite zones are recognized by the appearance of *Muderongia* sp. A Davey 1979 and the extinction of *Dingodinium tuberosum* and *Occisucysta balia*, respectively. The youngest rocks exposed are assigned to the *Progalbanites albani* Zone of the lower Portlandian. An increase in the ratio of microplankton to miospores in the upper Kimmeridgian and lower Portlandian is interpreted as indicating increasing distance from the terrestrial source due to fault-controlled widening of the submarine shelf.

INTRODUCTION

The discovery and development of hydrocarbon reservoirs in the North Sea Basin and adjacent areas during the past two decades have stimulated research on coeval strata onshore. In 1976, hydrocarbons in commercial quantities were discovered in the Moray Firth, 22 km east of Helmsdale (Fig. 1) and were subsequently developed as the Beatrice Field. This discovery renewed the interest in the onshore Jurassic exposures of the Moray Firth because the nearby surface geology can be directly related to the subsurface structure of the Beatrice Field.

Exposed on the coast between Brora and

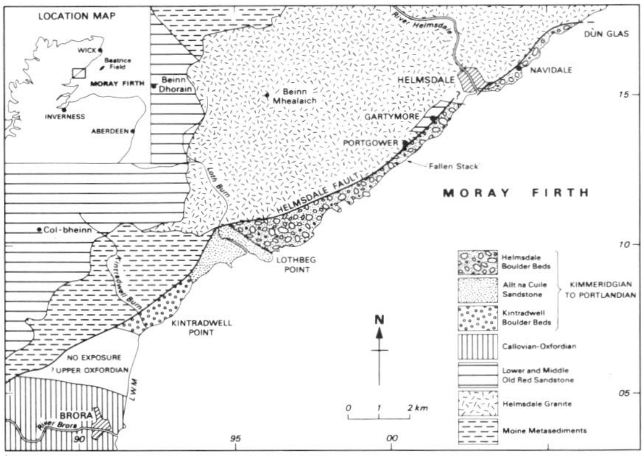

Fig. 1 — Geological map of the Brora–Helmsdale area. (Modified after Pickering (1984), MacDonald (1985) and the British Geological Survey's 1 inch sheet 103).

Helmsdale are Kimmeridgian and basal Portlandian sediments (Fig. 1), which form part of a thick Jurassic succession that represents the northwest margin of a fault-controlled sedimentary basin underlying the present Moray Firth. Lithologically the section consists of an interbedded sequence of sandstones, bituminous shales and boulder beds deposited on the downthrow side of the Helmsdale Fault. Many of the oilfields in the northern North Sea, particularly in the South Viking Graben, occur in similar structural settings (MacDonald 1985).

Palynomorphs extracted from the bituminous shales consist mainly of miospores and dinoflagellate cysts (henceforth termed dinocysts). The latter have been selected for biostratigraphical analysis in this study because of their more restricted ranges in the Upper Jurassic by comparison with miospores. The ratio of dinocysts to miospores can be used in palaeoenvironmental analyses especially when integrated with sedimentological studies.

The literature on the Kimmeridgian of the Helmsdale region is extensive; for comprehensive reviews of previous work see Bailey & Weir (1932) and Pickering (1984). Recent studies include palaeobotanical work on the fossil flora from Kintradwell and Lothbeg Point (van der Burgh & van Konijnenburg-van Cittert 1984), clay mineralogy (Hurst 1985) and sedimentological and facies analysis (MacDonald 1985).

GEOLOGICAL BACKGROUND

The geology of the Sutherland coast consists of a basement of metamorphosed Moinian sediments intruded by the Helmsdale Granite and unconformably overlain by Devonian Lower and Middle Old Red Sandstone deposits of the Orcadian Basin (Fig. 1). The Helmsdale Fault separates these pre-Mesozoic rocks from a narrow coastal strip of Triassic and Jurassic strata, representing the northwest margin of a large Mesozoic sedimentary basin in which a total of 2300 m of Jurassic sediments are present at the depocentre, 25 km east of Helmsdale (Chesher & Lawson 1983). This is one of the thickest Jurassic sections in the North Sea Basin (Brooks & Chesher 1975).

The Helmsdale Fault, active during Late Jurassic sedimentation, is one of a number of similar faults resulting from extensive rifting during the Late Jurassic and Early Cretaceous in this area. This rifting gave rise to the tilted horst blocks in the North Sea, evidence for which can be seen in angular unconformities in seismic sections (Selley 1976).

During the Mesozoic, the structure of the Moray Firth Basin seems to have broadly coincided with the Orcadian Basin that existed during Old Red Sandstone times (Chesher & Lawson 1983). Carboniferous sediments are not found either onshore or offshore, although past deposition and subsequent erosion are inferred from the presence of Carboniferous spores within the Jurassic succession (Lam & Porter 1977, Windle 1979, Riley 1980) and clay mineral provenance (Hurst 1985). Basic volcanics of Carboniferous age are, however, found in well 12/23-1, northeast of the Beatrice Field (Linsley *et al.* 1980).

The Triassic continental clastics exposed around Golspie (9 km to the southwest of Brora) are derived from Devonian and Caledonide sources, rejuvenated by Hercynian movements

(Linsley *et al.* 1980). This continental deposition is succeeded from the Hettangian to the Pliensbachian by fluvial to lagoonal, and then shallow marine siltstones, sandstones and shales (Batten *et al.* 1986). Overlying this is a broadly transgressive sequence of sandstones, carbonaceous shales and coal of Bathonian age. The succeeding Callovian to middle Oxfordian strata exposed around Brora represent stable shallow-water marine shelf sedimentation. Increased activity of the Helmsdale Fault from the late Oxfordian to the Portlandian led to deepening of the basin, resulting in the deposition of the thick fault-controlled Kimmeridgian to basal Portlandian succession of boulder beds, shales and sandstones.

STRATIGRAPHY

Upper Jurassic rocks are exposed on the foreshore between Kintradwell and Dùn Glas. They form a narrow coastal strip about 19 km long and up to 1 km wide (Figs. 1 and 2), with a total estimated thickness of about 630 m (MacDonald 1985). The strata reveal a complex array of boulder beds interdigitated with bituminous shales and sandstones which generally become younger to the northeast. Minor faults of up to several metres displacement are common and the whole succession has been deformed by simple open, gently plunging folds (Pickering 1984).

The lithostratigraphical nomenclature of Pickering (1984) is adopted here. This consists of

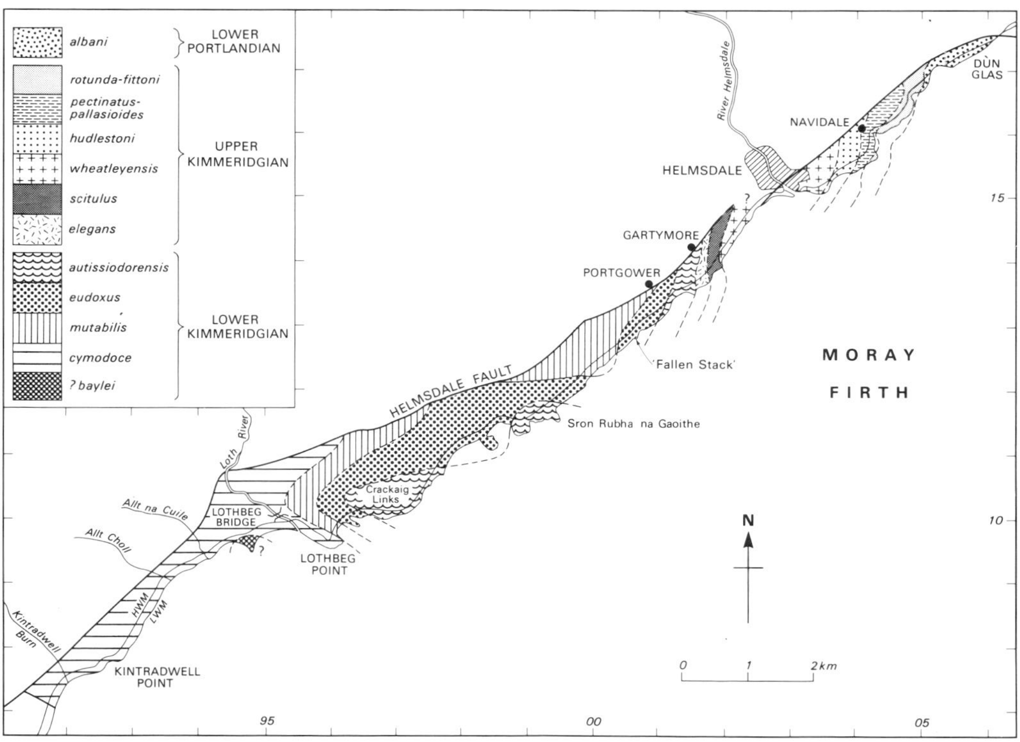

Fig. 2 — Ammonite zonation of the Upper Jurassic of east Surtherland. The *cymodoce* to the base of the *elegans* Zones are based on ammonite data of Linsley (1972), while the base and top of the *scitulus* and *hudlestoni* Zones are based on structural reinterpretation of Linsley's (1972) ammonite zones by MacDonald (1985). The base of the *rotunda* and the top of the *fittoni* Zones are newly determined in this paper.

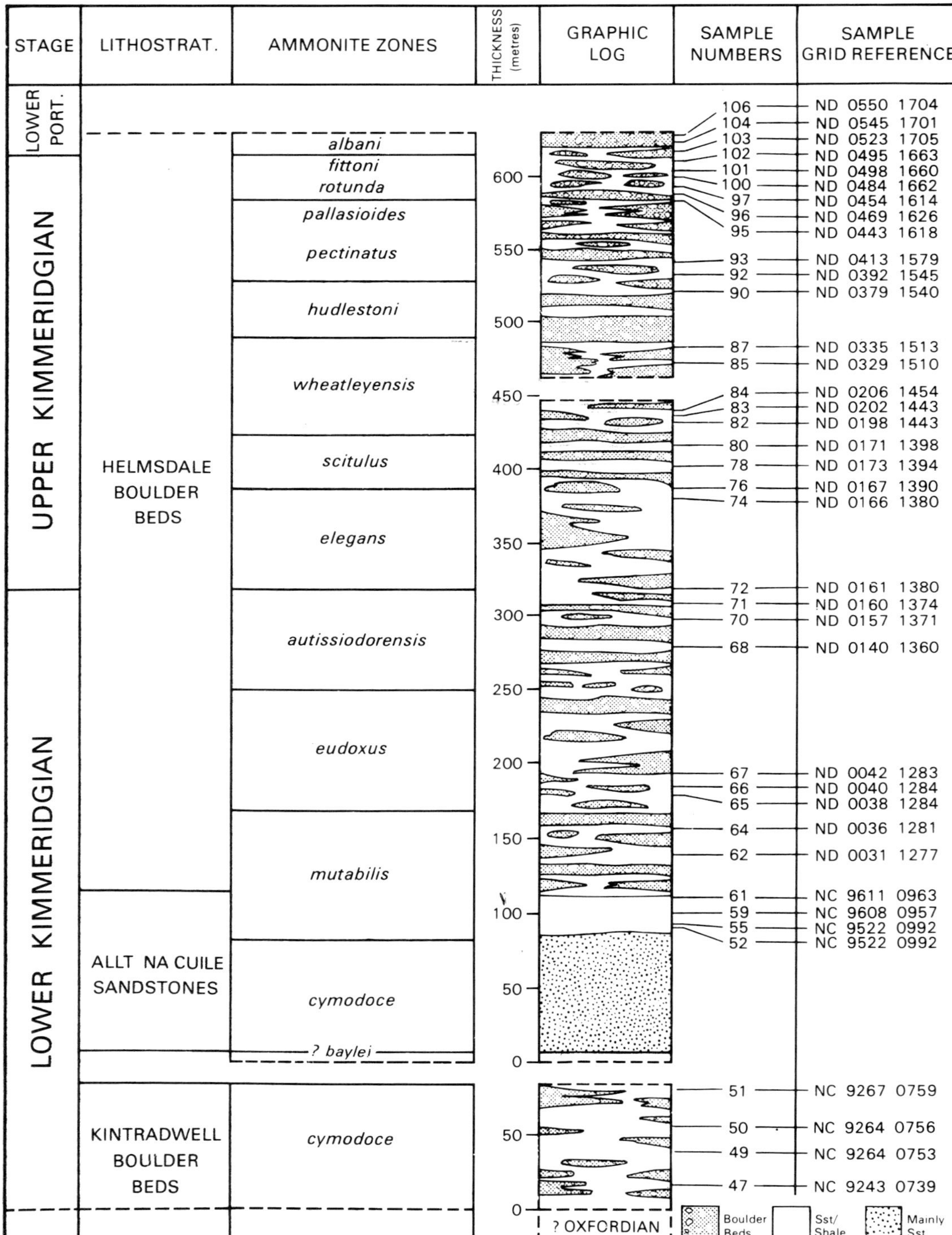

Fig. 3 — Lithostratigraphy and biostratigraphy of the Upper Jurassic section. Derivation of ammonite zonation as for Fig. 2.

the Kintradwell Boulder Beds, the Allt na Cùile Sandstones and the Helmsdale Boulder Beds. The section has been dated using ammonites (Bailey & Weir 1932, Linsley 1972, Brookfield 1976) and, with additional palynological data from Riley (1980) and the present study, a total age range of *Rasenia cymodoce* to *Progalbanites albani* Zones is proposed (Fig. 3). Age relations from structural studies by MacDonald (1985) generally confirm the biostratigraphy of earlier workers.

Exposed on the foreshore at Kintradwell are the Kintradwell Boulder Beds (Fig. 1), which consist of black shales with lesser amounts of sandstones and several boulder beds. Bedding is irregular with common evidence of synsedimentary deformation, such as slides, faulting and folding. They are assigned to the *cymodoce* Zone (Linsley 1972) and the maximum thickness is estimated to be 85 m (Pickering 1984).

A sequence of bioturbated mudstones and sandstones exposed on the foreshore between the Loth River and the Allt na Cùile (Fig. 2) is interpreted by Macdonald (1985) to be referable to the *cymodoce* zone or the *Pictonia baylei* Zone as they underlie the Allt na Cùile Sandstones. Deposition in a well-oxygenated shelf environment, similar to that of the Callovian and Oxfordian strata of the Brora region, is suggested by the bioturbation and general lack of sapropelic kerogen residues in palynological preparations (MacDonald 1985). These conditions probably represent the closing stages of Jurassic shallow water, marginal marine sedimentation, prior to major basin margin faulting and subsidence (MacDonald 1985).

Outcrops of the Allt na Cùile Sandstones are present at Lothbeg Bridge, Lothbeg Point and in the gorges of the Allt Choll and Allt na Cùile (Fig. 2). The succession consists of sandstones that vary from massive to cross-bedded and laminated carbonaceous, and black shales with limestone nodules. Its total thickness is estimated to be 120 m (Pickering 1984, MacDonald 1985). As defined by Pickering (1984), the Allt na Cùile Sandstones include the Loth River Shales of Brookfield (1976) and Lam & Porter (1977) and range in age from at least the basal *cymodoce* to the basal *Aulacostephanoides mutabilis* Zones. The contact between the Allt na Cùile Sandstones and the Kintradwell Boulder Beds is not exposed. Neves & Selley (1975) consider the sandstones to underlie the boulder beds, whereas Lam & Porter (1977) and Pickering (1984) place them above. From field relations and palaeontological evidence, MacDonald (1985) regards the two sequences as approximate lateral equivalents.

The laminated carbonaceous sandstones at Lothbeg Bridge are very similar to the 'tiger-stripe' facies described from Upper Jurassic sands in the Viking Graben of the northern North Sea Basin by Stow *et al.* (1982). In the Brae oilfield, these sands are interpreted as a series of coalescing submarine fans with associated fault scarp boulder beds. Pickering (1984) interpreted the Allt na Cùile Sandstones as submarine channel deposits whereas MacDonald (pers. comm.) believes they are subwave base, but relatively shallow-water deposits.

The Helmsdale Boulder Beds exposed on the foreshore between Lothbeg Point and Dùn Glas (Fig. 1) are estimated to be 500 m thick (MacDonald 1985), although considerable lateral facies variation makes this difficult to calculate accurately. They range in age from the *mutabilis* to the *albani* Zones (Linsley 1972, Riley 1980, this work). Linsley (1972) and Lam & Porter (1977) had suggested the *Pectinatites scitulus* Zone to be faulted out at Gartymore and the *Pectinatites hudlestoni* Zone to be present in the Helmsdale area but not exposed. From structural and sedimentological data, MacDonald (pers. comm.) assumed the section to be almost continuous, with only part of the *Pectinatites wheatleyensis* Zone unexposed at Helmsdale (Fig. 2) and no evidence of faulting at Gartymore, although angular disconformities are present on the Helmsdale to Navidale and Gartymore foreshores.

Dominant lithologies are boulder beds, carbonaceous shales and sandstones in varying proportions throughout the section. At Portgower and Gartymore, boulder beds from 1 m to nearly 10 m in thickness interdigitate with dark-grey laminated sandy shales and numerous bedded sandstones. Contained within the boulder beds are randomly orientated blocks of Old Red Sandstone which range in size from several centimetres to the Portgower 'fallen stack' (Bailey & Weir 1932) which measures 45 m × 27 m × 9 m.

Between Helmsdale and Dùn Glas, the boulder beds become more numerous and are associated with reduced amounts of sandstone and shale. On the Dùn Glas foreshore, the Jurassic rocks form a breccia with the Helmsdale Granite as a result of shearing associated with the Helmsdale Fault (Pickering 1984).

Clast composition within the boulder beds changes from Kintradwell to Dùn Glas. In the *cymodoce* and *mutabilis* Zones the clasts are all composed of Jurassic sandstone while, from the *Aulacostephanus eudoxus* to the *Aulacostephanus autissiodorensis* Zones, they are both Jurassic and Old Red Sandstone and, in the *Pectinatites elegans* to the *albani* Zones only Old Red Sandstone clasts are present (MacDonald 1985).

SAMPLES AND PROCESSING TECHNIQUES

38 samples were collected from shales and carbonaceous sandstones along the foreshore between Kintradwell and Dùn Glas and subjected to standard palynological processing techniques (Batten & Morrison 1983). The stratigraphical positions and eight-figure national grid references of the sample locations are shown in Fig. 3. Palynological preparations and figured specimens are deposited in the Marischal College Palynology Collection (MCP) of the Department of Geology and Petroleum Geology at the University of Aberdeen.

PALYNOMORPH PRESERVATION

Palynomorph preservation in these samples is variable. Miospores are generally moderately well preserved, probably owing to their more robust nature compared with the dominantly thinner-walled dinocysts, most of which show evidence of physical degradation. Skolochorate cysts especially are rarely well preserved, most specimens having broken and tangled processes, often obscured by adhering amorphous organic matter. Other forms often show surface corrosion and occasionally contain framboidal pyrite. Dinocyst preservation tends to be poorest in the lower Kimmeridgian and improves markedly in the younger strata examined, especially in the *Pavlovia rotunda* to *albani* Zones. Only samples 52, 104 and 106 were found to be barren of palynomorphs. Sample 52 is a fine sandstone, while 104 and 106 were collected from an indurated shale in a shear zone associated with the Helmsdale Fault; the preparations consist mainly of amorphous organic matter. Lam & Porter (1977) and Riley (1980) have also commented on the poor preservation of the dinocysts from the Helmsdale Boulder Beds. Dinocyst assemblages from the Kimmeridgian of Europe are often poorly preserved (Riding 1984).

SYSTEMATIC PALYNOLOGY

All the dinocyst species or groups encountered during the study are alphabetically listed below, together with author citations, full reference to which can be found in Lentin & Williams (1985). Species or groups which merit comment or discussion are marked with an asterisk (*). The plate and figure numbers are given for the species that are illustrated.

Dinocyst taxa

***Adnatosphaeridium caulleryi* (Deflandre 1938) Williams & Downie 1969 group
Apteodinium nuciforme (Deflandre 1938) Stover & Evitt 1978
Cassiculosphaeridia magna Davey 1974
Chlamydophorella sp.
Chytroeisphaeridia chytroeides (Sarjeant 1962) Downie & Sarjeant 1965 emend. Davey 1979.
Cleistosphaeridium ehrenbergii (Deflandre 1947) Davey *et al.* 1969
C. polyacanthum Gitmez 1970

C. tribuliferum (Sarjeant 1962) Davey *et al.* 1969
Cribroperidinium globatum (Gitmez & Sarjeant 1972) Helenes 1984
C. granuligerum (Klement 1960) Stover & Evitt 1978
C. longicorne (Downie 1957) Lentin & Williams 1985, Plate 1, Fig. 1
**Cribroperidinium* sp. A Davey 1982a, Plate 1, Figs. 2 and 3
**Cribroperidinium* spp., Plate 1, Figs. 4 and 5
Cyclonephelium hystrix (Eisenack 1958) Davey 1978, Plate 1, Fig. 7
Dichadogonyaulax? pannea (Norris 1965) Sarjeant 1969, Plate 1, Fig. 6
Dingodinium tuberosum (Gitmez 1970) Riley, in Fisher & Riley 1980, Plate 1, Fig. 8
Egmontodinium expiratum Davey 1982a, Plate 1, Fig. 9
E. polyplacophorum Gitmez & Sarjeant 1972, Plate 1, Fig. 10
Ellipsoidictyum cinctum Klement 1960
Escharisphaeridia pocockii (Sarjeant 1968) Erkmen & Sarjeant 1980
Glossodinium dimorphum Ioannides *et al.* 1977, Plate 2, Fig. 1
Gochteodinia mutabilis (Riley, in Fisher & Riley 1980) Davey 1982a, Plate 2, Fig. 3
Gonyaulacysta eisenackii (Deflandre 1938) Dodekova 1967 emend. Sarjeant 1982
G. jurassica (Deflandre 1938) Norris & Sarjeant 1965 emend. Sarjeant 1982
Hystrichodinium pulchrum Deflandre 1935
Hystrichosphaeridium petilum Gitmez 1970
Kleithriasphaeridium porosispinum Davey 1982a
Leptodinium antigonium Ioannides *et al.* 1977, Plate 2, Fig. 4
L. eumorphum (Cookson & Eisenack 1960) Sarjeant 1969
Meiourogonyaulax pila Gitmez & Sarjeant 1972
Mendicodinium groenlandicum (Pocock & Sarjeant 1972) Davey 1979
Muderongia simplex Alberti 1961, Plate 2, Figs. 8 and 9
**Muderongia* sp. A Davey 1979, Plate 2, Fig. 5
Occisucysta balia Gitmez 1970, Plate 2, Fig. 7
**Oligosphaeridium pulcherrimum* (Deflandre & Cookson 1955) Davey & Williams 1966 *sensu* Ioannides *et al.* 1977, Plate 2, Fig. 2
Pareodinia ceratophora Deflandre 1947 emend. Gocht 1970
Perisseiasphaeridium sp., Plate 2, Fig. 10
Prolixosphaeridium capitatum (Cookson & Eisenack 1960) Singh 1971
P. granulosum (Deflandre 1937) Davey *et al.* 1966
P. parvispinum (Deflandre 1937) Davey *et al.* 1969
Rhynchodiniopsis cladophora (Deflandre 1938) Below 1981
R. nealei (Sarjeant 1962) Jan du Chêne *et al.* 1985
Senoniasphaera jurassica (Gitmez & Sarjeant 1972) Lentin & Williams 1976, Plate 2, Fig. 6
Sentusidinium echinatum (Gitmez & Sarjeant 1972) Sarjeant & Stover 1978
S. pilosum (Ehrenberg 1854) Sarjeant & Stover 1978 emend. Erkmen & Sarjeant 1980
S. sparsibarbatum Erkmen & Sarjeant 1980
Sirmiodinium grossii Alberti 1961 emend. Warren 1973
Subtilisphaera? inaffecta (Drugg 1978) Bujak & Davies 1983
Systematophora areolata Klement 1960
S. penicillata (Ehrenberg 1843) Sarjeant 1980
Tubotuberella apatela (Cookson & Eisenack 1960) Ioannides *et al.* 1977 emend. Sarjeant 1982

Genus *Adnatosphaeridium* Williams & Downie 1966 *Adnatosphaeridium caulleryi* (Deflandre 1938) Williams & Downie 1969 group

Remarks. Included in this group are forms attributed to *Adnatosphaeridium caulleryi* (Deflandre 1938) Williams & Downie 1969, *Hystrichosphaerina orbifera* (Klement 1960) Stover & Evitt 1978 and *Systematophora* sp. 1 Davey 1982a. The close similarity and process complexity of these species precludes their separation in the present study, especially as all the specimens found are damaged and poorly preserved. Davey (1982a) stated that *Systematophora* sp. 1 may fall within the range of variation acceptable for *A. caulleryi*. *Hystrichosphaerina orbifera* is similar to *Systematophora* sp. 1 but has more complexly branched processes.
Dimensions. Central body diameter,

54(57)64 μm; process length, 20(23)28 μm; six specimens measured.
Occurrence. *cymodoce* to *albani* Zones.

Genus *Cribroperidinium* Neale & Sarjeant 1962 emend. Helenes 1984
Cribroperidinium sp. A Davey 1982a
(Plate 1, Figs. 2 and 3)

Remarks. Davey (1982a) noted that *Cribroperidinium* sp. A is morphologically similar to *C. longicorne* (Downie 1957) Lentin & Williams 1985 but is less spherical and thicker walled and has better indications of paratabulation. The specimens of *Cribroperidinium* sp. A observed in this study conform to Davey's description, although the wall is often not as densely intraperforate and the spines at the base of the apical horn are not usually as well developed.
Dimenstion. Length, 97(119)136 μm; breadth, 53(59)81 μm; horn length, 20(40)49 μm; five specimens measured.
Occurrence. *scitulus* to *albani* Zones.

Cribroperidinium spp.
(Plate 1, Figs. 4 and 5)

Remarks. This group of species includes forms in which there is a continuous gradation in morphological features, thus making speciation difficult. The poor preservation and damaged nature of many of the specimens added to the problems involved. Helenes (1984) regarded the genus *Acanthaulax* Sarjeant 1968 as a possible junior synonym of *Cribroperidinium* and forms similar to *Acanthaulax* Sp. A Ioannides *et al.* 1977 have been included in *Cribroperidinium* spp.
Occurrence. *cymodoce* to *albani* Zones. Particularly abundant from the *scitulus* Zone upwards.

Genus *Muderongia* Cookson & Eisenack 1958
Muderongia sp. A Davey 1979
(Plate 2, Fig. 5)

Remarks. The specimens encountered in this study have large well-developed apical, lateral and antapical horns with the operculum normally remaining in place. They are similar to those illustrated by Davey (1979) in having an angular endocyst which normally extends into the proximal parts of the antapical horns. According to Davey (1979), *Muderongia simplex* Alberti 1961 has a characteristic subspherical rather than an angular endocyst. This differs from the original description which states (translated from the German) 'The inner body lies close to the outer margin of the cyst, often it stretches out somewhat into the horns' (Alberti 1961, p. 12). Thus it seems that *Muderongia* sp. A conforms to the original diagnosis of *Muderongia simplex*, although there appear to be two separate species involved.
Dimensions. Overall length with apex, 145(165)177 μm; overall breadth, 93(111)128 μm; 10 specimens measured.
Occurrence. *rotunda* to *albani* Zones.

Genus *Oligosphaeridium* Davey & Williams 1966 emend. Davey 1982a
Oligosphaeridium pulcherrimum (Deflandre & Cookson 1955) Davey & Williams 1966 *sensu* Ioannides *et al.* 1977
(Plate 2, Fig. 2)

Remarks. Riley (1979) considered the specimens illustrated by Ioannides *et al.* (1977) and Gitmez (1970) to have been wrongly assigned to *O. pulcherrimum* and considered them probably to represent a new species. The specimens found in this study are all poorly preserved with much adhering amorphous organic matter and many of their processes are broken or damaged. However, enough detail was seen to enable the specimens to be referred to *O. pulcherrimum sensu* Ioannides *et al.* 1977.
Dimensions. Central body diameter, 46(49)55 μm; process length, 29(33)41 μm; seven specimens measured.
Occurrence. *eudoxus* to *Pectinatites pectinatus* Zones.

BIOSTRATIGRAPHY

By the end of the Jurassic, several invertebrate groups (particularly the ammonites) had become separated geographically into a southern Tethyan Realm and a Boreal Realm lying to the north (Arkell 1956, Casey 1971, Hallam 1971). In addition, there were local provinces within these two realms, most notably in the Boreal Realm where shallow continental seas had prevailed (Casey 1973). The difficulties encountered in correlating between the provinces has resulted in the erection of several different nomenclatural schemes for the Upper Jurassic stages (Fig. 4).

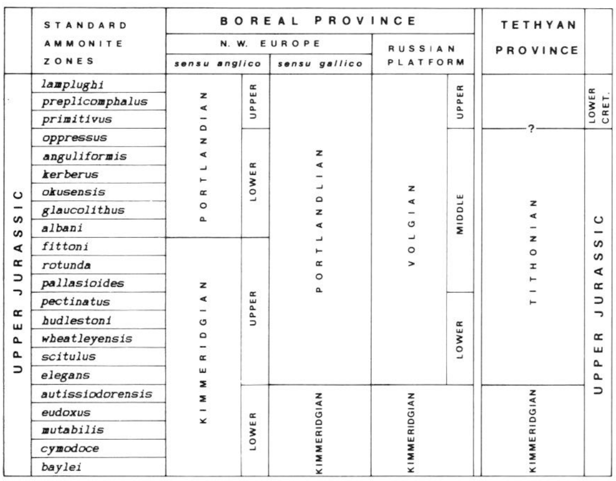

Fig. 4 — Upper Jurassic stage nomenclature. (Adapted from Riley (1977), Davey (1979) and Cope *et al.* (1980)).

Wimbledon (in: Cope *et al.* 1980) advocated the use of the Kimmeridgian and Portlandian *sensu anglico* in Britain and this recommendation is followed here. Most of the oil companies operating in the North Sea area, however, use the Kimmeridgian and Volgian stages, the latter based on condensed sequences on the Russian Platform.

Dinocyst assemblages in the Jurassic also show some degree of provincialism, although to a lesser extent than the invertebrate groups. Similar microplankton assemblages have been recorded from within the Boreal and Tethyan realms from the Hettangian to Oxfordian but, in the Kimmeridgian and Portlandian, several species are confined to one or other of these two regions (Riding & Sarjeant 1985).

In many publications on European dinocyst stratigraphy the ranges of taxa are defined in terms of ammonite zones (Riding 1984). While reviewing the correlation of Jurassic rocks in Britain, Torrens (in: Cope *et al.* 1980) argued against this practice and recommended (p. 12) '. . . a record of zonally significant forms plotted against a series of lithological divisions'. However, Woollam & Riding (1983) stated that the very refined biostratigraphy established from the study of ammonite faunas in the British Isles provides an ideal reference for comparison with any micropalaeontological zonation scheme. They also argued that ammonite zones have a very strong chronostratigraphic connotation and strata which contain no ammonite fauna are nevertheless frequently assigned to zones. This latter approach is followed here, the ammonite zones being treated as informal chronozones. Dinocyst biostratigraphy is used to allocate ammonite 'chronozones' to strata where ammonites are poorly preserved or lacking, e.g. in the section from Helmsdale to Dùn Glas.

The stratigraphical distribution of the 51 dinocyst species recorded (Fig. 5) reveals that approximately one third range through the whole or most of the section and consequently are of little use for biostratigraphical purposes. *Cribroperidinium* spp., *Cyclonephelium hystrix* and members of the *Adnatosphaeridium caulleryi* group are the most abundant forms. Other common long-ranging taxa include *Chytroeisphaeridia chytroeides*, *Escharisphaeridia pocockii*, *Pareodinia ceratophora*, *Sirmiodinium grossii*, *Systematophora areolata* and *Tubotuberella apatela*. Less common long-ranging taxa include *Chlamydophorella* sp., *Cleistosphaeridium polyacanthum* and *Glossodinium dimorphum*. The rarity of *G. dimorphum* is unusual considering that it is one of the commoner elements in most European assemblages (Riding 1984). Among the short-ranging, biostratigraphically useful dinocysts are *Cribroperidinium longicorne*, *Dichadogonyaulax*? *pannea*, *Dingodinium tuberosum*, *Egmontodinium*

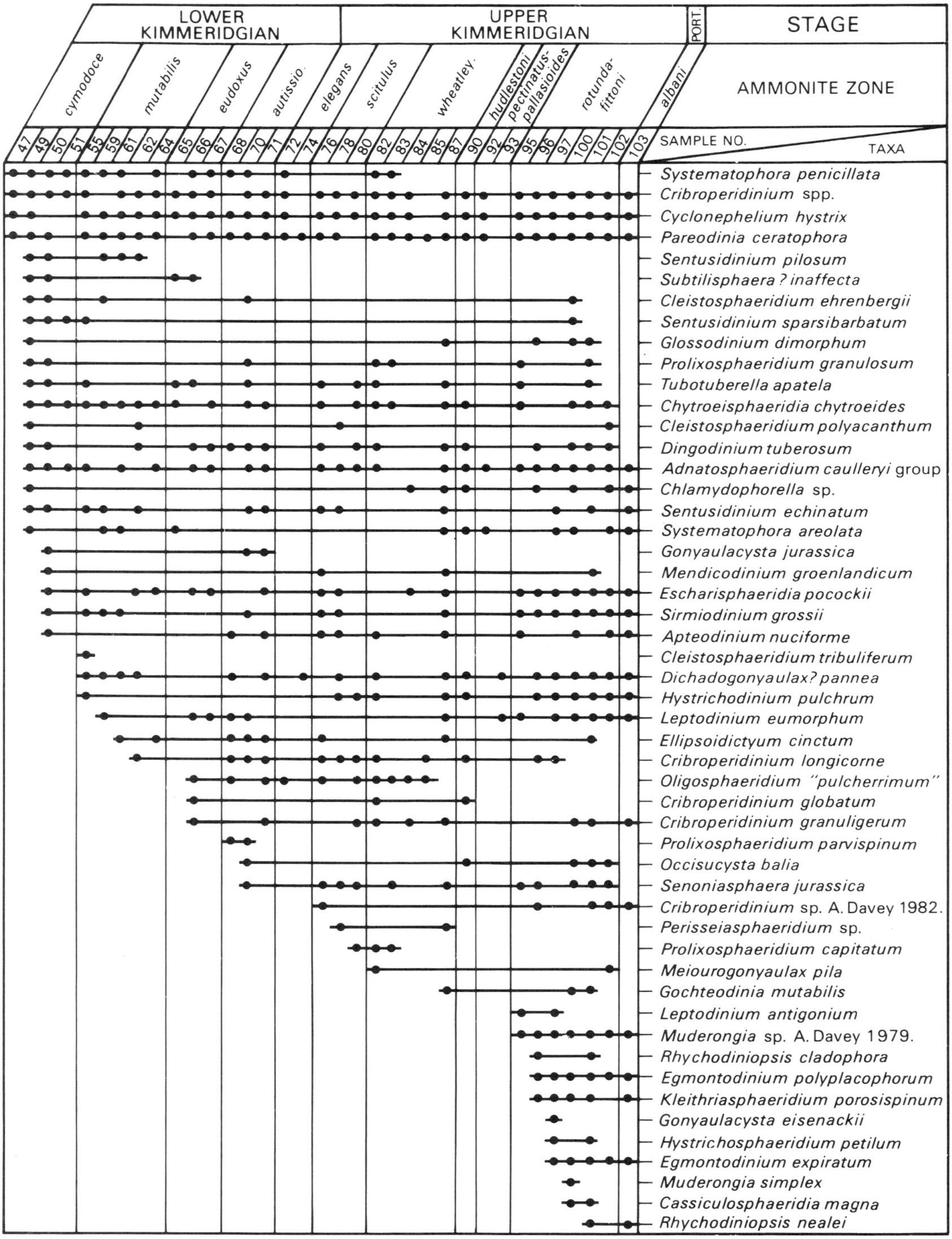

Fig. 5 — Stratigraphic ranges of dinocysts in the Upper Jurassic of east Sutherland. Derivation of ammonite zonation as for Fig. 2.

expiratum, *Kleithriasphaeridium porosispinum*, *Gochteodinia mutabilis*, *Muderongia* sp. A Davey 1979, *Occisucysta balia* and *Senoniasphaera jurassica*. These taxa have range bases or tops within the Kimmeridgian and are readily identifiable.

***cymodoce* and *mutabilis* Zones**

Common species within these zones include *Sentusidinium echinatum*, *S. sparsibarbatum*, *Systematophora areolata*, *S. penicillata* and *Tubotuberella apatela*. Taxa appearing in the *mutabilis* Zone include *Dichadogonyaulax? pannea*, *Hystrichodinium pulchrum*, *Leptodinium eumorphum* and *Cribroperidinium longicorne*. Lam & Porter (1977) also recorded the inception of *D.? pannea* within this zone, although it is not seen until the *wheatleyensis* Zone in the type Kimmeridgian section in Dorset (Riding, unpublished data). *Cleistosphaeridium tribuliferum* and *Sentusidinium pilosum* are found only within these two zones in the present study.

***eudoxus* and *autissiodorensis* Zones**

Systematophora penicillata is still abundant and *Dingodinium tuberosum* becomes increasingly common upwards within these two zones. Numbers of *Cribroperidinium longicorne* increase in the *autissiodorensis* Zone. Incoming taxa include *Cribroperidinium globatum*, *C. granuligerum*, *Oligosphaeridium pulcherrimum sensu* Ioannides *et al.* 1977 in the *eudoxus* Zones and *Occisucysta balia* and *Senoniasphaera jurassica* in the *autissiodorensis* Zone. A *eudoxus* Zone inception for *O. pulcherrimum sensu* Ioannides *et al.* 1977 was also noted by Riding (unpublished data) in the type Kimmeridgian section. *Subtilisphaera? inaffecta* is not seen above the *eudoxus* Zone and *Gonyaulacysta jurassica* last appears in the *autissiodorensis* Zone. Both species are very rare in this section.

***elegans* and *scitulus* Zones**

Dinocyst assemblages in these zones are similar to those in the *eudoxus* and *autissiodorensis* Zones. *Cribroperidinium* spp. become increasingly abundant in the *scitulus* Zone and *Systematophora penicillata* is rare in both zones. No species inceptions or extinctions are recorded from the *elegans* Zone, but first appearances of *Cribroperidinium* sp. A Davey 1982a, *Perisseiasphaeridium* sp. and *Prolixosphaeridium capitatum* were noted in the *scitulus* Zone. Davey (1982a) reported that *Cribroperidinium* sp. A was most abundant in the lower Portlandian of his sections; the occurrence here in the late Kimmeridgian represents an extension to that range. *Cribroperidinium longicorne* reaches its acme in the *scitulus* Zone, with a flood of specimens occurring in sample 76, near the base of the zone.

***wheatleyensis* and *hudlestoni* Zones**

Species of the *Adnatosphaeridium caulleryi* group are more common here than in the preceding zones, as are *Chytroeisphaeridia chytroeides*, *Dingodinium tuberosum* and *Escharisphaeridia pocockii*. *Gochteodinia mutabilis* first appears in the *wheatleyensis* Zone, although Riley (in: Fisher & Riley 1980) noted a *pectinatus* Zone inception for this species in the sections he examined in northwest Europe. *Systematophora penicillata*, *Oligosphaeridium pulcherrimum sensu* Ioannides *et al.* 1977 and *Perisseiasphaeridium* sp. have their range tops in the *wheatleyensis* Zone. A flood of *Senoniasphaera jurassica* specimens was noted in sample 87 at the top of the *wheatleyensis* Zone.

***pectinatus* to *albani* Zones**

Using published ranges for several dinocyst taxa, an attempt is made here to define ammonite zonal boundaries within the *pectinatus* to *albani* Zones. Linsley (1972) was unable to allocate this section to specific ammonite zones because of the lack of well-preserved material. As the number of known dinocyst species range bases and tops within these five ammonite zones is limited, only two zonal boundaries could be recognized for this part of the section, namely those of *Pavlovia pallasioides–rotunda* and *Virgatopavlovia fittoni–albani*, the latter also coinciding with the Kimmeridgian–Portlandian stage boundary. Riding (1984) placed the range base of *Muderongia* sp. A Davey 1979 at the base of the *rotunda* Zone, which in this study is between samples 93 and 95.

Dingodinium tuberosum and *Occisucysta balia* become extinct at the top of the *fittoni* Zone according to Woollam & Riding (1983), which is taken to be between samples 102 and 103 at Dùn Glas. This confirms Riley's (1980) suggestion that the youngest sediments exposed are early Portlandian in age.

The richest assemblages of dinocysts are encountered in the *rotunda* to *fittoni* Zones, where abundant species referable to the *Adnatosphaeridium caulleryi* group, *Cribroperidinium* spp., *Cyclonephelium hystrix*, *Dichadogonyaulax? pannea*, *Hystrichodinium pulchrum* and *Muderongia* sp. A Davey 1979 are found. Less abundant, although still common, are *Escharisphaeridia pocockii* and *Sirmiodinium grossii*. New taxa appearing for the first time in the *rotunda* to *fittoni* Zones include *Cassiculosphaeridia magna*, *Egmontodinium expiratum*, *E. polyplacophorum*, *Hystrichosphaeridium petilum*, *Kleithriasphaeridium porosispinum* and *Leptodinium antigonium*. In the type Kimmeridgian section, *E. expiratum* first appears in the *hudlestoni* Zone (Riding, unpublished data) and Davey (1982a) recorded a *pectinatus* Zone inception for *K. porosispinum* in Dorset and eastern England. The apparent later appearance of both taxa at Helmsdale is probably due to their rarity in the present study. *Cribroperidinium longicorne* is last seen in the *rotunda* to *fittoni* Zones, which is unusual and may be due to reworking as *hudlestoni* and *pallasioides* Zone range tops were recorded by Riding (unpublished data) and Raynaud (1978), respectively. The occurrences of *Gonyaulacysta eisenackii*, *Rhynchodiniopsis cladophora* and *R. nealei* are also presumed to be due to reworking as all have predominantly Oxfordian ranges in the literature. One specimen of *Muderongia simplex* is recorded here; previous records indicate an early Cretaceous age (Davey 1982a, Woollam & Riding 1983).

The assemblage recorded from the *albani* Zone is relatively rich in dinocysts, although the number of species is reduced compared with those recorded in the *rotunda* to *fittoni* Zones. The *A. caulleryi* group, *Cribroperidinium* spp., *C. hystrix*. *D.? pannea* and *Muderongia* sp. A Davey 1979 are all abundant in the *albani* Zone. Similar assemblages are reported in the *Ctenidodinium culmulum–C. panneum* (Cc–Cp) dinocyst Zone of Woollam & Riding (1983) which is equivalent to the *albani* to *Glaucolithites glaucolithus* Zones.

BIOSTRATIGRAPHIC SUMMARY AND COMPARISON WITH PREVIOUS ZONATION SCHEMES

Dinocysts confined to the lower Kimmeridgian of Helmsdale include *Gonyaulacysta jurassica*, *Sentusidinium pilosum*, *Subtilisphaera? inaffecta* and *Systematophora penicillata*. Many of the taxa present in the lower Kimmeridgian also range into the upper Kimmeridgian and basal Portlandian, where *Cribroperidinium* sp. A Davey 1982a, *Dichadogonyaulax? pannea*, *Egmontodinium polyplacophorum*, *E. expiratum*, *Gochteodinia mutabilis*, *Hystrichodinium pulchrum*, *Muderongia* sp. A Davey 1979, *Senoniasphaera jurassica* and *Sirmiodinium grossii* are distinctive forms.

The flood of *Cribroperidinium longicorne* in the *scitulus* Zone and *S. jurassica* in the *wheatleyensis* Zone could prove useful stratigraphic markers in the local area and possibly further afield.

Although the majority of the dinocysts are damaged and poorly preserved, most are considered to be indigenous. There is, however, some evidence of reworking from the records of *G. eisenackii*, *R. cladophora*, *R. nealei*, the extended range of *C. longicorne* and the presence of Devonian and Carboniferous miospores (Lam & Porter 1977, Riley 1980).

The most comprehensive Jurassic dinocyst zonation scheme to date is that of Woollam & Riding (1983) in which 16 dinocyst zones and 29 subzones were recognized, all of which were related to the standard ammonite zonation. The top of their *Scriniodinium luridum* dinocyst Zone is based on the extinction of *S. luridum*, a species not found in the present study. The overlying zone, the *Glossodinium dimorphum–Dingodinium tuberosum* (Gd–Dt) dinocyst Zone is divided into three subzones, based on the appearance of *E.*

polyplacophorum and the extinctions of *G. jurassica* and *D. tuberosum.* Owing to the lack of overlap between the ranges of *E. polyplacophorum* and *G. jurassica* in the present study, none of these subzones could be distinguished. However, the extinction of *D. tuberosum* and *Occisucysta balia* enabled the recognition of the boundary between the Gd–Dt and the Cc–Cp dinocyst Zones of Woollam & Riding (1983), which also marks the *fittoni–albani* Zone boundary. The assemblage from sample 103 probably represents subzone *a* of the Cc–Cp dinocyst Zone, despite the absence of *S. jurassica*, the extinction of which marks the top of this subzone. This confirms the suggestion of Riley (1980) of an early Portlandian age for the youngest samples exposed at Dùn Glas. The *pallasioides–rotunda* Zone boundary is marked by the appearance of *Muderongia* sp. A Davey 1979 in sample 95.

PALYNOMORPH DISTRIBUTION AND PALAEOENVIRONMENTS

The palynomorph assemblages of the early Kimmeridgian age are dominated by miospores, including trilete spores, bisaccate pollen and other gymnospermous pollen such as *Callialasporites*, *Cerebropollenites* and *Classopollis*, which account for between 58% and 94% of the total palynomorphs recorded, with a mean of 73% (Fig. 6). Terrestrially derived organic matter such as inertinite, vitrinite, 'cellular' tissue and cuticle is more abundant than marine algal 'sapropel'.

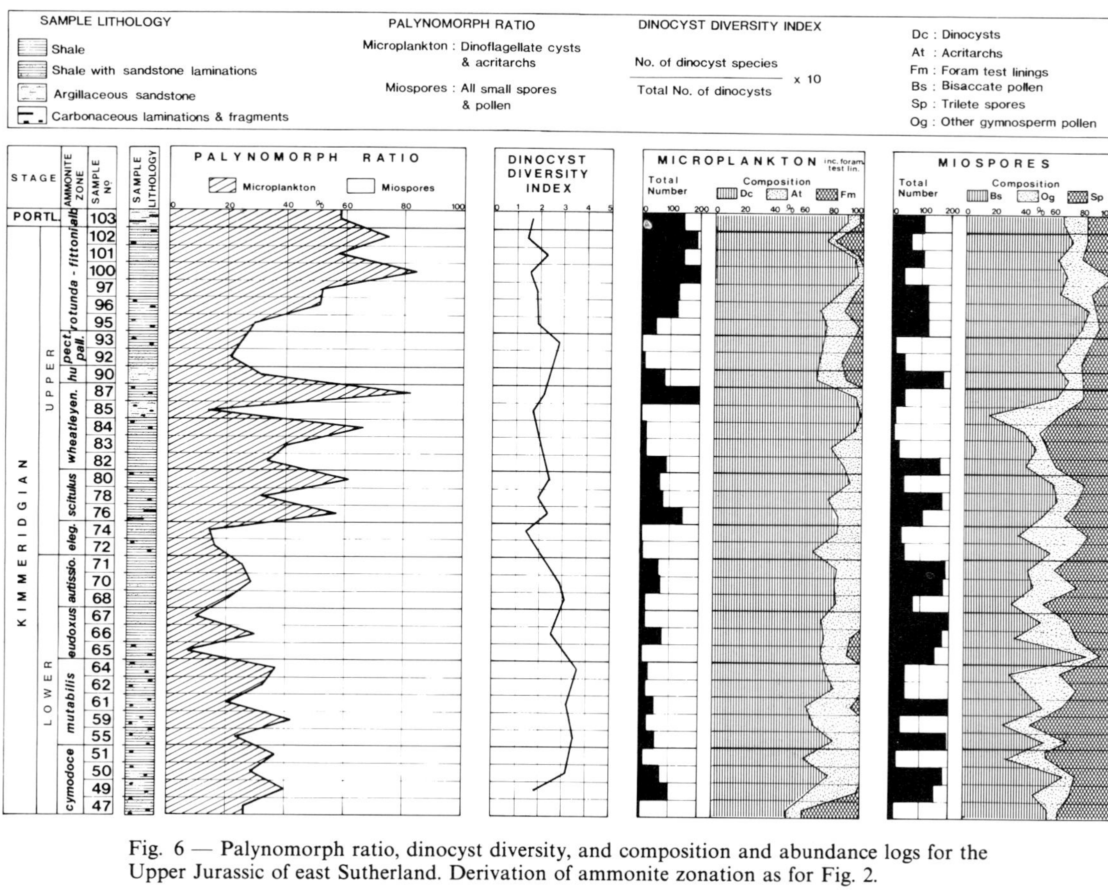

Fig. 6 — Palynomorph ratio, dinocyst diversity, and composition and abundance logs for the Upper Jurassic of east Sutherland. Derivation of ammonite zonation as for Fig. 2.

The removal of much of the fine detritus and amorphous matter during oxidation and ultrasonic treatment further supports this observation as terrigenous organic matter tends to be more easily removed by these means than that of marine origin (Batten 1981, 1983).

The upper Kimmeridgian and lowermost Portlandian assemblages show greater fluctuations in the microplankton:miospore ratio, although there is an increasing dominance of dinocysts and acritarchs in the *scitulus*, *wheatleyensis* and *rotunda* to *albani* Zones. Included with the acritarchs are probable prasinophycean algae such as *Crassosphaera*, *Pterospermella* and *Tasmanites*. Miospores, although less abundant, still average 55% of the total palynomorphs. Kerogen compositions are similar to those of the lower Kimmeridgian, although a decline in plant-derived detritus is observed with a corresponding increase in amorphous material of presumed marine algal origin, especially in the *rotunda* to *albani* Zones.

The presence of both microplankton and miospores in the preparations studied indicates deposition in a marine environment relatively close to a landmass from which fluvial discharge acted as a source of terrigenous material. Most of the samples collected came from black shales, which were probably deposited by suspension fallout sedimentation in a hemipelagic environment. Slow rates of deposition are suggested by the well preserved ammonites and plant fragments (Pickering 1984).

Dinocyst diversity (Fig. 6) is at a maximum in the *mutabilis* Zone and declines to about half in the *rotunda* to *albani* Zones, despite the increase in dinocyst abundance. This may be a reflection of the decline in total number of dinocyst species reported from the Kimmeridgian to the Portlandian worldwide (Bujak & Williams 1979).

Reactivation of the Helmsdale Fault, initiated at the beginning of the early Kimmeridgian, caused increased subsidence and basin deepening to occur (MacDonald 1985). The resultant changes in the width and composition of the submarine shelf affected palynomorph composition throughout the Kimmeridgian and basal Portlandian (Fig. 6). During the early Kimmeridgian this basin deepening was associated with the transition from relatively oxic to probable anoxic bottom conditions, leading to decreased microbial activity and thus allowing the better preservation of dinocysts observed in the upper part of the section. Restricted water circulation and rises in sea level caused widespread stagnant conditions throughout northwest Europe during the Kimmeridgian (Tyson *et al.* 1979).

During *cymodoce* and *mutabilis* Zone times the submarine shelf was narrow and relatively sandy, with evidence of limited wave reworking on the slope in the *cymodoce* Zone, indicating that the basin was still relatively shallow and above wave base at times (MacDonald 1985). The general lack of sapropelic kerogen and the poor dinocyst preservation suggests oxic, relatively high-energy bottom conditions. By *mutabilis* Zone times the basin was entirely below wave base. Abundant synsedimentary deformation at Kintradwell suggests that the sedimentary pile was oversteepened by rapid basin subsidence. MacDonald (1985) suggested that by the *eudoxus* Zone a wider, less sandy shelf was present, and from the *autissiodorensis* Zone to the early Portlandian, a wide sediment-starved shelf existed. The reduction in the sediment content on the shallow shelf allowed the development of coral reefs, the remains of which are seen as clasts in the boulder beds (Neves & Selley 1975, MacDonald 1985).

The composition of the palynomorph assemblages and kerogen types broadly supports MacDonald's (1985) palaeoenvironmental interpretation, with the abundance of miospores and terrigenous debris suggesting that a shoreline with sources of fluvial discharge was close to the depositional site located on the basin side of the fault (Fig. 7). Decreasing numbers of miospores and amounts of terrigenous detritus from the late Kimmeridgian supports the contention of a widening shelf area and increasing distance of the site from the shoreline. The larger recovery of dinocysts in the upper Kimmeridgian and lowermost Portlandian probably reflects a slower rate

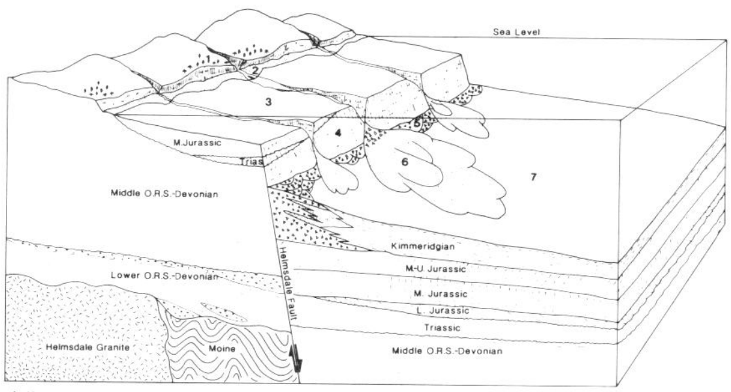

Fig. 7 — Depositional model for the Upper Jurassic boulder beds and associated deposits around Helmsdale. (Adapted from Trewin (1984) and Pickering (1984)).

of deposition (Davey 1982b), resulting from the fact that the decreased sediment load is deposited on the slope after transportation over the wider shelf, and the greater likelihood of preservation in a more anoxic environment.

Above the *wheatleyensis* Zone the miospore composition of the assemblages is seen to change with an increase in gymnospermous pollen and a corresponding decrease in trilete spore numbers. This could be interpreted as a reflection of an increase in the gymnosperm and a decrease in the pteridophyte component of the flora, due possibly to a climatic change, but there is no geological evidence to support this conclusion. An alternative explanation is that bisaccate pollen were generally lighter and more easily transported, and thus more likely to be found further from the shoreline or river effluence (Davey 1982b). This would also apply to small thin-walled spores (Batten 1973).

Rawson & Riley (1982) proposed that the widespread regression which occurred during the middle of the late Kimmeridgian of the North Sea area was eustatically controlled and never completely masked by local tectonics. MacDonald (pers. comm.) found no sedimentological evidence at Helmsdale for this regression and assumed that movements of the Helmsdale Fault completely masked the regional sea level trend. The microplankton:miospore ratio during the *pectinatus* and *pallasioides* Zones (Fig. 6), however, shows a marked decline in actual and relative dinocyst abundance, possibly reflecting a regression and hence providing support for Rawson & Riley's (1982) suggestion. If this is correct, Kimmeridgian palynomorph assemblage trends of the Helmsdale region would appear to be more sensitive indicators of sea-level fluctuations than is the sedimentological evidence.

ACKNOWLEDGEMENTS

This paper is based on the author's M. Sc. Thesis at the Department of Geology and Mineralogy (now Petroleum Geology), University of Aberdeen. Sincere thanks go to David Batten for his supervision and guidance and to Robertson Research International Ltd for financial support. I am grateful for help and advice from Jim Fenton, Dave Ewen, Tim Varney and Roger Davey of Robertson Research, and to Alister MacDonald of IKU, Trondheim, for valuable discussion on the stratigraphy and sedimentology. I am also indebted to Lorraine Morrison for help with palynological processing of the samples collected, to Debbie Daley (British Geological Survey, Keyworth) for draughting some of the figures and to Bernard Owens, Jim Riding and Beris Cox (British Geological Survey, Keyworth) for comments on the manuscript. This paper is published with the permission of the Director, British Geological Survey (Natural Environmental Research Council).

Plate 1 — All specimens are from the Helmsdale Boulder Beds.

Fig. 1 — *Cribroperidinium longicorne* (Downie 1957) Lentin & Williams 1985. Slide MCP 3818/3, sample 97, *rotunda* to *fittoni* Zones. Note the thin wall and faint paratabulation.

Figs. 2 and 3 — *Cribroperidinium* sp. A Davey 1982a. Slide MCP 3701/7, sample 76, *scitulus* Zone. Fig. 2. Detail of apical horn, which is hollow for approximately half of its length; note antapical ornament. Fig. 3. Median view, showing paratabulation.

Fig. 4 — *Cribroperidinium* sp. Slide MCP 3713/6, sample 102, *rotunda* to *fittoni* Zones. Specimen with prominent apical horn and antapical spines; composite photomicrograph.

Fig. 5 — *Cribroperidinium* sp. Slide MCP 3713/4, sample 102, *rotunda* to *fittoni* Zones. Specimen with prominent irregular spines.

Fig. 6 — *Dichadogonyaulax? pannea* (Norris 1965) Sarjeant 1969. Slide 3712/5, sample 100, *rotunda* to *fittoni* Zones. Oblique antapical view.

Fig. 7 — *Cyclonephelium hystrix* (Eisenack 1958) Davey 1978. Slide MCP 3712/5, sample 100, *rotunda* to *fittoni* Zones. Note distinct offset antapical bulge and bald central portion; operculum still attached.

Fig. 8 — *Dingodinium tuberosum* (Gitmez 1970) Riley in Fisher & Riley 1980. Slide MCP 3713/6, sample 102, *rotunda* to *fittoni* Zones. Apical archaeopyle, operculum in place.

Fig. 9 — *Egmontodinium expiratum* Davey 1982a. Slide MCP 3819/3, sample 103, *albani* Zone. Processes formed from extensions of the parasutural crests; composite photomicrograph.

Fig. 10 — *Egmontodinium polyplacophorum* Gitmez & Sarjeant 1972. Slide MCP 3819/3, sample 103, *albani* Zone. Specimen with prominent parasutural crests.

All magnifications × 500.

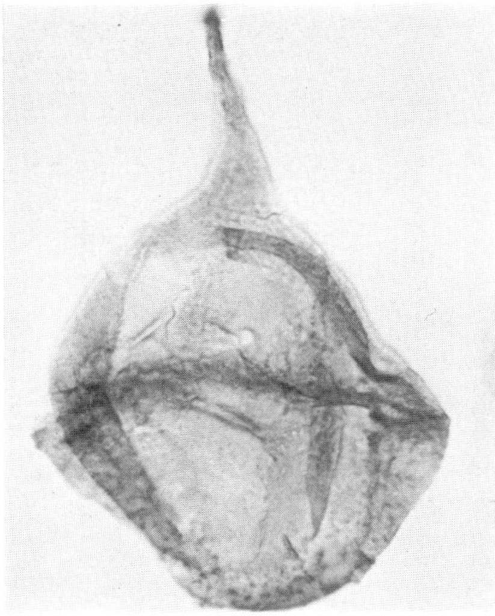
1

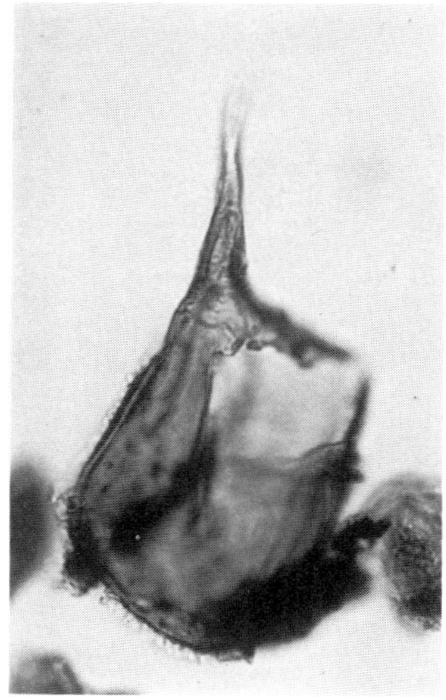
2

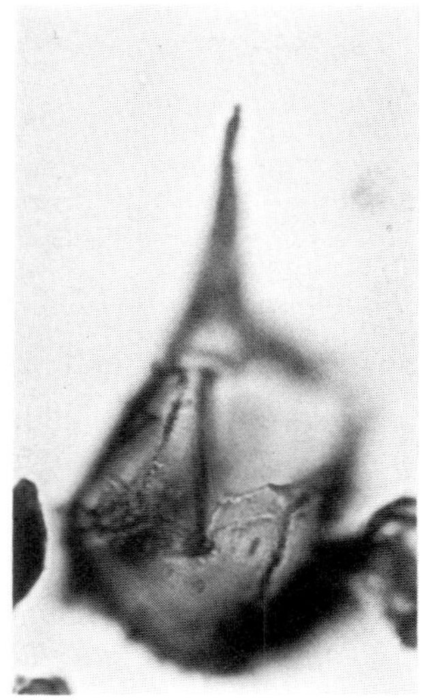
3

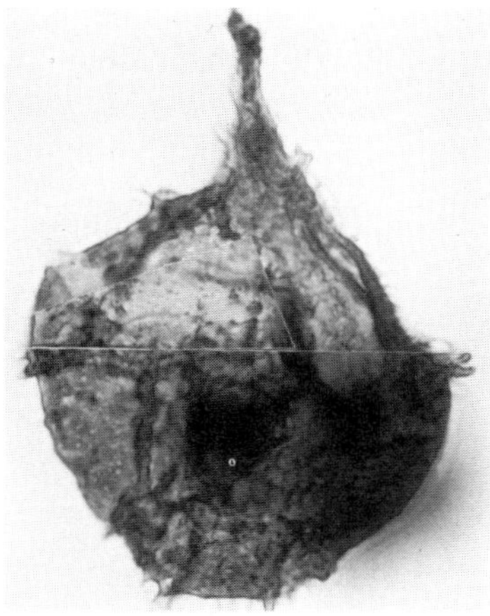
4

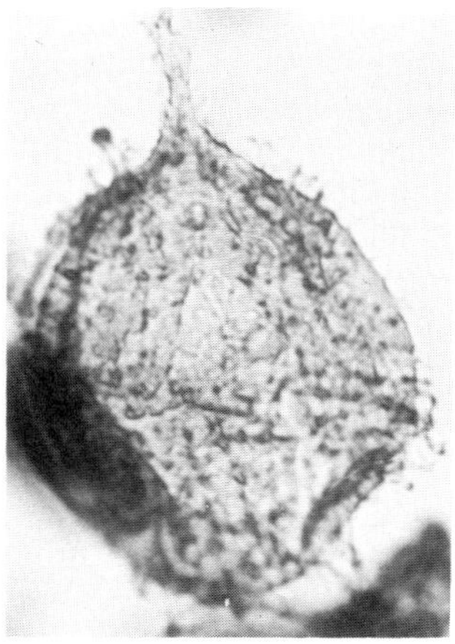
5

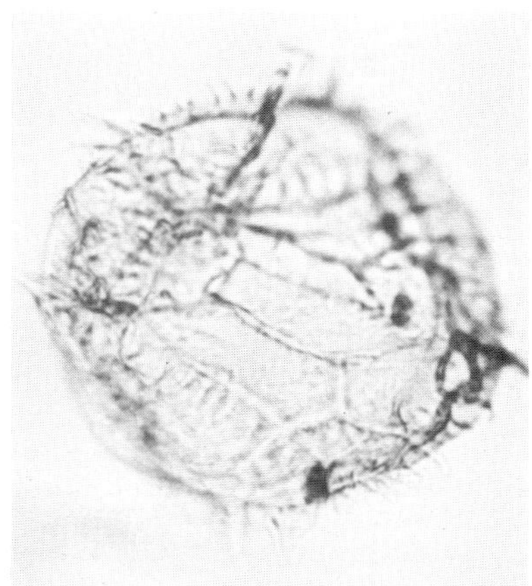
6

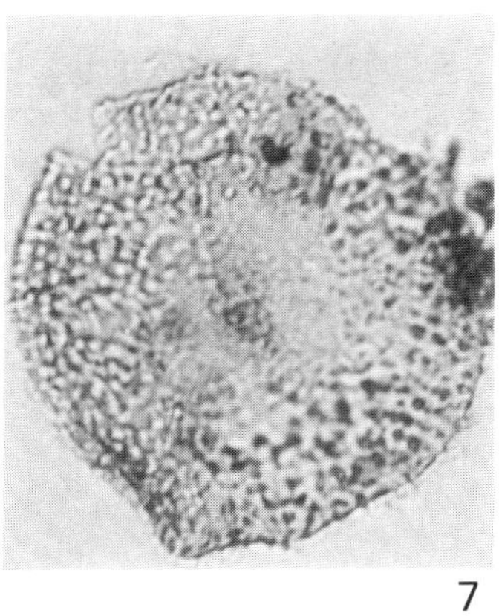
7

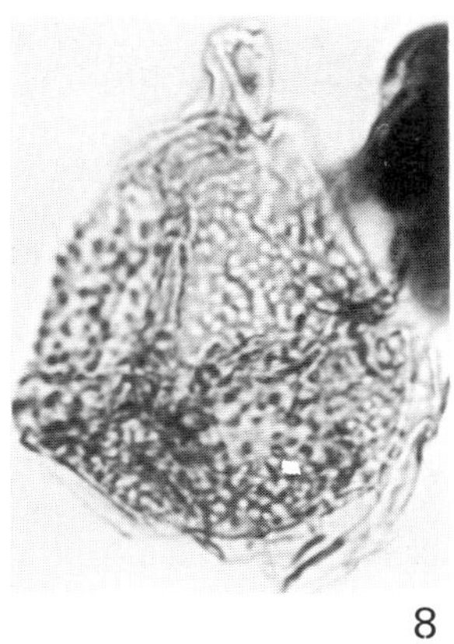
8

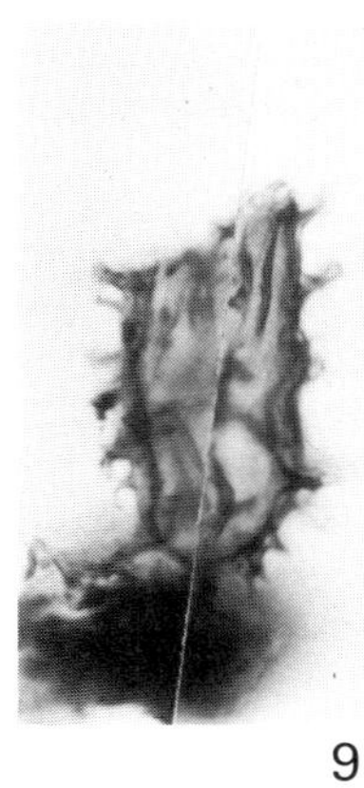
9

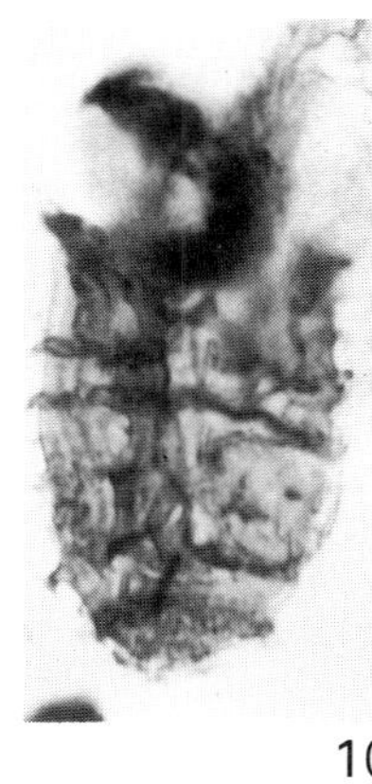
10

Plate 2 — All specimens are from the Helmsdale Boulder Beds.

Fig. 1 — *Glossodinium dimorphum* Ioannides *et al.* 1977. Slide MCP 3818/5, sample 97, *rotunda* to *fittoni* Zones. Note prominent digitate antapical process.

Fig. 2 — *Oligosphaeridium pulcherrimum* (Deflandre & Cookson 1955) Davey & Williams 1966 *sensu* Ioannides *et al.* 1977. Slide MCP 3703/4, sample 80, *scitulus* Zone. Note the distally expanded fenestrate processes; composite photomicrograph.

Fig. 3 — *Gochteodinia mutabilis* (Riley, in Fisher & Riley 1980) Davey 1982a. Slide MCP 3712/5, sample 100, *rotunda* to *fittoni* Zones. Note apicular structure.

Fig. 4 — *Leptodinium antigonium* Ioannides *et al.* 1977. Slide MCP 3818/4, sample 97, *rotunda* to *fittoni* Zones. Note denticulate antapical crest.

Fig. 5 — *Muderongia* sp. A Davey 1979. Slide MCP 3818/4, sample 97, *rotunda* to *fittoni* Zones. Angular endocyst extends into proximal regions at antapical horns.

Fig. 6 — *Senoniasphaera jurassica* (Gitmez & Sarjeant 1972) Lentin & Williams 1976. Slide MCP 3712/5, sample 100, *rotunda* to *fittoni* Zones. Specimen with partly detached operculum.

Fig. 7 — *Occisucysta balia* Gitmez 1970. Slide MCP 3713/4, sample 102, *rotunda* to *fittoni* Zones.

Figs. 8 and 9 — *Muderongia simplex* Alberti 1961. Slide MCP 3712/5, sample 100, *rotunda* to *fittoni* Zones. Specimen showing subspherical endocyst and detached operculum. Fig. 8. Low focus. Fig. 9. High focus.

Fig. 10. — *Perisseiasphaeridium* sp. Davey 1982a. Slide MCP 3706/5, sample 87, *wheatleyensis* Zone.

All magnifications × 500.

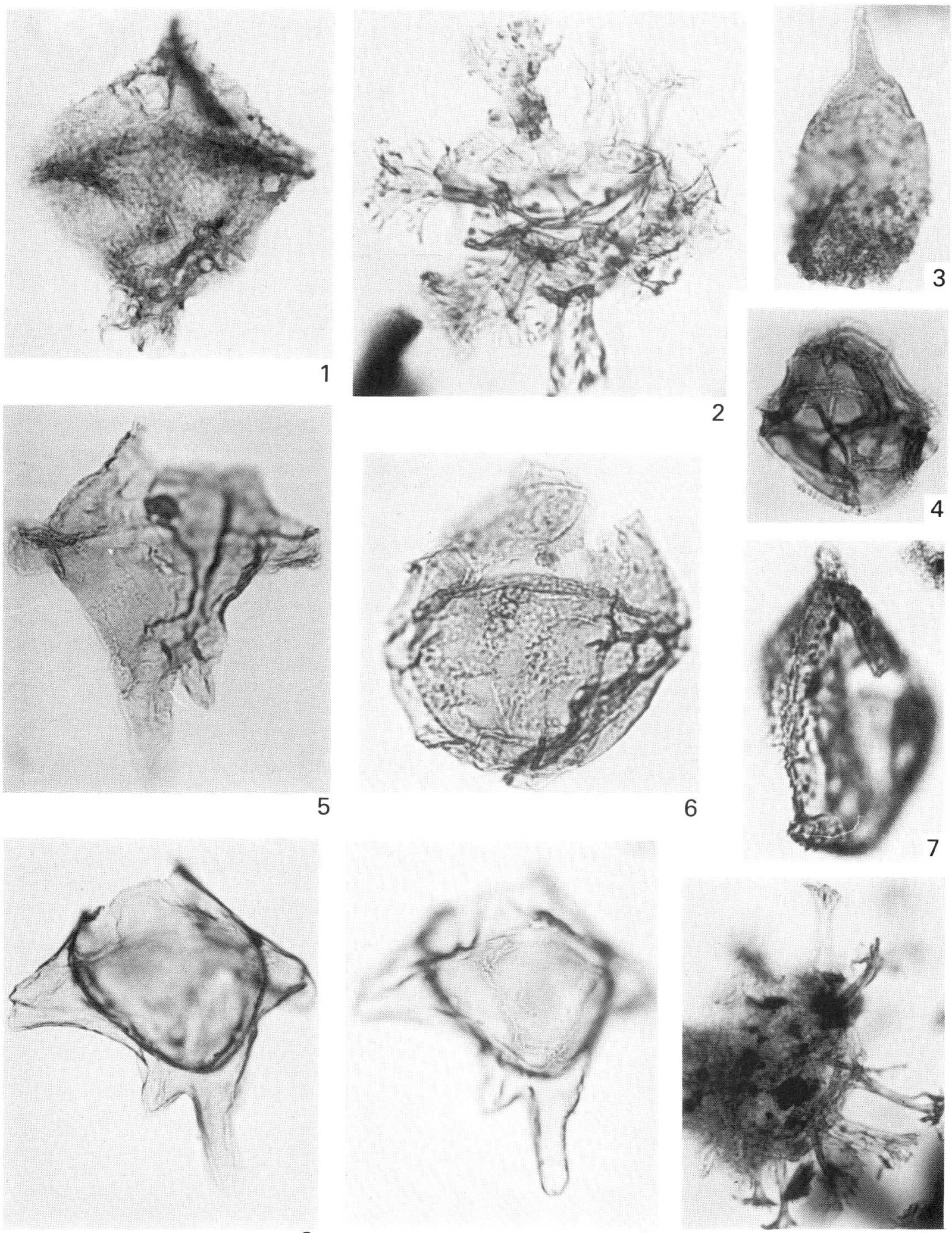
1
2
3
4
5
6
7
8
9
10

REFERENCES

Alberti, G. (1961). Zur Kenntnis mesozoischer und alttertiärer Dinoflagellaten und Hystrichosphaerideen von Nord- und Mitteldeutschland sowie einegen anderen europäischen Gebeiten. *Palaeontographica A* **116** 1–58.

Arkell, W. J. (1956). *Jurassic geology of the world.* Oliver and Boyd, Edinburgh, pp. 1–806.

Bailey, E. B. & Weir, J. (1932). Submarine faulting in Kimmeridgian times: east Sutherland. *Transactions of the Royal Society of Edinburgh* **57** 429–466.

Batten, D. J. (1973). Use of palynologic assemblage-types in Wealden correlation. *Palaeontology* **16** 1–40.

Batten, D. J. (1981). Palynofacies, organic maturation and source potential for petroleum. In: Brooks, J. (ed.), *Organic maturation studies and fossil fuel exploration.* Academic Press, London, pp. 201–223.

Batten, D. J. (1983). Identification of amorphous sedimentary organic matter by transmitted light microscopy. In: Brooks, J. (ed.), *Petroleum geochemistry and exploration of Europe. Special Publication, Geological Society, London* **12** 275–287.

Batten, D. J. & Morrison, L. (1983). Methods of palynological preparation for palaeoenvironmental, source potential and organic maturation studies. *Norwegian Petroleum Directorate Bulletin* **2** 35–53.

Batten, D. J., Trewin, N. H. & Tudhope, A. W. (1986). The Triassic–Jurassic junction at Golspie, inner Moray Firth Basin. *Scottish Journal of Geology* **22** 85–98.

Brookfield, M. E. (1976). The age of the Allt na Cùile Sandstones (Upper Jurassic, Sutherland). *Scottish Journal of Geology* **12** 181–186.

Brooks, J. R. V. & Chesher, J. A. (1975). Review of the offshore Jurassic of the UK northern North Sea. In: Finstad, K. G. & Selley, R.C. (eds), *Proceedings of the Jurassic Northern North Sea Symposium, Stavanger, September 1975. Norsk Petroleumforening* **JNNSS/2** 1–24.

Bujak, J. P. & Williams, G. L. (1979). Dinoflagellate diversity through time. *Marine Micropaleontology* **4** 1–12.

Casey, R. (1971). Facies, faunas and tectonics in the Late Jurassic–Early Cretaceous Britain. In: Middlemiss, F. A., Rawson, P. F. & Newall, G. (eds), *Faunal provinces in space and time. Geological Journal, Special Issue* **4** 153–168.

Casey, R. (1973). The ammonite succession at the Jurassic–Cretaceous boundary in eastern England. In: Casey, R. & Rawson, P. F. (eds), *The Boreal Lower Cretaceous. Geological Journal, Special Issue* **5** 193–266.

Chesher, J. A. & Lawson, D. (1983) . The geology of the Moray Firth. *Report, Institute of Geological Sciences, London* **83/5** 1–32.

Cope, J. C. W., Duff, K. L., Parsons, C. F., Torrens, H. S., Wimbledon, W. A. & Wright, J. K. (1980). A correlation of Jurassic rocks in the British Isles. Part two: Middle and Upper Jurassic. *Special Reports, Geological Society, London* **15** 1–109.

Davey, R. J. (1979). The stratigraphic distribution of dinocysts in the Portlandian (latest Jurassic) to Barremian (Early Cretaceous) of northwest Europe. *American Association of Stratigraphic Palynologists, Contributions Series* **5B** 49–81.

Davey, R. J. (1982a). Dinocyst stratigraphy of the latest Jurassic to Early Cretaceous of the Haldager No. 1 borehole, Denmark. *Danmarks Geologiske Undersøgelse, Series B* **6** 1–57.

Davey, R. J. (1982b). Die Verbreitung der palynomorphen im späten Apt und frühen Alb Nordwestdeutschlands. *Geologische Jährbuch, Hannover* **A65** 365–403.

Fisher, M. J. & Riley, L. A. (1980). The stratigraphic distribution of dinoflagellate cysts at the Boreal Jurassic–Cretaceous boundary. *Proceedings of the 4th International Palynological Conference, Lucknow, 1976–7,* Vol. 2, pp. 313–329.

Gitmez, G. U. (1970). Dinoflagellate cysts and acritarchs from the basal Kimmeridgian (Upper Jurassic) of England, Scotland and France. *Bulletin of the British Museum of Natural History, Geology* **18** 231–331.

Hallam, A. (1971). Provinciality in Jurassic faunas in relation to facies and palaeogeography. In: Middlemiss, F. A., Rawson, P. F. & Newall, G. (eds), *Faunal provinces in space and time. Geological Journal, Special Issue* **4** 129–152.

Helenes, J. (1984). Morphological analysis of Mesozoic–Cenozoic *Cribroperidinium* (Dinophyceae) and taxonomic implications. *Palynology* **8** 107–138.

Hurst, A. (1985). The implications of clay mineralogy to palaeoclimate and provenance during the Jurassic in NE Scotland. *Scottish Journal of Geology* **21** 143–160.

Ioannides, N. S., Stavrinos, G. N. & Downie, C. (1977). Kimmeridgian microplankton from Clavell's Hard, Dorset, England. *Micropaleontology* **22** 443–478.

Lam, K. & Porter, R. (1977). The distribution of palynomorphs in the Jurassic rocks of the Brora Outlier, NE Scotland. *Journal of the Geological Society, London* **134** 45–55.

Lentin, J. K. & Williams, G. L. (1985). Fossil and living dinoflagellates: index to genera and species, 1985 edition. *Canadian Technical Report of Hydrography and Ocean Sciences* **60** 1–451.

Linsley, P. N. (1972). The stratigraphy and sedimentology of the Kimmeridgian deposits of Sutherland, Scotland *Ph.D. Thesis.* University of London, pp. 1–214.

Linsley, P. N., Potter, H. C., McNab, G. & Racher, D. (1980). The Beatrice Field, Inner Moray Firth, UK North Sea. In: Halbouty, M. T. (ed.), *Giant oil and gas fields of the decade 1968–1978. American Association of Petroleum Geologists, Memoir* **30** 117–129.

MacDonald, A. C. (1985). Kimmeridgian and Volgian fault-margin sedimentation in the northern North Sea area. *Ph.D. Thesis.* University of Strathclyde, pp. 1–255.

Neves, R. & Selley, R. C. (1975). A review of the Jurassic rocks of north-east Scotland. In: Finstad, K. G. & Selley, R. C. (eds), *Proceedings of the Jurassic Northern North Sea Symposium, Stavanger, September 1975. Norsk Petroleumforening* **JNNSS/5** 1–29.

Pickering, K. T. (1984). The Upper Jurassic 'Boulder Beds' and related deposits: a fault controlled-submarine slope, NE Scotland. *Journal of the Geological Society, London* **141** 357–374.

Rawson, P. F. & Riley, L. A. (1982). Latest Jurassic–Early Cretaceous events and the 'Late Cimmerian Unconformity' in the North Sea area. *Bulletin of the American Association of Petroleum Geologists* **66** 2628–2648.

Raynaud, J. F. (1978). Principaux dinoflagellés caractéristiques du Jurassiques Supérior d'Europe du Nord. *Palinologia, Número Extraordinario* **1** 387–405.

Riding, J. B. (1984). Dinoflagellate cyst range-top biostratigraphy of the uppermost Triassic to lowermost Cretaceous of northwest Europe. *Palynology* **8** 195–210.

Riding, J. B. & Sarjeant, W. A. S. (1985). The role is dinoflagellate cysts in the biostratigraphic subdivision of the Jurassic System. *Newsletters on Stratigraphy* **14** 96–109.

Riley, L. A. (1977). Stage nomenclature at the Jurassic–Cretaceous boundary, North Sea basin. In: Finstad, K. G. & Selley, R. C. (eds), *Proceedings of the Mesozoic Northern North Sea Symposium, Oslo, 1977. Norsk Petroleumsforening* **MNNSS/4** 1–11.

Riley, L. A. (1979). Dinocysts from the upper Kimmeridgian (*pectinatus* Zone) of Marton, Yorkshire. *Mercian Geologist* **7** 219–222.

Riley, L. A. (1980). Palynological evidence of an early Portlandian age for the uppermost Helmsdale Boulder Beds, Sutherland. *Scottish Journal of Geology* **16** 29–31.

Selley, R. C. (1976). The habitat of North Sea oil. *Proceedings of the Geologists Association* **87** 359–387.

Stow, D. A. V., Bishop, C. D. & Mills, S. J. (1982). Sedimentology of the Brae oilfield, North Sea: fan models and controls. *Journal of Petroleum Geology* **5** 129–148.

Trewin, N. H. (1984). *Petroleum geology. M.Sc. course Excursion Guide.* University of Aberdeen.

Tyson, R.V., Wilson, R. C. L. & Downie, C. (1979). A stratified water column environmental model for the type Kimmeridge Clay. Nature **277** 377–380.

Van der Burgh, J. & Van Konijnenburg-van Cittert, J. H. A. (1984). A drifted flora from the Kimmeridgian (Upper Jurassic) of Lothbeg Point, Sutherland, Scotland. *Review of Palaeobotany and Palynology* **43** 359–396.

Windle, T. M. F. 1979. Reworked Carboniferous spores: an example from the Lower Jurassic of northeast Scotland. *Review of Palaeobotany and Palynology* **27** 173–184.

Woollam, R. & Riding, J. B. (1983). Dinoflagellate cyst zonation of the English Jurassic. *Report, Institute of Geological Sciences, London,* **83/2** 1–42.

9

A review of Late Jurassic to earliest Cretaceous radiolaria and their biostratigraphic potential to petroleum exploration in the North Sea

R. Dyer and **P. Copestake**

ABSTRACT

The nature and biostratigraphic potential of Late Jurassic and earliest Cretaceous radiolaria are reviewed in order to assess their role for age dating, zonation and correlation of the Kimmeridge Clay Formation of the northern North Sea. 13 radiolarian events are recognized, including species 'tops', 'acmes' and 'bases' which are distributed throughout the Kimmeridge Clay Formation. The correlation of these events is discussed in two wells from the northern North Sea. This suggests that the group can provide an important contribution towards stratigraphic evaluation of the formation in offshore exploration wells. The taxonomy of certain stratigraphically important species is briefly reviewed.

INTRODUCTION

The Kimmeridgian (Late Jurassic) to Ryazanian (Early Cretaceous) period is represented in the northern North Sea by the Humber Group (pars) (Deegan & Scull 1977). This unit contains the predominantly argillaceous Kimmeridge Clay Formation, which includes organic shales thought to be the main source rock for North Sea oil (Barnard & Cooper 1981). Tectonic activity along graben bounding faults and variation in sea level during this time (Vail & Todd 1981) were probably responsible for the deposition of several important clastic units, including the Fulmar, Ula, Magnus, Brae and Claymore sands (Deegan & Scull 1977, Johnson *et al.* 1986, Harms *et al.* 1981, De'Ath & Schuyleman 1981, Turner *et al.* 1984, Vollset & Doré 1984). The many clastic units, which vary in facies from shallow marine sands to deeper-water mass flow sands and conglomerates (Johnson & Stewart 1985), act as reservoirs for hydrocarbon accumulations in many fields currently under appraisal and development, including Claymore, Tartan, Fulmar, Clyde, Ula, Brae, Miller, Toni, Thelma, Tiffany, Magnus, Ettrick, Highlander, Galley and Petronella (Turner *et al.* 1984, Johnson *et al.* 1986, Gibbs 1984, Smith 1987, Bailey *et al.* 1981, Johnson & Stewart 1985, De'Ath & Schuylemann 1981, Harker *et al.* 1987) (Fig. 1). It is likely that many other accumulations occur in clastic units developed within the Kimmeridge Clay Formation of the northern North Sea,

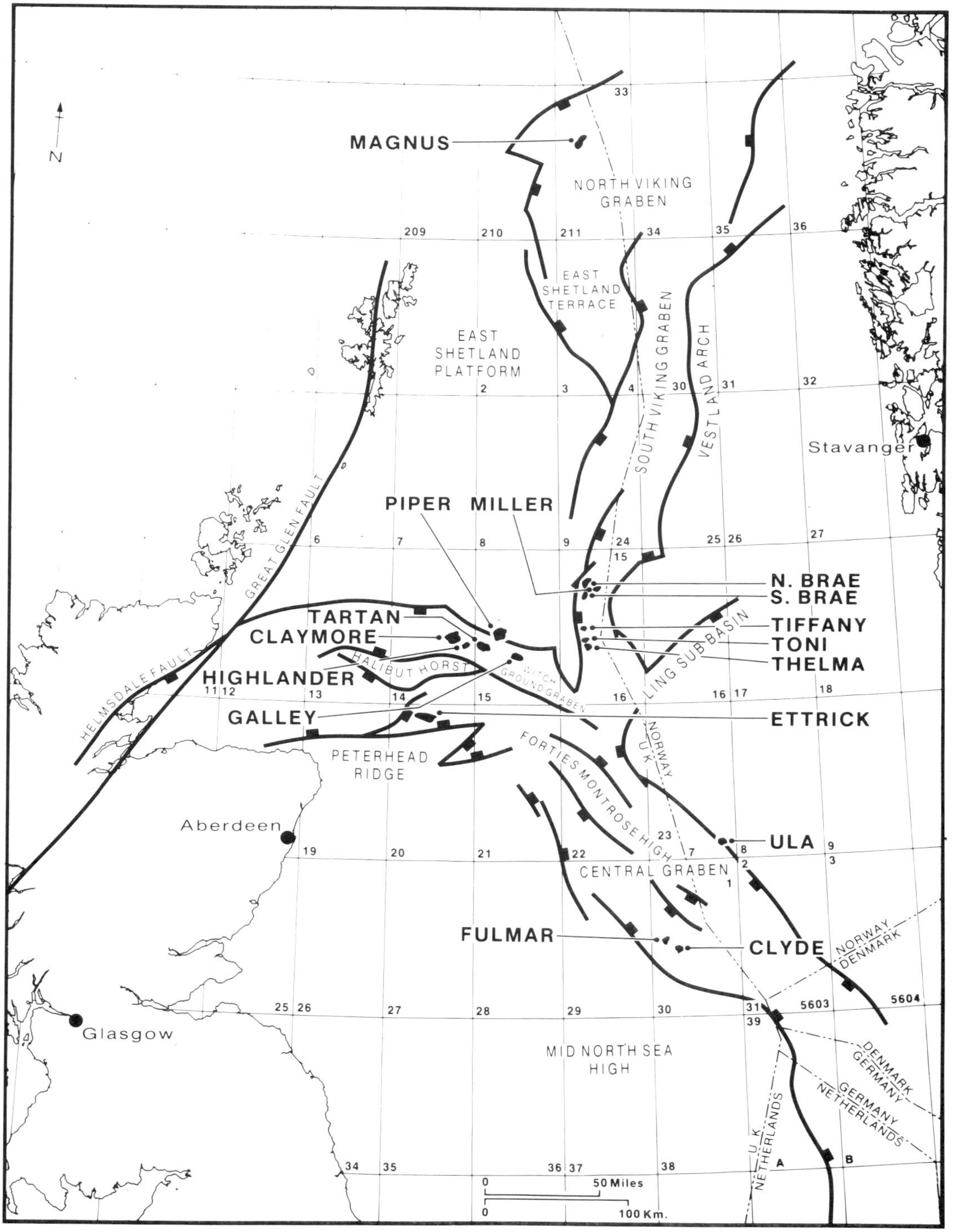

Fig. 1 — The location of oil fields within the Kimmeridge Clay Formation of the northern North Sea.

awaiting discovery, appraisal and development.

Biostratigraphy, in particular the application of palynology, has made an important contribution to the age dating, zonation, correlation and hence geological understanding of many of the reservoirs. While the use of palynomorphs such as dinocysts (and to a lesser extent other planktonic algae, pollen and spores) provides a high-resolution biostratigraphic framework with which to develop geological models, experience in industrial application has indicated that the utilization of additional zonation schemes can create significant improvements to the successful use of biostratigraphy in petroleum geology. The provision and utilization of a micropalaeontological scheme (radiolaria, foraminifera and ostracods), in conjunction with a palynological zonation, generate increased biostratigraphic resolution and also constrain stratigraphic interpretations based on single biostratigraphic disciplines alone. The stratigraphic and geographic distribution and physical recovery of the several micropalaeontological and palynological fossil groups are generally governed by different factors and it is therefore important to utilize a variety of microfossils in the event that specific geological or drilling factors may prevent the successful use of individual groups.

This paper aims to review the nature and biostratigraphic value of radiolaria from the Kimmeridge Clay of the northern North Sea (Kimmeridgian to Ryazanian) (i.e. north of the mid North Sea high). The degree of chronostratigraphic resolution attainable using basinwide changes in radiolarian assemblages is discussed. An example is provided of two sand-bearing wells 140 km apart in the North Sea which are correlatable on the basis of a set of 13 radiolarian events. The role of palaeoecological factors in tempering this application is also discussed. The taxonomy of certain stratigraphically important species is briefly considered and some of these forms are also figured.

The definition of the 13 radiolarian events and details of species distribution cannot be indicated in this paper because the data are considered confidential by Britoil plc and partners. The identification of one of the two wells has to be withheld for similar reasons.

Radiolaria are a planktonic group of microfossils whose use for biostratigraphic purposes in the hydrocarbon exploration industry of the North Sea has not been fully exploited, because of several factors. Firstly, micropalaeontologists with no formal research background in the study of radiolaria may find the group difficult to work with, to the extent that taxonomic 'lumping' is common. Secondly, published research on Jurassic and Cretaceous radiolaria has not been very productive until recently and has been limited to areas outside northwest Europe, such as California, Japan and the Tethyan area of Europe; the existing zonations cannot be applied to the North Sea, probably owing to faunal provincialism. Thirdly, Kimmeridge Clay radiolaria from the North Sea can be badly preserved and the group requires careful processing methods for successful extraction from rock samples. No major work has been published involving radiolaria from onshore UK Jurassic successions, probably because of a lack of abundant radiolaria in most of the marine facies exposed, thus providing no comparison of offshore faunas with well-dated onshore assemblages. Finally, there has been an over-reliance on palynology as the sole means for biozonation of the Kimmeridge Clay Formation, so that the research required to develop a radiolarian zonation scheme has not always been given significant priority.

There are several advantages of including radiolaria in a micropalaeontological zonation scheme for the Kimmeridge Clay Formation of the North Sea. Being planktonic microfossils, their geographic and stratigraphic distribution are less facies-controlled than benthonic microfossils, particularly during the deposition of the organic-rich Kimmeridge Clay Formation, when dysaerobic to anaerobic bottom conditions prevented a rich benthonic colonization of the sea floor. As a result, it is possible to extract abundant radiolaria from sediments devoid of foraminifera and ostracods and also to use stratigraphic top ranges and acmes confidently in the knowledge that planktonic events have great correlative value with limited influence from local facies control. Correlatable changes in radiolarian assemblages can thus be independent of sea bed sand deposition. Secondly, because Jurassic and Cretaceous radiolaria exhibited a preference for outer shelf to oceanic environments, variation in their abundance and diversity can be a good environmental indicator, particularly for recognizing transgressions and regressions. It is therefore possible to utilize the group for sequence stratigraphy analysis. Finally, when dealing with sediments from deeply buried graben axes, it is possible to extract a radiolarian assemblage from lithologies in instances where many microfossil groups such as dinocysts have been adversely affected by high thermal maturity or diamond bit drilling.

Unfortunately it is not possible to indicate details of radiolarian events and the stratigraphic distribution of individual taxa, because of reasons of confidentiality involving Britoil and its partners. For the same reasons, we are not able to provide wells and depths for the figured specimens. Despite these drawbacks, it is nevertheless hoped that this paper will contribute towards a standardized taxonomy for North Sea radiolaria and provide impetus for further research on the biostratigraphic application of the group.

PREVIOUS WORK ON JURASSIC TO EARLY CRETACEOUS RADIOLARIA

This review explains problems of relating radiolaria assemblages and published zonation schemes from various parts of the world, including the North Sea. The study of Mesozoic radiolaria has occurred mainly in three areas: western North America, the Alpine–Tethyan area of Europe, and Japan. Most research has focused on these regions because they contain sedimentary sequences rich in radiolarian remains. The sediments are specifically deep water in origin and form part of major orogenic belts such as subduction zones associated with island arc systems and oceanic spreading centres. Radiolaria are often the only fossils preserved in these sequences, particularly those deposited below the calcite compensation depth. The rocks consist of sand-shale flysch, siliceous mudstones and limestones and chert, associated with tuffs and/or pillow lavas.

The first significant work involving Mesozoic radiolaria was on German, Austrian, Swiss, French and Italian sections by Rust (1885, 1898), Parona (1890), Cayeux (1896), Vinassa (1899) and Squinabol (1913). These workers produced taxonomic monographs in which Mesozoic taxa were studied within a classificatory framework previously erected for Cenozoic radiolaria by Ehrenberg (1854) and Haeckel (1881, 1887). This tendency wrongly devalued the stratigraphic potential of Mesozoic radiolaria because it obscured true phylogenetic relationships and portrayed most genera as being extremely long ranging.

The study of Mesozoic radiolaria for stratigraphic purposes was resurrected during the 1960s and 1970s owing to the discovery of new processing methods, the development of the scanning electron microscope and the realization that radiolaria, being planktonic in nature, must hold significant stratigraphic potential. By successfully extracting radiolaria from siliceous sediments, Pessagno & Newport (1972) were able to revive the study of Mesozoic radiolaria, culminating in a series of papers describing the taxonomy and biozonation of radiolaria for the Late Jurassic and Cretaceous of the California Coast Ranges (the Great Valley Sequence and the Franciscan Complex) (Pessagno 1976, 1977a, b, c). These works represented the first attempt to develop a detailed system of radiolarian zonations for the Late Jurassic and Cretaceous. Pessagno discovered that radiolaria were valuable in dating the Coast Range sequences which were previously impossible to subdivide into mappable lithostratigraphic units because of their tectonic complexity and absence of other fossil types such as ammonites. Pessagno (1977a, c) erected six zones and two subzones for the Late Jurassic, and four zones and seven subzones for the Early Cretaceous. Pessagno (1977a, c) had difficulty in establishing an absolute age for his zones in terms of standard ammonite zones, however, because of the absence of reliable age diagnostic ammonites, although a bivalve scheme (*Buchia* zones) was available. Consequently it is difficult to correlate the Coast Range sequences with the European standard sequences. This problem is not confined to the Californian sections since it is often difficult to find abundant well-preserved ammonites in sediments containing rich radiolarian faunas. The zonation schemes of Pessagno were based on a major taxonomic review of radiolaria.

Further taxonomic and biostratigraphic work was subsequently carried out on the Early and Middle Jurassic radiolaria of north west North America by Pessagno & Blome (1980, 1982), Pessagno & Whalen (1982), Blome (1984), Pessagno *et al.* (1984) and Pessagno *et al.* (1986a).

The zonation scheme of Pessagno (1977a) was revised by Pessagno *et al.* (1984) as a result of additional work which extended the known ranges of many zonal markers. The work attempts to relate the zones to a standard chronostratigraphic framework. Pessagno *et al.* (1986a) have recently completed a preliminary zonation for the entire Jurassic of North America, providing seven zonal units for the Late Jurassic, eight for the Middle Jurassic and seven for the Early Jurassic. An attempt is made to calibrate the zones to the standard ammonite and calpionellid zones. Although Pessagno and his colleagues are to be congratulated on their excellent taxonomic and biostratigraphic studies of Mesozoic radiolaria of northwest North America, the problem of reliably and precisely relating radiolarian ranges and zones to an independent ammonite scheme must be considered a drawback to the application of this work outside the study region.

Recent studies of Mesozoic radiolaria from the Tethyan area of Europe and the central Atlantic have been undertaken by Baumgartner *et al.* (1980) and Baumgartner (1984). Baumgartner (1984) erected a series of radiolarian 'biochronozones', based on a computerized analysis of the first and last appearances of 110 species from the Bathonian to Hauterivian. The sequences were mainly siliceous sediments which accumulated on Tethyan and Atlantic oceanic basement. As in the studies by Pessagno and co-workers, Baumgartner (1984) discovered that the dating of siliceous oceanic sediments has greatly increased the understanding of tectonically complex areas, particularly the subsidence history of the various basins. Baumgartner (1984) concluded that some species ranges in the European Tethyan region were different to those documented by Pessagno (1977a, c) from North America.

Several recently published studies by Japanese workers have attempted to unravel the complicated Japanese stratigraphy by means of a radiolarian biostratigraphic framework, e.g. Nakaseko (1982), Nakaseko & Nishimura (1981), Mizutani *et al.* (1982), Matsuoka (1983), Ishida (1983), Aita & Okada (1986) and Matsuoka & Yao (1986). According to Baumgartner (1984) there is a striking similarity between the radiolarian morphotypes of Japan and the Tethyan region. Whilst the Japanese workers have erected locally based radiolarian zonation schemes, it is likely that correlation will eventually be achieved with other parts of the Tethyan province.

Very little work has been performed on radiolaria from Boreal Jurassic sediments deposited in continental shelf or epicontinental seas, including those of onshore UK, although such sequences could allow the comparison of radiolarian events against ammonite schemes. In a rare attempt to document Jurassic radiolarian faunas from onshore UK, Gregory (1986) examined the Kimmeridgian on the Moray Firth coast of east Scotland. In addition to foraminiferal microfaunas, Gregory (1986) recovered three radiolarian species identified herein as *Orbiculiforma mclaughlini* Pessagno, *Parvicingula blowi* Pessagno and *Hsuum* sp. 1

An ongoing study of radiolaria from Kimmeridgian to Volgian onshore sediments of the Inner Moray Firth by the present authors has revealed a similarity with offshore faunas and it is hoped that this work may allow a direct comparison of radiolarian ranges against ammonite and palynological zones.

RADIOLARIAN PALAEOECOLOGY AND BIOSTRATIGRAPHY OF THE KIMMERIDGE CLAY FORMATION OF THE NORTHERN NORTH SEA

Mesozoic radiolaria are a diverse group of planktonic organisms which rapidly evolved and were widely distributed throughout the world's oceans, thereby providing good potential for biostratigraphic application. This potential is tempered, however, by two interrelated limiting factors which controlled the distribution of taxa: provincialism and ecology.

As with all planktonic organisms, radiolaria display some degree of provincialism. Pessagno *et al.* (1984) recognized four latitudinal provinces during the Jurassic: central Tethyan, northern Tethyan, southern Boreal and northern Boreal, with highest diversity occurring in the Tethyan realms. Despite the provincialism, however, Pessagno *et al.* (1984) claimed that sufficient taxa are common to the various provinces to allow some degree of correlation, e.g. between the Boreal Californian assemblages and the Tethyan faunas of Europe. Nevertheless, it will be necessary to erect parallel zonal schemes for Boreal and Tethyan areas.

Modern and fossil radiolaria appear to be oceanic in their ecological preferences (Casey 1977, Pessagno 1977b). Jurassic radiolaria specifically exhibit greatest abundance and diversity in deep-water flysch and other sedimentary facies deposited in subduction zones and around spreading centres. They are particularly abundant in Jurassic rocks on the margins of the Pacific Ocean, including the west coast of North America and southwest Japan and the basinal sequences of the Tethys Ocean. In contrast, radiolaria are rare in the onshore shallow marine sequences of northwest Europe.

In the northern North Sea basin, tectonic rifting occurred between the Callovian and Kimmeridgian. This caused rapid subsidence which outpaced sedimentation to create basinal troughs in the Central Graben, Moray Firth Basin and Viking Graben (Ziegler 1982, Badley *et al.* 1988). This tectonic phase represented an increase in intensity of rifting which had begun earlier in the Middle Jurassic. Subsidence occurred by listric movement on graben margin faults so that the graben margins (Fig. 1) represented a tectonically controlled break. On the upthrown side of the graben margins, shallow marine shelf facies were deposited (e.g. the type Kimmeridge Clay Formation of Dorset and the southern North Sea) with calcareous benthonic foraminifera and ostracods, but no radiolaria. On the downthrown side of the graben margins occurred flanking zones of gently dipping fault block terraces ('shelf' areas) passing into more basinal regions in the graben axes.

The shallow marine sands of the Piper and Fulmar Formations were deposited on these terraces during the Late Oxfordian to 'mid'-Kimmeridgian, while submarine fan sands and basinal organic rich shales accumulated from 'mid'-Kimmeridgian to Ryazanian times (Harker *et al.* 1987, Turner *et al.* 1984) both on the terraces and in the graben axes. The submarine fan sands form the reservoirs and the organic-rich shales the source rocks for the hydrocarbons of fields such as Brae and Claymore (Turner *et al.* 1984). It is in these terrace and graben axis locations, which together constitute the graben areas shown in Fig. 1, that the radiolaria assemblages described in this paper occur, thus matching their known preference for deep-water flysch-type environments in the fossil record (see above). No volcanic, silica-rich sediments, which are usually associated with radiolaria, are known in the North Sea Upper Jurassic, however.

North Sea Kimmeridgian–Ryazanian radiolarian assemblages are typically northern Boreal in character (*sensu* Pessagno *et al.* 1984), in the lack of Pantanellidae and the presence of *Parvicingula*. Spumellarian taxa are dominant in both numbers and diversity, mainly represented by species of *Cenosphaera* and *Cenodiscus*, with subordinate development of *Orbiculiforma* and *Praeconocaryomma*. Many of these taxa are previously undescribed. Nasselarians are much less diverse but are nevertheless abundant at certain levels; *Parvicingula* and *Stichocapsa* are the most prominent genera, accompanied by *Tricolocapsa* and rare *Hsuum* and *Archaeodictyomitra*.

The low diversity of the radiolaria assemblages may be due to several factors, including water depth, the northerly palaeogeography (and colder climate), the possible reduced salinity because of proximity to nearby land areas, the lack of volcanogenic sediments or the restricted marine circulation. Lack of true oceanic conditions (and associated silica-rich sediments) in the North Sea Jurassic is considered to arise because, despite intense rifting in response to basin extension and crustal stretching in the Late Jurassic–Early Cretaceous, an oceanic spreading centre was never created. The North Sea remained, as it still is today, an essentially epicontinental sea. The northern North Sea developed therefore as a restricted deep-water arm of the sea, probably connected to an oceanic area north of the north Viking Graben (Ziegler 1982, Enclosure 20). Stratigraphic variation in the abundance and diversity of radiolaria suggests that water depths and the degree of oceanic influence fluctuated throughout the Kimmeridgian to Ryazanian, probably as a result of tectonically and eustatically influenced sea level changes.

The direct correlation of North Sea radiolarian events with published zones of true oceanic sequences is very difficult because of the limiting factors of provincialism and ecology, either owing to the absence of diagnostic zonal markers or because of differences in stratigraphic ranges in different geological settings. This is considered to arise because the North Sea assemblages are northern Boreal in character; such assemblages have not previously been described in detail and are less diverse than more southerly faunas (Pessagno *et al.* 1986b). However, the detailed study of numerous well sequences throughout the basin has allowed the establishment of a biozonation based on a succession of at least 13 radiolarian assemblage changes (Fig. 2). These include species inceptions ('bases'), extinctions ('tops') and the upper and lower limits of species abundances. Most of the events of biozone potential are abundance changes of longer-ranging species.

The age of the North Sea radiolarian events discussed herein can be determined indirectly by calibration to palynological events, which can be tied to ammonite zones (e.g. Woollam & Riding 1983) and by reference to onshore sections in which ammonite control is available, e.g. the coastal exposures of the Moray Firth. The events are widespread and occur throughout the northern North Sea (i.e. north of the mid North Sea High) and west of Shetland. The radiolarian biozones allow the recognition of most stage and substage boundaries in the Kimmeridgian–Ryazanian interval, namely the tops of the Kimmeridgian, early Volgian, middle Volgian, late Volgian and early Ryazanian intervals, and near base Kimmeridgian and intra-late Ryazanian horizons. In addition, several of these units can be subdivided using radiolarian events. The age terminology utilized is Boreal rather than Tethyan as applied by Pessagno and co-workers. Volgian and Ryazanian are used in preference to Portlandian–Berriasian in accordance with customary North Sea terminology (see Fig. 2).

Sands are commonly developed in the Kimmeridgian–Ryazanian of the North Sea, particularly in the middle Volgian when synsedimentary tectonism led to the widespread deposition of submarine fan mass flow–turbidite sands. These are interbedded with Kimmeridge Clay Formation organic-rich claystones and siltstones of oil prone source rock potential, often rich in radiolaria. The reservoir sands of several known fields are of this age and type (Claymore, Brae, Toni, Thelma, Magnus and Ettrick among others). Such sands are, however, often laterally impersistent, rapidly passing into claystone away from the clastic source. Furthermore, tectonically induced unconformities are frequent, cutting out varying amounts of section from well to well. Such features can be identified and quantified by microbiostratigraphy and are important as they are mechanisms by which hydrocarbons may be trapped at sand–shale contacts. In addition,

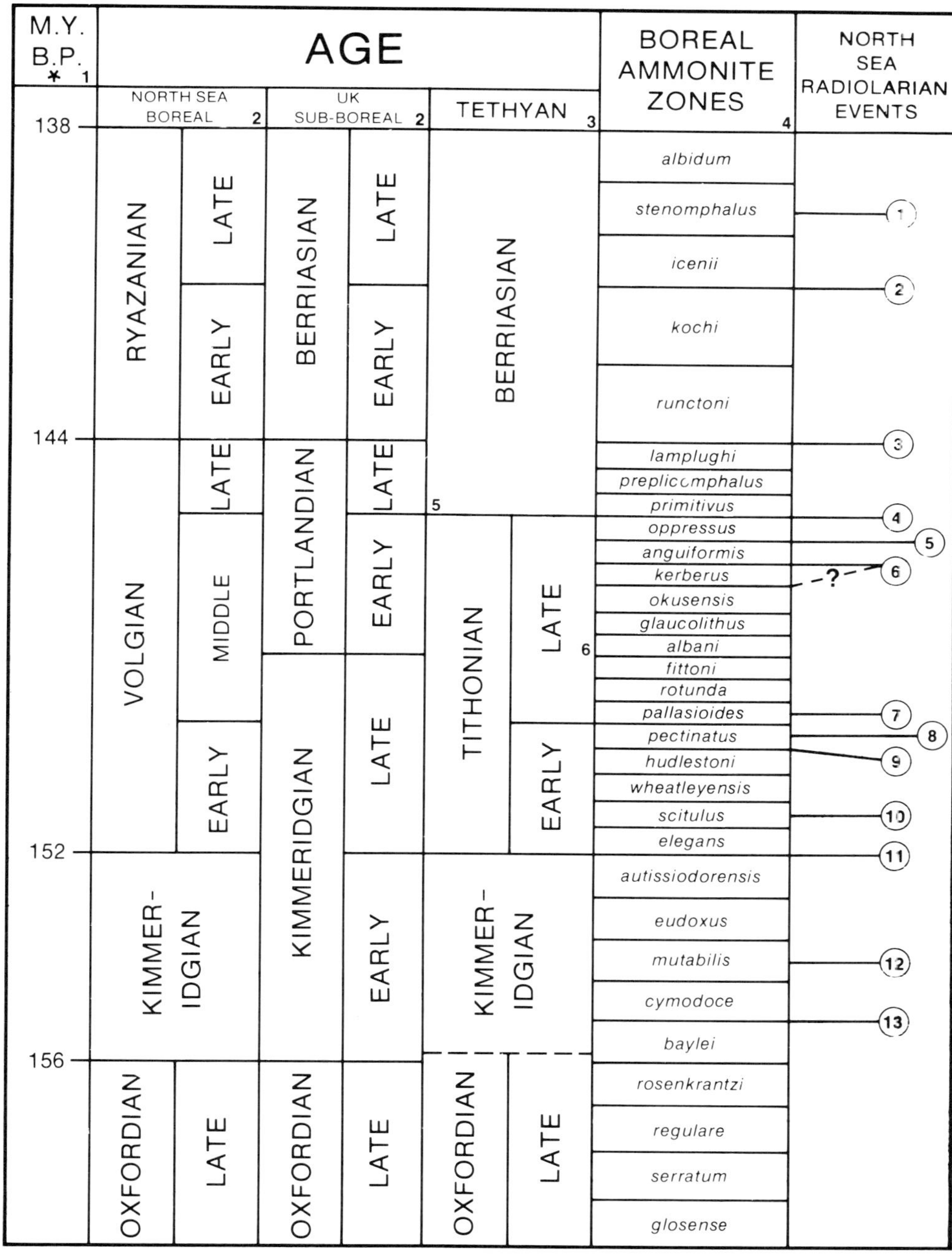

Fig. 2 — Comparison of Boreal–Tethyan Late Jurassic–earliest Cretaceous age terminologies and Boreal ammonite zones, showing the stratigraphic position of North Sea radiolarian events discussed in this paper: 1, after Kent and Gradstein (1985); 2, after Rawson *et al.* (1978), Cope *et al.* (1980) and Rawson & Riley (1982); 3, after Harland *et al.* (1982) and Cope (1985); 4, after Cope *et al.* (1980) and Rawson *et al.* (1978); 5, Tithonian–Berriasian boundary as placed by Pessagno *et al.* (1984) and Wimbledon (1985); 6, base of the late Tithonian as placed by Pessagno *et al.* (1984), following Imlay (1980).

amounts of missing reservoir must be evaluated correctly to allow accurate reserves calculations to be made.

The application of changes in radiolarian assemblages to well correlation in the North Sea is illustrated in Fig. 3. The two wells–well 12/21-2 and a quadrant 20 well—are located approximately 140 km apart. They can be zoned and correlated using a basic set of 13 basinwide assemblage changes. The Kimmeridge Clay Formation sections shown span the Kimmeridgian to intra-late Ryazanian. All stage and substages in this interval are recognizable and the most sand-prone interval, the middle Volgian, is subdivisible into four units using radiolarian events. Sands are also developed in the early Volgian, late Volgian, early Ryazanian and late Ryazanian.

The actual abundance variations in radiolaria are thought to reflect synchronous, basinwide palaeoenvironmental changes. The consistent recognition of species acmes, in alternation with zones of radiolarian absence or drastically reduced abundance and diversity, are

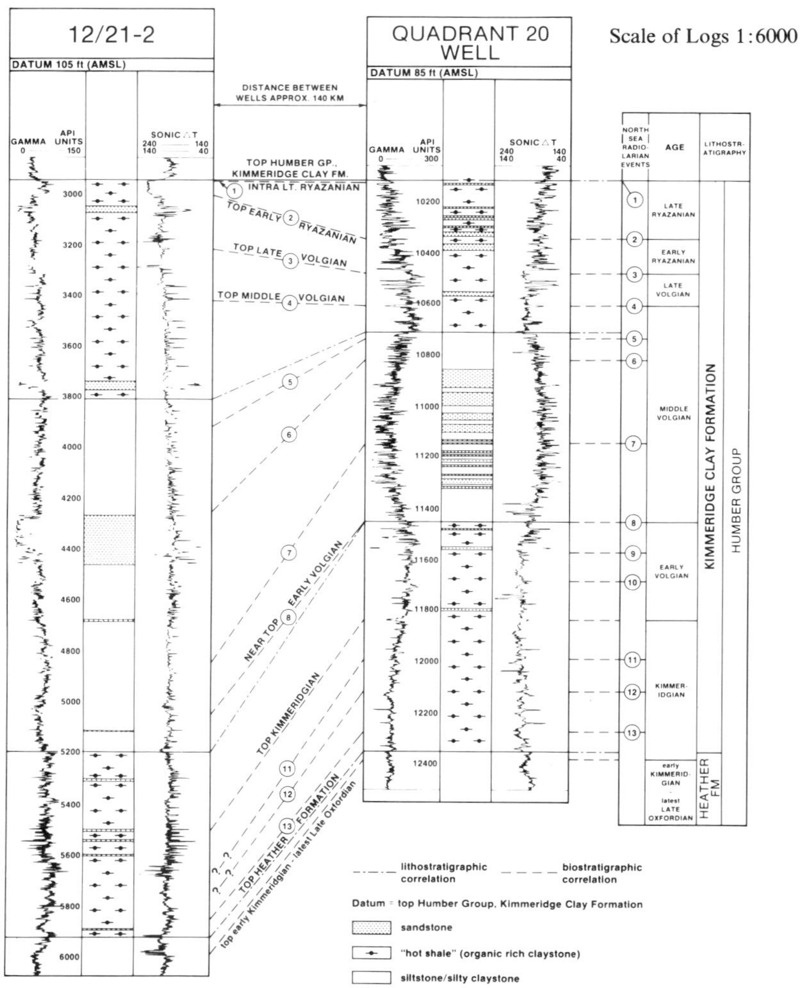

Fig. 3 — The age breakdown, zonation and correlation of the Kimmeridge Clay Formation and contained sands in 12/21-2 and a quadrant 20 well from the northern North Sea.

considered to be due to increases and decreases, respectively, in oceanic influence. These changes are caused by either eustatic or tectonically controlled sea level changes, or by a combination of both. Recognition of a consistent sequence of such events is now possible across the North Sea, and these are thought to be causally linked to phases of sand generation and provide a basis for sequence stratigraphy analysis. Some of these events agree in their timing with those of Rawson & Riley (1982) but others provide a refinement of their scheme. A detailed comparison with the proposed transgressive–regressive cycles of Vail & Todd (1981) and Haq *et al.* (1987) is beyond the scope of this paper.

METHODS OF STUDY

The majority of radiolarian tests recovered from the North Sea Kimmeridge Clay Formation are not preserved in their original state, the opaline silica having been replaced by calcite or pyrite. According to Sanfilippo & Riedel (1985), the original opaline silica changes with time to chalcedony through loss of water and may subsequently be replaced by other minerals. Pessagno *et al.* (1984) report that the opaline silica of most Mesozoic radiolarian tests of North America is replaced by quartz, chalcedony, smectite, pyrite or calcite. The diagenetic replacement of opaline silica by some other mineral faithfully maintains the external structure and appearance of the test in most instances, although the fine detail of surface ornament can be lost when tests are replaced by calcite. Because of these preservational changes to the test, Mesozoic radiolaria cannot be readily or routinely studied using a transmitted light microscope in the same manner as Cenozoic forms. Routine investigations can be performed using a reflected light stereomicroscope. Detailed observations and illustrations are best carried out with a scanning electron microscope although most diagnostic features can be recognized using a high-quality stereomicroscope.

Standard commercial micropalaeontological processing techniques (e.g. hydrogen peroxide treatment) can be used to extract radiolaria from most clastic and carbonate rocks of the North Sea with a wide range of success. In order to recover radiolaria successfully through standard micropalaeontological processing techniques, it is necessary to pick carefully the 63 μm and 125 μm residues. The failure to appreciate the value of radiolaria for stratigraphic purposes in the North Sea Kimmeridge Clay Formation has probably been largely due to inadequate picking of the fine fractions. Most published studies of Mesozoic radiolaria adopt a specialized technique, however, using hydrofluoric acid (Pessagno 1977b, Sanfilippo & Riedel 1985). The technique involves immersing a crushed rock sample (clay, shale, argillaceous siltstone or argillaceous sandstone) into dilute hydrofluoric acid (5–10%) until all the clays are broken down. The residue is then washed, sieved using a 63 μm mesh, dried and picked. Those tests preserved as calcite are changed to calcium fluorite while pyritized tests remain unaltered. Quartz tests undergo little corrosion, providing that the time spent in the hydrofluoric acid is not too long, since the clays break down much faster than pure silica. The advantage of this method over standard micropalaeontological processing techniques is that all the argillaceous component is removed, thereby concentrating the radiolaria in the insoluble residue.

THE TAXONOMY OF RADIOLARIA FROM THE KIMMERIDGE CLAY FORMATION

The early taxonomic studies of Mesozoic radiolaria (Rust 1885, 1898, Parona 1890) tended to follow the existing supraspecific taxonomic nomenclature previously erected by Ehrenberg (1854) and Haeckel (1862, 1881, 1887) for Cenozoic forms. This tendency unfortunately introduced two major setbacks to the study and stratigraphic appreciation of Mesozoic radiolaria. The classification schemes of Ehrenberg (1854) and Haeckel (1862, 1881, 1887) for Cenozoic radiolaria were artificial in the sense that they were based purely on morphological features such as shape which does not reflect natural phylogenetic relationships. Secondly, the use of Cenozoic genera for Mesozoic species gave rise to a long-held belief that radiolarian taxa were very long ranging and were therefore of little value for stratigraphic purposes. This fallacy was perpetuated even in fairly recent times (Campbell 1954).

The works by Pessagno (1977a, b, c), Pessagno & Whalen (1982) and Blome (1984) stress the need for a taxonomic classification of Mesozoic and Palaeozoic radiolaria based on internal structure and test construction, especially at suprageneric level, in order to understand fully their phylogenetic relationships and to optimize their use in biostratigraphy. This approach has led to the erection and description of many new families, genera and species of Mesozoic radiolaria during the last 15 years. The classification of radiolaria in this paper follows Pessagno (1977a, b, c). Descriptive terminology follows Pessagno (1977a, c, see appendices to both papers).

Many of the species encountered in the Kimmeridge Clay Formation of the North Sea are not known from previous work, reflecting some immaturity in the study of Mesozoic radiolaria. According to Pessagno *et al.* (1986a), 85% of Jurassic radiolarian species are currently undescribed. In particular, the spumellaria, which are dominant in the North Sea assemblages, have received less attention from researchers than have the nassellaria. 21 of the stratigraphically most important North Sea Late Jurassic–earliest Cretaceous taxa are briefly described in this paper. This is by no means an exhaustive coverage of all the species that are known to occur. The consistent recognition of new forms suggests that the species list will eventually be much greater.

SYSTEMATIC DESCRIPTIONS

Suborder Spumellariina Ehrenberg 1875
Superfamily Spongodiscacea Haeckel 1881, emend. Pessagno, 1973
Subsuperfamily Pseudoaulophacilae Riedel 1971
Family Orbiculiformidae Pessagno 1973
Genus *Orbiculiforma* Pessagno 1973

Orbiculiforma lowreyensis Pessagno 1977
(Plate 1, Figs. 1 and 2)

1977a *Orbiculiforma lowreyensis* Pessagno, n.sp.; p. 74, Plate 3, Figs. 15 and 16, Plate 4, Figs. 1–3.
1984 *Orbiculiforma lowreyensis* Pessagno; Pessagno *et al.* p. 22, Plate 1, Figs. 9–11.

Description. Test compressed with a small shallow central cavity; width of central cavity about one third test diameter. Sides of test sloping from central cavity to circumference, resulting in angled periphery. Short spines may be developed on the periphery of some specimens. Meshwork with tetragonal to pentagonal pore frames (Pessagno 1977a).
Size. Average diameter, approximately 250 μm.
Published range. Pessagno (1977a) and Pessagno *et al.* (1984) cite the North American range as late Tithonian–Berriasian (middle Volgian–Ryazanian in North Sea nomenclature).
Remarks. The sharp margin and shallow central cavity differentiates this species from others of the genus. A circumferential notch is usually developed. A rare species in the North Sea.

Orbiculiforma mclaughlini Pessagno 1977
(Plate 1, Figs. 3 and 4)

1977a *Orbiculiforma mclaughlini* Pessagno, n.sp. p. 74, Plate 4, Figs. 4–7.

Description. Test thick with large deep central cavity; central cavity commonly three quarters the diameter of the test. Meshwork consisting of coarse hexagonal and pentagonal pore frames arranged in a concentric fashion. Short spines are developed on the periphery of some specimens. Margin of test rounded although nearly vertical (Pessagno 1977a).
Size. Average diameter, approximately 250 μm.

Published range. Pessagno (1977a) gives the

North American range as late Kimmeridgian–early Tithonian (Kimmeridgian–early Volgian in North Sea nomenclature).
Remarks. Most specimens found do not reveal the spongy nature of the test because of poor preservation. A well preserved form is illustrated in Plate 1, Fig. 3. A circumferential notch is a common feature. No North Sea specimens have been found bearing peripheral spines, although such a feature could be easily lost during preservational changes. Rare to moderate in abundance within the Kimmeridge Clay Formation.

Subsuperfamily Spongodiscilae Haeckel 1881
Family Spongodiscidae Haeckel 1862
Genus *Spongodiscus* Ehrenberg 1854

Spongodiscus sp. 1

Remarks. A flattened discoidal spongodiscid with a subacute margin in edge view. Cross-sectional outline rhomboidal. Average diameter, approximately 200 μm.

Spongodiscus sp. 2

Remarks. A flattened spongodiscid with a rounded margin in edge view. Average diameter, approximately 200 μm.

Spongodiscus sp. 3
(Plate 1, Figs. 14 and 15)

Remarks. A moderately compressed spongodiscid with a straight margin in edge view. Average diameter, approximately 160 μm.

Spongodiscus sp. 4
(Plate 1, Fig. 16)

Remarks. A discoidal spongodiscid with a shallow circumferential notch. Several narrow and shallow concentric grooves are visible on the surface of the test. Cross-sectional shape generally rhomboidal. Average diameter, approximately 320 μm.

Superfamily Liosphaeracea Haeckel 1881, emend. Pessagno 1984
Family Praeconocaryommidae Pessagno 1976
Genus *Praeconocaryomma* Pessagno 1976

Praeconocaryomma hexagona (Rust 1898)
(Plate 1, Figs. 5 and 6)

1898 *Acanthosphaera hexagona* Rust, n.sp.; p. 12, Plate 3, Fig. 10.
1977a *Praeconocaryomma hexagona* Rust; Pessagno, p. 77, Plate 5, Fig. 13.

Description. Spherical test with surface ornamented with numerous, prominent mammae. The mammae are hemispherical and arranged regularly over the surface.
Size. Average diameter, approximately 150 μm.
Published range. Late Tithonian–Berriasian of North America (equivalent to middle Volgian–Ryazanian in North Sea terminology) (Pessagno 1977a).
Remarks. *P. hexagona* differs from *Praeconocaryomma* sp. 1 by possessing larger, more pronounced mammae.

Praeconocaryomma sp. 1
(Plate 1, Fig. 7)

Description. Spherical test with surface ornamented with numerous mammae. The mammae are very numerous, small and variable in size and shape, providing an irregularly ornamented surface.
Size. Average diameter, approximately 150 μm.
Remarks. The species is distinguished from *P. hexagona* in possessing more numerous, irregular, sharper and more raised mammae.

Praeconocaryomma(?) sp. 2
(Plate 1, Figs. 8 and 12)

Description. A small, spherical species with

numerous tiny, closely spaced mammae distributed uniformly throughout the test surface. The mammae have a stellate outline with a small central pore. The pores are often filled with secondary deposits.
Size. Average diameter, approximately 130 μm.
Remarks. The small test size and tiny numerous mammae are diagnostic features of this species. The taxon has a hispid appearance using a standard reflected light microscope. An inner medullary shell is often revealed when specimens are viewed in transmitted light. This species is tentatively placed in the genus *Praeconocaryomma* owing to the presence of numerous tiny mammae on the test surface. These are far more indistinct than those of *P. hexagona* and *Praeconocaryomma* sp. 1.

Liosphaeracea *Incertae Sedis*
Genus *Cenosphaera* Ehrenberg 1854

The genus *Cenosphaera* Ehrenberg should be restricted to Cenozoic species. Because there appears to be no demonstrated phylogenetic link between the Mesozoic and Cenozoic forms, it would be more appropriate to erect a new generic designation for Jurassic and Cretaceous species (Pessagno 1977c). In this paper the genus *Cenosphaera* includes all spherical spumellarians with a latticed, porous meshwork.

Cenosphaera sp. 1
(Plate 1, Fig. 11)

Remarks. A *Cenosphaera* with a smooth, slightly pitted surface. Average diameter, approximately 160 μm.

Cenosphaera sp. 2
(Plate 1, Figs. 9 and 10)

Description. A *Cenosphaera* with a coarse, open meshwork of hexagonal pore frames centred by large circular to subcircular pores. The pore frames possess well-developed nodes at each junction, providing a nodose surface ornamentation between the pores.
Size. Average diameter, approximately 150 μm.
Remarks. *Cenosphaera* sp. 2 is often found preserved in pyrite with much of the original surface ornament lost. These forms can be identified owing to the faithful preservation of the strong pore frames and nodes which impart a cancellate texture to the test surface (compare Plate 1, Figs. 9 and 10).

Genus *Cenodiscus* Haeckel 1887

As with *Cenosphaera*, the genus *Cenodiscus* should be restricted to Cenozoic species since there is little phylogenetic link between Mesozoic and Cenozoic forms. It would therefore be more appropriate to erect a new generic designation for Jurassic and Cretaceous forms (Pessagno 1977c). In this paper the genus *Cenodiscus* includes all discoidal spumellarians with a latticed, porous meshwork.

Cenodiscus sp. 1
(Plate 1, Fig. 13)

Remarks. A uniformly discoidal species with a smooth, pitted test surface. Average diameter, approximately 200 μm. Crushed specimens of *Cenosphaera* sp. 1 can often resemble *Cenodiscus* sp. 1 although they lack the regular geometric outline. The two taxa possess a similar surface ornament and often occur in association.

Cenodiscus sp. 2

Remarks. A biconvex, uniformly discoidal species with a very smooth surface and sharp, angled periphery. Average diameter, approximately 230 μm. The smooth, biconvex test surface separates this species from other discoidal radiolaria in the Kimmeridge Clay Formation.

Cenodiscus sp. 3

Remarks. A small, uniformly discoidal species with a smooth pitted test surface. Average diameter, approximately 100 μm.

Suborder Nassellariina Ehrenberg 1875
Superfamily Cyrtoidea Haeckel 1862
Family Hsuidae Pessagno & Whalen 1982
Genus *Hsuum* Pessagno 1977

Hsuum sp. 1
(Plate 2, Figs. 1 and 2)

Description. Conical test with massive discontinuous costae arranged in regular, parallel rows. A short horn is usually discernible but may be broken. The costae are developed over chamber joints. Three transverse rows of circular pores are present per chamber. The longitudinal costae are separated by a single vertical row of pores.
Size. Average length, approximately 200 μm; width, 90 μm.
Published range. Gregory (1986) recovered the species (as *Wrangellium* sp. cf. *W. thurstonense* Pessagno & Whalen 1982) from the Kimmeridgian (*cymodoce* Zone) of Sutherland, eastern Scotland.
Remarks. Several species of *Hsuum* have been described from the Jurassic of North America by Pessagno (1977a), Pessagno & Whalen (1982) and Blome (1984) although all differ from the present form by a greater number of vertical pore rows between costae.

Family Parvicingulidae Pessagno 1977, emend. Pessagno & Whalen 1982
Genus *Parvicingula* Pessagno 1977

Parvicingula blowi Pessagno 1977
(Plate 2, Figs. 3 and 4)

1977a *Parvicingula blowi* Pessagno, n.sp.; p. 85, Plate 8, Figs. 11–14.
1984 *Parvicingula blowi* Pessagno; Pessagno *et al.*, p. 26, Plate 2, Figs. 14 and 15.

Description. Test cylindrical with tapering proximal and distal portions. Cephalis dome-shaped with short massive horn and irregular, massive polygonal pore frames. Irregular pore frames persist to second post-abdominal chamber. Circumferential ridges are absent between first four chambers. Remaining post-abdominal chambers with regularly shaped polygonal pore frames arranged in three rows between circumferential ridges.
Size. Average length, approximately 300 μm; width, 150 μm.
Published range. Pessagno *et al.* (1984) give the range as late Kimmeridgian(?)–Tithonian (Kimmeridgian–Middle Volgian in North Sea terminology) for North America. Gregory (1986) recovered the species from the Kimmeridgian (*cymodoce* to *mutabilis* Zones) of Sutherland, eastern Scotland.
Remarks. The pores in the North Sea individuals are coarser than those of the holotype of *P. blowi*. This species is distinguished from *P. jonesi* and *Parvicingula* sp. 1 by its irregular, massive pore frames which persist to the second post-abdominal chamber. The North Sea individuals are generally shorter than the holotype.

Parvicingula blowi Pessagno ssp. A
(Plate 2, Figs. 5 and 6)

Description. A form distinct from *P. blowi* in which the cephalis, thorax and abdominal chambers each possess massive, regular polygonal pore frames and circular pores, with no circumferential ridges. The pores of the cephalis are somewhat irregularly arranged although those of the thorax to post-abdominal chambers possess regularly shaped polygonal pore frames arranged in three rows. Circumferential ridges are developed on the post-abdominal chambers.
Size. Average length, approximately 260 μm; width, 120 μm.
Remarks. This subspecies differs from *P. blowi* in its more numerous regular pore frames. This taxon differs from other species of the genus by the lack of circumferential ridges in the proximal

chambers. It is possible that this difference may be partly influenced or enhanced by preservational factors, particularly involving replacement by pyrite.

Parvicingula jonesi Pessagno 1977
(Plate 2, Figs. 7 and 8)

1977c *Parvicingula jonesi* Pessagno, n.sp.; p. 48, Plate 8, Fig. 14.
1984 *Parvicingula jonesi* Pessagno; Pessagno *et al.*, p. 28, Plate 3, Figs. 5–7, 14, 19, 24 and 25.

Description. Test conical with dome-shaped cephalis bearing short, massive horn. All subsequent chambers trapezoidal in outline. Cephalis and thorax with small polygonal pore frames. Abdomen and post-abdominal chambers separated by well-developed circumferential ridges and having three rows of large polygonal pore frames. Post-abdominal chambers eight to twelve in number (Pessagno 1977c).
Size. Average length, approximately 325 μm; width, 150 μm.
Published range. Late Tithonian–Berriasian (middle Volgian–Ryazanian in North Sea terminology) (Pessagno 1977c, Pessagno *et al.* 1984).

Parvicingula sp. 1
(Plate 2, Fig. 12)

Description. Test subcylindrical with spherical cephalis bearing a short, massive horn. Thorax, abdomen and first post-abdominal segment trapezoidal in outline. All subsequent chambers are parallel sided and impart an overall subcylindrical test outline with a blunt, rounded proximal end. Each chamber possesses three rows of polygonal pore frames and strong circumferential ridges.
Size. Average length, approximately 330 μm; width, 150 μm.
Remarks. This species is closely related to *P. jonesi* but differs in its more cylindrical, less tapering test outline. Individuals with complete horns are rare.

Cyrtoidea *Incertae Sedis*
Genus *Stichocapsa* Haeckel 1881

Stichocapsa devorata Rust 1885
(Plate 2, Figs. 9–11)

1885 *Stichocapsa devorata* Rust, n.sp.; p. 318, Plate 41, Figs. 7 and 8.
1898 *Stichocapsa devorata* Rust; Rust, p. 65.

Description. Test an elongate ellipsoid with a sharp pointed proximal end and blunt, gently rounded distal end. Chambers broad and low, enlarging gradually in width, although maintaining a constant height. Final two post-abdominal chambers decrease in width to provide an elliptical test outline. Chambers usually seven in number. Three transverse rows of circular pores are irregularly arranged on each chamber, bordered by strongly developed polygonal pore frames. Sutures flush and usually obscured by pore frames. Large aperture present at base of final chamber.
Size. Average length, approximately 210 μm; width, 140 μm.
Published range. No useful distributional data are given by Rust (1885, 1898).
Remarks. *Stichocapsa devorata* is consistently present throughout the Kimmeridge Clay of the North Sea and its absence in the wide range of recently published radiolarian literature is enigmatic, suggesting that the species has a narrow geographic distribution, perhaps largely restricted to the northern Boreal province. The species is often represented by internal casts which reveal the chamber arrangement (see Plate 2, Fig. 11), not usually discernible externally.

Genus *Tricolocapsa* Haeckel 1881

Tricolocapsa sp. 1
(Plate 2, Figs. 13–15)

Description. A small species consisting of a small hemispherical cephalis and trapezoidal thorax and a large inflated spherical abdomen. The three chambers of the test are well defined in outline view. Cephalis smooth; thorax and abdomen

possess strongly developed hexagonal pore frames, each enclosing a circular depression centred by a small circular pore. Nodes are developed at the junctions of pore frames. The base of the abdomen possesses a large circular aperture which imposes a slightly truncated outline to the distal edge of the test.
Size. Average length, approximately 115 µm; width, 85 µm.
Remarks. This small valuable index species is retrieved from the smallest lithological residues (63 µm and 125 µm) examined for micropalaeontological analysis. The species differs from *T.* cf. *elongata* described by Ozvoldova (1975) from the Late Jurassic–Early Cretaceous of the Carpathians in having a smaller, less prominent thorax with a less distinct surface ornament.

Tricolocapsa sp. 2
(Plate 2, Figs. 16 and 17)

Description. A moderately sized species (approximately 200 µm in length) consisting of a small hemispherical cephalis, an intermediate-sized trapezoidal thorax and a large, inflated spherical abdomen. The three chambers of the test are well defined in outline view. Cephalis smooth; thorax and abdomen possess strongly developed hexagonal pore frames, each enclosing a circular depression centred by a small circular pore. Nodes are developed at the junctions of pore frames. The base of the abdomen either possesses a small circular aperture, or may be closed.
Size. Average length, approximately 200 µm; width, 160 µm.
Remarks. *Tricolocapsa* sp. 2 is larger than *Tricolocapsa* sp. 1 and possesses a significantly larger thorax in proportion to the cephalis and abdomen. Surface ornament is also much coarser in *Tricolocapsa sp. 2.*

ACKNOWLEDGMENTS

The authors are grateful for constructive comments made by C. MacLean, A. Parsley, D. Skevington, P. Watson and G. Williams. The following companies are thanked for providing permission to publish well details: Agip, B. P. Exploration, Britoil, Deminex, Elf, Fina, Hunt Oil, Kerr McGee, Premier, Swedish Petroleum, Tricentrol, and U.K. Oil and Gas. This paper is published with the permission of Britoil plc and GeoStrat Ltd. Financial and other assistance provided by these companies is gratefully acknowledged. Thanks are due to J. Williams and H. A. Quigley for typing the manuscript.

Plate 1 —

Fig. 1 — *Orbiculiforma lowreyensis* Pessagno.
Fig. 2 — *Orbiculiforma lowreyensis* Pessagno.
Fig. 3 — *Orbiculiforma mclaughlini* Pessagno.
Fig. 4 — *Orbiculiforma mclaughlini* Pessagno.
Fig. 5 — *Praeconocaryomma hexagona* (Rust, reflected light).
Fig. 6 — *Praeconocaryomma hexagona* (Rust).
Fig. 7 — *Praeconocaryomma* sp. 1.
Fig. 8 — *Praeconocaryomma* (?) sp. 2.
Fig. 9 — *Cenosphaera* sp. 2.
Fig. 10 — *Cenosphaera* sp. 2.
Fig. 11 — *Cenosphaera* sp. 1.
Fig. 12 — *Praeconocaryomma* (?) sp. 2.
Fig. 13 — *Cenodiscus* sp. 1, edge view
Fig. 14 — *Spongodiscus* sp. 3.
Fig. 15 — *Spongodiscus* sp. 3 (edge view).
Fig. 16 — *Spongodiscus* sp. 4.

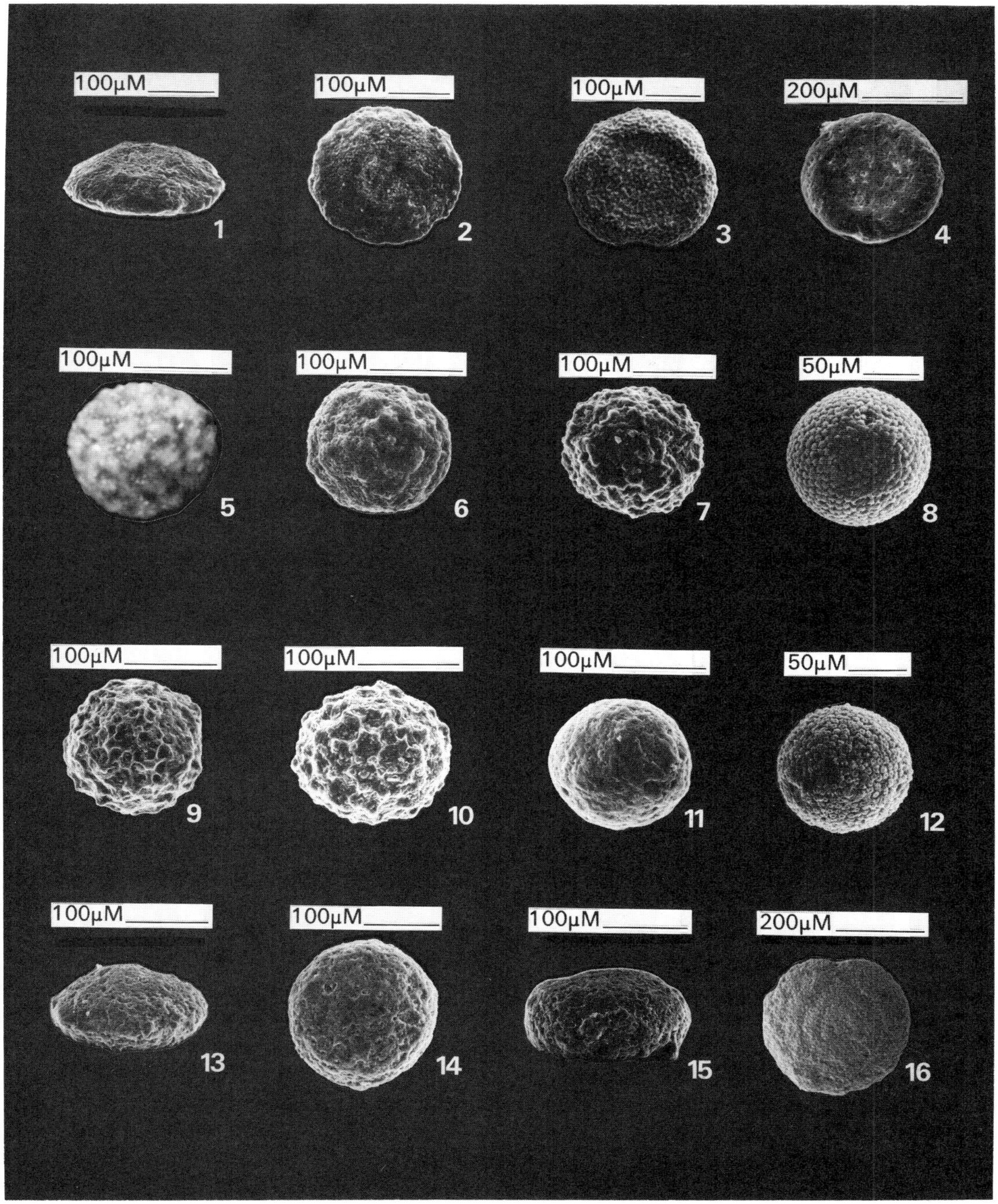
100μM
1
100μM
2
100μM
3
200μM
4
100μM
5
100μM
6
100μM
7
50μM
8
100μM
9
100μM
10
100μM
11
50μM
12
100μM
13
100μM
14
100μM
15
200μM
16

Plate 2 —

Fig. 1 — *Hsuum* sp. 1.
Fig. 2 — *Hsuum* sp. 1.
Fig. 3 — *Parvicingula blowi* Pessagno.
Fig. 4 — *Parvicingula blowi* Pessagno.
Fig. 5 — *Parvicingula blowi* Pessagno ssp. A.
Fig. 6 — *Parvicingula blowi* Pessagno ssp. A.
Fig. 7 — *Parvicingula jonesi* Pessagno.
Fig. 8 — *Parvicingula jonesi* Pessagno.
Fig. 9 — *Stichocapsa devorata* Rust.
Fig. 10 — *Stichocapsa devorata* Rust.
Fig. 11 — *Stichocapsa devorata* Rust (internal cast).
Fig. 12 — *Parvicingula* sp. 1.
Fig. 13 — *Tricolocapsa* sp. 1.
Fig. 14 — *Tricolocapsa* sp. 1.
Fig. 15 — *Tricolocapsa* sp. 1.
Fig. 16 — *Tricolocapsa* sp. 2.
Fig. 17 — *Tricolocapsa* sp. 2.

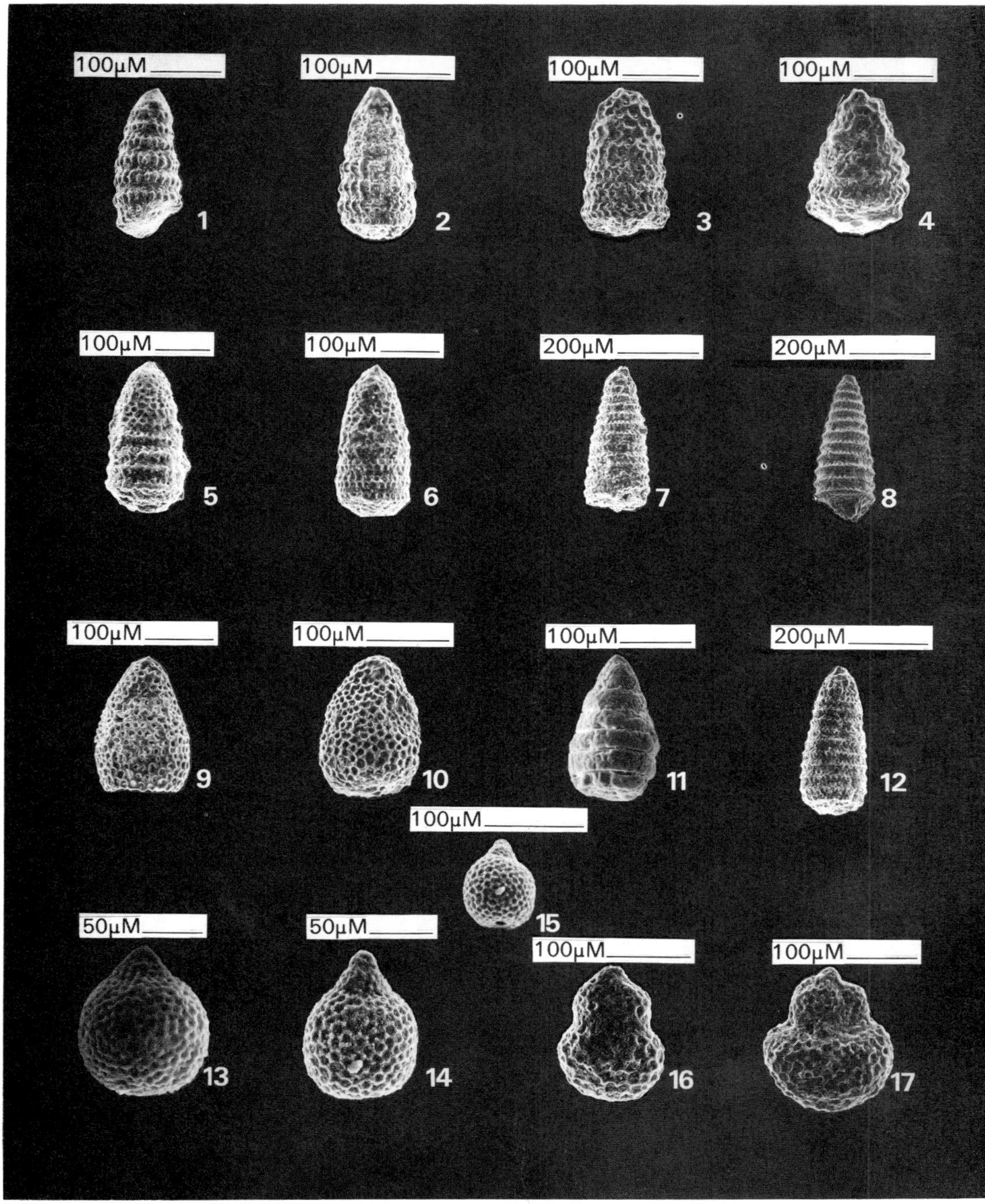
100μM
100μM
100μM
100μM
1
2
3
4
100μM
100μM
200μM
200μM
5
6
7
8
100μM
100μM
100μM
200μM
9
10
11
12
100μM
15
50μM
50μM
100μM
100μM
13
14
16
17

REFERENCES

Aita, Y. & Okada, H. (1986). Radiolaria and calcareous nannofossils from the uppermost Jurassic and Lower Cretaceous strata of Japan and Tethyan regions. *Micropaleontology* **32** 97–128.

Badley, M. E., Price, J. D., Rambech Dahl, C. & Agdestein, T. (1988). The structural evolution of the northern Viking Graben and its bearing upon extensional modes of basin formation. *Journal of the Geological Society, London* **145** 455–472.

Bailey, C. C., Price, I. & Spencer, A. M. (1981). The Ula oil field, block 7/12, Norway. In: *The sedimentation of the North Sea reservoir rocks. Norkse Petroleum Forening.*

Barnard, P. C. & Cooper, B. S. (1981). Oils and source rocks of the North Sea area. In: Illing, L. V. & Hobson, G. D. (eds), *Petroleum geology of the continental shelf of NW Europe.* Institute of Petroleum, London, pp. 169–175

Baumgartner, P. O. (1984). A Middle Jurassic–Early Cretaceous low-latitude radiolarian zonation based on Unitary Association and age of Tethyan radiolarites. *Eclogae Geologicae Helveticae* **77** 729–837.

Baumgartner, P. O., De Wever, P. & Kocher, R. (1980). Correlation of Tethyan Late Jurassic–Early Cretaceous radiolarian events. *Cahiers Micropaleontologie* **2** 23–72.

Blome, C. D. (1984). Middle Jurassic (Callovian) radiolarians from carbonate concretions, Alaska and Oregon. *Micropaleontology* **30** 343–389.

Campbell, A. S. (1954). Radiolaria. In: Moore, R. C. (ed.), *Treatise on invertebrate paleontology, Part D, Protista 3,* Geological Society of America, University of Kansas Press, pp. Dll–D195.

Casey, R. E. (1977). The ecology and distribution of recent Radiolaria. In: Ramsey, A. T. S. (ed.), *Oceanic micropalaeontology*, pp. 809–846.

Cayeux, L. (1896). De l'existence de nombreux radiolaires dans le Tithonique superieur de l'Ardeche. *Compte Rendu de l'Academie des Sciences* **122** 343.

Cope, J. C. W. (1985). Report of the Terminal Jurassic Stage Working Group. *Proceedings of the First International Symposium on Jurassic Stratigraphy.* Geological Survey of Denmark, pp. 111–116.

Cope, J. C. W., Duff, K. L., Parsons, C. F., Torrens, H. S., Wimbledon, W. A. & Wright, J. K., (1980). A correlation of Jurassic rocks of the British Isles. Part Two: Middle and Upper Jurassic., Special Report, *Geological Society, London* **15** 1–109.

De' Ath, N. G. & Schuyleman, S. F. (1981). The geology of the Magnus Oilfield. In: Illing, L. V. & Hobson, G. D. (eds), *Petroleum geology of the continental shelf of NW Europe.* Institute of Petroleum, London, pp. 342–351.

Deegan, C. E. & Scull, B. J. (compilers) (1977). A proposed standard lithostratigraphic nomenclature for the central and northern North Sea. *Report, Institute of Geological Sciences, London* **77/25**; *Bulletin of the Norwegian Petroleum Directorate* **1.**

Dumitrica, P. (1970). Cryptocephalic and cryptothoracic Nassellaria in some Mesozoic deposits of Romania. *Revue Roumaine de Geologie, Geophysique et Geographie, Serie de Geologie* **14** 1–124.

Ehrenberg, C. G. (1854). *Mikrogeologie.* Voss, Leipzig, pp. 1–374.

Ehrenberg, C. G. (1875). Fortsetzung der mikrogeologischen Studien als Gesammt-Uebersicht der mikroskopischen Palaontologie gleichartig analysirter Gebirgsarten der Erde, mit specieller Rucksicht auf den Polycystinen-Mergel von Barbados. *Konigliche Akademie der Wissenschaften zu Berlin, Abhandlungen*, 1–225.

Foreman, H. P. (1973). Radiolaria from D.S.D.P. Leg 20. *Initial Reports of the Deep Sea Drilling Project* **20** 249–305.

Gibbs, A. D. (1984). Clyde field growth fault secondary detachment above basement faults in North Sea. *American Association of Petroleum Geologists Bulletin* **68** 1029–1039.

Gregory, J. (1986). A study of the foraminifera and radiolaria from the Kimmeridgian (*cymodoce* Zone–*eudoxus* Zone) of the Brora/Helmsdale outlier, east Sutherland. *M. Sc. Thesis.* University of Hull.

Haeckel, E. (1862). *Die Radiolarien (Rhizopoda Radiolaria)* **14** 1–572.

Haeckel, E. (1881). Entwurf eines Radiolarien-Systems auf Grund von Studien der Challenger-Radiolarien. *Jenaische Zeitschrift fur Naturwissenschaft, New Series 8* **15** 418–472.

Haeckel, E. (1887). Report on the Radiolaria collected by HMS Challenger during the years 1873–1876. In: *Report on the Scientific results of the voyage of the HMS Challenger, Zoology*, **18** 1–1803.

Haq, B. U., Hardenbol, J. & Vail, P. R. (1987). Chronology of fluctuating sea levels since the Triassic. *Science* **235** 1156–1166.

Harker, S. D., Gustav, S. H. & Riley, L. A. (1987). Triassic to Cenomanian Stratigraphy of the Witch Ground Graben. In: Brooks, J. & Glennie, K. W. (eds), *Petroleum geology of northwest Europe.* Vol. 2. Graham & Trotman, London pp. 809–819.

Harland, W.B., Cox, A. V., Llewellyn, P. G., Pickton, C. A. G., Smith, A. G. & Walters, R. (1982). *A geologic time scale.* Cambridge University Press, Cambridge, pp. 1–128.

Harms, J. C., Tackenberg, P., Pickles, E. & Pollack, R. E. (1981). The Brae oilfield area. In: Illing, L. V. & Hobson, G. D. (eds), *Petroleum geology of the continental shelf of NW Europe.* Institute of Petroleum, London, pp. 352–357.

Imlay, R. W. (1980). Jurassic paleobiogeography of the conterminous United States in its continental setting. *United States Geological Survey Professional Paper* **1062** 1–134.

Ishida, H. (1983). Stratigraphy and radiolarian assemblages of the Triassic and Jurassic siliceous sedimentary rocks in Konose Valley, Tokushima Prefecture, southwest Japan. *Journal of Science, University of Tokushima* **16** 111–141.

Johnson, H. D., Mackay, T. A. & Stewart, D. J. (1986). The Fulmar oilfield (central North Sea): geological aspects of its discovery, appraisal and development. *Marine Petroleum Geology* **3** 99–125.

Johnson, H. D., & Stewart, D. J. (1985). The role of clastic sedimentology in the exploration and production of oil in the North Sea. In: Williams, B. P. J. & Brenchley, P. J. (eds), *Sedimentology: recent developments and applied aspects.* Blackwell Scientific, Oxford.

Kent, D. V. & Gradstein, F. M. (1985). A Cretaceous and Jurassic geochronology. *Geological Society of America Bulletin* **69** 1419–1427.

Maher, C. E. (1981). The Piper oilfield. In: Illing, L. V. &

Hobson, G. D. (eds). *Petroleum geology of the continental shelf of NW Europe.* Institute of Petroleum, London, pp. 358–370.

Matsuoka, A. (1983). Middle and Late Jurassic radiolarian biostratigraphy in the Sakawa and adjacent areas, Shikoku, southwest Japan. *Journal of Geosciences, Osaka City University* **26** 1–48.

Matsuoka, A. & Yao, A. (1986). A newly proposed radiolarian zonation for the Jurassic of Japan. *Marine Micropaleontology* **11** 91–106.

Mizutani, S., Nishiyama, H. & Ito, T. (1982). Radiolarian biostratigraphic study of the Shimanto Group in the Nanto–Nansei area, Mie Prefecture, Kii Peninsula, central Japan. *Journal of Earth Sciences, Nagoya University* **30** 31–107.

Nakaseko, K. (ed.) (1982). *Proceedings of the 1st Japanese Radiolarian Symposium. Special Volume, News Osaka Micropaleontology* **5** 1–485.

Nakaseko, K. & Nishimura, A. (1981). Upper Jurassic and Cretaceous radiolaria from the Shimanto Group in southwest Japan. *Science Reports, College of General Education, Osaka University* **30** 133–203.

Ozvoldova, L. (1975). Upper Jurassic radiolarians from the Kisuca Series in the Klippen Belt. *Zapadne Karpaty, Seria Paleontologia* **1** 73–86.

Parona, C. F. (1890). Radiolarie nei noduli selciosi del calcare giurese di Cittiglio presso Laveno. *Bollettino della Societa Geologica Italiana* **9** 132–175.

Pessagno, E. A. (1973). Upper Cretaceous Spumellariina from the Great Valley Sequence, California Coast Ranges. *Bulletin of American Paleontology* **63** 49–102.

Pessagno, E. A. (1976). Radiolarian zonation and stratigraphy of the Upper Cretaceous portion of the Great Valley Sequence, California Coast Ranges. *Micropaleontology, Special Publication* **2** 1–95.

Pessagno, E. A. (1977a). Upper Jurassic Radiolaria and radiolarian biostratigraphy of the California Coast Ranges. *Micropaleontology* **23** 56–113.

Pessagno, E. A. (1977b). Radiolaria in Mesozoic stratigraphy. In: Ramsay, A. T. S. (ed), *Oceanic Micropalaeontology*, Vol. 2. Academic Press, London, pp. 913–950.

Pessagno, E. A. (1977c). Lower Cretaceous radiolarian biostratigraphy of the Great Valley Sequence and Franciscan Complex, California Coast Ranges. *Cushman Foundation for Foraminiferal Research, Special Publication* **15** 1–87.

Pessagno, E. A. & Blome, C. D. (1980). Upper Triassic and Jurassic Pantanellidae from California, Oregon and British Columbia, *Micropaleontology* **26** 225–273.

Pessagno, E. A. & Blome, C. D. (1982). Bizarre Nassellariina (Radiolaria) from the Middle and Upper Jurassic of North America. *Micropaleontology* **28** 289–318.

Pessagno, E. A., Blome, C. D., Carter, E. S., MacLeod, N., Whalen, P. A. & Yeh Kuei-Yu (1986a). Preliminary radiolarian zonation for the Jurassic of North America. *Cushman Foundation For Foraminiferal Research, Special Publication* **23** 1–18.

Pessagno, E. A., Blome, C. D. and Longoria, J. F. (1984). A revised radiolarian zonation for the Upper Jurassic of western North America. *Bulletin of American Paleontology* **87** 1–51.

Pessagno, E. A. & Newport, R. L. (1972). A technique for extracting radiolaria from radiolarian cherts. *Micropaleontology* **18** 231–234.

Pessagno, E. A. & Whalen, P. A. (1982). Lower Jurassic and Middle Jurassic Radiolaria from California, east-central Oregon and the Queen Charlotte Islands, B. C. *Micropaleontology* **28** 111–169.

Pessagno, E. A., Whalen, P. A. & Yeh, K.-Y. (1986b). Jurassic Nassellariina (Radiolaria) from North American Geologic Terranes. *Bulletin of American Paleontology* **91** 1–75.

Rawson, P. F., Curry, D., Dilley, F. C., Hancock, J. M., Kennedy, W. J., Neale, J. W., Wood, C. J. & Worssam, B. C. (1978). A correlation of Cretaceous rocks in the British Isles. *Special Report, Geological Society, London.* **9** 1–70.

Rawson, P. F. & Riley, L. A. (1982). Latest Jurassic–Early Cretaceous events and the 'Late Cimmerian Unconformity' in North Sea area. *American Association of Petroleum Geologists Bulletin* **66** 2628–2648.

Riedel, W. R. (1971). Systematic classification of polycystine Radiolaria. In: Funnell, B. M. & Riedel, W. R. (eds), *Micropalaeontology of oceans.* Cambridge University Press, Cambridge, pp. 649–661.

Rust, D. (1885). Beitrage zur Kenntnis der Fossilien Radiolarien aus Gesteinen des Jura. *Palaeontographica, Ser. 3* **31** 273–321.

Rust, D. (1898). Neue Beitrage zur Kenntnis der Fossilien Radiolarien aus Gesteinen des Jura und der Kreide. *Palaeontographica* **45** 1–67.

Sanfilippo, A. & Riedel, W. R. (1985). Cretaceous Radiolaria. In: Bolli, H. M., Saunders, J. B. & Perch-Nielsen, K. (eds) *Plankton stratigraphy.* pp. 573–630.

Smith, R. L. (1987). The structural development of the Clyde Field. In: Brooks, J. & Glennie, K. W. (eds), *Petroleum geology of northwest Europe.* Vol. 1. Graham & Trotman, London, pp. 523–531.

Squinabol, S. (1913). Radiolari della strada nazionale al Monginevro. *Bolletino del R. Comitato Geologico d'Italia* **43** 281–289.

Turner, C. C., Richards, P. C., Swallow, J. L. & Grimshaw, S. P. (1984). Upper Jurassic stratigraphy and sedimentary facies in the central Outer Moray Firth Basin, North Sea. *Marine Petroleum Geology* **1** 105–177.

Vail, P. R. & Todd, R. G. (1981). North Sea unconformities, chronostratigraphy and sea-level changes from seismic stratigraphy. In: Illing, L. V. & Hobson, G. D. (eds), *Petroleum geology of the continental shelf of NW Europe.* Institute of Petroleum, London, pp. 216–235.

Vinassa de Regny, P. E. (1899). I radiolari delle ftaniti titoniane di Carpena (Spezia). *Palaeontographia Italica* **4** 217–238.

Vollset, J. & Doré, A. G. (1985). A revised Triassic and Jurassic lithostratigraphical nomenclature for the Norwegian North Sea. *Bulletin of the Norwegian Petroleum Directorate* **3.**

Wimbledon, W. A. (1985). The Portlandian, the terminal Jurassic stage in the Boreal realm. In: *Proceedings of the First International Symposium on Jurassic Stratigraphy.* Geological Survey of Denmark, pp. 533–549.

Woollam, R. & Riding, J. B. (1983). Dinoflagellate cyst zonation of the English Jurassic. *Report, Institute of Geological Sciences, London* **83/2.**

Ziegler, P. A. (1982). *Geological atlas of Western and Central Europe.* Shell International Petroleum Maatschappij B. V., Elsevier, Amsterdam, pp. 1–130.

10

Upper Cretaceous microbiostratigraphy of Bornholm, Denmark

S. Packer, M. B. Hart, B. A. Tocher and S. Braley

ABSTRACT

The Upper Cretaceous succession of the Island of Bornholm (Denmark) has been investigated as part of a larger study of northwest European Cretaceous microbiostratigraphy. The strata have been logged in detail and micropalaeontological samples collected at approximately 1 m intervals. The Upper Cretaceous succession (Arnager Greensand, Arnager Limestone and Bavnodde Greensand) is all of marine origin and contains rich assemblages of dinoflagellate cysts, benthonic and planktonic foraminifera. The foraminifera can be used for both international and inter-regional correlation and indicate a mid-Cenomanian to ?late Turonian–early Coniacian hiatus in the succession.

INTRODUCTION

The Danish island of Bornholm lies in the central southern Baltic (55 °N, 15 °E) 65 km southeast of Ystad, on the Swedish coast (Fig. 1). The greater part of the island is formed of Precambrian gneiss, migmatite and granite with sediments of Mesozoic age exposed only in the southwest corner, in the area immediately north and east of Rønne. Running northwest–southeast through the island are a series of major faults which are associated with the Fennoscandian Border Zone (Christensen 1984).

Between Rønne and Hasle, on the west coast of the island, is the very edge of a north–south trending graben. The intersection of the two trends (northwest–southeast and north–south) provides the main geological structure, as well as the overall shape of the island. During the Mesozoic the Fennoscandian Border Zone acted as a boundary between the stable Fennoscandian Shield to the north and east and the subsiding Danish sub-basin to the south and west.

The present investigation is part of a larger integrated study involving the biostratigraphy of the major macrofaunal groups (ammonites, belemnites, inoceramids, etc.) that have been recorded from this succession (Christensen 1984, Kennedy *et al.* 1981). Work on the foraminifera, dinoflagellate cysts, ostracods and calcareous nannofossils is still in progress.

THE CRETACEOUS SUCCESSION

The Mesozoic sedimentary succession of Bornholm (Surlyk 1980) consists of only intermittent

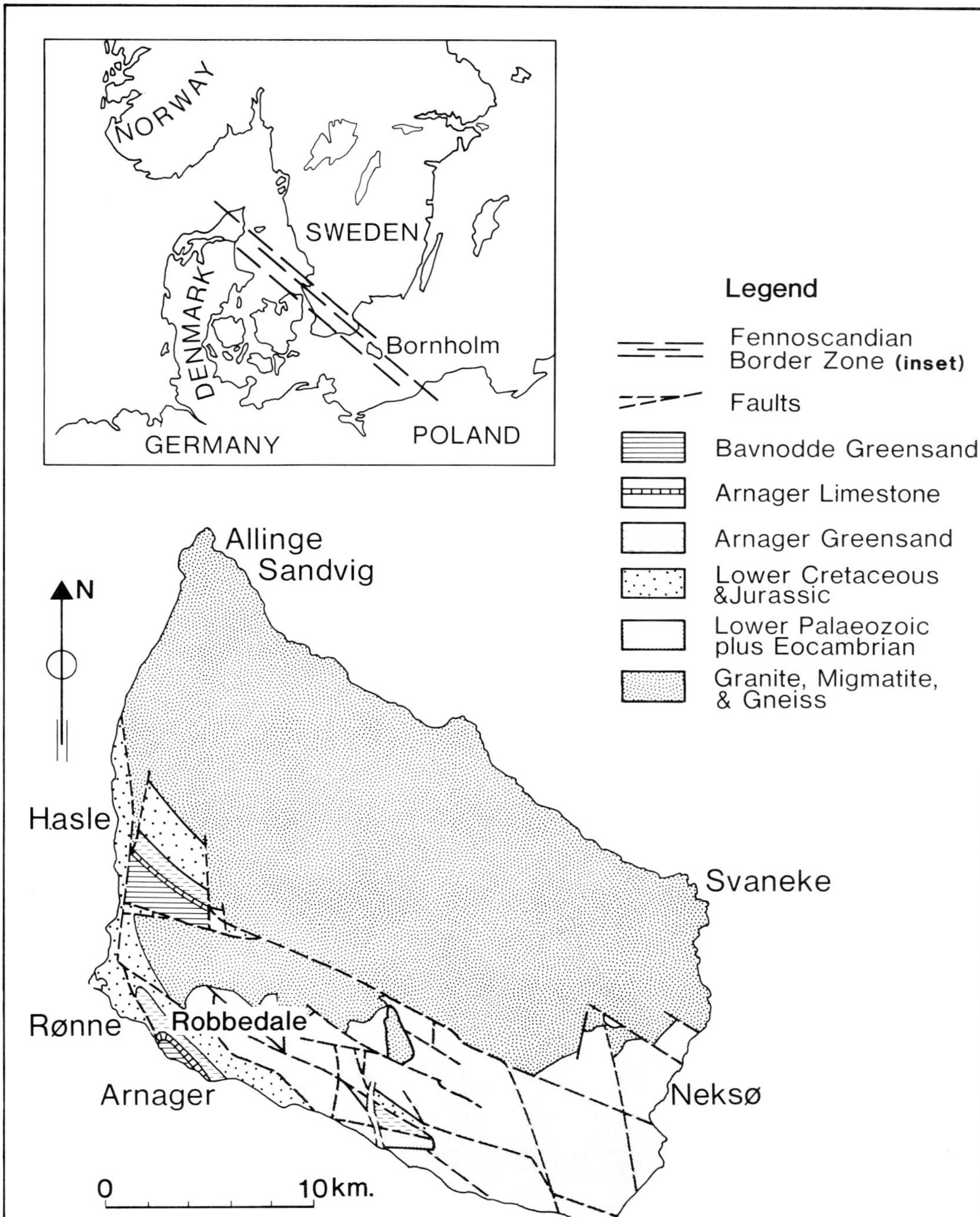

Fig. 1 — Outline geological map of the Island of Bornholm.

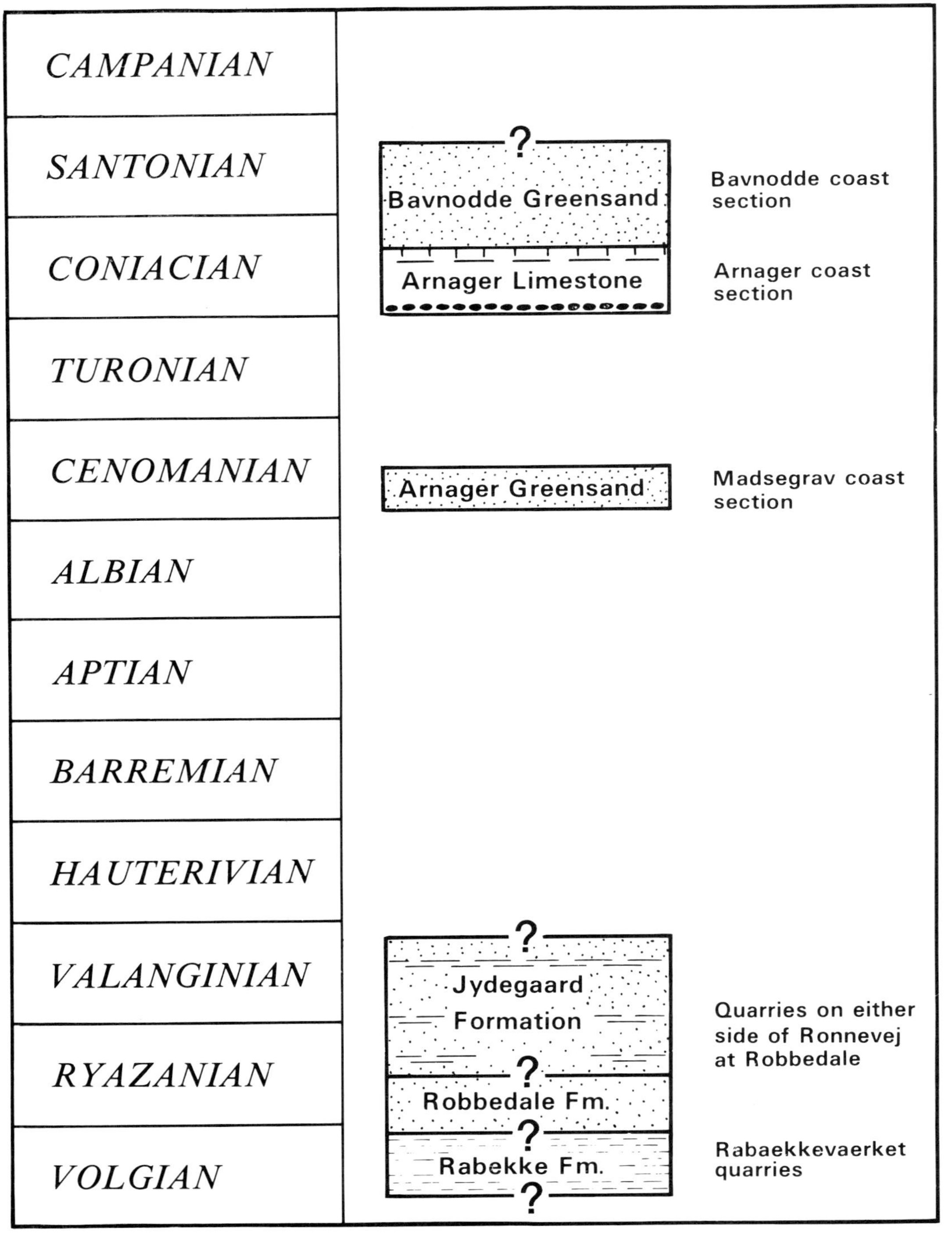

Fig. 2 — The latest Jurassic and Cretaceous succession of Bornholm. The principal localities for each of the intervals investigated are indicated in the right-hand column.

packages of sediment between which there are major intervals of non-deposition and/or erosion. Following deposition of the early Jurassic Sose Beds, Rønne Beds, Hasle Beds and Baga Beds, there is a major hiatus through to the latest Jurassic. Following Cimmerian movements on the Fennoscandian Border Zone, deposition began in the late Volgian. These sediments, which constitute the Rabekke Formation (Fig. 2), rest unconformably on the Jurassic sediments of the Robbedale–Sose Block and also overlap onto the Palaeozoic strata and the Pre-Cambrian basement. The Robbedale Formation of ? latest Volgian–Ryazanian age overlies the Rabekke Formation.

On the south coast at Madsegrav (east of Arnager) the Ryazanian–Valanginian Jydegaard Formation is overlain by glauconitic sandstones of Cenomanian age, thus indicating a major hiatus. The marine sediments of Late Cretaceous age include, from bottom to top (Fig. 2), the Arnager Greensand, Arnager Limestone and Bavnodde Greensand. This succession is exposed on the south coast of the island between Bavnodde (2 km west of Arnager Point) and Madsegrav (1 km east of Arnager Point).

The lower part of the Arnager Greensand at Madsegrav is marked by a distinct conglomerate (Fig. 3) with phosphatic pebbles and boulders of glauconitic sandstone. There is evidence of an early Albian ammonite fauna (Kennedy *et al.* 1981, Christensen 1984) in pebbles which are enclosed in nodules that yield a late Albian to early Cenomanian fauna. The Arnager Greensand itself yields a distinctive early–middle Cenomanian macrofauna and microfauna (Hart 1979). The soft, generally uncemented quartz–glauconitic sands are heavily bioturbated and contain intermittent pebble beds. One major erosion surface has been identified and there may well be other structures present. This was impossible to determine, however, because of the poorly exposed nature of the succession.

The top of the Arnager Greensand and its contact with the Arnager Limestone is exposed at Arnager. The intensely bioturbated horizon (Fig. 4) is overlain by the basal bed (20–25 cm) of the limestone and contains several generations of phosphatized and glauconitized pebbles (Fig. 5). The Arnager Limestone is approximately 15–20 m thick and is a white to grey splintery brittle calcilutite or calcisiltite. Noe-Nygaard & Surlyk (1985) have described mound bedding within these limestones. Above this the limestone becomes blue-grey and contains a greater proportion of clay, fine sand and silica (Christensen 1984). The contact with the overlying Bavnodde Greensand was seen by one of us (M.B.H.) in 1975 and has recently been excavated by Solakius (1986 pers. comm.).

The Bavnodde Greensand is a thick succession of soft glauconitic sandstone; coarse levels within it may well represent storm sands (Fig. 6).

MICROPALAEONTOLOGICAL ANALYSIS

Samples for micropalaeontological analysis have been collected at approximately 1 m intervals throughout the Arnager Greensand, Arnager Limestone and Bavnodde Greensand, although more detailed sampling was carried out at the complex horizon shown in Figs. 4 and 5. All samples were broken into subsets for foraminiferal, palynological and nannofossil investigation.

The foraminiferal fauna is abundant throughout the marine succession and, apart from in the Arnager Limestone, is well preserved. Planktonic foraminifera are relatively abundant and, surprisingly for such a high palaeolatitude, include many keeled taxa. The lower part of the Arnager Greensand (at Madsegrav) yields *Praeglobotruncana delrioensis* (Plummer), *Hedbergella delrioensis* (Carsey), *H. brittonensis* (Loeblich & Tappan), *H. planispira* (Tappan), *Heterohelix moremani* (Cushman) and *Guembelitria harrisi* (Tappan). In the Arnager Greensand immediately below the contact with the Arnager Limestone this assemblage is joined by large numbers of *Rotalipora reicheli* Mornod (Hart 1979). The geographical distribution of this taxon (Hart 1979, Fig. 3) is clearly controlled

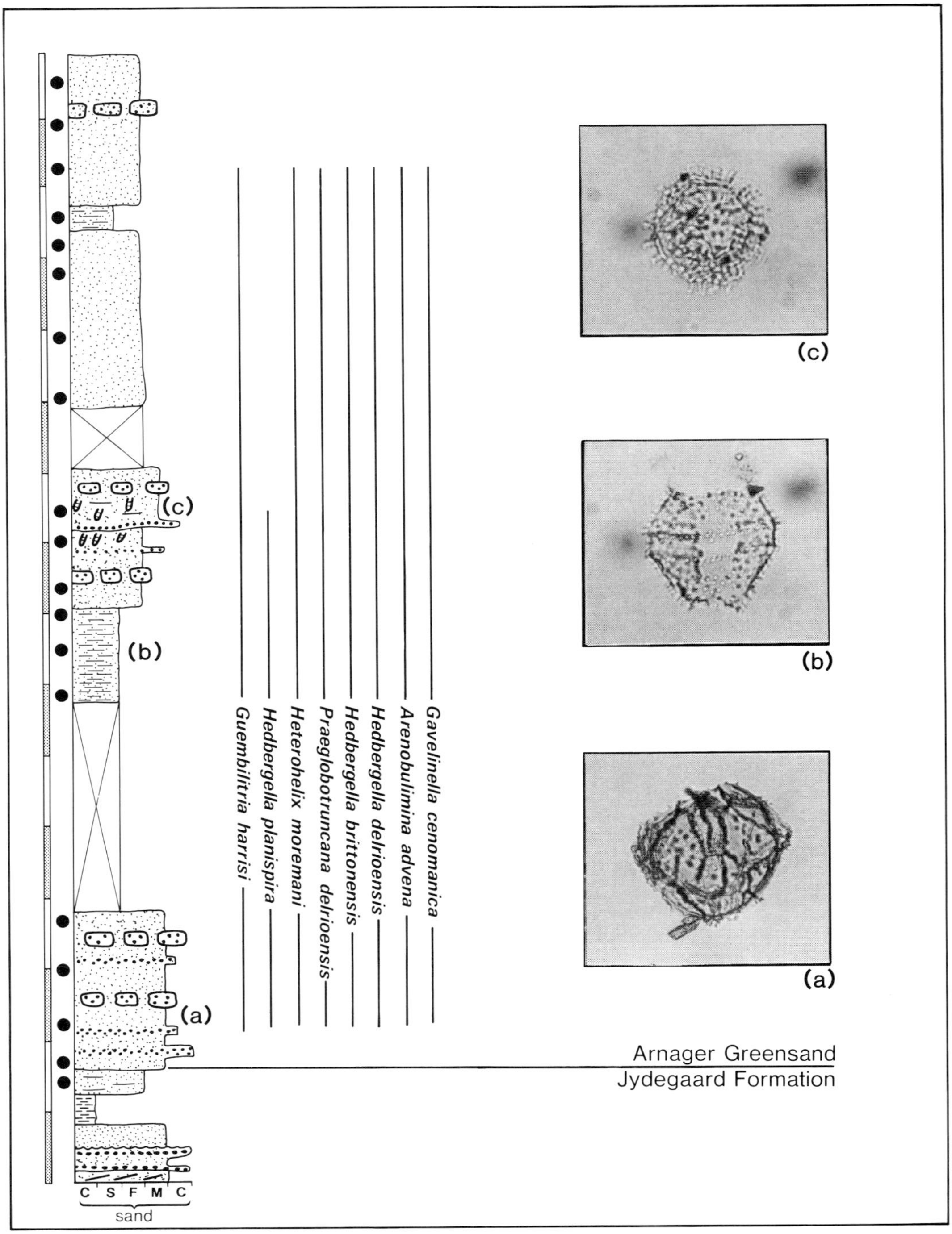

Fig. 3 — Composite sedimentological log of the lower part of the Arnager Greensand exposed at, and just west of, Madsegrav. This log was assembled from six very limited exposures and as such, must be treated with caution. The remainder of the Arnager Greensand succession (some 70 m) is overgrown and covered with sand dunes. Representative dinoflagellate cysts: (a) *Cribroperidinium edwardsii*; (b) *Epelidosphaeridia spinosa*; (c) *Cleistophaeridium? huguoniotii*.

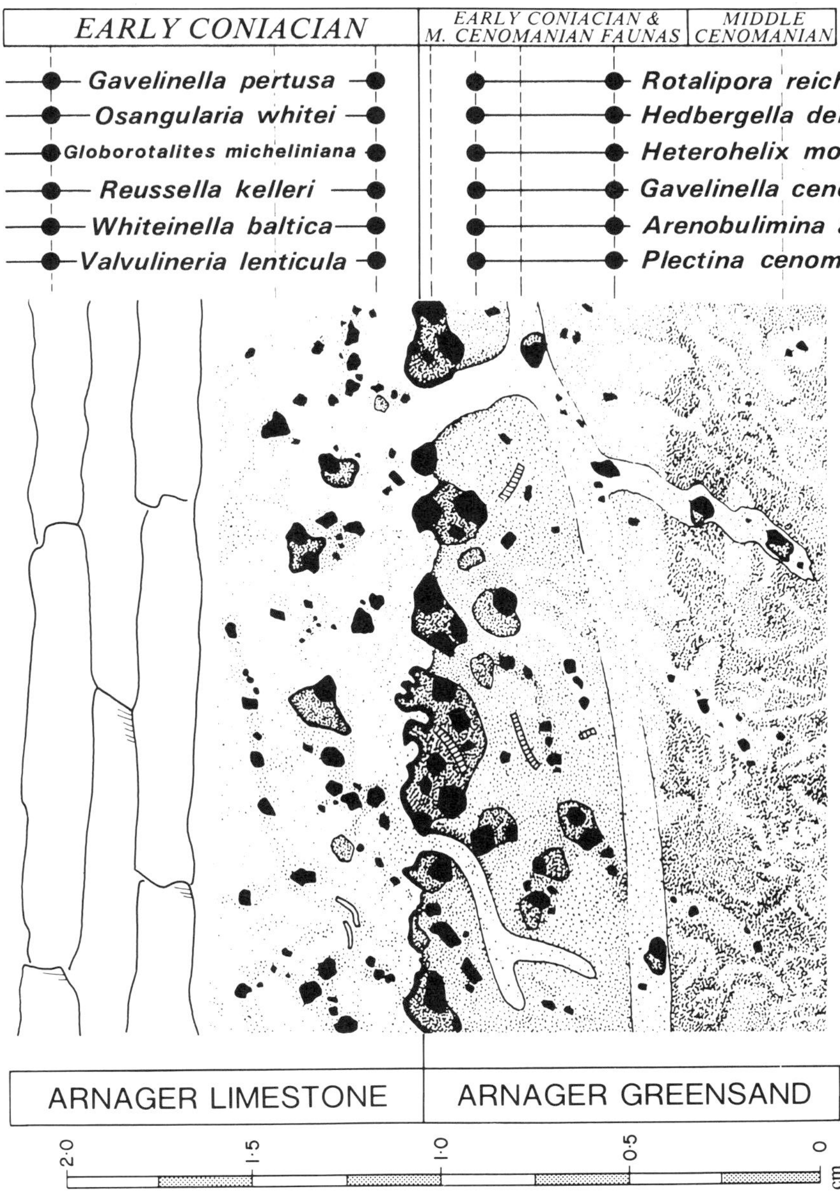

Fig. 4 — The Arnager Greensand–Arnager Limestone boundry as it is exposed in the low sea cliff immediately west of the sea front car park at Arnager. The sketch is based on field work by the present authors and that of Bromley (1979). Immediately overlying the Arnager Greensand is a buff-coloured, faintly glauconitic sandstone that contains a mixed Cenomanian–Coniacian fauna and flora.

a

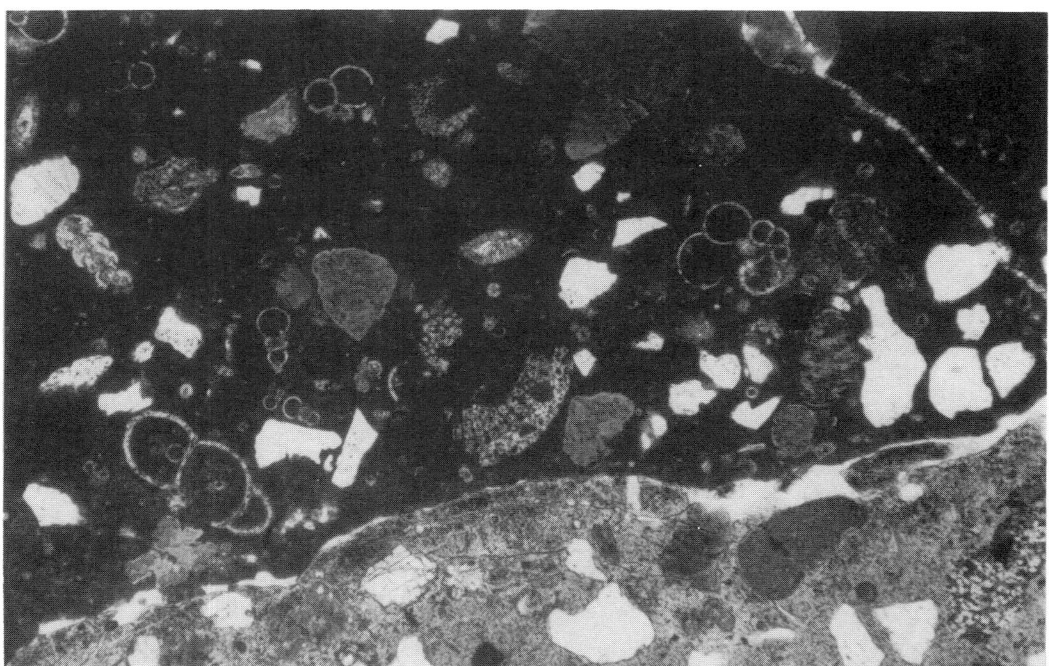

b

Fig. 5 — (a) Thin section of the contact between the phosphatized pebbles shown in Fig. 6 and the overlying limestone. The specimen in the lower left-hand corner is *Whiteinella baltica* Douglas & Rankin. (b) Thin section of the edge of a phosphatized pebble and the adjacent Arnager Limestone. Assemblage dominated by *Whiteinella* spp., *Hedbergella* sp. and *Archaeoglobigerina* sp.

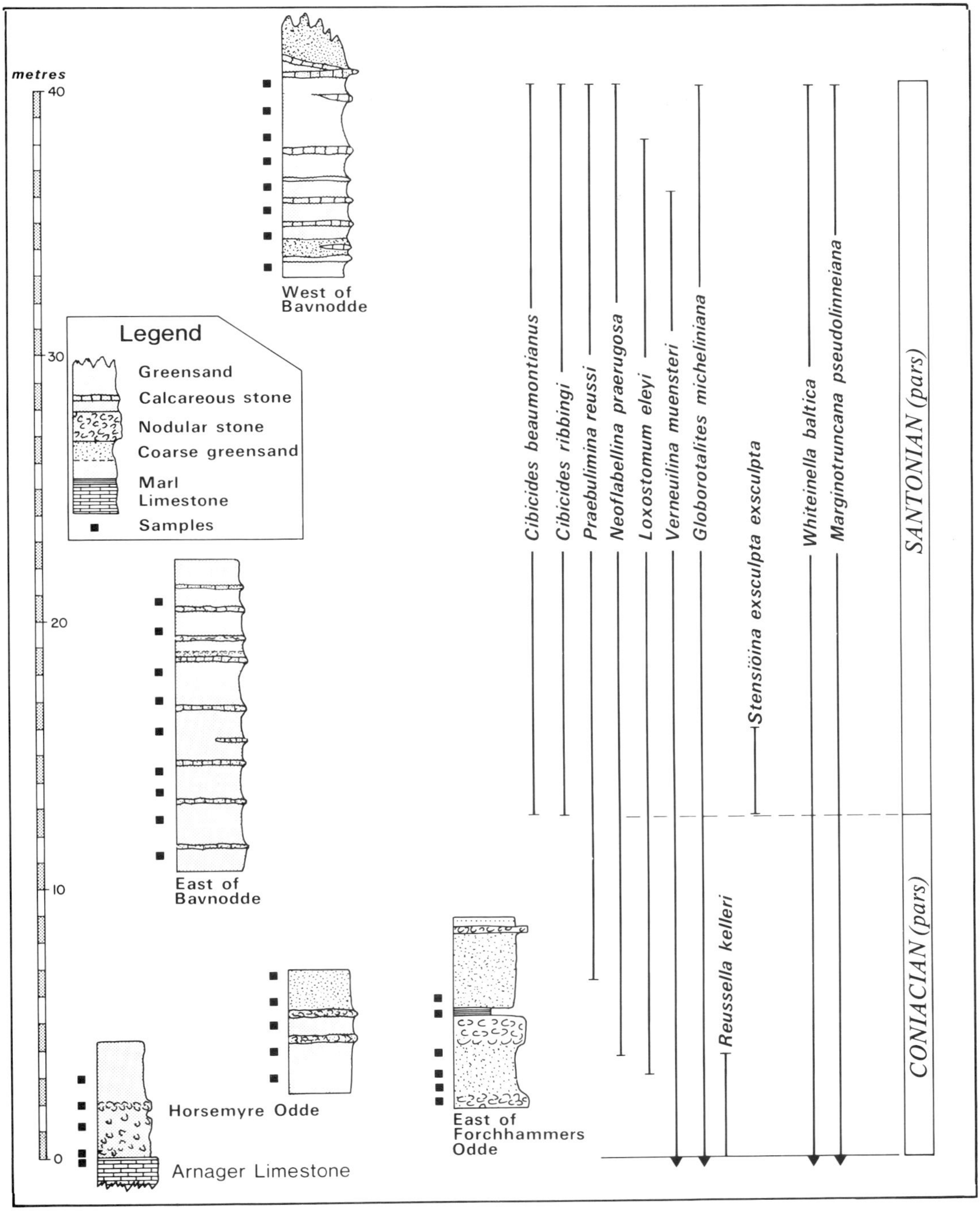

Fig. 6 — Sedimentological logs of the Bavnodde Greensand which is exposed on the south coast between Horsemyre Odde and West Bavnodde (immediately adjacent to the airport perimeter fence), and ranges of some stratigraphically diagnostic Foraminifera (the contact with the Arnager Limestone is no longer exposed). The tentative placing of the Coniacian–Santonian boundary based on the Foraminifera agrees well with the known distribution of belemnites (W. K. Christensen, pers. comm.).

by the increased water depths available in the Danish Sub-basin (Danish Polish Furrow of Hart (1979)). This species probably indicates that the uppermost Arnager Greensand is of earliest middle Cenomanian age.

The Arnager Limestone yields an abundant planktonic fauna although the diversity is low. *Marginotruncana pseudolinneiana* (Pessagno) and *Whiteinella baltica* Douglas & Rankin dominate the fauna while *M. marginata* (Reuss) and *M. coronata* (Bolli) are also recorded. We have not found (as yet) specimens referable to the *Dicarinella primitava/concavata* groups as recorded by Solakius & Larsson (1986). In the Bavnodde Greensand, *Globotruncana bulloides* Vogler appears, together with *Archaeoglobigerina bosquensis* Pessagno and *Heterohelix* spp. Within the planktonic fauna there is a distinct change in the proportion of keeled to non-keeled species in the transition from the Arnager Limestone to the Bavnodde Greensand. Within the limestone, keeled taxa predominate and this is also true within the lowest level of the Bavnodde Greensand. Higher in the Greensand, the proportion of keeled taxa reduces markedly and the fauna gradually becomes dominated by *Whiteinella* and *Archaeoglobigerina.* If Hart & Bailey (1979) are correct, this would imply a gradual reduction of effective water depth during this interval; this was probably caused by sedimentation, there being no sharp dramatic change.

Investigation of the benthonic foraminifera (Hart 1979, Bailey & Hart 1979, Solakius & Larsson 1986) indicate that, throughout the succession, stratigraphically important taxa are present. In the Arnager Greensand the majority of the benthonic taxa—*Arenobulimina advena* (Cushman), *A. depressa* (Perner), *Marssonnella trochus* (d'Orbigny), *Plectina mariae* (Franke), *Quinqueloculina antiqua* (Franke), *Gavelinella cenomanica* (Brotzen) and *G. intermedia* (Berthelin)—are well-known long-ranging species characteristic of the Cenomanian; *Gavelinella belorussica* (Akimez), *Cibicides gorbenkoi* Akimez and *Lingulogavelinella* sp. cf. *L. ornatissima* are little-known eastern European species (Gawor-Biedowa 1972), with general Cenomanian ranges. The overlapping ranges of *Plectina cenomana* (Carter & Hart) and *Lingulogavelinella jarzevae* (Vasilenko) within the Arnager Greensand succession indicate a proximity to the early–middle Cenomanian boundary, which agrees with the occurrence of the *R. reicheli* fauna already reported.

The fauna of the Arnager Limestone has been described by Bailey & Hart (1979) and in more detail by Solakius & Larsson (1986). It is dominated by *Reussella kelleri* Vasilenko, *Verneuilina muensteri* Reuss, *Gavelinella pertusa* Marsson, *G. thalmanni* (Brotzen), *Osangularia whitei* (Brotzen) and *Stensiöeina* sp. The latter are important (Bailey *et al.* 1983, Hart *et al.* 1981, Koch 1977) for stratigraphic work but our present material is not well enough preserved to identify the subspecies used in the various zonal schemes for this interval.

The overlying Bavnodde Greensand fauna shows that many of the above taxa continue (except *R. kelleri*) and are joined by *Gavelinella stelligera* (Marie), *Cibicides ribbingi* Brotzen, *C. beaumontianus* d'Orbigny, *Loxostomum eleyi* (Cushman), *Neoflabellina praerugosa* Hiltermann and *Praebulimina reussi* (Morrow). Bailey & Hart (1979) tentatively placed the Arnager Limestone within the Coniacian, there being no evidence of any Turonian indicators. The placing of the Coniacian–Santonian boundary at that time was most difficult and we would now locate this at a level within the lower part of the Bavnodde Greensand. There may be a hiatus at the Arnager Limestone–Bavnodde Greensand boundary as suggested by Solakius & Larsson (1986) although at the present time we have no indication of its relative importance or magnitude.

Samples processed for palynomorphs from the base of this formation yield *Cribroperidinium exilicristatum* (Davey) Stover & Evitt, *Cyclonephelium hughesii* Clarke & Verdier, *Exochosphaeridium bifidum* (Clarke & Verdier) Clarke *et al.*, *Florentinia deanei* (Davey & Williams) Davey & Verdier, *Odontochitina costata* Alberti and *Palaeohystrichophora infusorioides* Deflandre, which have previously been recorded (Davey & Verdier 1973) as first appearing in the late Albian in northwest Europe. Furthermore, the presence of *C. exilicristatum* and *Epelidosphaeridia spinosa* (Cookson & Hughes) Davey is thought to indicate that the succession is no younger than middle Cemomanian (Davey 1969, Tocher 1984). These two forms in association with *Litosphaeridium siphoniphorum glabrum* Lucas-Clark, which has not been recorded in sediments younger than later Cenomanian (Lucas-Clark 1984), are also found in the middle part of the Arnager Greensand succession.

Samples from the top of the Arnager Greensand and its contact with the overlying Arnager Limestone yield a complicated assemblage. Previously useful forms such as *C. exilicristatum*, *E. spinosa* and *L. siphoniphorum glabrum* are absent, although the presence of *Litosphaeridium siphoniphorum siphoniphorum* (Cookson & Eisenack) Davey & Williams, *Florentinia mantellii* (Davey & Williams) Davey & Verdier, *Microdinium ornatum* Cookson & Eisenack and *Stephodinium coronatum* Deflandre suggest an age no younger than late Turonian (Foucher 1979, 1981, Lucas-Clark 1984). However, these species are here found to occur in association with specimens of *Chatangiella tripartita* (Cookson & Eisenack) Lentin & Williams, *Dinogymnium microgranulosum* Clarke & Verdier, *Florentinia deanei*, *Senoniasphaera rotundata* Clarke & Verdier and *Spiniferites? velatus* (Clarke & Verdier) Stover & Evitt. The presence of *F. deanei* indicates an age no younger than late Coniacian, while *D. microgranulosum* and *S.? velatus* have previously been recorded from the Anglo-Paris Basin (Clarke & Verdier 1967, Foucher 1979) as first appearing in the Coniacian and early Santonian, respectively. On the basis of the foraminiferal analysis, however, it is thought the former age is more likely. Thus, at the junction between the Arnager Greensand and the Arnager Limestone there appears to be a mixed assemblage of ?late Cenomanian–late Coniacian forms.

REGIONAL IMPLICATIONS

The Robbedale and Jydegaard Formations represent depositional environments close to a shoreline that appears to be trending WNW–ESE across the island. This shoreline was probably fault induced by the late Cimmerian movements that affected the Border Zone. Palaeocurrent trends produced by one of us (S.B.) indicate a net transport direction off the Pre-Cambrian basement towards the south, with subordinate trends towards the southeast and southwest. Offshore these sands may thicken to the south and provide some reservoir potential. To the west of Bornholm, in the Rønne Graben, sands of Early Cretaceous age are probably present, although only sediments of Jurassic age are exposed onshore between Rønne and Hasle. To the north, in Hano Bay (Christensen 1984, Norling 1981) sands of mid-Cretaceous age are extensively recorded, including the glauconitic sandstones and siltstones recorded in the Åhus borehole (Christensen 1975). In the central part of Hano Bay, the Barremian–Aptian is reportedly 50–100 m thick with the Albian–Cenomanian attaining a thickness of 75 m. Clearly these sediment thicknesses may be greater in the downfaulted centre of the graben, now well known in the area around Bornholm, and as such could provide suitable reservoir sandstones. Knowledge of their stratigraphy is therefore of paramount importance when considering futher offshore exploration.

Recently, Christensen (1984) has provided an intepretation of the sea-level changes represented in the Bornholm succession. He records a maximum level in the middle Cenomanian followed by a gradual regressive phase throughout the Turonian. This minimum in the middle–late Turonian is followed by a transgressive phase (and renewed sedimentation) in the Coniacian to Santonian.

This interpretation appears to be at variance with most views of Cretaceous sea levels (Hancock 1976, Hancock & Kauffman 1979, Hart & Bailey 1979, Hart 1980). Our present data suggest that the hiatus in the middle Cenomanian was produced by a marked increase in water depth (Carter & Hart 1977), thereby isolating the horst (later to become Bornholm) from sediment input. The sea-level high, existing through the late Cenomanian and early–middle Turonian, left Bornholm isolated, receiving little sediment and creating a phosphate conglomerate (Figs. 4 and 5). Lowering of worldwide sea levels in the late Turonian brought Bornholm back into a depositional domain and sedimentation began in the late Turonian–Coniacian and Santonian. This scenario would agree well with the foraminiferal and palynological data presented here and very nearly comply with the predictions for areas such as south England where the succession is relatively complete (Hart & Bailey 1979, Hart 1980). This interpretation is at some variance with that suggested by Christensen (1984, Fig. 6) and our future research must be directed towards a solution of this problem.

ACKNOWLEDGEMENTS

The authors acknowledge the assistance given by our Scandinavian colleagues: Dr W. K. Christensen, Professor H. J. Hansen, Professor F. Surlyk, Dr R. Bromley, Dr E. Norling, Dr E. Stenestad, J. Gabrielsen and N. Solakius. Professor V. Poulsen is thanked for allowing us to use the University of Copenhagen Field Centre on Bornholm. Our research in southern Scandinavia is funded by the Research Initiative of the National Advisory Body for Public Sector Higher Education.

This paper is dedicated to the memory of Professor Tove Birkelund (University of Copenhagen), who died in the summer of 1986, and who until shortly before that time had been Chairman of the International Subcommission on Cretaceous Stratigraphy. Her advice and encouragement in the earlier stages of our work were much appreciated.

REFERENCES

Bailey, H. W., Gale, A. G., Mortimore, R. N., Swiecicki, A. & Wood, C. J. (1983). The Coniacian–Maastrichtian Stages of the UK with particular reference to southern England. *Newsletters in Stratigraphy* **12** 29–42.

Bailey, H. W. & Hart, M. B. (1979). The correlation of the early Senonian in western Europe using foraminiferida. *Aspekte der Kreide Europas, IUGS, Series A* **6** 159–169.

Bromley, R. G. (1979). Chalk and Bryozoan limestone facies, sediments and depositional environments. In: Birkelund, T. & Bromley, R. G. (eds), *Cretaceous–Tertiary Boundary Events Symposium*, I. *The Maastrichtian and Danian of Denmark*. Univesity of Copenhagen, Copenhagen, pp. 16–32.

Carter, D. J. & Hart, M. B. (1977). Aspects of mid-Cretaceous stratigraphical micropalaeontology. *Bulletin of the British Museum (Natural History), Geology* **29** 1–135.

Christensen, W. K. (1975). Upper Cretaceous belemnites from the Kristiansand area in Scania. *Fossils and Strata* **7** 1–69.

Christensen, W. K. (1984). The Albian to Maastrichtian of southern Sweden and Bornholm, Denmark: a review. *Cretaceous Research* **5** 313–327.

Clarke, R. F. A. & Verdier, J.-P. (1967). An investigation of the microplankton assemblages from the Chalk of the Isle of Wight. *Verhandelingen der Konninklijke Nederlands Akademie van Wetenschappen, Afdeeling Natuurkunde, Eerste Reeks* **24** 1–96.

Davey, R. F. (1969). Non-calcareous microplankton from the Cenomanian of England, northern France and North America. Part 1. *Bulletin of the British Museum (Natural History)* **17** 103–180.

Davey, R. J. & Verdier, J.-P. (1971). An investigation of microplankton assemblages from the Albian of the Paris Basin. *Verhandelingen der Koninklijke Nederlandsche Akademie van Wetenschappen, Afdeeling Natuurkunde, Eerste Reeks* **26** 1–58.

Davey, R. J. & Verdier, J.-P. (1973). An investigation of microplankton assemblages from latest Albian (Vraconian) sediments. *Revista Española Micropaleontologia* **5** 173–212.

Fauconnier, D. (1979). Les dinoflagelles de l'Albien et du Cenomanien inferieur du Bassin de Paris. Repartition stratigraphique et relations avec le nature du dépôt. *Thèse de Doctorat d'Université*. Orleans. *Serie 'Documents' du Bureau de Recherches Geologiques et Minières* **5** 1–186.

Foucher, J.-C. (1979). Distribution stratigraphique des kystes de dinoflagelles et des acritarches dans la Crétacé supérieur du Bassin de Paris et de l'Europe Septentrionale. *Palaeontographica B* **169** 78–105.

Foucher, J.-C. (1981). Kystes de Dinoflagelles du Cretace Moyen Européen: Proposition d'une Echelle Biostratigraphique pour de Domain Nord-Occidental. *Cretaceous Research* **2** 331–338.

Gawor-Biedowa, E. (1972). The Albian, Cenomanian and Turonian Foraminifers of Poland and their stratigraphic importance. *Acta Palaeontologica Polonica* **17** 3–151.

Gravesen, P., Rolle, F. & Surlyk, F. (1982). Lithostratigraphy and sedimentary evolution of the Triassic, Jurassic and Lower Cretaceous of Bornholm, Denmark. *Danmarks Geologiske Undersogelse* **137** 1–51.

Hancock, J. M. (1976). The petrology of the Chalk. *Proceedings of the Geologists' Association* **86** 499–536.

Hancock, J. M. & Kauffman, E. G. (1979). The great transgressions of the Late Cretaceous. *Journal of the Geological Society, London* **136** 175–186.

Hart, M. B. (1979). Biostratigraphy and palaeozoology of planktonic Foraminifera from the Cenomanian of Bornholm, Denmark. *Newsletters in Stratigraphy* **8** 83–96.

Hart, M. B. (1980). The recognition of Mid-Cretaceous sea-level changes by means of foraminifera. *Cretaceous Research* **1** 289–297.

Hart, M. B. & Bailey, H. W. (1979). The distribution of Planktonic Foraminiferida in the Mid-Cretaceous of

NW Europe. *Aspekte der Kreide Europas, IUGS Series A* **6** 527–542.

Hart, M. B., Bailey, H. W., Fletcher, B., Price, R. & Swiecicki, A. (1981). Cretaceous. In: Jenkins, D. G. & Murray, J. W. (eds), *Stratigraphical atlas of fossil Foraminifera.* Ellis Horwood, Chichester, pp. 149–227.

Kennedy, W. J., Hancock, J. M. & Christensen, W. K. (1981). Albian and Cenomanian ammonites from the Island of Bornholm (Denmark). *Bulletin of the Geological Survey of Denmark* **29** 203–244.

Koch, W. (1977). Biostratigraphie in der Oberkreide und Taxonomie von Foraminiferen. *Geologisches Jarbuch A* **38** 1–128.

Lucas-Clark, J. (1984). Morphology of species of *Litosphaeridium* (Cretaceous, Dinophyceae). *Palynology* **8** 165–193.

Neo-Nygaard, N. & Surlyk, F. (1985). Mound bedding in a sponge-rich Coniacian Chalk, Bornholm, Denmark. *Bulletin of the Geological Survey of Denmark* **34** 237–249.

Norling, E. (1981). Upper Jurassic and Lower Cretaceous geology of Sweden. *Geologiska Foreningens i Stockholm Forhandlingar* **103** 253–269.

Solakius, N. & Larsson, K. (1986). Foraminifera and biostratigraphy of the Arnager Limestone, Bornholm, Denmark. *Danmarks Geologiske Undersogelse Series C* **5** 1–41.

Surlyk, F. (1980). Denmark. In: *Geology of the European Countries, Denmark, Finland, Iceland, Norway, Sweden. In association with 26th International Geological Congress, Paris.* CNFG, Paris.

Tocher, B. A. (1984). Palynostratigraphy of uppermost Albian to basal Coniacian (Cretaceous) sediments of the western Anglo-Paris Basin. *Ph.D. Thesis.* City of London Polytechnic, pp. 1–228.

Verdier, J.-P. (1975). Les kystes de Dinoflagelles de la section de Wissant et leur distribution stratigraphique au Crétacé moyen. *Revue de Micropalaeontologie* **17** 191–197.

11

Oligocene ostracod biofacies from onshore areas of the North Sea Basin

Michael C. Keen

ABSTRACT

Six benthonic ostracod biofacies have been recognized in the onshore Oligocene of the North Sea Basin. Biofacies 1 represents a fauna from a coastal complex; four subfacies can be distinguished, indicating shallow and deep freshwater lakes, brackish lagoons and polyhaline bays. Biofacies 2 is recognized from shallow coastal waters above wave base, predominantly from a high-energy environment with arenaceous deposition depth 0–25 m. Biofacies 3 is from a more protected lower-energy environment, of depth 0–25 m, with more argillaceous deposition. Biofacies 4, 5 and 6 represent intergrading low-energy muddy environments of successively deeper water: shallow infralittoral, 25–75 m; deep infralittoral to circalittoral, 75–150 m; circalittoral and upper bathyal, deeper than 150 m.

INTRODUCTION

Ostracods are a largely benthonic group of organisms, whose distribution is closely related to the type of substrate. They are sensitive to variations in substrate conditions, such as sediment, vegetation, water depth, temperature, and water and sediment chemistry, and therefore make excellent environmental markers. Living marine ostracods frequently show a patchy distribution related to microenvironments, such as the occurrence of areas of shelter provided by plants, bryozoans or epifaunal bivalves; however, on death the empty carapaces tend to become evenly distributed within an environment. Much work has been carried out in recent years on their ecology and palaeoecology, so the many methods used have become standard practice. It is not necessary to describe the techniques used; for general accounts of these and the underlying philosphy see Carbonel (1980), Keen (1971, 1977), Neale (1983) and Whatley (1983). The biofacies described below are believed to represent autochthonous assemblages, members of meiofaunal communities.

The material examined in this study comes mainly from the classical onshore areas of Germany, Belgium, France and the Isle of Wight (Fig. 1). Although some of the biofacies described can be recognized in offshore areas of the North Sea, they have been determined from onshore material.

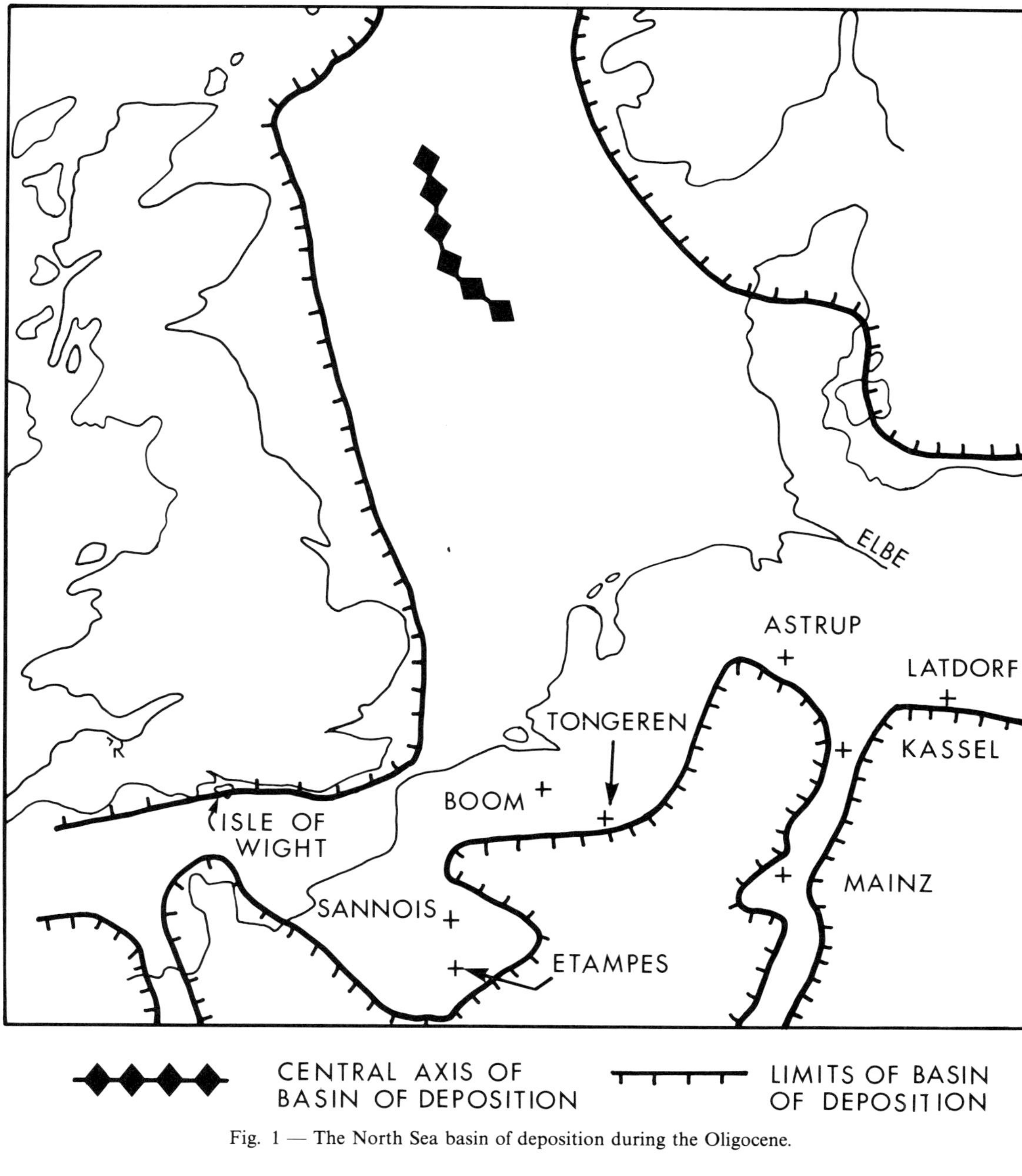

Fig. 1 — The North Sea basin of deposition during the Oligocene.

THE OLIGOCENE OF THE NORTH SEA BASIN

The precise definition of the Oligocene is subject to continuing controversy and, until agreement is reached on stratotypes, especially to define the base, there is little hope that any consensus will be reached. The Oligocene is traditionally divided into three stages: the Lattorfian at the base, Rupelian, and finally Chattian at the top. A recent paper in support of this view is that by Martini *et al.* (1986). Other workers, however, have argued that the Lattorfian should be regarded as Eocene and suggest a twofold division of the epoch (e.g. Cavelier 1975, Cavelier *et al.* 1981, Berggren *et al.* 1985). The problem is not entirely one of definition, because correlation has proved to be difficult. A threefold division can be recognized on the basis of lithology and biofacies in the onshore areas of the North Sea Basin and in some offshore areas. These facies are diachronous and difficult to correlate and provide a dubious basis for stage classification. At the base are either freshwater or brackish deposits of the Sannoisian facies, or shallow marine sands and clays. These are succeeded by the typical Oligocene facies of the North Sea Basin, the thick marine clay sequence of the Boom Clay; more rarely, sandy marine deposits are present as seen in the Paris and Mainz Basins. Finally, the top of the succession witnesses the reappearance of shallow marine sediments, which are normally sands.

The Oligocene of the North Sea Basin is well documented for the onshore areas of France, Belgium, Germany and Denmark, but the offshore areas have not been described in any detail. The most recent published account is in the results of the International Geological Correlation Programme (IGCP) No. 124 (Vinken 1988) where full details can be found. The isopachytes and palaeogeography reflect those of the Palaeogene in general, with the thickest sequences and deepest water located in the Viking and Central Grabens, thinning and shallowing towards the margins of the basin (Fig. 1). The greatest thickness known is 1200 m in the Central Graben; in most offshore areas it is usually between 100 and 500 m thick. Onshore thicknesses are comparable with those of the Southern North Sea, e.g. 100–200 m in the Boom area of Belgium, with over 60 m for the Chattian of Astrup. Rapid subsidence in The Netherlands' Central Graben allowed 500 m of late Oligocene sediments to accumulate near the present Dutch–German border. The marginal deposits show the greatest variation in lateral and vertical facies changes, with more uniform deposits in the centre of the basin. The marginal deposits include lagoonal and shallow marine sediments in Belgium and Germany and around the southern limits of the North Sea Basin, with westerly prograding sedimentary wedges in the north in the Viking Graben, Outer Moray Firth and Central Graben. In general, during the Palaeogene the basinal deposits were argillaceous and the marginal deposits arenaceous. During the Rupelian transgression the deeper-water muddy facies became widespread over Belgium, The Netherlands and Germany.

The biostratigraphy of the North Sea area is complex, with numerous publications dealing with the onshore areas. The most comprehensive scheme for the Tertiary of the North Sea itself is that of King (1982), where 16 planktonic zones and 16 benthonic foraminiferal zones have been described. The Oligocene corresponds to the planktonic zone NSP 9 and the benthonic zones NSB 6 (uppermost part), NSB 7 and NSB 8. Detailed biostratigraphy, with correlation between various fossil zonations have recently been published in the results of IGCP No. 124 (Vinkers 1988).

OSTRACOD BIOFACIES

A great deal of work has been published in recent years on the palaeoecology of Tertiary ostracods, together with the recognition of assemblages or biofacies and their environmental interpretation. Examples dealing with Oligocene faunas can be seen in the works of Keen and Uffenorde cited below. Uffenorde in particular has recognized the presence of recurring ecozones in the Middle Oligocene–Miocene of northern Germany, which he related to depth of water and position of storm wave base (Uffenorde 1979, 1980, 1981, 1986). Many of the species used in defining the ecozones have fairly long stratigraphical ranges with little sign of any evolutionary change. This contrasts with those ostracod species used for biostratigraphical zonation. The recognition of distinct ecozones allows the possibility of correlating shallow-water biostratigraphic zones with those present in deeper water where different species are used for zonation. The biofacies described here are mostly related to water depth which enables them to be used in recognizing transgressive and regressive events. The shallow marine biofacies show provincial differences due to climate and geography. The shores of the Oligocene North Sea (Fig. 1) had several partially isolated areas of deposition such as the Paris and Mainz Basins, which enabled semi-indigenous species to evolve. The more westerly areas had a closer connection with warmer regions to the south so that a 'southern' influence is seen in the Paris Basin and sometimes in Belgium. In contrast with this, northern Germany and much of the North Sea area was situated at the western end of a great boreal sea spreading over Poland and the USSR, giving rise to a strong 'boreal' influence. These differences are discussed under the relevant biofacies. The deeper-water biofacies do not show this provincialism and many species are widespread in Europe and even further afield.

The biofacies are described in a transect from the coastal complex of lakes, bays and lagoons of Biofacies 1, through shallow marine Biofacies 2 and 3 to offshore Biofacies 4, 5 and 6. The two shallow marine biofacies are unlikely to have occurred together because the arenaceous deposits of Biofacies 2 indicates that high-energy conditions were present at times, while the argillaceous deposits of Biofacies 3 indicate quieter deposition in more sheltered environments. Both were present in shallow waters above the storm wave base, but only Biofacies 2 suffered frequent agitation.

Four intergrading biofacies can be recognized for muddy low-energy environments (Biofacies 3–6 (see Figs. 4 and 5)). The fauna shows a gradual change, with no sharp breaks. The biofacies are distinguished on the basis of diversity and on appearance or disappearance of diagnostic species. The thick clays of these biofacies form the typical Oligocene sediment of the North Sea Basin, but they are far from being monotonous in character. They may be more, or less, silty; some layers are rich in pyrite, others in organic matter and yet others in carbonate; septarian nodules occur and their horizons may be used for correlation. Macrofossils are only abundant in the shallow water facies, although *Nucula* is commonly present. Trace fossils such as *Thalassanoides* and small worm tubes are present. Planktonic foraminifera became common during the deposition of Biofacies 4–6, reflecting open contact with the Atlantic Ocean. Some horizons are barren of calcareous fossils; this may reflect original conditions but is probably more usually caused by post-mortem dissolution.

The biofacies described below have been recognized on a semiquantitative basis. They are distinguished by total composition, diversity, abundance, and presence or absence of key species (see Figs. 3–5). The samples examined range from closely spaced and accurately logged collections, to single-sample locality collections and localized material supplied by numerous colleagues over the past 22 years. The sample size is extremely variable; richly fossiliferous Meeressand requires much less sample than the Boom Clay, the latter frequently requiring several kilograms of sediment to yield a few specimens. North Sea material is mainly of single unlocalized specimens. Representative species of the biofacies are illustrated in Plates 1 and 2, and lists, by no means exhaustive, of characteristic species are given in Table 1. Specimens with catalogue numbers are deposited either in the British Museum (Natural History) (Io) or with the British Geological Survey, Keyworth (MPK); uncatalogued specimens are at present in the author's collection. The general interpretation of the biofacies is shown in Fig. 2.

COASTAL PLAIN BIOFACIES

Biofacies 1

This indicates a coastal complex of bays, lagoons, rivers, lakes and marshes. The ostracods reflect these varying environments where salinity is the most obvious controlling factor. In this study they are regarded as a single biofacies which can be subdivided into four subfacies; these have been

Table 1 — Representative species of the Ostracod biofacies.

Biofacies 1
1a: shallow freswhater
Sannoisian
Candona fertilis Triebel 1963, *Cypridopsis soyeri* (Margerie 1961), *Ilyocypris boehli* Triebel 1941, *Virgatocypris tenuistriata* (Dollfus 1877)

1b: deeper freshwater
Sannoisian
Moenocypris forbesi (Jones 1856);
Chattian
M. francofurtana Triebel 1959

1c: brackish water
Sannoisian
Cladarocythere apostolescui (Margerie 1961), *Hemicyprideis montosa* (Jones & Sherborn 1889), *Neocyprideis williamsoniana* (Bosquet 1852)
Chattian
Hemicyprideis clivosa Malz 1978, *H. rhenana* (Lienenklaus 1905)

1d: polyhaline
Sannoisian
Cyamocytheridea punctatella producta (Margerie 1961), *Cytheridea pernota* Oertli & Keij 1955, *Cytheromorpha zinndorfi* (Liennenklaus 1905), *Hemicyprideis elongata* Keen 1972, *Loxoconcha nystiana* (Bosquet 1852)

Biofacies 2
Rupelian
Cyamocytheridea punctatella punctatella (Bosquet 1852) (P), *Cytherelloidea jonesiana* (Bosquet 1852) (P), *Cytheretta tenuistriata* (Reuss 1853) (P, M), *C. tenuipunctata* (Bosquet 1852) (P, B, M), *C. minor* (Lienenklaus 1905) (P. M), *C. rhenana* (Triebel 1952) (P, M), *Cytheridea pernota* (P, B), *Hammatocythere hebertiana* (Bosquet 1852) (P, B), *Leguminocythereis scrobiculata* (von Muenster 1830) (P, B, M), *Pokornyella limbata* (Bosquet 1852) (P), *Pterygocythereis fimbriata fimbriata* (von Muenster 1830) (P, B, M), *Quadracythere macropora* (Bosquet 1852) (P, B, M)
Chattian
Cytherella beyrichi (Reuss 1851), *Cytherella semipunctata* (Bornemann 1855), *Cytheridea muelleri* (von Muenster 1830), *C. pernota* Oertli & Keij 1955, *Echinocythereis hispida* (Speyer 1863), *E. scabra* (von. Muenster 1830), *Eucytheridea fissodentata* (Lienenklaus 1894), *Flexus plicatus* (von Muenster 1830), *Leguminocythereis scrobiculata* (von Muenster 1830), *Muellerina latimarginata* (Speyer 1863), *Quadracythere confluens* (Reuss 1856), *Schuleridea* gp. *perforata* (Roemer 1838)

Biofacies 3
Rupelian
Cyamocytheridea punctatella punctatella (P), *Cytheretta tenuipunctata* (Bosquet 1852) (P, B, E), *Cytheridea pernota* (P, B, E), *Echinocythereis? hamsteadensis* Keen 1972 (P, B, E), *Eocytheropteron plicatoreticulatum* Margerie 1963 (P), *Eucytheridea lienenklausi* (Kuiper 1918) (B), *Hammatocythere hebertiana* (P, B, E), *Loxoconcha kuiperi* Keij 1957 (B), *L. nystiana* (P, B, E), *Paracypris bouldnorensis* Keen 1978 (P, B), *Pokornyella limbata* (P, B), *Pterygocythereis fimbriata fimbriata* (B)
Chattian
Cytheridea pernota, Echinocythereis hispida, Eucytheridea fissodentata

Biofacies 4

Cytherella beyrichi, C. compressa, Cytheretta semipunctata, Cytheridea pernota, Echinocythereis hispida, Eucytheridea rostrata Bassiouni 1962, *Paracyprideis rarefistulosa* (Lienenklaus 1905), *Pterygocythereis fimbriata fimbriata, Trachyleberidea hollandica* (Kuiper 1918)

Biofacies 5

Cytherella compressa, C. transversa Speyer 1863, *Cytheropteron tricorne* (Bornemann 1855), *Henryhowella asperrima* (Reuss 1850), *Krithe pernoides* (Bornemann 1855), *Pterygocythereis fimbriata fimbriata, Trachyleberidea hollandica*

For Biofacies 2 and 3: P, Paris Basin; M, Mainz Basin; B, Belgium; E, Isle of Wight.

considered in detail in Keen (1971, 1972, 1975, 1977). Two freshwater subfacies can be recognized; one (Biofacies 1a) representing shallow lakes or lake edges, with a diverse fauna including *Cypridopsis, Candona* and *Virgatocypris*; the other (Biofacies 1b) is much less diverse, characterized by *Moenocypris*, and representing deep-water lakes. The genus *Ilyocypris*, which became widespread during the Oligocene, can be present in either subfacies. Typical brackish-water oligohaline and mesohaline environments are characterized by the third subfacies (Biofacies 1c), with often monospecific assemblages of abundant *Hemicyprideis*, and less frequent *Cladarocythere* and *Neocyprideis*. The fourth subfacies (Biofacies 1d) indicates polyhaline conditions of open bays with *Cyamocytheridea, Cytheromorpha, Cytheridea, Hemicyprideis* and *Loxoconcha*. This polyhaline subfacies grades into the more brackish-water subfacies on the one hand and into the

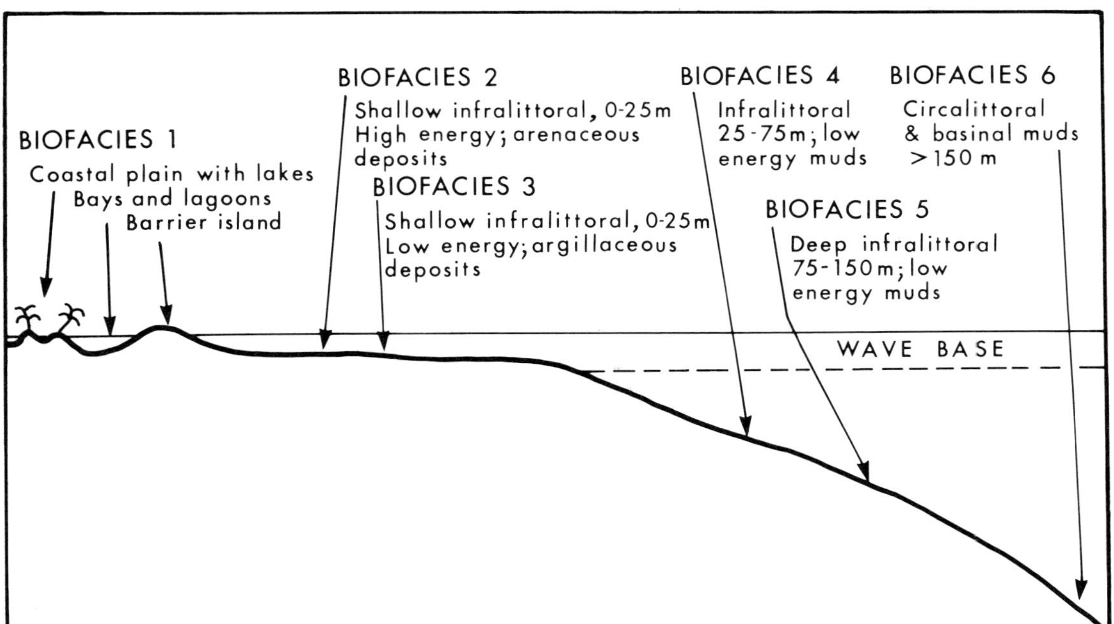

Fig. 2 — Oligocene ostracod biofacies of the North Sea Basin.

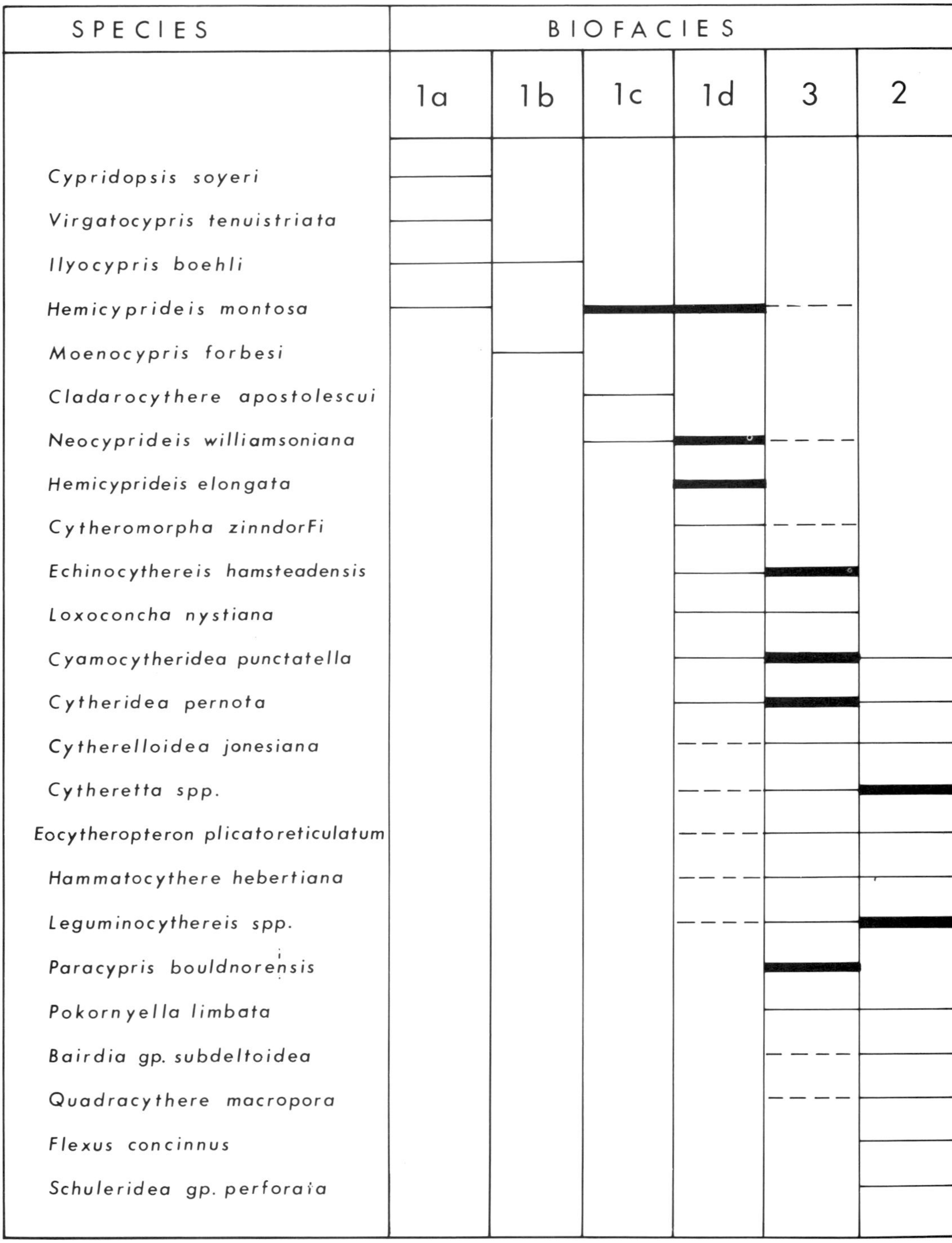

Fig. 3 — The range of typical species between the ostracod biofacies of the Rupelian of the Paris Basin and Isle of Wight. See Fig. 4 for key.

shallower marine facies on the other hand. It is distinguished from the former by its much greater diversity and the presence of the more typically marine genera, and from the latter by the presence of substantial numbers of typical brackish-water genera (Fig. 3). *Hemicyprideis, Neocyprideis* and *Loxoconcha* develop phenotypic characters related to reduced salinity; the first two become tuberculate, and the third loses its reticulate ornamentation (Keen 1982).

The sediments associated with this biofacies are mainly argillaceous, often sandy clays, with some freshwater limestones. A rich molluscan fauna of gastropods and bivalves is usually present. The constantly shifting geography gave rise to complex lateral and vertical facies changes. The various subfacies can only be revealed by accurate and closely spaced sampling; coarse sampling is likely to lead to mixing. The brackish-water species can be recognized in small samples, although they may be mixed with species from other biofacies. The presence of these species at least suggests the proximity of Biofacies 1 which is characteristic of the Sannoisian of the Isle of Wight, Paris Basin, Belgium, the Rhine Graben, the Latdorfian of the Mainz Basin, the Hessian Depression (Keen 1982, Sonne 1982, Ritzkowski 1986), and the late Chattian of the Mainz Basin.

SHALLOW MARINE BIOFACIES

The sediments and faunas formed in the shallow marine environment are highly variable and present many difficulties in biofacies determination. The two biofacies described here are very generalized and not always as distinct as this report may suggest. While it is easy to generalize on the sedimentology of a particular locality, in detail both sandy and argillaceous deposits are frequently present. These may alternate, or the sediment may be homogeneous but consist of a sandy–clay mixture, or they may be very fine sands. Shell beds range from thin simple low-diversity beds to thick complex beds indicative of several different environments of accumulation; life position, varying amounts of post-mortem movement, and winnowing are evident. The designation of a particular sample to either biofacies is essentially determined by the faunal composition; this may result in some sandy samples being placed into Biofacies 3, which is usually argillaceous. The second problem is that of provincialism, with considerable faunal differences between the Paris Basin on the one hand and Belgium and Germany on the other hand. The distribution of some characteristic species of the Paris Basin and Isle of Wight are shown in Fig. 3, where the differences and similarities between Biofacies 1d, 2 and 3 can be seen.

Several species are present in all the shallower marine biofacies, e.g. *Cytherella beyrichi, Cytheridea pernota* and *Pterygocythereis fimbriata* (Figs. 3–5). It might be assumed that they were independent of the sediment forming the substrate. However, this is unlikely when living members of these genera are considered. These generally prefer a substrate of mud or sandy mud, and these Oligocene species are all commoner in the muddy sediments of Biofacies 3 and 4 compared with the sandy deposits of Biofacies 2. *Cytheridea* and related genera are typical of sheltered coastal waters, which agrees well with its maximum abundance in Biofacies 3. *Cytherella* species are also characteristic of fine-grained substrates and, if the living species *C. abyssorum* Sars is taken as a modern analogue for *C. beyrichi*, then the latter would have been adapted for living on offshore sandy and clayey muds (Uffenorde 1988, pers. comm.). Living species of *Pterygocythereis* avoid agitated environments and are typical of stable sandy or clayey muds below storm wave base, which supports the interpretation of Biofacies 4 and 5, where it is most abundant. Therefore the presence of these species in coarse-grained sediments is either because they are allochthonous or because they were present during calmer periods, perhaps living on thin layers of mud or organic detritus; these thin layers, together with their microfauna, would have become mixed in with the more normal sediment during subsequent more agitated conditons giving time-averaged microfaunas. These species were unlikely to have been

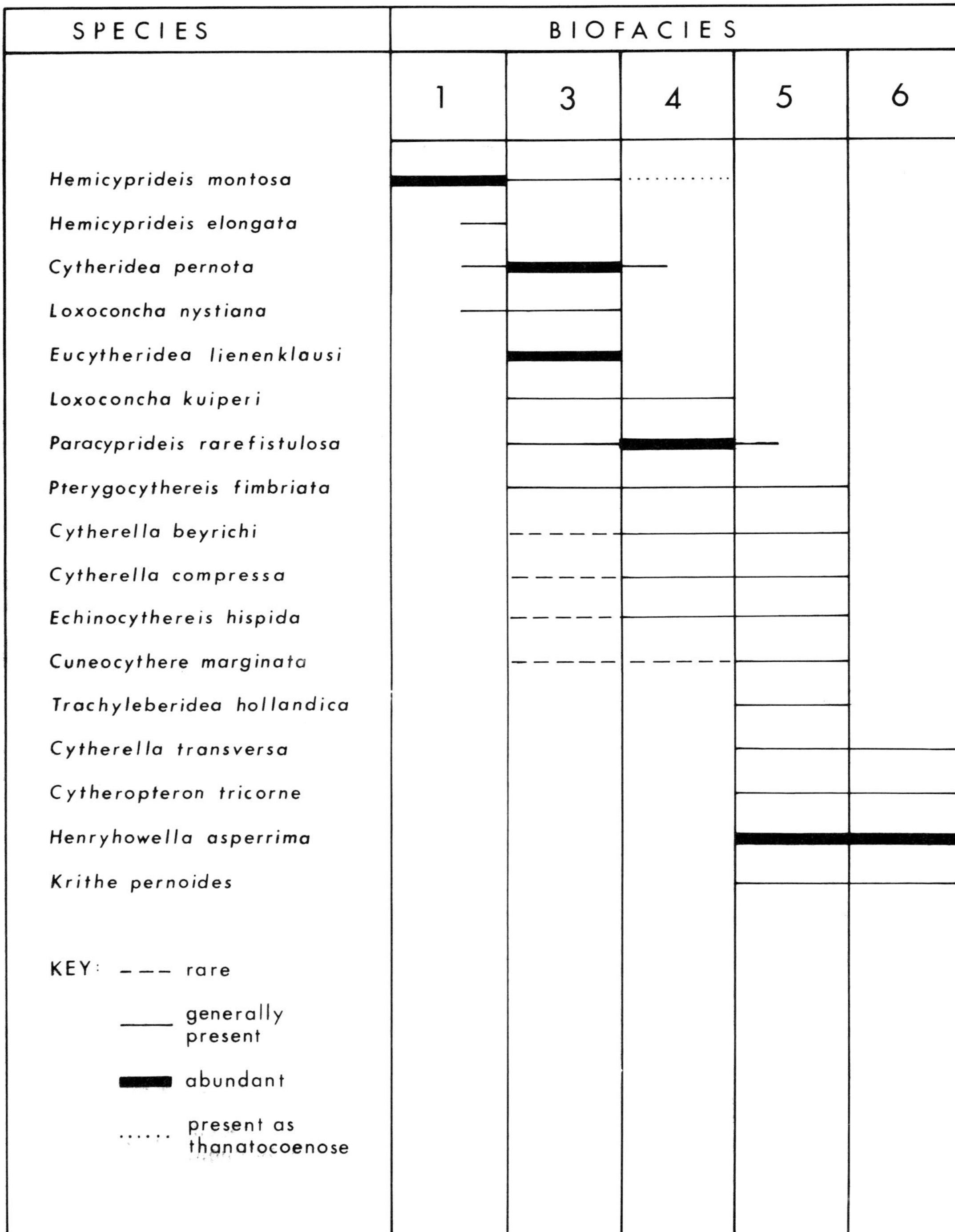

Fig. 4—The range of typical species between the ostracod biofacies of the Rupelian of Belgium and northern Germany.

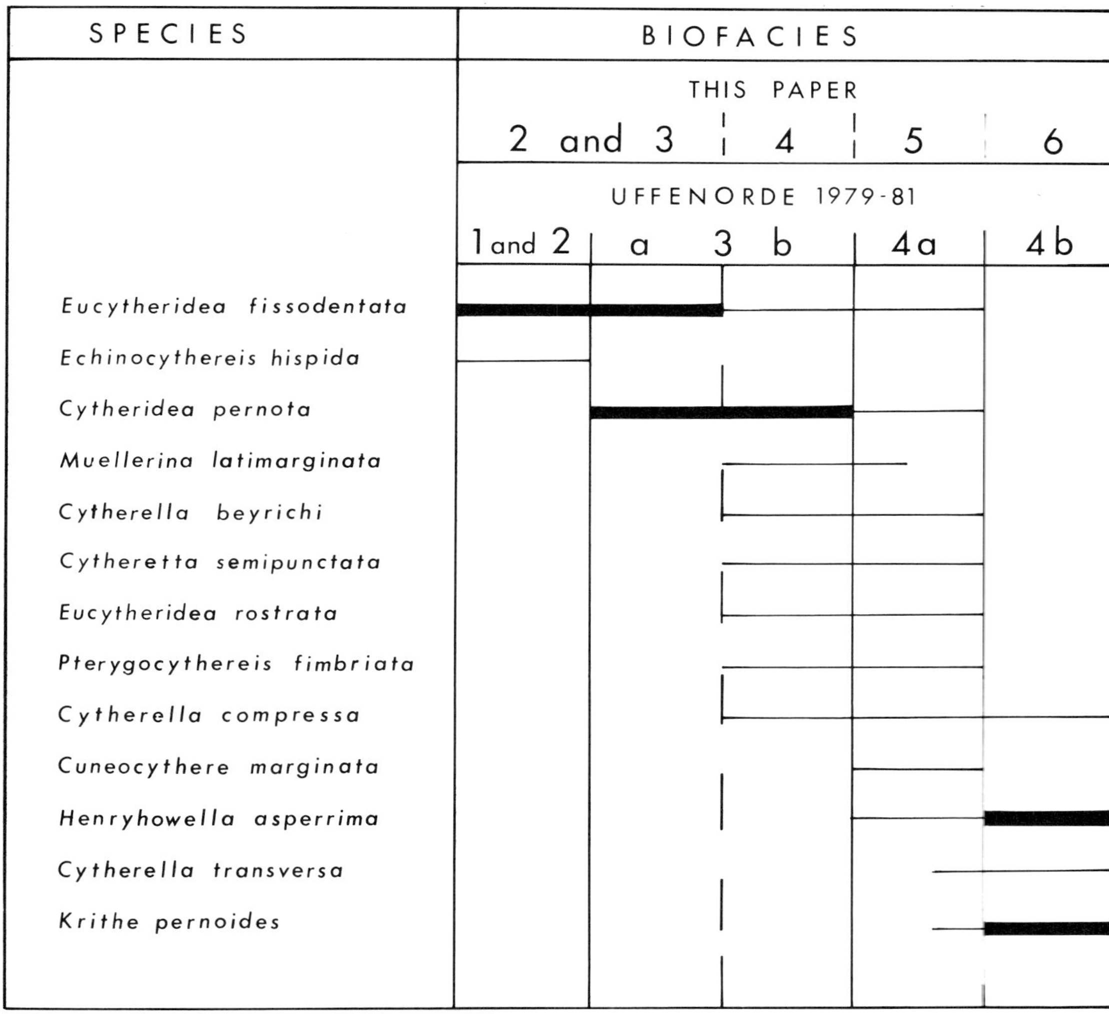

Fig. 5 — The range of typical species between the ostracod biofacies of the Upper Oligocene of northern Germany. See Fig. 4 for key.

phytal and thus independent of the sediment, by comparison with living relatives, and also because macroalgae are rare on muddy bottoms.

Biofacies 2

This is represented by shallow marine sands deposited above storm wave base where conditions were agitated at times. The sediments were frequently reworked in these high-energy conditions. However, the presence of well-preserved macrofossils in various types of shell bed suggests a variable regime, with quiet interludes allowing their accumulation. The microfauna is frequently well preserved, giving support to this interpretation. Evidence of periods of emersion can be seen in the Paris Basin where it is also likely that sea grass meadows flourished in the shallow marine waters. This biofacies is seen in the Middle

Oligocene Stampian sands of the Paris Basin, the Rupelian Unterer Meeressand of the Mainz Basin, parts of the Chattian Kasseler Meeressand around Kassel, and in the sandy sediments of the Outer Moray Firth. The richest and most diverse ostracod faunas are found in this biofacies, samples often yielding 50–70 species. The common genera are *Cyamocytheridea*, *Cytheretta*, *Cytheridea*, *Leguminocythereis* and *Quadracythere*. Provincial differences are best seen in this biofacies. Northern Europe lay in the so-called 'Boreal Province' during the Tertiary; this refers to an area of relatively cold water compared with Tethys, but with temperatures comparable with the modern-day Mediterranean. A warmer southerly influence is indicated by the presence of *Hammatocythere* and *Pokornyella*, which had their centres of development in Aquitaine and are commonly found in the Paris Basin but are rare elsewhere. *Cytherelloidea* is a well-known warm-water genus and is rare in the Oligocene of the North Sea Basin outside the Paris Basin during the Oligocene. *Eucytheridea* and the species referred to *Leguminocythereis* have a well-defined boreal distribution in the Tertiary of Europe. *Cytheridea*, although widespread throughout the North Sea area, was also most abundant in the boreal areas during the Tertiary. Some of the genera of Biofacies 2 have robust shells which could possibly indicate adaptation to more agitated conditions, e.g. *Leguminocythereis*, *Quadracythere* and some *Cytheretta* species. Provincialism was also controlled by the geography; the semi-isolation of various regions allowed the evolution of distinct species or subspecies. This is well illustrated in the case of *Cytheretta tenuipunctata* and *Cytheretta tenuistriata* which show distinct geographical subspecies (Keen 1982).

Biofacies 3

This was deposited in relatively quiet and protected near-shore areas with a water depth of 0–25 m. Typical deposits are the Marnes à Huîtres of the Paris Basin and the Argiles à *Nucula comta* of Belgium. These both indicate a strong influence from nearby polyhaline lagoonal areas and were presumably deposited in fully marine open bays. The sediment is a fossiliferous clay, sandy clay or fine sand, often quite variable when seen in exposures. Uffenorde's Ecozones 1 and 2, otherwise described as the *Eucytheridea fissodentata* Ecozone, is also included here, although Uffenorde (1979, 1981, 1986) considered this is to be subject to storm wave action; the sediments consist of silty sands, silts and clays, with a high-diversity molluscan fauna. The characteristic species of the Rupelian of Belgium and of the Upper Oligocene of northern Germany are shown in Figs. 4 and 5. Typical genera in the Paris Basin are *Cyamocytheridea*, *Cytheretta*, *Cytheridea*, *Hammatocythere* and *Paracypris*, and in Belgium *Eucytheridea* and *Cytheridea*. In northern Germany, Uffenorde recognizes a shallower Ecozone 1 with a low-diversity ostracod fauna with *Eucytheridea*, and an Ecozone 2 with similar but more abundant ostracods; these two ecozones are better distinguished in Miocene sediments.

OFFSHORE BIOFACIES

Biofacies 4

This represents the accumulation of muds in shallower infralittoral depths (25–75 m) in quiet conditions generally below storm wave base. It has a diverse fauna; ostracods may be abundant, are always well preserved and have common moult stages preserved. The biofacies is characteristic of some of the Boom Clay of Belgium and The Netherlands, the Rupelton of the Mainz Basin, and some of the Rupelian of northern Germany. Characteristic genera are *Cytherella*, *Eucytheridea*, *Loxoconcha*, *Paracyprideis* and *Pterygocythereis*. *Cytheridea pernota* and *Cytheretta* spp. may be common in the shallower water parts of this biofacies, with *Trachyleberidea* becoming common in the deeper parts. This corresponds to Uffenorde's Ecozone 3b and in the late Oligocene Chattian stage includes the deposits of Astrup and some of the Kasseler Meeressand.

Biofacies 5

This is distinguished from Biofacies 4 in having a lower abundance of ostracods, in losing the shallower-water species of Biofacies 4, and in the appearance of *Henryhowella* and *Krithe*. It is interpreted as deep infralittoral to circalittoral (75–150 m) and is present in the Boom Clay of Belgium and the Rupelian clays of northern Germany and Denmark. It corresponds to Uffenorde's Ecozone 4a.

Biofacies 6

This is recognized by the disappearance of the shallower-water species of Biofacies 4 and 5, leaving a much lower-diversity fauna of *Argilloecia*, *Cytheropteron*, *Henryhowella* and *Krithe*. The biofacies is recorded in the Rupelian of northern Germany and Denmark and the offshore Oligocene of the southern North Sea but has not been recognized so far in Belgium. No truly psychrospheric ostracods have been reported; so it is interpreted as outer shelf, circalittoral to shallow bathyal (Uffenorde 1986). Uffenorde (1980) described this fauna from the Wursterheide Borehole of northwest Germany, regarding it as an outer shelf to slope fauna which spread over large areas of the southern North Sea Basin during the course of the Rupelian transgression. He analysed the morphology of *Krithe pernoides*, concluding that the bottom waters were well oxygenated in the Rupel 4, becoming less oxygenated at the close of the Rupelian.

DISCUSSION

The Oligocene biofacies described above allow a unique traverse from coastal plain to deep basin for the onshore Palaeogene rocks of northern Europe. Several of the biofacies can be recognized in other parts of Europe, but the facies changes associated with the Rupelian transgression give a more coherent picture than at other times. The closest comparison with these biofacies from outside the area is the series of biofacies described from the Oligocene of Aquitaine by Ducasse (1974). She recognized three ostracod biofacies spanning coastal to bathyal environments; her Littoral Facies A corresponds to Biofacies 2 and 3, Littoral Facies B to Biofacies 4, and her Bathyal Facies to Biofacies 5 and 6. Freshwater ostracod faunas of Biofacies 1a are known from the Palaeocene and Eocene of the Paris Basin (e.g. Guernet 1981), the late Eocene of the Isle of Wight (Keen 1975) and the Miocene of Germany (e.g. Malz & Moayedpour 1973). Biofacies 1b is only known from the late Eocene and Oligocene of the Isle of Wight and Oligo-Miocene of Germany. Brackish-water faunas have been extensively described and are well developed in the Palaeogene of southern England, northern France, the Paris Basin, the Pyrenees and as far afield as Turkey. *Neocyprideis* is ubiquitous in these deposits and ranges throughout the Palaeogene; relatives of *Hemicyprideis*, such as *Vetustocytheridea*, are other important constituents of this biofacies in the Palaeocene. Brackish-water faunas appear sporadically throughout the Eocene of the Paris Basin but are best seen in the late Eocene deposits of the Isle of Wight (Keen 1977). The shallow marine faunas dominated by genera such as *Cyamocytheridea*, *Cytheretta*, *Cytheridea*, *Leguminocythereis*, *Loxoconcha*, *Pokornyella*, *Pterygocythereis* and *Quadracythere* are typical of most of the onshore Palaeogene and are widespread throughout the Palaeogene of the northern hemisphere from Kazakstan in the USSR through Europe to the Gulf coast of the USA. The deeper-water Biofacies 5 and 6 are rarely developed in onshore areas of northern Europe except during the Rupelian. In this region, basinal facies are only seen outside the Oligocene in the Ypresian, where equivalents of Biofacies 5 are present in northern Germany and the eastern part of the London Basin. *Henryhowella* and associated *Krithe* species are recorded from many localities throughout the world in Oligocene–Pliocene sediments. *Henryhowella* is usually indicative of deeper marine waters, and blind members of the *H. asperrima* group are important members of the shallower part of the psychrosphere, being recorded from many Deep Sea Drilling Project sites (e.g. Benson & Peypouquet 1983). *Krithe* is probably the most characteristic bathyal–abyssal genus of ostracods.

ACKNOWLEDGEMENTS

This paper was extensively reviewed by Henning Uffenorde and I should like to acknowledge his valuable help in preparing it.

Plate 1 — **Biofacies 1a** (Figs. 1 and 2), **Biofacies 1b** (Fig. 3), **Biofacies 1c** and **1d** (Figs. 4–9) and **Biofacies 2** (Figs. 10–18).

Fig. 1 — *Cypridopsis soyeri* (Margerie 1961), left valve, MPK 1575, length 0.49 mm, × 85, Rupelian, Middle Hamstead Beds, Bouldnor Cliff, Isle of Wight.

Fig. 2 — *Virgatocypris edwardsi* Keen 1977, right valve, Io 6741, length 1.16 mm, × 35, Bartonian, Upper Headon Beds, Colwell Bay, Isle of Wight.

Fig. 3 — *Moencypris reidi* Keen 1972, carapace left view, MPK 1577, length 1.29 mm, × 32, Bartonian, Upper Headon Beds, Headon Hill, Isle of Wight.

Fig. 4 — *Cladarocythere apostolescui* (Margeri 1961), female right valve, Io 3709, length 0.70 mm, × 60, Bartonian, Bembridge Limestone, Bouldnor Cliff, Isle of Wight.

Fig. 5 — *Hemicyprideis montosa* (Jones & Sherborn 1889), female right valve, Io 3724, length 0.70 mm, × 55, Sannoisian, Glaises à Cyrènes, Montfermeil, Paris Basin.

Fig. 6 — *Neocyprideis williamsoniana* (Bosquet 1852), right valve, Io 3810, length 0.75 mm, × 63, Sannoisian, Upper Hamstead Beds, Bouldnor Cliff, Isle of Wight.

Fig. 7 — *Loxoconcha nystiana* (Bosquet 1852), female left valve, Io 3787, length 0.71 mm, × 60, Rupelian, Sables de Berg, Bilzen Belgium.

Fig. 8 — *Hemicyprideis elongate* Keen 1972, female right valve, Io 3745, length 0.66 mm, × 55, Sannoisian, Upper Hamstead Beds, Bouldnor Cliff, Isle of Wight.

Fig. 9 — *Cytheromorpha zinndorfi* (Leinenklaus 1905), female right valve, Io 3789, length 0.34 mm, × 117, Sannoisian, Upper Hamstead Beds, Bouldnor Cliff, Isle of Wight.

Fig. 10 — *Pokornyella limbata* (Bosquet 1852), female left valve, length 0.62 mm, × 68, Rupelian, Marnes à Huitres, Cormeilles, Paris Basin.

Fig. 11 — *Quadracythere macropora macropora* (Bosquet 1852), female left valve, length 0.79 mm, × 52, Stampian, Falun de Jeurre, Auvers St Georges, Paris Basin.

Fig. 12 — *Cyamocytheridea punctatella punctatella* (Bosquet 1852), female left valve, length 0.69 mm, × 52, Stampian, Falun de Jeurre, Auvers St Georges, Paris Basin.

Fig. 13 — *Cytherelloidea jonesiana jonesiana* (Bosquet 1852), female right valve, Io 3663, length 0.77 mm, × 58, Stampian, Falun de Jeurre, Auvers St Georges, Paris Basin.

Fig. 14 — *Hammatocythere hebertiana hebertiana* (Bosquet 1852), female right valve, length 0.77, × 50 Rupelian, Marnes a Huitres, Cormeilles, Paris Basin.

Fig. 15 — *Cytheretta tenuipunctata* (Bosquet 1852) *absoluta* Keen 1972, female left valve, Io 3884, length 0.84 mm, × 53, Stampian, Marnes a Huitres, Cormeilles, Paris Basin.

Fig. 16 — *Echinocythereis hispida* (Speyer 1863), female left valve, length 1.05 mm, × 40, Rupelian, Boom Clay, St Niklaas, Belgium.

Fig. 17 — *Leguminocythereis scrobiculata* (von Munster 1830), female right valve, length 1.31 mm, × 33, Chattian, Astrup, Germany.

Fig. 18 — *Muellerina latimarginata latimarginata* (Speyer 1863), female left valve, length 0.82 mm, × 50, Chattian, Voort Formation, Terhaegan, Belgium.

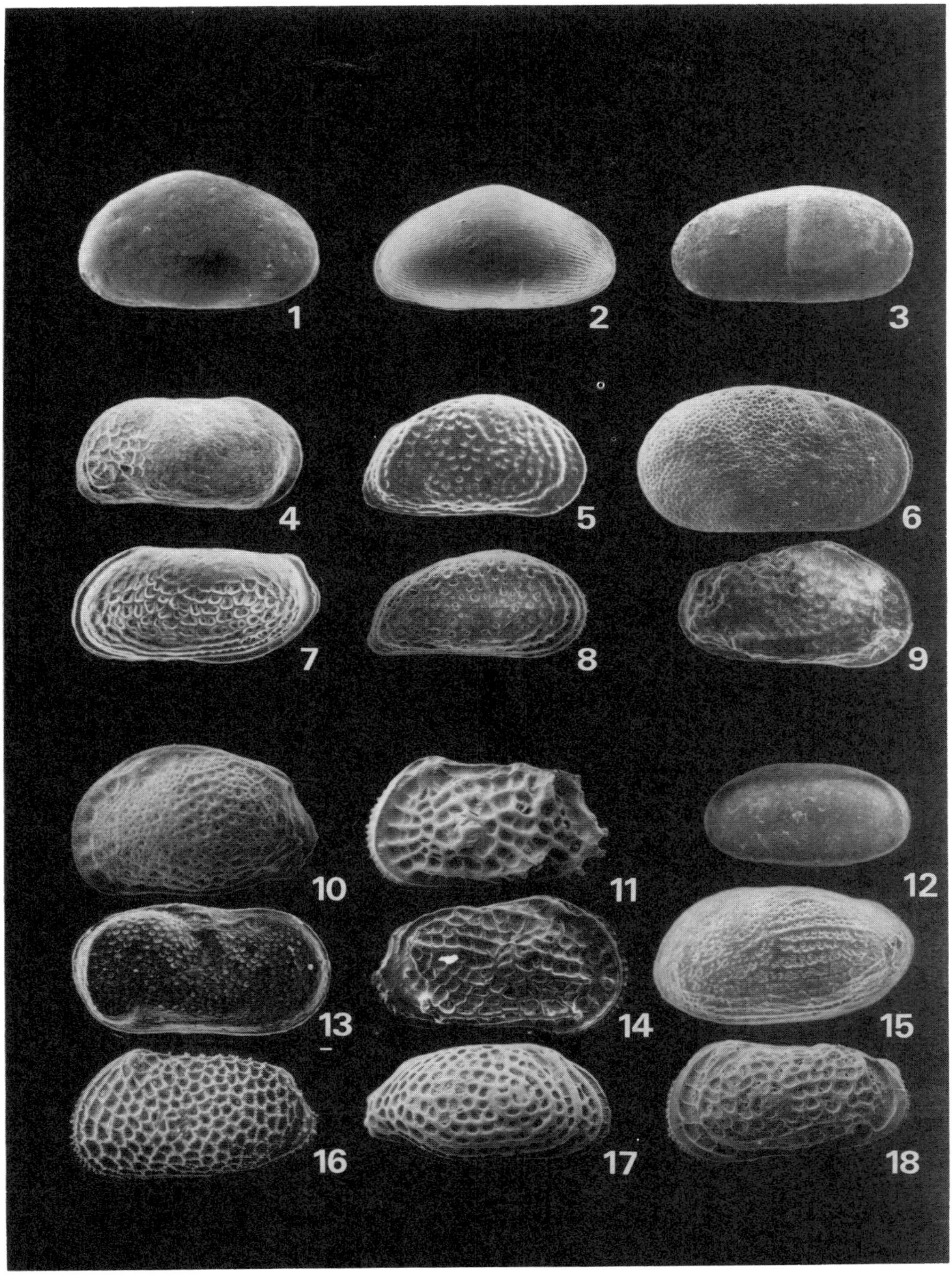
1
2
3
4
5
6
7
8
9
10
11
12
13
14
15
16
17
18

Plate 2 — Biofacies 3 and **4** (Fig. 1–5) and **Biofacies 5** and **6** (Figs. 6–11). Figs. 6–10 are from the Rupelian Boom Clay of St Niklass, Belgium.

Fig. 1 — *Cytheridea pernota* Oertli & Keij 1955, female right valve, length 0.77 mm, × 60, Rupelian, Argile à *N. comta*, Bilzen, Belgium.

Fig. 2 — *Eucytheridea rostrata* Bassiouni 1962, female right valve, length 0.95 mm, × 45, Chattian Voort Formation, Terhaegan, Belgium.

Fig. 3 — *Cytheretta semipunctata* (Bornemann 1855), female left valve, length 0.85 mm, × 50, Chattian, Kasseler Meeresand, Kassel, Germany.

Fig. 4 — *Paracyprideis rarefistulosa* (Lienenklaus 1905), female left valve, length 0.87 mm, × 50, Rupelian Rupelton, Wollstein, Mainz Basin.

Fig. 5 — *Pterygocythereis fimbriata fimbriata* (von Munster 1830), female left valve, length 1.00 mm, × 43, Rupelian Rupelton, Neumuhle bei Weinheim, Mainz Basin.

Fig. 6 — *Cytherella beyrichi* (Reuss 1851), left valve, length 0.82 mm, × 50.

Fig. 7 — *Cytherella transversa* Speyer 1863, left valve, length 0.87 mm, × 48.

Fig. 8 — *Trachyleberidea hollandica* Kuiper (1918), female left valve, length 0.79 mm, × 55.

Fig. 9 — *Krithe pernoides* (Bornemann 1855), male left valve, length 0.71 mm, × 60.

Fig. 10 — *Henryhowella asperrima* (Reuss 1850), female right valve, length 0.82 mm, × 50.

Fig. 11 — *Cytheropteron tricorne* (Bornemann 1855), right valve, length 0.44 mm, × 98, Rupelian, Boom Clay, Kruibeke, Belgium.

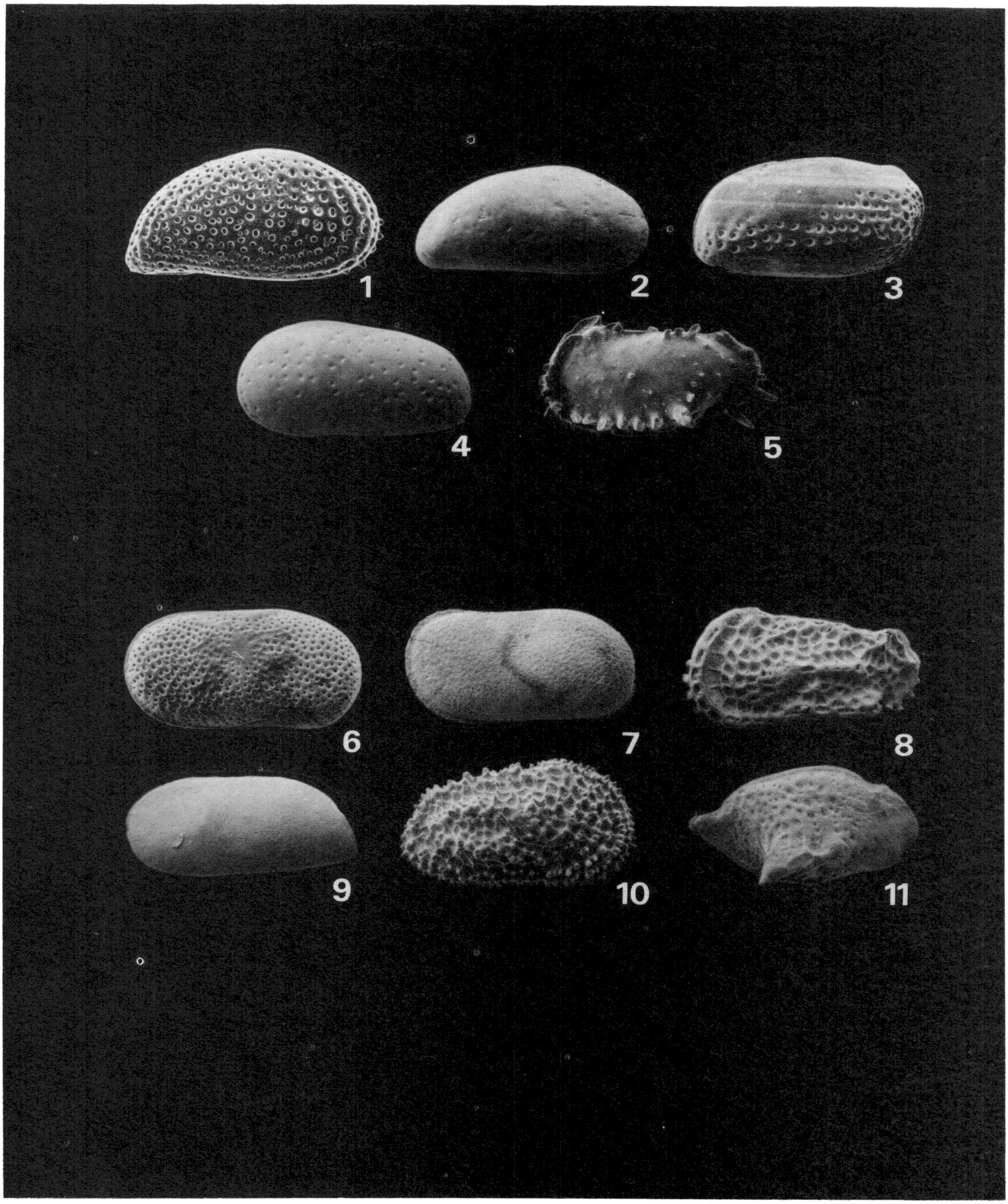
1
2
3
4
5
6
7
8
9
10
11

REFERENCES

Benson, R. H. & Peypouquet, J. P. (1983). The upper and mid-bathyal Cenozoic ostracode fauna of the Rio Grande rise found on leg 72 DSDP. *Initial Reports of the Deep Sea Drilling Project* **72** 805–818.

Berggren, W. A., Kent, D. V., Flynn, J. J. & Ven Convering, J. A. (1985). Cenozoic geochronology. *Geological Society of America Bulletin* **96** 1407–1418.

Carbonel, P. (1980). Les ostracodes et leur interet dans la definition des ecosystemes estuariens et de plateforme continentale. Essais d'application à des domains anciens. *Memoires de l'Institut de Geologie du Bassin d'Aquitaine* **11** 1–350.

Carbonnel, G. & Jacobzone, M. (1977). L'ornamentation chez les Loxoconchidae: un moyen de prediction (paleo)ecologique. In: Loffler, H. & Danielopol, D. (eds), *Aspects of ecology and zoogeography of Recent and fossil Ostracoda.* Junk Publications, The Hague, pp. 121–132.

Cavalier, C. (1975). La diachronisme de la zone a *Ericsonia subdisticha* (Nannoplankton) et la position de la limite Eocene–Oligocene en Europe et en Amerique du Nord. *Bulletin du Bureau de Recherches Geologiques et Minières (France), Section IV* **3** 201–225.

Cavalier, C., Chateauneuf, J.-J. Pomerol, Ch., Rabussier, D., Renard, M. & Vergnaud-Grazzini, C. (1981). The geological events at the Eocene/Oligocene boundary. *Palaeogeography, Palaeoclimatology, Palaeoecology* **36** 223–248.

Ducasse, O. (1974). La faune d'Ostracodes des differents domaines marines de l'Oligocene en Aquitaine meridionale. *Comptes Rendues Sommaires Société Geologique de France* **1** 7–9.

Gramann, F. & Spiegler, D. (1986). The northwest German Tertiary Basin–Oligocene. In Tobien, H. (coord.), *Nordwestdeutschland in Tertiaer. Beitraege regionelen Geologie Erde* **18** 669–678.

Guernet, C. (1981). Ostracodes Sparnaciens du Bassin de Paris (France). *Revue de Micropaleontologie* **24** 51–66.

Keen, M. C. (1971). A palaeoecological study of the ostracod *Hemicyprideis montosa* (Jones & Sherborn) from the Sannoisian of north-west Europe. *Bulletin Centre Recherche Pau-S.N.P.A., suppl.* **5** 523–543.

Keen, M. C. (1972). The Sannoisian and some other Upper Palaeogene Ostracoda from north west Europe. *Palaeontology* **15** 267–325.

Keen, M. C. (1975). The palaeobiology of some Upper Palaeogene freshwater ostracodes. *Bulletin American Palaeontology* **65** 271–283.

Keen, M. C. (1977). Ostracod assemblages and the depositional environments of the Headon, Osborne and Bembridge Beds (Upper Eocene) of the Hampshire Basin. *Palaeontology* **20** 405–445.

Keen, M. C. (1982). Intraspecific variation in Tertiary ostracods. In: Bate, R. H., Robinson, E. & Sheppard, L. M. (eds), *Fossil and Recent ostracods.* Ellis Horwood, Chichester, pp. 381–405.

King, C. (1982). Cainozoic micropalaeontological biostratigraphy of the North Sea. *Report, Institute of Geological Sciences, London* **82/7** 1–40.

Malz, H. & Moayedpour, E. (1973). Miozane Susswater Ostracoden aus der Rhon. *Senckenbergiana Lethaea* **54** 281–309.

Martini, E., Fahlbusch, V. & Hagn, H. (1986). The Eocene/Oligocene boundary and the Latdorfian (Lower Oligocene). *Newsletter in Stratigraphy* **17** 37–43.

Neale, J. W. (1983). The Ostracoda and Uniformitarianism. 1. The later record: Recent Pleistocene and Tertiary. *Proceedings of the Yorkshire Geological Society* **44** 305–326.

Pomeral, C. & Premoli-Silvan, I. (eds) (1986). *Terminal Eocene events. Developments in palaeontology and stratigraphy.* Elsevier, Amsterdam, pp. 1–414.

Ritzkowski, S. (1986). Ostracods from the Tertiary of the Hessian Depression. In: H. Tobien (coord.), *Nordwestdeutschland in Tertiaer. Beitraege regionelen Geologie Erde* **18** 440–445.

Sonne, V. (1982). Bestandaufnalume der Foraminiferin und Ostradkoden im 'prae-aquitanen' Tertiaer des Mainzer Bechens. *Mainzer geowissenschaftliche Mitteilungen* **10** 37–82.

Uffenorde, H. (1979). On the bio- and ecostratigraphical distribution of Ostracoda in the Oligo-/Miocene of the southern North Sea Basin. A preliminary report. In Vinken, R. (ed.), *Project 124: The Northwest European Tertiary Basin, Report* **5** 19–38.

Uffenorde, H. (1980). Ostracode biostratigraphy and palaeoecology of the boring Wursterheide (N.W. Germany): first results from the Middle Oligocene to the Upper Miocene Section. In: Vinken, R. (ed.) *Project 124: The Northwest European Tertiary Basin, Report* **6** 119–134.

Uffenorde, H. (1981). Ostracoden aus dem Oberoligozaen und Miozaen des unteren Elbe-Gebietes (Niedersachsen und Hamburg, NW-deutsches Tertiaerbecken). *Palaeontographica A* **172** 103–198.

Uffenorde, H. (1986). Stratigraphical and palaeoecological aspects of Upper Oligocene and Miocene Ostracoda of Lower Saxony. In: H. Tobin (coord.), *Nordwestdeutschland im Tertiaer. Beitraege regionelen Geologie Erde* **18** 422–436.

Vinken, R. (ed.) (1988). The Northwest European Tertiary Basin. *Geologisches Jahrbuch, Series A* **100** 1–508.

Whatley, R. (1983). The application of Ostracoda to palaeoenvironmental analysis. In: R. F. Maddocks (ed.), *Applications of Ostracoda.* University of Houston, Geosciences, pp. 51–77.

12

Palynomorph and palynodebris distributions in modern British and Irish estuarine sediments

Katherine M. Farr

ABSTRACT

Intertidal sediments of modern British and Irish estuaries and tidally influenced embayments have been analysed to investigate the degree to which their deposits are characterized by their palynomorph assemblages. At the head of a typical estuary there is a sharp and conspicuous change in the palynomorph assemblages; brackish and marine palynomorphs are generally abundant throughout the intertidal reaches, whereas they are absent upstream of the tidal limit. In general, marine and brackish types show few distinguishable trends in distribution between the head and mouth of the estuary, being evenly spread by tidal flushing. Dinoflagellate cysts, for example, are as abundant in the upper reaches as they are downstream. The marine–estuarine transition is poorly characterized by changes in palynomorph assemblages. Relative proportions of most of the terrestrially derived palynomorphs increase upstream although woody fragments have a greater relative abundance in the lower reaches, possibly because they are more resistant and can remain in suspension longer than most other types of plant debris, Biogenic pyrite and invertebrate faecal material both occur more commonly in estuarine than in fluvial deposits. These observations should provide criteria for distinguishing estuarine from fully marine or freshwater facies in Quaternary and perhaps older sediments. Additionally, the trends observed within estuaries may aid in the interpretation of palaeosalinities.

INTRODUCTION

Previous work on the acid-resistant organic content of intertidal sediments includes that of the broad geographical pattern of distribution of palynomorphs around the British Isles by Reid (1972a, 1972b, 1974, 1977). The present study is an extension of this work in which I have investigated distribution patterns within individual estuaries and intertidal embayments. The sites chosen for investigation were the Ythan Estuary and Montrose Basin in northeast Scotland, the Shannon and Clarinbridge estuaries in western Ireland, the Tresillian (Falmouth) and Hayle estuaries in southwest England (Fig. 1) and numerous estuaries feeding into the Solway Firth (Fig. 2). This paper is a preliminary report with results based mainly on data from the Ythan

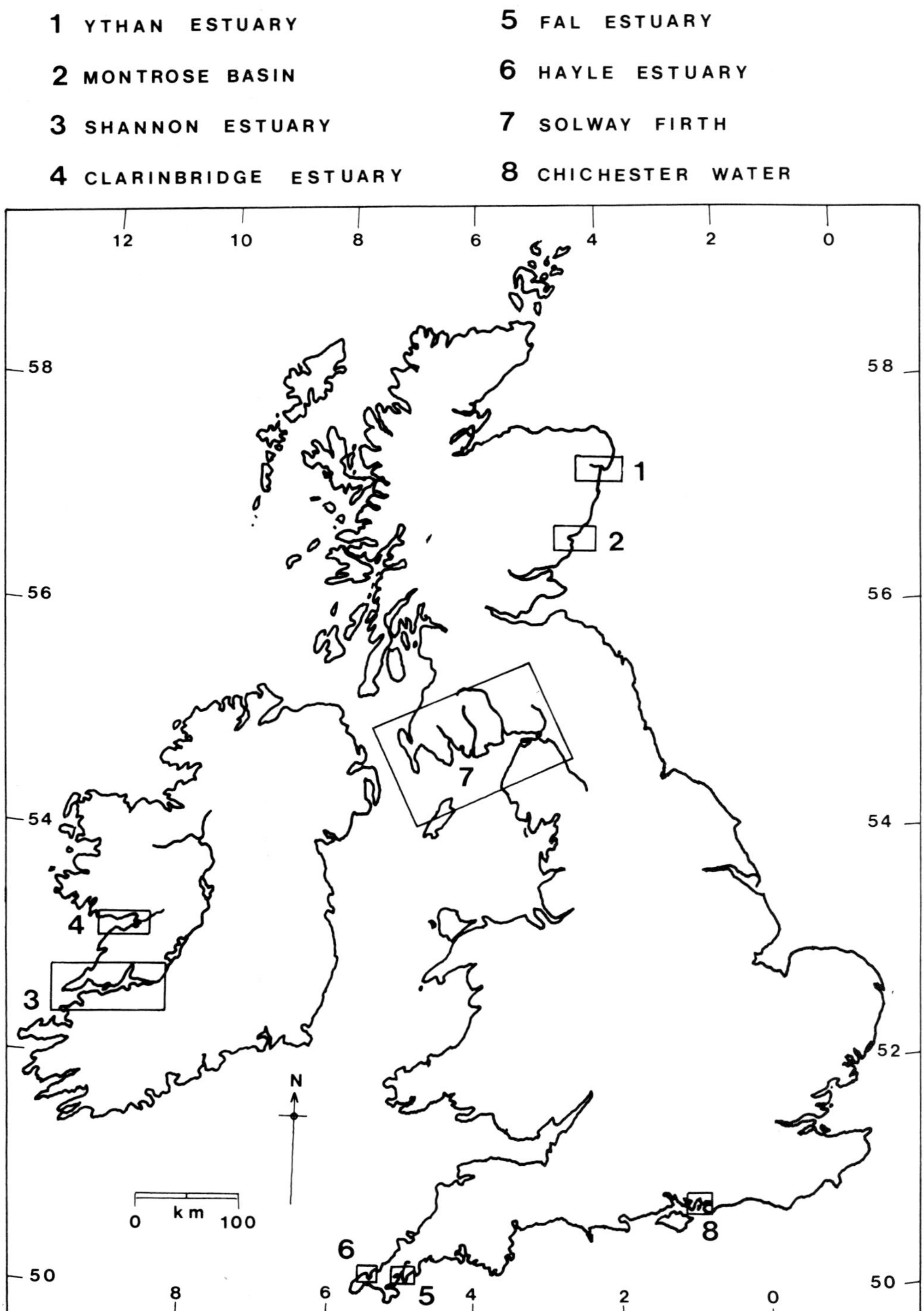

Fig. 1 — Map of the British Isles showing sampling localities.

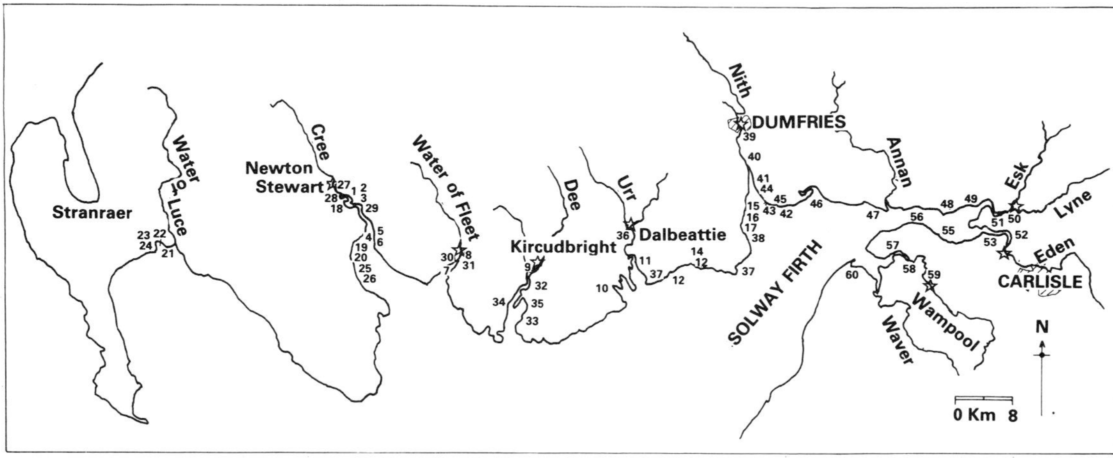

Fig. 2 — Map of the Solway Firth showing feeder estuaries. The numbers represent sample localities and the stars indicate spring tidal limits.

Estuary in northeast Scotland, which was studied in greater detail than the other areas. It has not been possible to present general conclusions covering all estuarine situations; much more work is required to gain an understanding of the local factors which affect the species of palynomorphs present and their distributions.

CHOICE OF SAMPLE LOCALITIES

Estuaries were chosen which were relatively unpolluted (those with little industry nearby and away from large centres of population) showed minimum modification by man and, as far as could be ascertained from maps and personal knowledge, had silty intertidal areas. They were selected from regions spaced well apart to obviate local climatic effects and to include tidal flushing by different marine water bodies so that the full range of palynomorphs occurring in estuarine sediment samples from around the British coasts could be established. Where possible, samples from two neighbouring tidal inlets from each region were examined to compare palynomorph assemblages deposited under similar environmental conditions. Most of the selected estuaries are small; the Shannon is the longest. The Ythan Estuary was studied in more detail than the others because its sediments contained the richest and most diverse palynomorph assemblages and because its proximity to Aberdeen enabled frequent visits to be made to the site. Consequently, this paper is biased towards results obtained from this locality and the way in which they compare with data obtained from elsewhere.

COLLECTION PROCEDURES AND PROCESSING TECHNIQUES

Samples of dark organic-rich silty mud were collected from intertidal areas of net deposition at regular intervals along the estuaries from the mouth to about 2 miles beyond the spring tidal limit. In general, sediment was taken at low tide from a depth of 2–6 cm. This commonly represented anaerobic material below an oxidized surface layer. Occasionally, the deposits were too sandy to be worth sampling for particulate organic matter beneath the surface; in these instances, material was instead skimmed from surficial mud drapes. The intertidal flats were generally bioturbated to a maximum depth of about 30 cm, so that the palynomorphs were well mixed in the upper layers. In saltmarshes and non-tidal stretches of the rivers, sedimentary laminations were poorly developed, suggesting disruption by infaunal organisms although burrows were scarce or poorly preserved.

10 g of wet sediment from each sample were treated with 20% HCl and 60% HF using standard techniques (Batten & Morrison 1983) to remove carbonates and silicates and then with further 20% HCl to remove any insoluble fluorides and colloidal SiO_2. After the residue had been neutralized and filtered, four drops of residue were mounted in glycerine jelly on a microscope slide and covered with a glass cover slip.

Further treatment methods included ultrasonic vibration for 30–60 s to disaggregate amorphous organic matter (AOM), and oxidation with fuming nitric acid to remove faecal and other debris surrounding and adhering to the palynomorphs. However, these treatments tended to damage or change the appearance of the more delicate palynomorphs. After ultrasonic vibration, for example, the yellow 'Palynomorph Y' (Plate 1 Fig. 14) became colourless whereas on oxidation it darkened to orange–brown while tintinnid loricae, some prasinophycean algal cysts and certain invertebrate eggs disappeared completely. To prevent loss or damage to palynomorphs, many samples were simply washed using a 10 μm nylon mesh sieve after the initial processing.

For consistency of comparison, counts were performed on only those preparations that had not undergone further treatment. However, all slides of additionally treated material were analysed microscopically to detect any rare species which were not present in the original slide and to select particularly photogenic specimens so that a record could be kept of all palynomorph types encountered. At least 200 palynomorphs were logged from each slide and the relative abundance of each was determined. The types recorded are listed in Table 1. Plant tissue and AOM occurred in abundance in most slides but were excluded from the counts since only discrete palynomorph entities were logged. However, a visual estimate of the relative abundance of these materials was made. Figure 3, 1–25, shows how the relative abundances of selected palynomorphs vary in an upstream direction in the estuaries. To supplement the study of palynomorph assemblages a brief survey of the absolute abundance of various types of palynodebris was also undertaken.

RESULTS

Dinoflagellate cysts are present in high numbers in many of the estuarine sediment samples, being particularly abundant in those from northeast Scotland. All palynomorph assemblages show low diversity. Figure 3, 1–6, shows how the Ythan Estuary dinoflagellate cysts comprise mainly *Operculodinium centrocarpum* and *Spiniferites* spp., with other species such as *Trinovantedinium capitatum*, *Bitectatodinium tepikiense*, *Polykrikos schwartzii* and 'round brown' peridinioid cysts each forming only about 1–5% of the total assemblages. The northeast Scottish assemblages are a little more diverse than those from elsewhere. *Operculodinium centrocarpum* and *Spiniferites* spp. dominate assemblages everywhere except in the west-facing estuaries of the Solway Firth where *Lingulodinium machaerophorum* is the commonest cyst (Fig. 3, 22; cf. Fig. 3, 23). The proportions of individual and total dinocysts (and most other marine-derived palynomorphs) generally remain relatively constant in nearly all the samples along the lengths of the estuaries, decreasing only near the tidal limit (Fig. 3, 7–11). Dinocysts are absent from almost all fluvial samples. A cyst of uncertain affinity (*Impletosphaeridium* sp. (Plate 1, Fig. 11) and several types of invertebrate eggs show different distribution trends (Fig. 3, 8 and 9). These palynomorphs are rare in samples from the lower reaches of the estuaries and reach a peak in relative abundance further upstream but they were not found in freshwater sediments. The only palynomorphs found in both fluvial and estuarine deposits were *Tristolidium leiozosterum* (Plate 1, Figs. 1 and 2) '*Tasmanites? gelidus*' (Plate 1, Fig. 4) and Palynomorph F (Plate 1, Fig. 6).

Table 1 — Palynomorph categories used in counts.

Aquatic forms	*Terrestrial forms*
Dinoflagellate cysts	Pollen grains
1. *Spiniferites membranaceus* (Rossignol) Sarjeant	32. *Pinus sylvestris* L.
2. *S. bulloideus* (Deflandre & Cookson) Sarjeant *sensu* Reid	33. *Picea* sp.
3. *S. elongatus* Reid	34. *Abies* sp.
4. *Other Spiniferites* spp.	35. *Betula* sp.
5. *Bitectatodinium tepikiense* Wilson	36. *Alnus glutinosa* (L.) Gaertn,
6. *Operculodinium centrocarpum* (Deflandre & Cookson) Wall	37. *Corylus avellana* L.
7. *Lingulodinium machaerophorum* (Deflandre & Cookson) Wall	38. *Quercus* sp.
8. *Nematosphaeropsis labyrinthea* (Ostenfeld) Reid	39. *Calluna vulgaris* (L.) Hull
9. *Trinovantedinium* spp.	40. Graminae
10. *Lejeunacysta* spp.	41. *Plantago* spp.
11. *Quinquecuspis concreta* (Reid) Harland	42. *Taraxacum officinale* Weber
12. *Polykrikos schwartzii* Bütschli	43. Other Compositae
13. *Brigantedinium* spp.	44. *Sucissa pratensis* Moench
14. 'Round brown' cysts	45. Chenopodiaceae
Incertae sedis cysts and tintinnids	
15. *Impletosphaeridium* sp. Morgenroth *sensu* Reid 1972a	Spores
16. 'Palynomorph G' (algal cyst?, Plate 1, Fig. 7)	46. *Sphagnum* sp.
17. *Pacillina arctica* (?prasinophycean cyst)	47. *Lycopodium* sp.
18. *Hexasterias problematica* (*incertae sedis*)	48. *Polypodium vulgare* L.
19. *Sternhaarstatoblast* sp. (*incertae sedis*)	49. Other trilete spores
20. *Saturni conceptaculi* (*incertae sedis*)	50. Monoporate pollen
21. *Parafavella* sp. Kofoid & Campbell (tintinnid lorica)	51. Filicales undifferentiated
22. *Stenosomella* sp. Jörgensen (tintinnid lorica)	
23. Other loricae (tintinnids and amoebae)	Other palynomorphs
24. Cyst P Reid (tintinnid cyst)	52. Fungal spores
25. Other vasiform cysts (tintinnid cysts)	53. Fungal hyphae
	54. Fungal sporangia
	55. *Tristolidium leiozosterum* Reid (algal cyst?)
Other palynomorphs	
26. 'Palynomorph Y' (egg case)	56. 'Palynomorph F' (?)
27. Turbellarian remains (marine, brackish and freshwater)	57. '*Tasmanites? gelidus*' *sensu* Reid (algal cyst?)
28. 'Palynomorph D' (*incertae sedis*)	
29. Cyst A Harland (copepod egg)	58. pollen and spores
30. Foraminiferal linings (eight morphotypes distinguished)	59. Other cysts (marine)
31. Invertebrate fragments; limbs, integument, eyes (marine, brackish and terrestrial)	60. Other material (marine, brackish and terrestrial)

Foraminiferal test linings generally occur in moderate proportions in almost all samples up to, but not beyond, the spring tidal limit (Fig. 3, 11). Their relative abundance decreases as the head of the estuary is approached. Species diversity is low: the four groups of estuaries studied (Shannon–Clarinbridge, Solway Firth, Ythan–Montrose and Falmouth–Hayle–Chichester) each contain a different dominant foraminiferan morphotype. Observation of material untreated

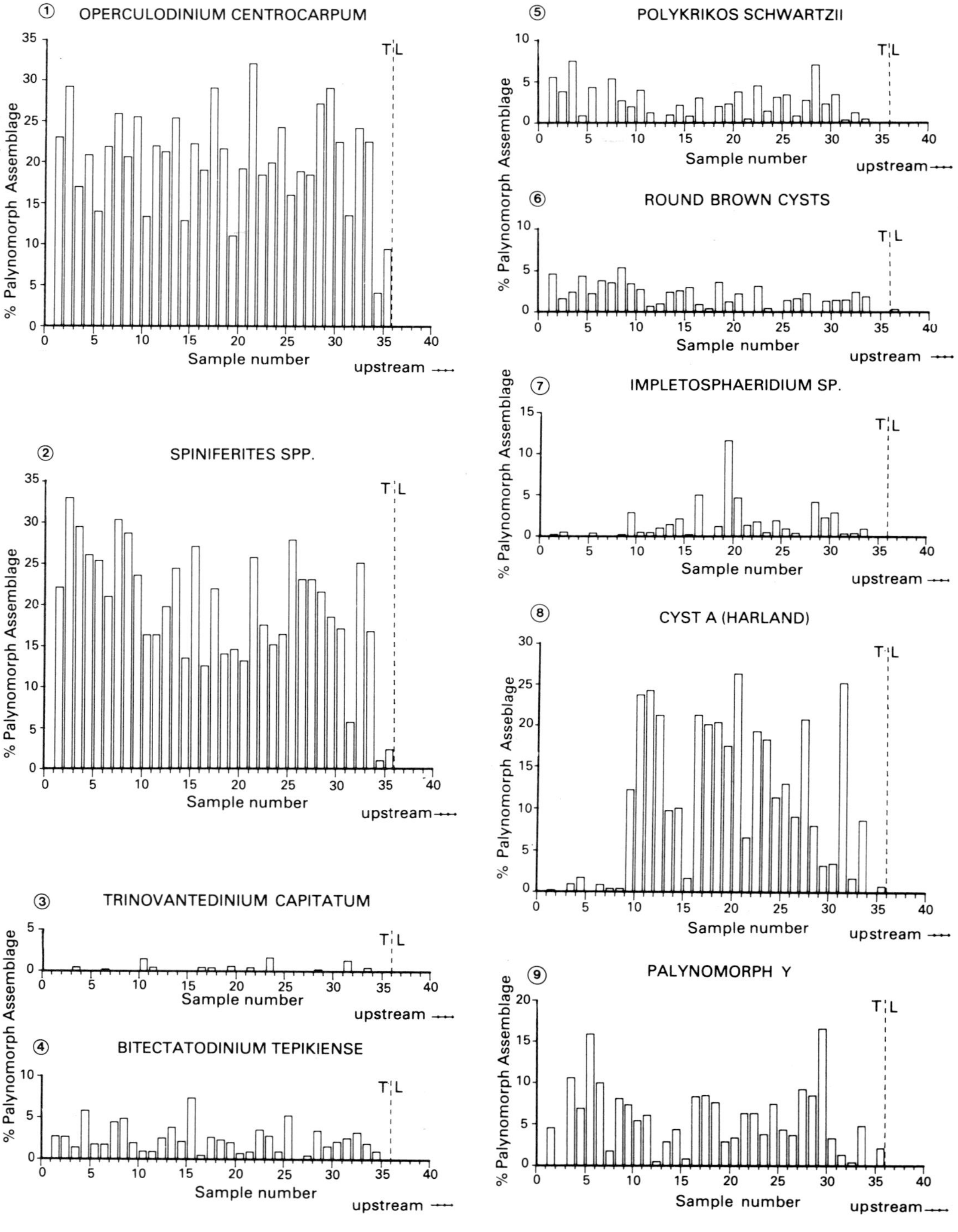
① OPERCULODINIUM CENTROCARPUM
② SPINIFERITES SPP.
③ TRINOVANTEDINIUM CAPITATUM
④ BITECTATODINIUM TEPIKIENSE
⑤ POLYKRIKOS SCHWARTZII
⑥ ROUND BROWN CYSTS
⑦ IMPLETOSPHAERIDIUM SP.
⑧ CYST A (HARLAND)
⑨ PALYNOMORPH Y
% Palynomorph Assemblage
% Palynomorph Asseblage
Sample number
upstream
T L

(10) TOTAL DINOCYSTS

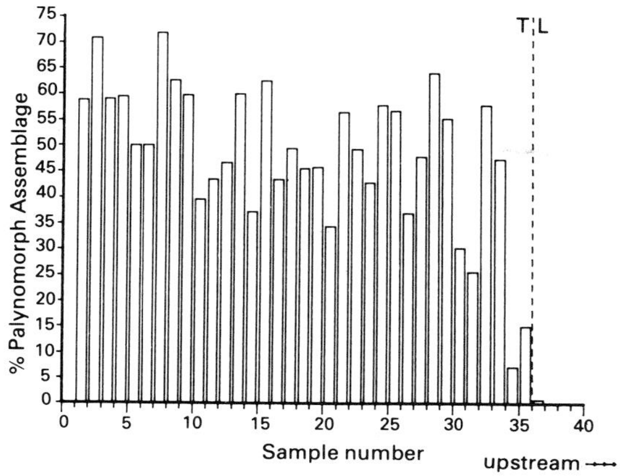

(11) TOTAL FORAM LININGS

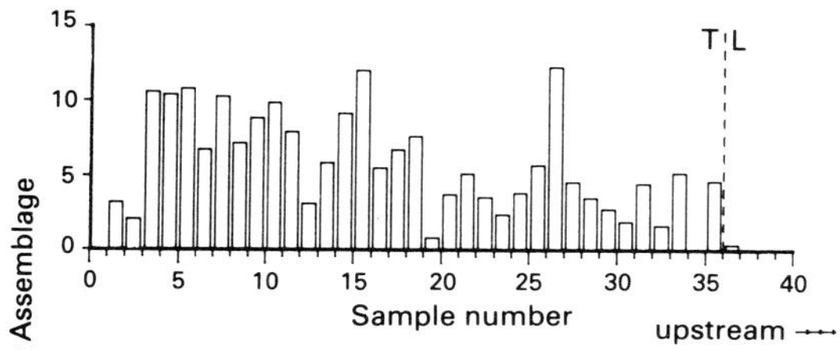

(12) PINUS SYLVESTRIS

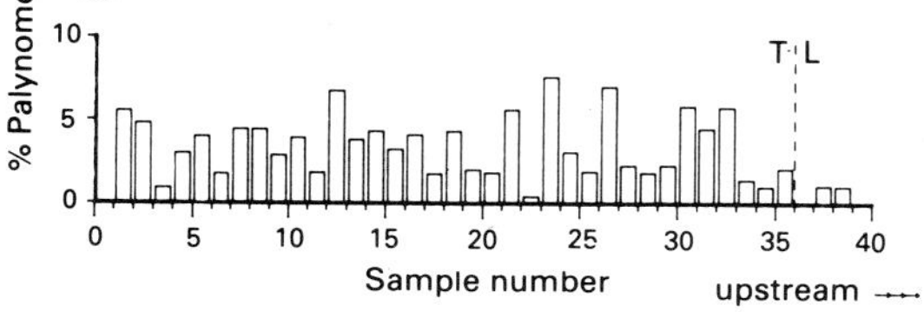

(13) TOTAL ARBOREAL POLLEN

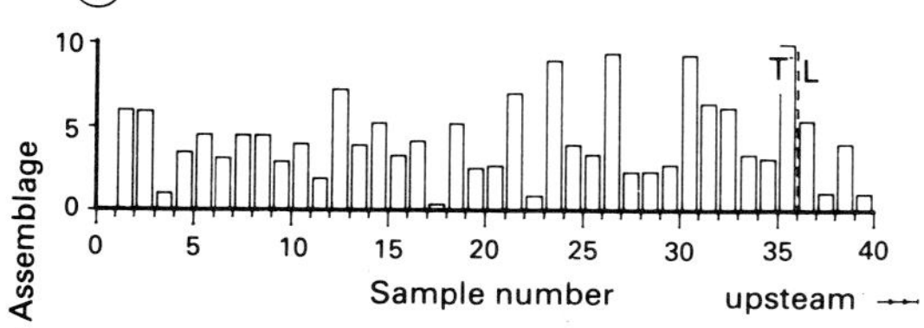

(14) TOTAL ARBOREAL POLLEN LESS PINUS

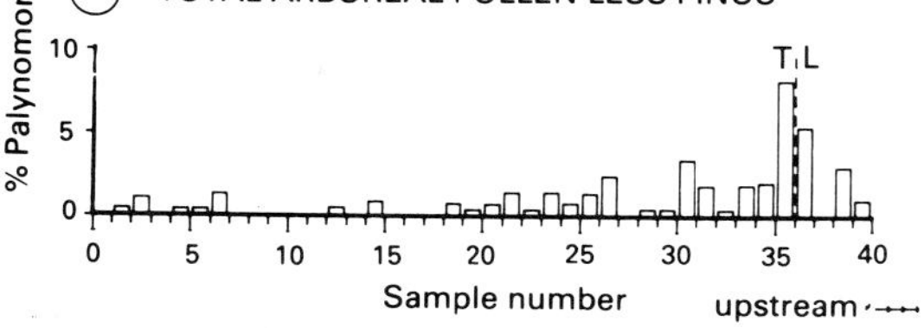

(15) TOTAL NON-ARBOREAL POLLEN

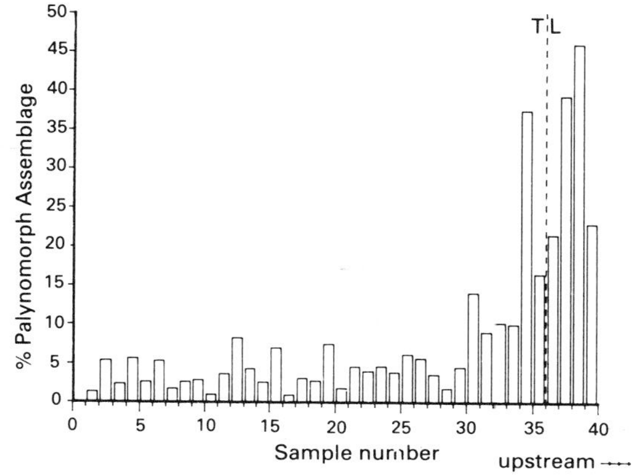

(16) TOTAL POLLEN AND SPORES

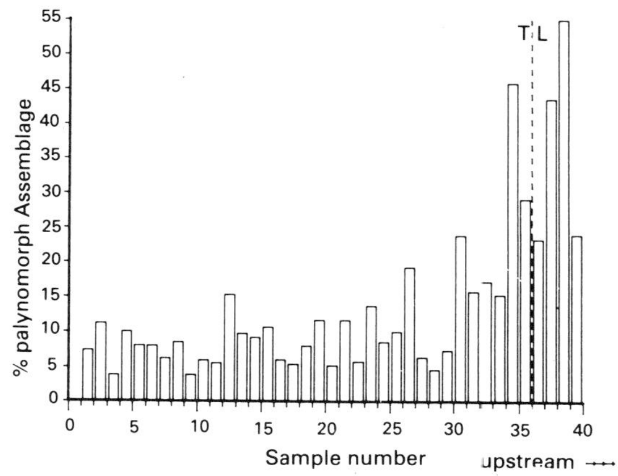

(17) TOTAL FUNGAL PALYNOMORPHS

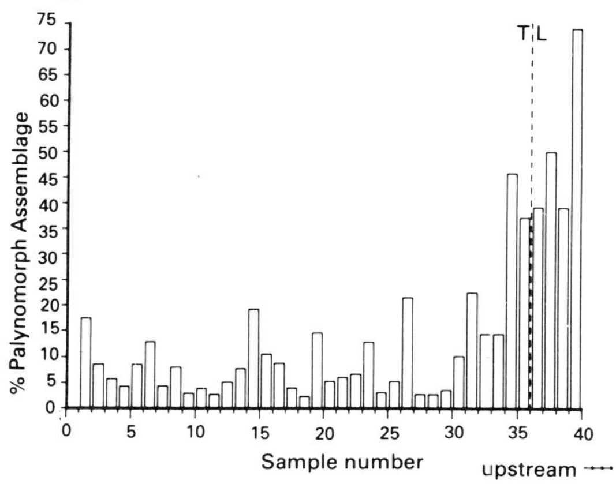

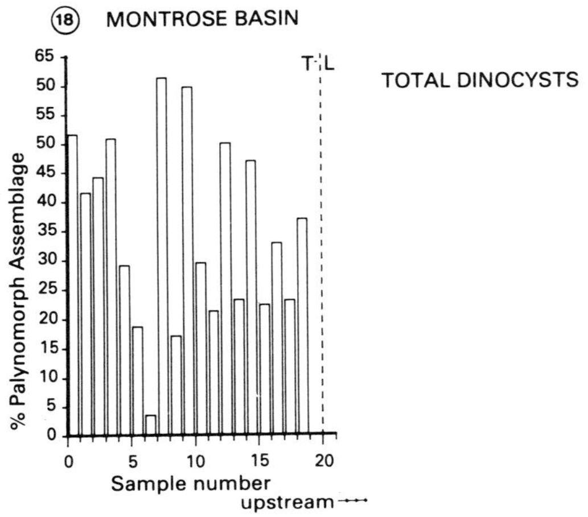
18 MONTROSE BASIN
TOTAL DINOCYSTS
% Palynomorph Assemblage
Sample number
upstream
T L

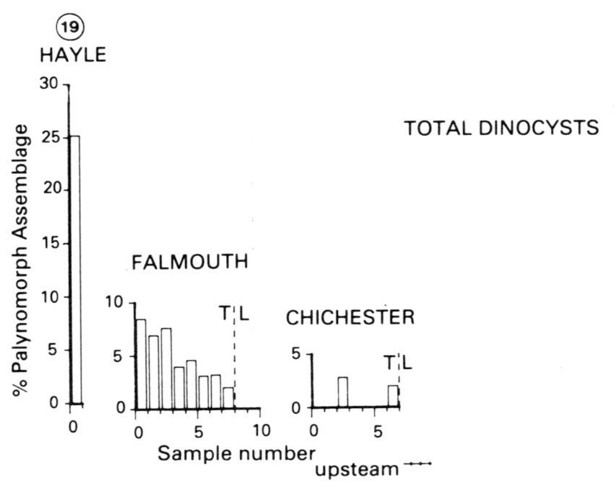
19 HAYLE
TOTAL DINOCYSTS
FALMOUTH
CHICHESTER
% Palynomorph Assemblage
Sample number
upsteam
T L

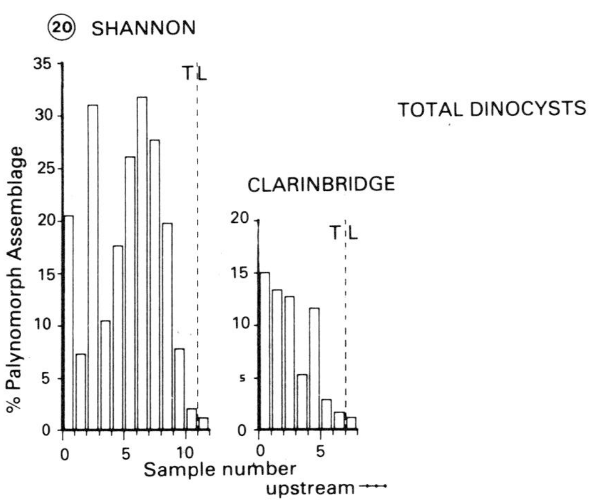
20 SHANNON
TOTAL DINOCYSTS
CLARINBRIDGE
% Palynomorph Assemblage
Sample number
upstream
T L

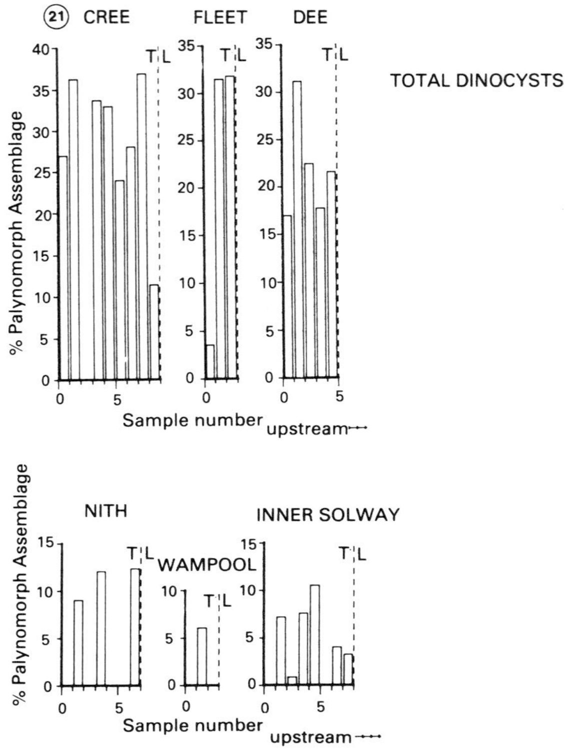
21 CREE
FLEET
DEE
TOTAL DINOCYSTS
% Palynomorph Assemblage
Sample number
upstream
NITH
WAMPOOL
INNER SOLWAY
T L

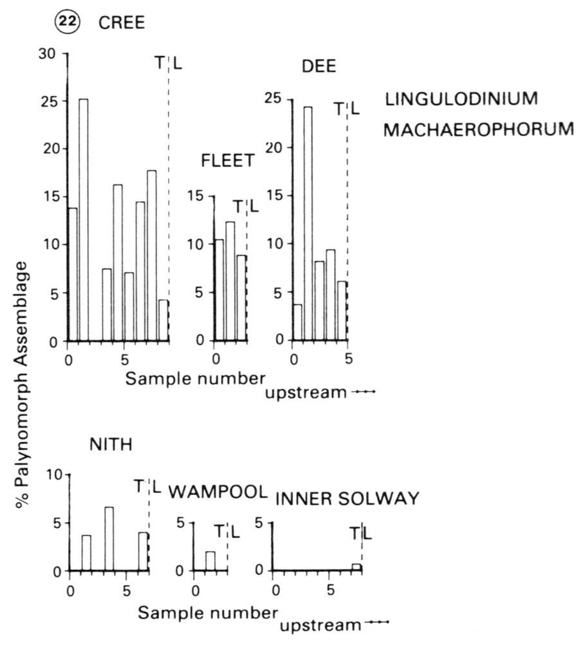
22 CREE
DEE
FLEET
LINGULODINIUM
MACHAEROPHORUM
% Palynomorph Assemblage
Sample number
upstream
NITH
WAMPOOL
INNER SOLWAY
T L

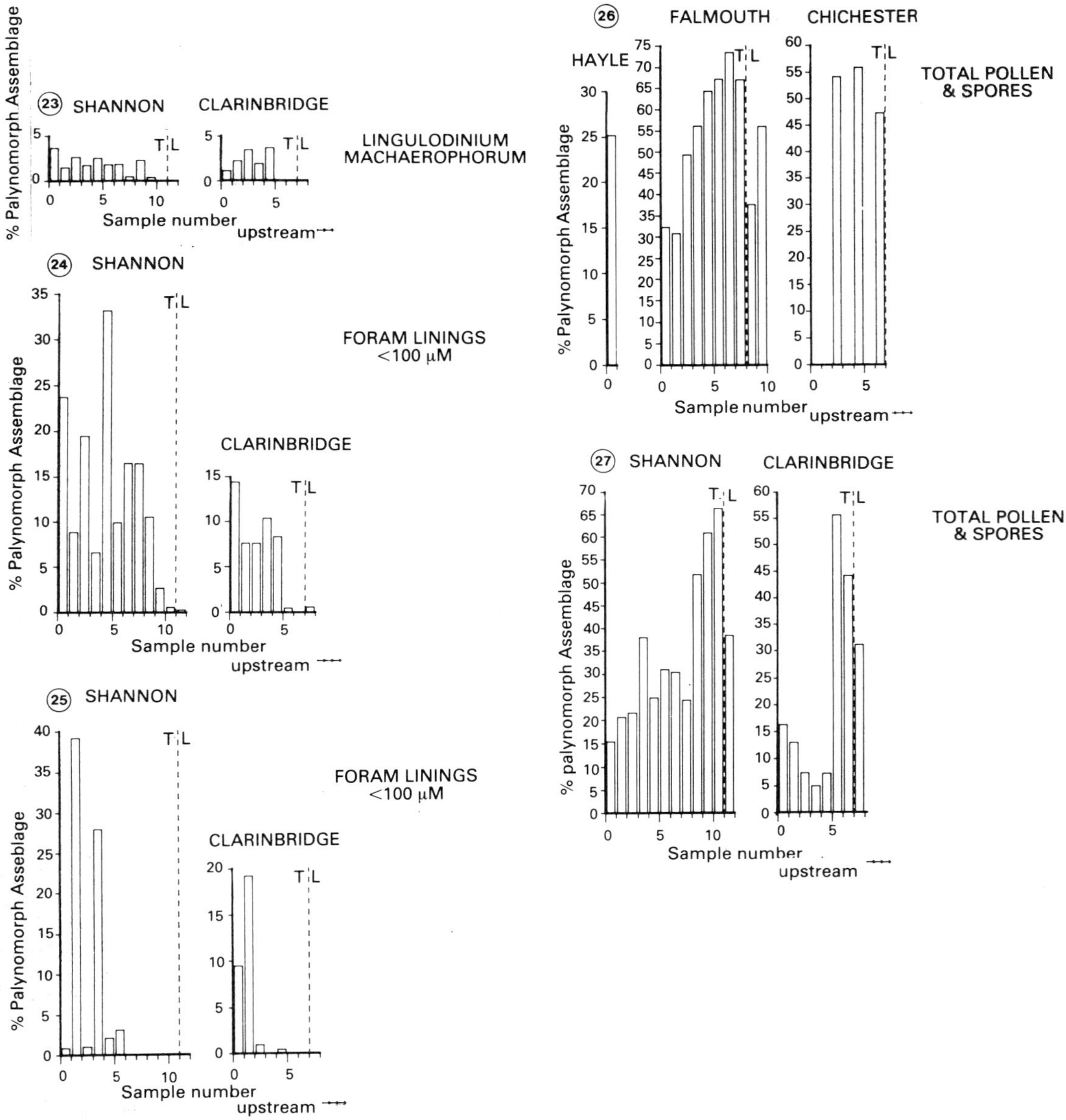

Fig. 3 — Histograms showing palynomorph distributions in intertidal mudflat samples along the length of the estuaries: 1–17, palynomorph distributions in the Ythan Estuary, Aberdeenshire; 18–21, distribution of total dinoflagellate cysts in the other estuaries; 22 and 23, distribution of *Lingulodinium machaerophorum* in the Solway Firth and Irish estuaries; 24 and 25, distribution of small and large foraminiferal linings in the Irish estuaries; 26 and 27, distribution of total pollen and spores in the estuaries of southern England and Ireland. The relative proportion of individual species or groups of palynomorphs as a percentage of total palynomorphs is plotted as ordinate. Six histograms are necessary to display data from the Solway Firth: the estuaries of the Cree, Fleet, Dee, Nith, Wampool and the Inner Solway Firth are arranged such that the estuary nearest the seaward end of the Firth (the Cree) is on the left-hand side of the figures. Estuaries which debouch further upstream are placed to the right and are divided into two groups of three. TL indicates the position of the mean spring tidal limit.

by mineral acids did not reveal the presence of other genera, although relatively diverse foraminiferan faunas, most of which appear to comprise empty tests of fully marine forms washed in by the tide, have been reported from other estuarine waters (Brasier 1981, Murray 1968, 1980).

Relative abundances of total pollen and spores remain approximately constant in the lower and middle reaches of the estuaries but in the upper reaches begin to show an increase upstream towards the source of the grains (Fig. 3, 16). Diversity also increases upstream. Relative abundance of terrestrial grains increases markedly beyond the tidal limit where marine palynomorphs are absent, although the rise in absolute abundance (calculated as a fraction of the total number of palynomorphs in four drops of residue) is much less pronounced. Individual pollen species show the same trend, except for bisaccate *Pinus sylvestris* grains which are distributed more evenly along the estuaries (Fig. 3, 12; cf. Fig. 3, 14). The patterns of distribution of total arboreal and total non-arboreal pollen are illustrated in Fig. 3, 13–15.

Total fungal material (i.e. total relative proportions of fungal spores, hyphae and sporangia) shows a random distribution in the estuarine sediments (Fig. 3, 17) but, along with the pollen and spores, becomes relatively much more abundant in the oligohaline waters just below the tidal limit and in the riverine deposits where marine forms are absent. Cuticular plant tissue is as abundant in fluvial sediments as in the estuarine sediments. Woody fragments tend to increase in abundance downstream.

In some estuaries (Ythan, Montrose, Fleet, Dee, Shannon and Clarinbridge), marine forms are dominant over terrestrial forms along almost the whole length of the intertidal reaches (Figs. 4(a)–4(d)). Terrestrial material becomes dominant near the tidal limit, and is the only acid-resistant organic matter present in the fluvial sediments. Conversely, the Cree, Nith and Inner Solway Firth sediments are dominated by land-derived palynomorphs, while the Chichester Water and Falmouth samples contain almost no marine palynomorphs other than foraminiferal test linings and turbellarian egg cases (Figs. 4(c) and 4(e)). In the Irish estuarine samples, forms with marine affinities outnumber terrestrial palynomorphs by about 2:1, whereas the north-east Scottish material has a much higher proportion of marine palynomorphs (a ratio of about 10:1). It is thus impossible to generalize on whether marine or terrestrial palynomorphs dominate estuarine intertidal sediments.

Species diversity of marine or brackish palynomorphs in the estuarine sediments is moderate to low and decreases further near the tidal limit; yet in many samples several species may be very abundant. This mirrors data for macrobiota in many estuarine environments (e.g. Khayrallah & Jones 1975, McLusky 1981). Similar observations were made by Muller (1959) on Recent palynomorphs from the Orinoco Delta, Trinidad. Offshore, Recent dinocyst and foraminiferan assemblages are more diverse (e.g. Harland 1973, 1977, 1978, 1981, 1983, Reid 1972b, 1974, 1975, Wall *et al.* 1977, Murray 1973). Terrestrial palynomorphs generally show moderate diversities and numbers, and both diversity and relative abundance increase upstream within the estuaries studied.

All of the dinocysts and papulifers (tintinnid and other vasiform cysts) found in the intertidal sediments have been recorded by other authors from Quaternary and Recent offshore marine deposits around the coasts of Britain (e.g. Harland 1973, 1977, 1978, 1981, 1983, Harland & Sharp 1986, Reid 1972a, b, 1974, 1975, 1977, Reid & John 1978). Most of the cyst species recorded in this study occur in low numbers in the estuarine intertidal sediments and lack opercula, and so are probably discarded rather than resting cysts.

Gonyaulacoid cysts (i.e. *Operculodinium*, *Spiniferites* and *Lingulodinium*) dominate the assemblages. Approximately equal numbers of gonyaulacoid and peridinoid cysts constitute the remainder of the cysts. About 10% of the dominant cysts contain partly degraded protoplasm. A similar number of round brown cysts also have cell contents although these are subordinate to the above-mentioned genera. Up to 70% of

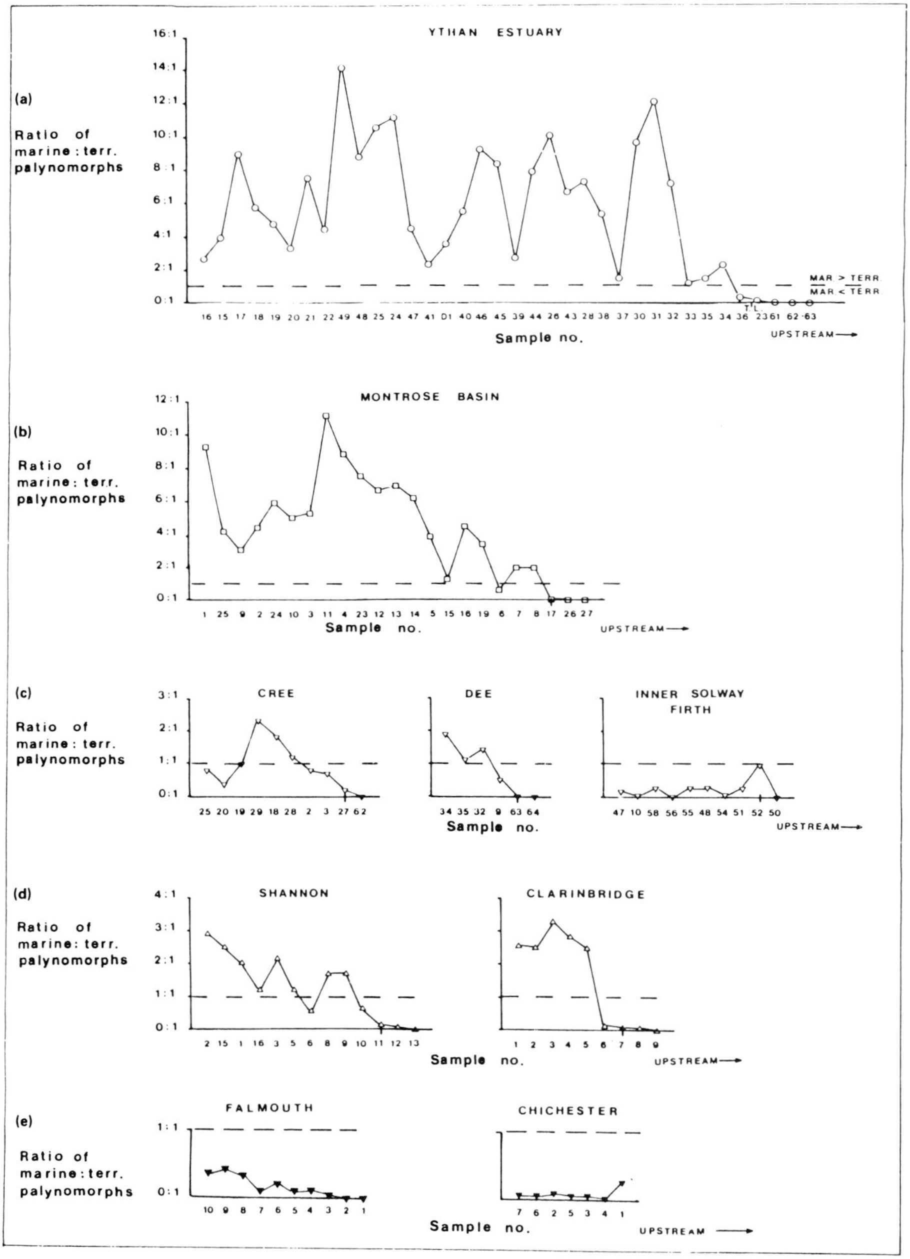

Fig. 4 — Graphs showing the ratio of marine and brackish palynomorphs to terrestrial palynomorphs: (a) Ythan Estuary; (b) Montrose Basin; (c) Cree Estuary, Dee Estuary and Inner Solway Firth; (d) Shannon Estuary and Clarinbridge Estuary; (e) Tresillian Estuary (Falmouth) and Chichester Water. Note change of scales between (a), (b) and (c), (d) and (e). Above the horizontal broken line, marine and brackish forms outnumber terrestrial forms. Below this line, terrestrial forms are dominant. Original sample numbers are plotted along the horizontal axis— those nearest the mouth of the estuaries to the left of the figures and those further upstream to the right. TL represents mean spring tidal limit.

Impletosphaeridium sp. cysts (*sensu* Reid 1972a) in the Ythan samples contain yellow–green degraded protoplasm. Most of the remaining cysts and eggs occasionally have cell contents (eg. Plate 1, Fig. 8). *Operculodinium*, *Spiniferites* and *Lingulodinium* dominate the dinocyst populations in offshore sediments (references above). They form an even higher percentage of the assemblages in the estuaries. No morphological differences are discernible between fully marine dinocyst species and those found in the estuaries, nor are morphological changes seen within the estuaries in successively more upstream samples.

The modern silty estuarine sediments studied are rich in faecal matter, plant epidermal tissue (Plate 1, Fig. 13), woody fragments (Plate 1, Fig. 12) and pyrite. In general, there is little variation in the absolute abundance (total fragments in four drops of residue released from a pipette) of these materials over most of the intertidal reaches, except that woody fragments tend to occur less commonly upstream. Cuticular plant debris occurs more frequently than lignitic fragments throughout the estuaries and in the lower part of the rivers, and ranges from well-preserved fragments to gelified masses.

Fungal spores, hyphae and sporangia are distributed irregularly throughout the estuarine sediments. Relative proportions of fungal material increase in the upper reaches of the estuarine and non-tidal river sediments, and also in reedswamp and saltmarsh samples. I am unaware of any work that describes fungal spore distributions in small temperate estuaries. However, distributions in the large Orinoco Delta in Trinidad (Muller 1959) are similar to those that I have found in British estuaries, i.e. the number of fungal spores and hyphae are generally high but erratic throughout the intertidal reaches, and decrease seawards. 15–35 km offshore, fungal debris is very rare (Muller 1959).

Faecal matter is abundant in the preparations but occurs less frequently in the upper reaches of the estuaries and becomes even less common in fluvial deposits. It is disseminated throughout the preparations and clings tightly to all palynomorphs (Plate 1, Fig. 15) except those with smooth hyaline walls and no processes, such as turbellarian cocoons. Faecal pellets are rare but when present they are approximately 50–100 µm × 30–70 µm with rounded ends (Plate 1, Fig. 5) and are probably produced by small invertebrates or possibly zooplankton. A sample of *Mytilus edulus* faecal mud from the Solway Firth contained no distinctive pellets after palynological processing; the pellets presumably disintegrate quickly to form disseminated faecal matter (Plate 1, Fig. 10).

Pyrite occurs in the preparations as framboids less than 10 µm in size. It tends to be concentrated within palynomorphs and cellular tissue of all types but also occurs scattered throughout the preparations (Plate 1, Figs. 9 and 16). The proportion of pyrite decreases in the oligohaline part of the estuaries studied and it is rare in fluvial deposits upstream of the tidal limit.

Offshore samples were not analysed in this study. The use of data amassed by other workers on palynomorph distributions in shelf sediments has proved unsatisfactory for comparison with my intertidal and fluvial samples since most workers have studied only one particular group of organisms whereas I have examined *all* microscopic acid-resistant organic remains. Also, many of the studies have been of pre-Flandrian rather than of Recent deposits, and sample preparation techniques have not been identical, so that a strict comparison is not possible.

DISCUSSION OF PALYNOMORPH DISTRIBUTIONS

The estuarine palynomorphs represent a mixture of indigenous and allocthonous elements. The main aim of this study was not to investigate their 'ecology' but to determine the composition of palynomorph assemblages preserved in intertidal sediments. Consequently, relatively little time was spent in attempting to ascertain whether the acid-resistant, biogenic remains were derived from living (brackish–euryhaline) or dead (brackish–euryhaline, freshwater, terrestrial and fully marine) organisms. To understand more fully the factors governing estuarine palynomorph distributions, more work must be done to discover which estuarine biota produce preservable tests and cysts.

The estuaries are all partially mixed (Leach 1971, Perkins 1974) (also noted from personal observations), having a saline wedge which flows upstream along the channel bed with the incoming of the tide and which ebbs as the tide recedes. Mixing of fresh and salt water takes place at their interface, where turbulence occurs. Marine palynomorphs carried into the estuary on the rising tide are deposited on the mudflats during a period of minimum flow velocity, usually but not always at the turnaround of the tide (Dyer 1973). Terrestrially derived palynomorphs are also forced out of the main channel and onto the mudflats as the tide rises. The histograms in Fig. 3, 1–27 indicate that most palynomorphs are carried up an estuary by the tide and settle with a roughly even distribution between the mouth and upper reaches. The strength of the flow decreases near the head of the estuary which may partly explain the reduction in palynomorph numbers there. Salinites near the tidal limit are also extremely low compared with coastal waters, and even estuarine organisms experience difficulty in adapting to such conditions (McLusky 1981). Marine-derived palynomorphs are generally absent from fluvial deposits because salt water does not penetrate beyond the spring tidal limit. The very few, poorly preserved specimens found in fluvial samples were probably transported by birds or other animals; analysis of faecal material from Eider ducks (*Somateria mollissima*) in the Ythan Estuary showed that the ducks ingest a wide range of palynomorphs with their food.

Certain benthic foraminiferans inhabit various niches in estuaries (Murray 1965, 1968, 1973, 1980). Tests of planktonic species present in intertidal sediments are thought to be washed in from offshore waters (Murray 1965, 1968, 1973, 1980, Brasier 1981). I detected no clear distribution patterns for test linings except to note that there is a predominance of large types in the lower reaches and smaller types upstream (Fig. 3, 24 and 25); whether they are derived from dead or living foraminiferans is unknown.

The intertidal reaches are characterized by macrofaunal zonations, both longitudinally and transversely, the faunal assemblages being controlled by both salinity and substrate (e.g. Perkins 1973, Trewin 1987). Most animals that contribute to the zonation are bottom-dwelling or actively swimming macrofauna and are able to remain in particular parts of the estuary, whereas most of the palynomorphs derive from living or dead, planktonic or nektonic organisms and are swept upstream and downstream by the tide and thus are probably unable to establish themselves in particular areas. Exceptions include certain types of egg which are produced by benthic or free-swimming brackish-water invertebrates; the eggs probably sink quickly after production, either directly or via a host estuarine animal. Ciliate and flagellate organisms are capable of migrating up and down within the water column but the extent to which they can regulate their position along the estuary is uncertain. Those dinocysts and other palynomorphs that occur in large numbers and dominate the assemblages may be produced by euryhaline and/or mesohaline species. The intact state of many of these palynomorphs and the presence of protoplasm in a high proportion of them may indicate that they have been produced by indigenous organisms.

All the dinoflagellate cysts recorded in the estuarine samples in this study have been reported from Quaternary and Recent shallow shelf deposits around the coasts of Britain; so none is confined in their habitat to brackish water. The overwhelming dominance of *Operculodinium centrocarpum*, *Spiniferites* spp. (especially *S. membranaceus*, *S. bulloideus* and *S. elongatus*) and *Lingulodinium machaerophorum* in the cyst assemblages suggests that the dinoflagellates which produced them are euryhaline and are exploiting the low-salinity environments. Other dinocysts occur in low proportions and are probably swept into the estuary by the tide from near-shore marine waters and resuspended sediments. If the presence of a large number of intact protoplasm-containing cysts is taken as an indication of the success of a particular species in adapting to estuarine conditions, then this is further support for the euryhaline nature of *Operculodinium*, *Spiniferites* and *Lingulodinium*.

The relatively high number of peridinioid round brown cysts containing protoplasm may indicate that some of them can retain viable contents in brackish environments and/or that some of the cysts are produced by brackish-water organisms.

The distribution patterns shown by *Impletosphaeridium* sp., Palynomorph Y and Cyst A suggest that they are probably produced by indigenous brackish-water algal and invertebrate organisms, such as prasinophycean algae, copepods, flatworms and trematodes (Van Geel *et al.* 1981, Kasahara *et al.* 1974). *Tristolidium leiozosterum*, previously believed to be marine (Reid 1972a), Palynomorph F and '*Tasmanites? gelidus*' (Reid 1972a) are interpreted as fluvial palynomorphs because of their presence in river sediments. The reason why there should be many more algal and invertebrate eggs in estuarine sediments from northeast Scotland than elsewhere around the British Isles is unclear.

The low diversity and occasional high abundance shown by foraminiferan test linings in the estuarine sediments suggests that most of them may be produced by brackish or euryhaline species. The tendency for large forms (diameter, greater than 100 μm) to be concentrated in the lower reaches, and for smaller types to be found throughout the estuaries could reflect a difference in sorting of dead tests or, if the linings are derived from indigenous species, it may indicate that the smaller foraminiferans tolerated lower salinities. It is also possible that low salinities cause stunting of the growth of otherwise larger species. No differences between assemblages from mudflat and saltmarsh environments were seen even though some workers have reported differences in the live populations (e.g. Phleger & Bradshaw 1966, Phleger 1970).

POSSIBLE FACTORS CONTROLLING PALYNOMORPH DISTRIBUTIONS

A brief review of the physicochemical and biological differences between the estuaries is necessary to identify possible factors governing the distribution of palynomorphs. Suggestions as to the main environmental controls can only be speculated upon at this stage.

The Ythan Estuary and Montrose Basin have relatively small drainage basins which may partly explain the low proportions and diversities of pollen and spores reaching the intertidal sediments. However, there is no direct relationship between the size of hinterland area and the proportion of terrestrial palynomorphs in the estuarine sediments, since those estuaries with the smallest areal extent of hinterland (Clarinbridge and Chichester) contain relatively high proportions of pollen and spores. The vegetation in the drainage basins, particularly in the lower river courses, is reflected in the terrestrial palynomorph content of the sediments. Diverse pollen floras are present in the Solway Firth and Shannon estuaries, both of which drain large hinterlands. Almost all the pollen grains and spores present in the samples can be attributed to wild plants and trees (native and introduced) rather than to cultivated and exotic varieties, despite the proximity of garden plants and fields of crops to the estuaries studied.

There are small differences in climate around the British Isles. Rainfall on the west coast is higher than in the east (Fullard 1972, Lee & Ramster 1981) and may wash pollen out of the air more thoroughly. This can enter the river systems by surface runoff. Surface and bottom temperatures in the North Sea are generally several degrees below those in western and southern waters (Lee & Ramster 1981). This may explain why the Ythan Estuary and Montrose Basin contain more 'Arctic' microfloras than west coast estuaries: for example, *Bitectatodinium tepikiense* and *Spiniferites elongatus* (both described by Harland (1983) as north temperate–Arctic species) are relatively common. Mean surface salinities also vary around Britain. The dominance of *Lingulodinium machaerophorum* in the Solway Firth estuaries may be associated with the below-average salinities (34‰) of the Irish Sea. The overall low numbers of marine-derived palynomorphs in the Solway Firth estuaries could be a result of low numbers in the Irish Sea itself caused by the lower salinities.

Palynomorphs settle out with the fine silt fraction of clastic sediments; so silty intertidal deposits are more likely to contain high proportions of acid-resistant organic remains than are coarser or finer sediments. The intertidal areas of the Ythan Estuary and Montrose Basin have, of all the sediments studied, a grain size most conducive to palynomorph deposition (3.0–8.0 φ) and contain rich assemblages of organic debris. The Solway Firth sediments (0.0–5.0 φ) are deficient in marine palynomorphs and I consider the 'sanding-up' of the Solway Firth and the lack of silt (Perkins & Williams 1966) to be a major cause of this. The kaolinitic clay-grade sediment at Falmouth (8.0–13.0 φ), which is probably derived from the waste products of the local China Clay industry, is severely limiting to the establishment of benthic communities and fish stocks (e.g. Herbert *et al.* 1961, Howell & Shelton 1970) and, from the results of the present study, apparently also to the establishment of micro-organisms. The rarity of marine-derived flora in the silty deposits of Chichester Water is not easily explained. Perhaps a local pollutant is responsible. Tributyl tin (TBT), used as an antifouling agent on boat hulls, is believed to have caused malformities and death in shellfish in numerous British estuaries and it is possible that it may have affected the microflora in the popular yachting harbour at Chichester. Diatoms are killed by a variety of antifouling paints (Perkins 1974) although, to my knowledge, no work has been published to date on the effects of TBT on other micro-organisms.

Protozoan cysts in the seas around Britain are more intimately associated with the distribution of surface currents and water masses than with any other factors, such as salinity, climate and nutrients (Williams 1971, Reid 1972b). Micro-organisms from different water masses are swept into the estuaries studied. This may account for the variations in palynomorph assemblages between the estuaries on the west and east coasts of the British Isles and may also explain the similarity of assemblages in groups of neighbouring estuaries (i.e. Ythan–Montrose, Shannon–Clarinbridge, Cree–Fleet–Dee, Nith–Wampool–Inner Solway Firth, Falmouth–Hayle–Chichester).

Marine current directions at the mouths of the estuaries are complex and can change in direction and intensity throughout the year. The estuaries richest (Ythan–Montrose) and poorest (Chichester) in marine palynomorphs experience currents approximately parallel to the shore, those with average numbers of marine palynomorphs (Shannon–Clarinbridge) experience onshore currents while there are offshore currents in the Solway Firth. There thus appears to be no direct correlation between directions of water movement at the mouths of the estuaries and the richness of the palynomorph content of the intertidal sediments. However, detailed current data may reveal differences in the general trends in the vicinity of the mouths which would allow more useful conclusions to be drawn.

All British estuaries experience prevailing southwesterly winds. Those in northeast Scotland also receive frequent southeasterlies. One might expect pollen and spores to be carried northeastwards, away from the mouths at Falmouth, Clarinbridge, the Shannon and the Solway Firth; yet these localities yielded higher pollen counts than Hayle, the Ythan and the Montrose Basin whose mouths face away from the prevailing winds. More frequent washout of pollen by rain in the west may explain this. It appears that long- and medium-range wind transport of terrestrial grains to the estuaries is insignificant. From the composition of pollen and spore assemblages it seems likely that most grains enter feeder streams and rivers from local sources (either directly or by being blown short distances) and are swept down into the river mouths in suspension and by flotation. The pollen spectra thus represent the local vegetation, particularly that around the middle and lower courses of the main river, since upland pollen and spores are relatively uncommon and non-arboreal pollen (especially Graminae) is dominant.

The distribution histograms show a sharp increase in the relative proportion of terrestrial palynomorphs upstream of the tidal limit, which is to be expected as there is no longer a contribution here from marine palynomorphs. Similarly, relative abundances of pollen and spores are high in those estuaries which contain few marine-derived palynomorphs. The gradual rise in the percentage of terrestrial palynomorphs seen throughout the lower and middle estuarine reaches probably continues at a similar rate further upstream, despite being masked by the more pronounced apparent increase. As most pollen grains probably enter the estuary via rivers and their tributaries, they may well be settling out of suspension progressively as they are carried downstream.

The scarcity of pollen and spores in a soil adjacent to the Ythan Estuary is probably the result of oxidation and the lack of daily flushing by palynomorph-bearing tidal and fluvial waters. The absence of aquatic palynomorphs distinguishes the soil palynofacies from those of the tidally influenced saltmarsh deposits. The latter contain palynomorph assemblages which are virtually indistinguishable from those of intertidal flat deposits except for a higher proportion of halophytic plant pollen. Thus the lateral margins (spring high-tide level) of an estuary can be traced according to the palynomorph content of its sediments. This has potential applications in pre-Quaternary environmental analyses. It should be possible to map ancient coastlines accurately and to trace sea-level oscillations, given the presence of transgressive–regressive marine sequences.

I have been unable to find sufficient data concerning nutrient and pollutant levels to ascertain whether there is any correlation with marine palynomorph abundance or diversity. There are high levels of radionucleides (Lee & Ramster 1981) and low numbers of marine palynomorphs in the Solway Firth, but no relationship has been established to link these two factors. Unlike some of the macrofauna (e.g. N. Pentreath 1987, pers. comm.), none of the palynomorphs shows any discernible morphological deformity.

DISCUSSION OF PALYNODEBRIS DISTRIBUTIONS

In many estuaries, the high concentration of nutrients resulting from the mixing of river and marine waters permits high organic productivity. Consequently, in the water column there are copious supplies of faecal pellets produced by zooplankton and other organisms (Schäfer 1972, Bayliss & Syvitski 1982). Once incorporated into the sediment, these tend to break down into irregular clumps of faecal debris which cling to palynomorphs. This debris is dispersed throughout those preparations which have not been oxidized by HNO_3. It has been suggested by P. C. Reid (1985, pers. comm.) that many dinocysts are able to attract faecal debris to themselves to afford protection to their cell walls when they pass through the digestive tracts of higher organisms; spines and granular ornamentation on the cyst trap the faecal material. U. Pollingher (1985, pers. comm.) has succeeded in culturing certain *Protoperidinium* cysts from the stomachs of fish, showing that some are still viable after being eaten by larger animals. Davey & Rogers (1975) have shown numbers of faecal pellets to be moderate to low in shelf sediments, except in areas of upwelling of nutrient-rich marine currents where organic productivity is raised.

Fungal spores are probably produced both in the hinterland and within estuarine soils in mats of decaying vegatation. They are not distributed far by the wind owing to the low stature of most of the spore-bearing plants but, once washed into the estuaries, they can be carried large distance owing to their small size (5–30 μm) (Muller 1959). Muller believed the majority of spores and hyphae to be liberated from saltmarsh soils and vegetation mats by water erosion and hence to enter the water column directly rather than floating on the surface before sinking as is the case for many pollen grains. Consequently, he argued that fungal spores are less likely than pollen grains to be carried out to sea.

Woody material, with lignified cell walls, is more resistant to abrasion and corrosion than cuticular plant debris (e.g. Batten 1974, Masran & Pocock 1981), particularly when it has been partly or wholly carbonized (Venkatachala 1981), and it is more likely than cuticular tissue to be reworked (Cope 1981). On moving seawards from the shore, it is selectively concentrated in shallow marine sediments (Muller 1959). This observation corresponds with the downstream increase in the proportion of woody fragments seen in the estuarine samples.

Pyrite is believed to be produced by sulphate-reducing bacteria in anoxic conditions in the water column and at, or below, the sediment–water interface (e.g. Love 1958, Love & Murray 1963;). Conditions for its formation are optimized in certain intertidal sediments, such as coastal lagoons, saltflats and estuaries, and in this study pyrite is more abundant in the estuarine intertidal reaches than in the fluvial sediments. Offshore, the availability of organic matter that can be metabolized by sulphate-reducing bacteria is a limiting factor on the formation of biogenic sulphide although, in areas of upwelling, organic production and consequently rates of pyrite accumulation can be high (Siesser & Rogers 1976). The conditions inside certain organic-walled tests and eggs as protoplasm begins to decay also appear to be suited to the formation of pyrite by sulphate-reducing bacteria.

CONCLUSIONS

The intertidal sediments of estuaries around the coasts of the British Isles show variations in the composition, diversity and abundance of palynomorphs and palynodebris. For example, the Ythan Estuary and Montrose Basin have particularly rich assemblages, the Solway Firth is dominated by terrestrial grains, while Chichester Water and the estuary at Falmouth contain very few palynomorphs. Regional and local physical, chemical and climatic factors are probably responsible for this variation. Within any individual estuary there is sufficient variation in the palynomorph assemblages of adjacent samples to necessitate collection of numerous samples to determine the 'average' assemblage in that particular part of the estuary. In general, the palynomorph assemblages show low divesity with a few species being abundant.

Within the estuaries, aquatic palynomorphs are dominated by dinoflagellate cysts and/or invertebrate eggs. Most cysts are marine species washed in by the tide. Some appear to be euryhaline and a few may be the products of brackish-water forms. The majority of these organisms are distributed evenly between the mouth and upper reaches of the estuary, their number and diversity falling only near the tidal limit. They are generally absent from fluvial samples. The few fluvial aquatic palynomorphs that were recognized are present in both fluvial and intertidal deposits. The cyst *Tristolidium leiozosterum*, hitherto believed to be marine (Reid 1972a), has been identified as a fluvial palynomorph. Terrestrially derived palynomorphs are dominated by pollen grains and spores. Long-distance wind transport of these grains to the estuaries is insignficant; most probably enter the river systems directly or by surface runoff and reflect local vegetation, mainly that bordering the lower river course.

Changes in palynomorph assemblages mark the inland extent (spring tidal limit) and lateral margins (spring high-tide level) of modern estuaries. Palynological investigation of Quaternary and older deposits should establish the extent to which their palynomorph and palynodebris assemblages mirror those of Recent temperature estuarine sediments. Tidally and non-tidally influenced facies should be distinguishable in older sediments if palynomorph assemblages similar to those in modern sediments are present. Pre-Quaternary estuarine palaeoenvironments can be difficult to recognize on the basis of microfaunal and macrofaunal, geochemical and sedimentary facies analyses. Use of data on the composition, diversity and abundance of palynomorphs and palynodebris should help to support or refute estuarine interpretations but on present evidence subenvironments, such as saltmarshes, tidal creeks, reedswamps and intertidal flats, are likely to be difficult to distinguish.

ACKNOWLEDGEMENTS

This study was undertaken as part of a Ph.D. project at Aberdeen University and I am grateful to Dr D. J. Batten for his advice, supervision and critical reading of the manuscript of this paper, and to Mrs L. Morrison for technical assistance. I thank Miss K. Hackney, Dr T. N. Harrison, Mr J. C. Rushton and Mr B. N. Young for fieldwork assistance and useful discussion, Dr G. E. Farrow for introducing me to the Solway Firth estuaries, Miss J. O'Mahony for cooperation with fieldwork in Eire and Mr B. N. Farr for field assistance in southern England. Mr A. A. Finch wrote the computer programs to produce the histograms (Fig. 3). I thank especially Dr D. A. Rogers for his advice and field assistance, and for critical comments on early drafts of this manuscript. The work was funded by a Natural Environment Research Council studentship.

Plate 1 — The locations of the specimens illustrated are given by the sample number, the east–west and north–south microscope coordinates and the preparation number (prefixed by MCP). The vernier readings are from the Leitz Dialux 20 EB microscope (020 452 008) in the Department of Geology and Mineralogy, University of Aberdeen, where all the slides and samples have been deposited .

Fig. 1 and 2 — *Tristolidium leiozosterum*, SOL 50/1, 46.2/107.4, MCP3524/1, × 400. Fig. 1, lower focus, granular ornament; Fig. 2, upper focus, vermicular ornament.

Fig. 3 — Palynomorph D, YTH 2/1, 33.0/100.0, MCP2967/1, × 1000. Palynomorph contains granular protoplasm.

Fig. 4 — '*Tasmanites? gelidus*', CHI 1/1, 36.0/102.0, MCP3656/1, × 1000. Operculum within palynomorph.

Fig. 5 — Invertebrate faecal pellet, MONT 9/1, 20.8/109.5, MCP3663/1, × 400.

Fig. 6 — Palynomorph F, CHI 1/1, 50.0/96.7, MCP3656/1, × 1000.

Fig. 7 — Palynomorph G, SOL 10/1, 23.0/96.4, MCP3430/1, × 400. Aperture in focus.

Fig. 8 — *Bitectatodinium tepikiense*, YTH 21/2, 33.0/99.0, MCP3127/2, × 400. Partly condensed protoplasm within cyst.

Fig. 9 — Pyrite within foraminiferal test lining, YTH 2/1, 32.0/99.0, MCP2967/1, × 400.

Fig. 10 — Disseminated faecal debris, SOL 5/1, 37.0/110.5, MCP3425/1, × 100.

Fig. 11 — *Impletosphaeridium* sp., YTH 39/1, 26.0/109.5, MCP3797/1, × 400. Cyst contains degraded protoplasm.

Fig. 12 — Brown wood, SOL 33/2, 27.0/98.7, MCP3506/2, × 250.

Fig. 13 — Cellular plant tissue, SOL 5/1, 39.2/101.0, MCP3425/1, × 100.

Fig. 14 — Palynomorph Y, YTH 9/1, 26.7/105.1, MCP2974/1, × 400. Membrane and ?eggs in focus.

Fig. 15 — Dinocysts bound by faecal debris, YTH 24/4, 23.0/109.1, MCP3680/4, × 400.

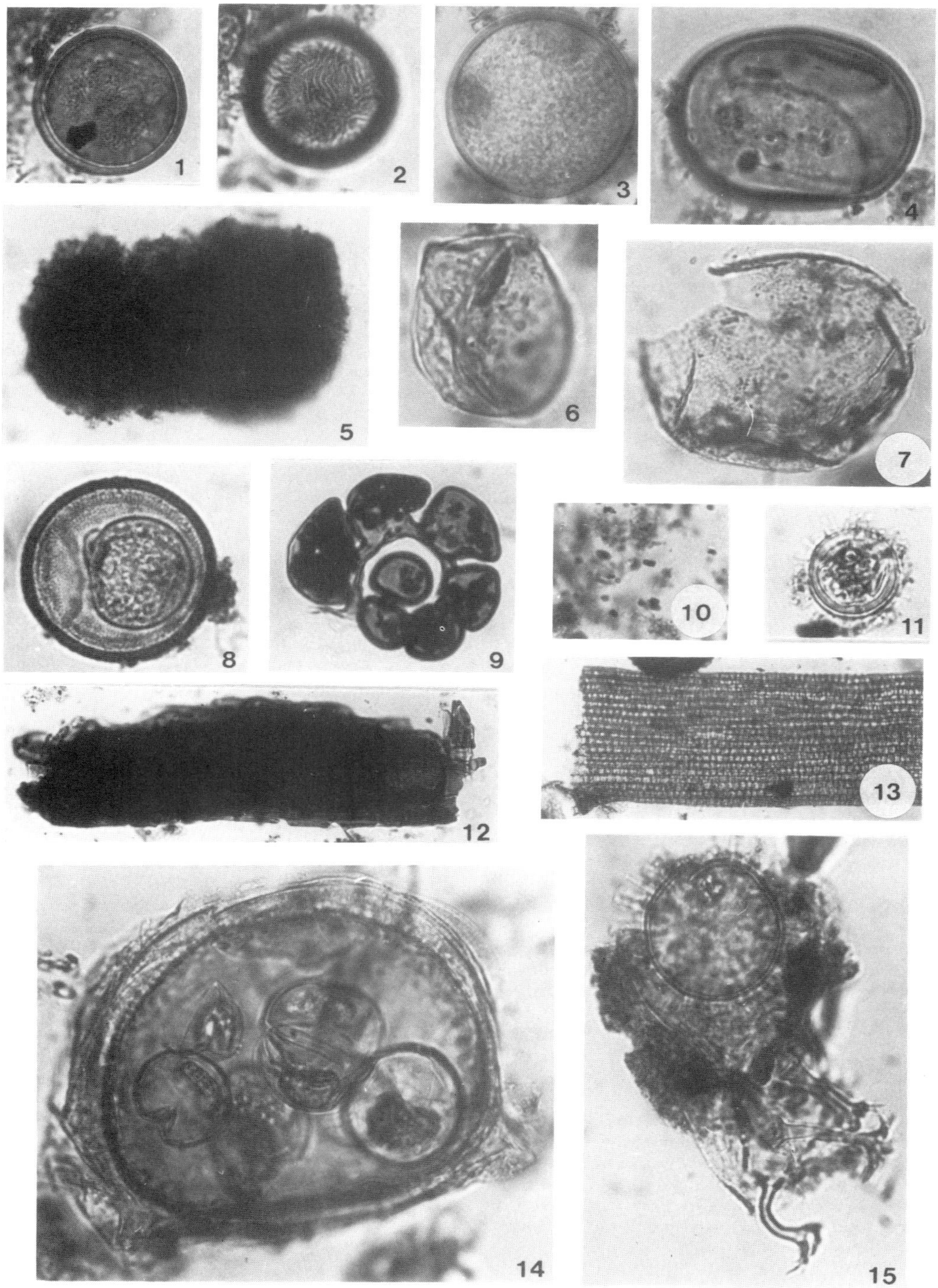
1
2
3
4
5
6
7
8
9
10
11
12
13
14
15

REFERENCES

Batten, D. J. (1974). Wealden palaeoecology from the distribution of plant fossils. *Proceedings of the Geologists' Association* **85** 433–458.

Batten, D. J. & Morrison, L. (1983). Methods of palynological preparation for palaeoenvironmental, source potential and organic maturation studies. *Norwegian Petroleum Directorate, Bulletin* **2** 35–53.

Bayliss, P. & Syvitski, J. P. M. (1982). Clay diagenesis in Recent marine faecal pellets. *Geo-Marine Letters* **2** 83–88.

Brasier, M. D. (1981). Microfossil transport in the tidal Humber Basin. In: Neale, J. W. & Brasier M. D. (eds), *Microfossils from Recent and fossil shelf seas.* Ellis Horwood, Chichester, pp. 314–322.

Cope, M. J. (1981). Products of natural burning as a component of the dispersed organic matter in sedimentary rocks. In: Brooks, J. (ed.), *Organic maturation studies and fossil fuel exploration.* Academic Press, New York, pp. 89–109.

Davey, R. J. & Rogers, J. (1975). Palynomorph distribution in Recent offshore sediments along two traverses off southwest Africa. *Marine Geology* **18** 213–225.

Dyer, K. R. (1973). *Estuaries: a physical introduction.* Wiley-Interscience, London, pp. 1–140.

Fullard, H. (1972). *Philip's modern school atlas*, 69th edition. George Philip & Son, London, pp. 1–132.

Harland, R. (1973). Microplankton from boreholes in the lower reaches of the Firth of Clyde. In: Deegan, C. E., Kirby, R., Rae, I. & Floyd R. (eds), The superficial deposits of the Firth of Clyde and its sea lochs. Appendix II. *Report, Institute of Geological Sciences, London* **73/9** 1–42.

Harland, R. (1977). Recent and Late Quaternary (Flandrian and Devensian) dinoflagellate cysts from marine continental shelf sediments around the British Isles. *Palaeontographica B* **164** 87–126.

Harland, R. (1978). Modern and Quaternary organic-walled microplankton from the north-east Irish Sea. In: Pantin, H. M. (ed.), *Quaternary sediments from the north-east Irish Sea: Isle of Man to Cumbria. Bulletin of the Geological Survey of Great Britain* **64** 41–43.

Harland, R. (1981). Cysts of the colonial dinoflagellate *Polykrikos schwartzii* Bütschli 1873 (Gymnodiniales) from Recent sediments, Firth of Forth, Scotland. Palynology **5** 65–79.

Harland, R. (1983). Distribution maps of Recent dinoflagellate cysts in bottom sediments from the North Atlantic Ocean and adjacent seas. *Palaeontology* **26** 321–387.

Harland, R. & Sharp, J. (1986). Elongated *Spiniferites* cysts from North Atlantic bottom sediments. *American Association of Stratigraphic Palynologists* **10** 25–34.

Herbert, D. W. M., Alabaster, J. S., Dart, M. C. & Lloyd, R. (1961). The effect of China Clay waste on trout streams. *International Journal of Air and Water Pollution* **5** 56–74.

Howell, B. R. & Shelton, R. G. J. (1970). The effect of china-clay on the bottom fauna of St Austell and Mevagissey bays. *Journal of the Marine Biological Association UK* **50** 593–607.

Kasahara, S., Uye, S. & Onbé, T. (1974). Calanoid copepod eggs in sea-bottom muds I. *Marine Biology* **26** 167–171.

Khayrallah, N. & Jones, A. M. (1975). A survey of the benthos of the Tay Estuary. *Proceedings of the Royal Society of Edinburgh B* **75** 113–135.

Leach, J. H. (1971). Hydrology of the Ythan Estuary with reference to distribution of major nutrients and detritus. *Journal of the Marine Biological Association (UK)* **51** 137–157.

Lee, A. J. & Ramster, J. W. (1981). *Atlas of the seas around the British Isles.* Ministry of Agriculture, Fisheries and Food, Lowestoft, pp. 1–88.

Love, L. G. (1958). Micro-organisms and the presence of syngenetic pyrite. *Quarterly Journal of the Geological Society, London* **113** 429–440.

Love, L. G. & Murray, J. W. (1963). Biogenic pyrite in Recent sediments of Christchurch Harbour, England. *American Journal of Science* **261** 433–448.

McLusky, D. S. (1981). *The estuarine ecosystem.* Blackie, Glasgow, pp. 1–150.

Masran, T. C. & Pocock, S. A. J. (1981). The classification of plant-derived particulate organic matter in sedimentary rocks. In: Brooks, J. (ed.), *Organic maturation studies and fossil fuel exploration.* Academic Press, New York, pp. 146–175.

Muller, J. (1959). Palynology of Recent Orinoco delta and shelf sediments: reports of the Orinoco Shelf Expedition; Volume 5. *Micropalaeontology* **5** 1–32.

Murray, J. W. (1965). On the foraminiferida of the Plymouth region. *Journal of the Marine Biological Association (UK)* **45** 481–505.

Murray, J. W. (1968). The living Foraminiferida of Christchurch Harbour, England. *Micropalaeontology* **14** 83–96.

Murray, J. W. (1973). *Distribution and ecology of living benthic foraminiferids.* Heinemann, London, pp. 1–274.

Murray, J. W. (1980). The Foraminifera of the Exe Estuary. In: *Essays on the Exe Estuary. Devonshire Association Special Number* **2** 89–115.

Perkins, E. J. (1973). *The marine fauna and flora of the Solway Firth.* The Dumfriesshire and Galloway Natural History and Antiquarian Society, pp. 1–112.

Perkins, E. J. (1974). *The biology of estuaries and coastal waters.* Academic Press, London, pp. 1–678.

Perkins, E. J. & Williams, B. R. H. (1966). The biology of the Solway Firth in relation to the movement and accumulation of radioactive materials. II.—The distribution of sediments and benthos. *UK Atomic Energy Authority PG Report* **587 (CC)**. HMSO, London, pp. 1–37.

Phleger, F. B. (1970). Foraminiferal populations and marine marsh processes. *Limnology and Oceanography* **15** 522–534.

Phleger, F. B. & Bradshaw, J. S. (1966). Sedimentary environments in a marine marsh. *Science* **154** 1551–1553.

Reid, P. C. (1972a). The distribution of dinoflagellate cysts, acritarchs and pollen in coastal sediments from the British Isles. *Ph.D. Thesis.* University of Sheffield, pp. 1–250.

Reid, P. C. (1972b). Dinoflagellate cyst distribution around the British Isles. *Journal of the Marine Biological Association (UK)* **52** 939–944.

Reid, P. C. (1974). Gonyaulacacean dinoflagellate cysts from the British Isles. *Nova Hedwigia* **25** 579–637.

Reid, P. C. (1975). A regional subdivision of dinoflagellate cysts around the British Isles. *New Phytologist* **75** 589–603.

Reid, P. C. (1977). Peridiniacean and Glenodiniacean dinoflagellate cysts from the British Isles. *Nova Hedwigia* **29** 429–463.

Reid, P. C. & John, A. W. G. (1978). Tintinnid cysts. *Journal of the Marine Biological Association* (*UK*) **58** 551–557.

Schäfer, W. (1972). *Ecology and palaeoecology of marine sediments*. Oliver & Boyd, Edinburgh, pp. 1–568.

Siesser, W. G. & Rogers, J. (1976). Authigenic pyrite and gypsum in south west African continental slope deposits. *Sedimentology* **23** 567–577.

Thia-Eng. C. (1973). An ecological study of the Ponggal Estuary, Singapore, *Hydrobiologia* **43** 505–533.

Trewin, N. H. (1987). Sedimentology of the Ythan estuary, Newburgh. In: Trewin, N. H., Kneller B. C. & Gillen, C. (eds), *Excursion guide to the geology of the Aberdeen area*. Geological Society of Aberdeen, Scottish Academic Press, Edinburgh, pp. 274–283.

Van Geel, B., Bohncke, S. J. F. & Dee, H. (1981). A palaeoecological study of an upper late glacial and Holocene sequence from 'De Borchert', the Netherlands. *Review of Palaeobotany and Palynology* **31** 367–448.

Venkatachala, B. S. (1981). Differentiation of amorphous organic matter types in sediments. In: Brooks, J. (ed.), *Organic maturation studies and fossil fuel exploration*. Academic Press, New York, pp. 177–200.

Wall, D., Lohmann, G. P. & Smith, W. K. (1977). The environmental and climatic distribution of dinoflagellate cysts in modern marine sediments from regions in the North and South Atlantic Oceans and adjacent seas. *Marine Micropalaeontology* **2** 121–200.

Williams, D. B. (1971). The distribution of marine dinoflagellates in relation to physical and chemical conditions. In: Funnell, B. M. & Riedel W. R. (eds), *The micropalaeontology of oceans*. Cambridge University Press, Cambridge, pp. 91–95.

General Index

Taxonomic Index

Bold numbers identify references to plates. The plate number is prefixed by its chapter number.